Marine Metapopulations

Marine Metapopulations

Editors
Jacob P. Kritzer
Peter F. Sale

AMSTERDAM • BOSTON • HEIDELBERG • LONDON
NEW YORK • OXFORD • PARIS • SAN DIEGO
SAN FRANCISCO • SINGAPORE • SYDNEY • TOKYO
Academic Press is an imprint of Elsevier

Senior Acquisitions Editor: Nancy Maragioglio
Associate Editor: Kelly Sonnack
Senior Publishing Editor: Andy Richford
Publisher: Dana Dreibelbis
Project Manager: Sarah Hajduk
Marketing Manager: Trevor Daul
Cover Design: Tyson Levasseur
Composition: SNP Best-set Typesetter Ltd., Hong Kong
Cover Printer: Phoenix Color
Interior Printer: The Maple-Vail Book Manufacturing Group

Elsevier Academic Press
30 Corporate Drive, Suite 400, Burlington, MA 01803, USA
525 B Street, Suite 1900, San Diego, California 92101-4495, USA
84 Theobald's Road, London WC1X 8RR, UK

This book is printed on acid-free paper. ∞

Copyright © 2006, Elsevier Inc. All rights reserved.

No part of this publication may be reproduced or transmitted in any form or by any means, electronic or mechanical, including photocopy, recording, or any information storage and retrieval system, without permission in writing from the publisher.

Permissions may be sought directly from Elsevier's Science & Technology Rights Department in Oxford, UK: phone: (+44) 1865 843830, fax: (+44) 1865 853333, e-mail: permissions@elsevier.com. You may also complete your request on-line via the Elsevier homepage (http://elsevier.com), by selecting "Customer Support" and then "Obtaining Permissions."

Library of Congress Cataloging-in-Publication Data
Marine metapopulations / editors, Jacob P. Kritzer, Peter F. Sale.
    p. cm.
    Includes bibliographical references and index.
    ISBN 0-12-088781-9 (hardcover : alk. paper)   1. Marine ecology. 2. Animal populations.
I. Kritzer, Jacob P. II. Sale, Peter F.
    QH541.5.S3M284 2006
    577.7—dc22

                                                                                                    2005028925

British Library Cataloguing in Publication Data
A catalogue record for this book is available from the British Library

ISBN 13: 978-0-12-088781-1
ISBN 10: 0-12-088781-9

For all information on all Elsevier Academic Press Publications,
visit our Web site at www.books.elsevier.com

Printed and bound by CPI Group (UK) Ltd, Croydon, CR0 4YY

Transferred to Digital Print 2012

**Working together to grow
libraries in developing countries**

www.elsevier.com | www.bookaid.org | www.sabre.org

ELSEVIER   BOOK AID International   Sabre Foundation

CONTENTS

Foreword — xvii
Preface — xxi
About the Editors — xxv
Contributors — xxvii

PART I
Introduction

1. The Merging of Metapopulation Theory and Marine Ecology: Establishing the Historical Context — 3
Peter F. Sale, Ilkka Hanski, and Jacob P. Kritzer

    I. Introduction — 3
   II. Differences and Similarities between Marine and Terrestrial Systems — 6
      A. Describing Actual Spatial Population Processes and Structures — 6
      B. A Framework for Asking Research Questions — 8
      C. A Paradigm for Population and Conservation Biology — 9
      D. Applications to Marine Populations — 10
  III. History and Effects of Predominant Research Questions in Marine Ecology — 12
  IV. Conservation and Fisheries Management Influences on Marine Ecology — 17
      A. Scale of Study — 17

B. Spatial Resolution	18
C. Marine Protected Areas	19
V. Summary	20
References	22

PART II
Fishes

2. The Metapopulation Ecology of Coral Reef Fishes — 31
Jacob P. Kritzer and Peter F. Sale

I. Introduction	31
II. Spatial Structure	34
A. Geographic Extent	34
B. Spatial Subdivision	35
C. Interpatch Space	41
III. Biology and Ecology of Coral Reef Fishes	42
A. Postsettlement Life Stages	42
B. Dispersal and Connectivity	51
C. Metapopulation Dynamics	51
IV. Factors Dissolving Metapopulation Structure	55
A. Spawning Aggregations	56
B. Nursery Habitats	57
V. Summary	60
References	61

3. Temperate Rocky Reef Fishes — 69
Donald R. Gunderson and Russell D. Vetter

I. Introduction	70
II. Geological Processes and the Types and Distribution of Temperate Reef Habitat	71
A. Subduction, Volcanism, and Faulting	74
B. Glaciation	76
C. Fluvial and Dynamic Submarine Erosive Processes	76
III. Temperate Reef Fish Communities	78
A. Biogeographic Provinces and Temperate Reef Fish Communities	78

	B. Depth as a Master Variable in Temperate Reef Fish Communities	79
	C. Typical Fish Fauna of Temperate Rocky Reef Communities	80
IV.	The Role of Oceanography in Metapopulation Structuring	80
	A. Major Oceanographic Domains	83
	B. Dispersal and Retention Mechanisms	84
V.	Climate, Climate Cycles, and Historical Metapopulation Structuring	85
VI.	The Role of Life History in Metapopulation Structuring	87
	A. Early Life History	87
	B. Juveniles	88
	C. Adults	90
	D. Longevity	92
VII.	Empirical Approaches to Measuring Dispersal and Metapopulation Structure	94
VIII.	Population Genetic Studies in North Pacific Rocky Reef Fishes	96
IX.	Human Impacts	103
X.	Future Directions for Metapopulation Studies of Temperate Reef Fishes	105
	References	108

4. Estuarine and Diadromous Fish Metapopulations   119
Cynthia M. Jones

	I. Introduction to Metapopulation Concepts in Estuarine and Diadromous Fish	119
II.	Mechanisms that Form Distinct Populations	125
III.	Tools to Quantify Migration in Diadromous Fish	127
	A. Genetics	127
	B. Artificial Tags	128
	C. Natural Tags with Emphasis on Otolith–Geochemical Tags	129
IV.	Diadromous Fish Exemplify Metapopulation Theory	130
	A. Salmonids	131
	B. Alosines	135
	C. Comparing Salmonids and Alosine Herrings	137
	D. Sciaenids	137
	E. Atherinids	138

V. Value of the Metapopulation Concept in Understanding and
   Managing Diadromous Fisheries   139
   A. Conservation of Local Populations   140
   B. Historical Management of Local Populations in
      a Fisheries Context   140
   C. Mixed-Stock Analysis   142
   D. Effect of Demography on Metapopulation Management   143
   E. Marine Protected Areas as a Spatial Management Tool   144
   F. Can Fisheries Be Managed as Metapopulations?   144
VI. Acknowledgments   146
   References   146

PART **III**
Invertebrates

5. **Metapopulation Dynamics of Hard Corals**   157
   Peter J. Mumby and Calvin Dytham

   I. Introduction   158
   II. Structure of this Chapter   160
   III. Summary of Model Structure and Parameterization   160
   IV. Existing Models of Dynamics on Coral Reefs   166
   V. Development of a Prototype Model of Spatially Structured
      Coral Reef Communities   167
      A. Scales   167
      B. Reproduction   169
      C. Connectivity   170
      D. Recruitment   171
      E. Growth   174
      F. Mortality   175
      G. Competition (and Modeling the Dynamics of Competitors)   178
      H. Herbivory   181
   VI. Testing the Model: Phase Shifts in Community Structure   185
   VII. Sensitivity of Model to Initial Conditions   186
   VIII. Exploration of Model Behavior   189
      A. Interactions between Initial Coral Cover, Algal Overgrowth
         Rate, Coral Growth Rate, and Herbivory   189
      B. Recruitment Scenario and Overfishing of Herbivores   190

C. Impact of Hurricane Frequency on Local Dynamics	193
D. Effect of Reduced Hurricane Frequency on a Reserve Network	194
IX. Summary	195
X. Acknowledgment	195
References	196

6. Population and Spatial Structure of Two Common Temperate Reef Herbivores: Abalone and Sea Urchins 205
Lance E. Morgan and Scoresby A. Shepherd

I. Introduction	205
II. Life History	208
A. Abalone	208
B. Sea Urchins	208
III. Larval Dispersal and Settlement	209
A. Abalone	209
B. Sea Urchins	212
C. Summary of Larval Dispersal	217
IV. Population Genetics	218
A. Abalone	218
B. Sea Urchins	219
C. Summary of Genetics	221
V. Spatial Variability in Adult Distributions and Demographics	221
A. Abalone: Adult Habitat and Spatial Structure	223
B. Sea Urchins: Adult Habitat and Spatial Structure	226
VI. Fishing and Management	228
A. Abalone	228
B. Sea Urchins	229
C. Optimal Harvesting of Invertebrate Metapopulations	231
VII. Summary	233
VIII. Acknowledgment	234
References	234

7. Rocky Intertidal Invertebrates: The Potential for Metapopulations within and among Shores 247
Mark P. Johnson

I. Introduction	247
II. Patch Models	249

III.	Within-Shore Metapopulations	251
IV.	Metapopulations at Different Scales	257
V.	Measured Scales of Variability	260
VI.	Summary	263
	References	266

8. Metapopulation Dynamics of Coastal Decapods 271
Michael J. Fogarty and Louis W. Botsford

I.	Introduction	271
II.	Decapod Life Histories	274
III.	Identifying Decapod Metapopulations	277
IV.	Case Studies	279
	A. American Lobster (*Homarus americanus*)	279
	B. Blue Crab (*Callinectes sapidus*)	285
	C. Dungeness Crab (*Cancer magister*)	293
	D. Pink Shrimp (*Pandalus borealis*)	300
V.	Discussion	305
VI.	Summary	310
VII.	Acknowledgment	311
	References	311

9. A Metapopulation Approach to Interpreting Diversity at Deep-Sea Hydrothermal Vents 321
Michael G. Neubert, Lauren S. Mullineaux, and M. Forrest Hill

I.	Introduction	321
II.	Vent Systems as Metapopulations	328
	A. Dynamics and Distribution of Vent Habitat	328
	B. Dispersal and Colonization	328
III.	Species Interactions	332
IV.	Biogeography and Diversity	333
V.	Metapopulation Models for Vent Faunal Diversity	335
	A. A Null Model	337
	B. Facilitation	341
VI.	Summary	343
	References	347

PART IV
Plants and Algae

10. **A Metapopulation Perspective on the Patch Dynamics of Giant Kelp in Southern California**   353
   Daniel C. Reed, Brian P. Kinlan, Peter T. Raimondi, Libe Washburn, Brian Gaylord, and Patrick T. Drake

   I. Introduction   353
   II. Dynamics of Giant Kelp Populations   355
   III. Factors Affecting Colonization   360
       A. Life History Constraints   360
       B. Modes of Colonization   363
       C. Spore Production, Release, and Competency   365
       D. Postsettlement Processes   366
   IV. Spore Dispersal   367
       A. Factors Affecting Colonization Distance   367
       B. Empirical Estimates of Spore Dispersal   368
       C. Modeled Estimates of Spore Dispersal   371
   V. Connectivity Among Local Populations   375
   VI. Summary   377
   VII. Acknowledgment   381
       References   381

11. **Seagrasses and the Metapopulation Concept: Developing a Regional Approach to the Study of Extinction, Colonization, and Dispersal**   387
    Susan S. Bell

    I. Introduction   388
    II. Seagrass Reproduction   389
    III. Patches: Colonization and Extinction   391
    IV. Examples of Potential Seagrass Metapopulations   394
        A. *Halophila decipiens* on the West Florida Shelf: A Local Population That Exhibits Patch Extinction and Regional Recruitment   396
        B. *Phyllospadix scouleri* on Exposed Pacific Shores: Regional Patterns of Suitable Sites Exist among a Matrix of Unsuitable Sites; Patch Extinction Documented   399

C. *Halophila johnsonii* in Southeastern Florida: A Well-defined Regional Population with Limited Dispersal between Patches; Extinction of Patches Documented 399
V. The Metapopulation Model and Seagrass Populations: A Useful Concept? 400
   A. Collection of Good Information on Spatial Organization of Seagrasses 401
   B. Collection of Genetic Information to Help Analyze Spatial Structure of a Population 402
   C. Seed Dormancy: A Special Problem for Plant Populations 402
VI. Summary 403
VII. Acknowledgments 403
References 404

PART V

Perspectives

12. Conservation Dynamics of Marine Metapopulations with Dispersing Larvae 411
Louis W. Botsford and Alan Hastings

   I. Introduction 411
   II. Single Population Persistence 413
   III. Metapopulation Persistence 416
      A. Consequences for the Distribution of Meroplanktonic Species 420
      B. Consequences for the Success of Marine Reserves 421
   IV. Role of Variability 421
   V. Discussion 422
   VI. Summary 426
   References 426

13. Genetic Approaches to Understanding Marine Metapopulation Dynamics 431
Michael Hellberg

   I. Introduction 431
      A. Subdivision in Marine Populations 432
      B. The Island Model and Its Limitations 432

II. Delineating Populations — 434
  III. Inferring Patterns of Connectivity — 438
  IV. Inferring Nonequilibrium Population Dynamics — 441
      A. Population Extinction and Recolonization — 442
      B. Source–Sink Relationships — 444
      C. Mixing in the Plankton and the Genetics of Larval Cohorts — 445
  V. Summary — 447
  VI. Acknowledgment — 449
      References — 449

14. Metapopulation Dynamics and Community Ecology of Marine Systems — 457
Ronald H. Karlson

  I. Introduction — 458
      A. Scale of Dispersal — 460
      B. Dispersal and Population Dynamics — 462
      C. Relevance to Marine Metacommunities — 465
  II. Metacommunities and Species–Area Relationships — 466
      A. Background — 466
      B. Relevance to Marine Metacommunities — 469
      C. Regional-Scale Differentiation — 470
      D. Summary — 471
  III. Metacommunities and Local–Regional Species Richness Relationships — 472
      A. Background — 472
      B. Relevance to Marine Metacommunities — 474
      C. Effects of Transport Processes and Relative Island Position — 476
      D. Summary — 477
  IV. Metacommunities and Relative Species Abundance Patterns — 477
      A. Background — 477
      B. Metacommunities and Relative Species Abundance Patterns — 478
      C. Relevance to Marine Metacommunities — 482
      D. Summary — 483
  V. Summary — 483
      References — 483

15. Metapopulation Ecology and Marine Conservation 491
Larry B. Crowder and Will F. Figueira

 I. Introduction 491
 II. Sources, Sinks, and Metapopulation Dynamics 494
 III. Case Studies 495
  A. Coral Reef Fishes 495
  B. Caribbean Spiny Lobster (*Panulirus pargus*) 499
  C. Red Sea Urchin (*Strongylocentrotus franciscanus*) 504
  D. Loggerhead Sea Turtles (*Caretta caretta*) 509
 IV. Summary 510
 References 511

16. The Future of Metapopulation Science in Marine Ecology 517
Jacob P. Kritzer and Peter F. Sale

 I. Introduction 517
 II. The Amphibiousness of Metapopulation Theory 518
  A. Commonalities between Marine and Terrestrial Ecology 518
  B. Differences between Marine and Terrestrial Systems 519
 III. Where in the Sea Is Metapopulation Theory Less Relevant? 521
  A. Highly Mobile Species 521
  B. Highly Isolated Populations 522
  C. Widely Dispersing Species 523
  D. The Special Case of Clonal Organisms 523
 IV. Outstanding Research Questions 524
  A. Knowledge of Dispersal Pathways and Mechanisms 524
  B. Knowledge of Rates and Extents of Connectivity 525
  C. Marine Metacommunities 525
 V. Spatially Explicit Management of Marine Fishery Resources 526
 VI. Summary 527
 References 528

Index *531*

Color plates can be found between pages 386 and 387.

In memory of Scott

*A biologist and geographer,
An environmentalist and outdoorsman,
My cousin and friend,*

A kind heart who left us far too soon.

—Jake

FOREWORD

This book marks one of the most important milestones in marine ecology during the last 50 years, and its content has been in the making for more than two decades. I hope you take the time to read through these lucidly and authoritatively written pages. If you are a professional in marine ecology, fisheries biology, conservation biology, or biological oceanography; if you are an educator in a secondary school or university, or a hobbyist into fishing or diving; in short, if you are anyone who cares about life in the ocean, this book has something for you. And if you enjoy the history of science, this book also offers a case study in how awkwardly and painfully, although inevitably, scientists revise their ideas.

I began writing about the metapopulation concept in marine ecology in the early 1980s, and may have been the first marine ecologist to use the word *metapopulation* specifically for marine populations, even though the idea had been widely used in terrestrial ecology since Richard Levins coined the term in the late 1960s. The editors observe in the book's first chapter how citations to the word *metapopulation* have spread slowly in marine ecology relative to terrestrial ecology, and present some conjectures as to why this has happened. My recollection is somewhat different, and revolves around hostility and continuing resistance to the idea.

When my first grant proposal to the biological oceanography panel of the National Science Foundation in the mid 1980s was rejected, reviewers made fun of the word *metapopulation*. I vividly remember the panel director scoffing at the idea by saying, "Where did you come up with that one?" I replied that Levins had coined the word more than 10 years before, and mentioned Doug Gill's (1978) study of the metapopulation structure of salamanders. I felt my work was merely extending the accepted metapopulation concept from aquatic systems into marine systems. Words that did resonate with marine scientists however were *recruitment*, which had long been discussed in fisheries biology; and *settlement*, in invertebrate zoology. One could propose and conduct research about recruitment that was immediately understood, work that today might be pitched as metapopulation research. Another phrase fisheries biologists used was *stock identification*, which meant finding where adult populations occur, which in turn leads to a metapopulation picture of the species population. If the literature search of Chapter 1 for citations to metapopulation were widened to include phrases like *recruitment*, *settlement*, *stock identification*, and *benthic/oceanic coupling*, then I suspect the number of marine papers would equal or exceed those in terrestrial systems.

Although fisheries biologists and oceanographers were resistant to metapopulation terminology, terrestrial ecologists were not particularly helpful either. In 1991, Gilpin and Hanski explicitly disqualified marine spatially structured populations from consideration as true metapopulations because marine dynamics involving pelagic larvae with benthic adults leads to an equation different from Levins'. I felt this was unfair. My sense of the concept of a metapopulation is that it is a "population of populations." Levins (1968) introduced a particular model for a metapopulation, that in 1998 I termed the *logistic weed*, which envisions a dynamics of within-patch extinction and between-patch colonization leading to a logistic equation for the patch-occupancy fraction. It seems foolish to restrict the metapopulation concept to systems with spatial dynamics that happen to produce a logistic equation. Nonetheless, denying marine populations the status of "true" metapopulations chilled research because one had to deal with the xenophobia of fisheries biologists without the benefit of terrestrial allies.

To make matters worse, the grand old men of intertidal ecology were especially threatened by the metapopulation concept and tried to prohibit its use. Intertidal research leaders had reputations invested in experimental protocols that relied on small-scale manipulative experiments, in food-web pictures presenting intertidal community ecology as self-contained within a site, and in explaining intertidal community structure as resulting from local dynamics and species interactions. Accepting a metapopulation conceptualization required rethinking and redoing intertidal research. Recruitment fluctuation had been regarded as noise impinging on the local system dynamics, partially obscuring what was really happening, which was taken to be hierarchical competition among space occupiers like barnacles and mussels tempered by top-down "keystone" regulation from predators like starfish. Recruitment fluctuation merely veiled this underlying dynamic. Instead, according to the metapopulation concept, recruitment was the signature of the local system's being coupled to a much larger system with regional dynamics that are largely responsible for what occurs in any local site. Recruitment fluctuation is, then, the waxing and waning of the coupling among various parts of a larger system, not noise. Hierarchical competition and keystone predators become insignificant coincidences that happen to occur in some local systems whenever their oceanographic coupling to the regional system winds up permitting high enough recruitment for intertidal space to become physically limiting there. This disenfranchising of local system dynamics from its dominant explanatory power was also anticipated by Peter Sale's 1977 critiques of niche theoretical explanations of coral reef fish community structure in his "lottery model" that has now, I suppose, been superseded by coupled benthic–oceanographic process models.

Outsiders cannot readily appreciate the angst that metapopulation thinking causes in marine ecology, an angst that continues to impede its acceptance. Generations of students have visited tide pools to be regaled by stories of the thadid snails drilling through barnacle shells, starfish everting their stomachs to gobble up hapless mussels, giant barnacles overgrowing and suffocating their tiny cousins, waves and logs demolishing huge swaths of gentle plantlike animals leaving scars of vacant space upon the rocks. Yet, according to the metapopulation concept, these dramatic and venerable stories are ecologically meaningless unless repeated at all other local sites spanned by the metapopulation's scale and thereby licensed by the metapopulation perspective. A visceral discomfort with this disempowering of local explanations for locally observed phenomena underlies continual efforts to resuscitate local thinking. One sees a continual stream of papers searching in vain for local entrainment of larvae, or evidence of local genetic differentiation taken as evidence that the larval pool across the entire metapopulation is not uniformly mixed. A need to misstate the marine metapopulation conceptualization is a red flag signaling continuing discomfort. No one, of course, has ever said or implied that a larval pool is uniformly mixed across its spatial extent. I have often stated and written that the larval pool mixing scale is determined by the larval life span, which in turn points to the appropriate fluid dynamic mixing scale. Tunicate larvae in the water for hours are mixed with tidal fluxes, barnacle larvae in the water for weeks are mixed with mesoscale eddy diffusion, and decapod larvae in the water for months are mixed by seasonal currents and basin scale events. An implausible older literature attributing sophisticated choice to larvae in where they settle is being continually revived in hopes of accounting for something, for anything, maybe vertical zonation, in local sites, thus preserving at least some explanatory power of local process for local pattern.

Yet, as this book testifies, the clock will not be turned back. The metapopulation concept is here to stay in marine ecology. Science demands it, fisheries management needs it, and it is the last hope for marine conservation. The convergence of marine and terrestrial ecologists present in this book ushers in a new era of unified approach to wet and dry ecosystems simultaneously, even as it marks the most important milestone of marine ecology in more than 50 years.

Joan Roughgarden
Department of Biological Sciences
Stanford University
Stanford, CA 94305
joan.roughgarden@stanford.edu
June 7, 2005

REFERENCES

Gill, D. (1978). The metapopulation ecology of the red-spotted newts, *Notophthalmus viridescens* (Rafinesque). *Ecol. Monogr.* **48**, 145–166.

Gilpin, M.E., and Hanski, I. (1991). *Metapopulation dynamics: Empirical and theoretical investigations.* London: Linnaean Society of London and Academic Press.

Levins, R. (1968). *Evolution in changing environments.* Princeton, NJ: Princeton University Press.

Roughgarden, J. (1998). *Primer of ecological theory.* Upper Saddle River, NJ: Prentice Hall.

Roughgarden, J., and Iwasa, Y. (1986). Dynamics of a metapopulation with space-limited subpopulations. *Theor. Popul. Biol.* **29**, 235–261.

Roughgarden, J., Iwasa, Y., and Baxter, C. (1985). Demographic theory for an open marine population with space-limited recruitment. *Ecology.* **66**, 54–67.

Sale, P. (1977). Maintenance of high diversity in coral reef fish communities. *Amer. Natur.* **111**, 337–359.

PREFACE

Marine ecology has been moving toward metapopulation thinking for some time. We track this progression along with Ilkka Hanski in the opening chapter of this book, and Joan Roughgarden offers some personal insights on how this came to be in her thought-provoking Foreword. Like Joan, we have our own stories about how we came to metapopulation ideas and their application to marine systems. For Peter Sale, a career spent thinking about the role of recruitment in structuring populations and communities of coral reef fishes led him to wonder about, if not work on, the larval phase and what happened before settlement. Beyond the edge of the reef, questions floated about where these newly arriving larvae had originated and, therefore, how distinct populations, once viewed largely in isolation, were interconnected. The flexibility of metapopulation dynamics also fitted well with his increasingly unstructured view of how populations functioned. Sale hinted at this transformation from a single population view to one focused on networks of populations in his own contribution to his 1991 reef fish ecology book. He continued to move in this direction, drawing attention, in his chapter in his 2002 book on coral reef fishes, to the need to quantify connectivity among populations if reef fishes were ever to be managed well using marine-protected areas.

Jake Kritzer came to metapopulation ideas from a different angle. His doctoral research under Howard Choat and Campbell Davies examined patterns of spatial variation in demographic traits of adult reef fish. Describing these patterns and speculating on their causes was only satisfying to a point, and there were more intriguing questions to be asked about the consequences of the variation. Larger body size in certain populations means production of more eggs, and the dispersive nature of reef fish larvae means that increased fecundity in one population can have replenishment benefits beyond that population. He stumbled upon Hanski and Gilpin's 1997 metapopulation book and began to think about reef fishes in the same context. So, while Sale came to think about metapopulations by moving back in the life cycle from recruitment, Kritzer got there by moving forward from spawning.

Contributors to Hanski and Gilpin's book applied the metapopulation concept to a broad range of taxa, from butterflies to small mammals to plants. But all the focal species were terrestrial. There seemed to be a role for metapopulation thinking in marine ecology as well, perhaps with some changes or at least with a different focus, and a need for more marine examples of the application of metapopulation concepts. Kritzer conceived the idea for this book sometime in the late 1990s while a graduate student, figuring it would be something to do years down the road after the weight of the dissertation was lifted, a "postdoc" or perhaps two were under the belt, and he was settled into a cozy professorship somewhere.

A conversation with Bruce Mapstone in 2001 revised that schedule. Bruce liked the idea, and urged moving on it sooner rather than later. The range of possible career paths for a marine ecologist is much broader today, and that cozy professorship might not come to pass (quite prescient, in hindsight!). Bruce argued that the topic was timely and might be taken up by others if not acted upon. Furthermore, while working in the Sale lab, Kritzer could draw upon the wisdom of a veteran of two edited volumes. Despite having just finished the 2002 reef fish book, Sale enthusiastically agreed to the project. The result is now in your hands.

Several papers have been written asking whether metapopulation concepts should even be applied to marine systems and, if so, how? Unlike our own contribution to that debate, some call for more limited application of the metapopulation concept in the marine realm. We will not take up that debate again here, except to say that the breadth of perspectives represented in this book (and, indeed, the enthusiasm with which the authors responded to our requests to contribute) underscore the momentum and readiness within the marine ecology community to think about marine populations as metapopulations.

But momentum and readiness do not mean that there is universal agreement on what the idea is all about and what to do with it in marine systems. And we have not tried to force a common viewpoint on our contributors. We did not feel it was our place to impose definitions, especially in a book that is taking the first in-depth look at how marine ecology and metapopulation ecology can interact. It was more important to present a range of perspectives and approaches to help stimulate debate and new thinking.

Consequently, some, such as Dan Reed and his colleagues in their chapter on kelps, adopt the "classic" metapopulation perspective and examine extinction–recolonization dynamics. Others, such as Mike Fogarty and Loo Botsford, in their chapter on decapods, focus much more on fluctuations in abundance with less consideration of local extinctions. Most chapters address single-species population ecology, but Pete Mumby and Calvin Dytham, and Michael Neubert et al. consider community ecology in their chapters on tropical corals and deep-sea vents respectively. Conservation and fisheries issues are prominent themes in some chapters, such as Lance Morgan and Scoresby Shepherd's contribution on urchins and abalone, but others are more purely ecological, such as Mark Johnson's look at the rocky intertidal. The final series of five chapters draws together some of the varied ideas raised in the preceding system-specific chapters in a more general context.

The marine systems addressed in this book do not represent the full spectrum of those to which metapopulation concepts can and should be applied. Previous papers have examined a range of other taxa, from isopods and infaunal microinvertebrates to groundfish and herring. There are other species for which a metapopulation approach might be useful as well, including large schooling pelagics such as tuna, and amphibious megafauna (i.e., pinnipeds and sea turtles) whose populations are defined by areas where they haul out to rest, breed, and nest. But we needed to draw the line somewhere, and feel good about the broad array of systems we have assembled.

As with any venture of this sort, this project could not have been completed without the efforts of a wide range of people. The 26 other authors contributing to this volume all added one more significant task to their already full workloads, and each managed to produce a stimulating chapter. We thank all the authors for their promptness (for the most part), for their responsiveness to our requests (even if the responses were slowed by other work), and, above all, for the chapters they produced.

Along the way, it has been a pleasure to work with the enthusiastic, supportive, gracious, and infinitely patient team at Academic Press, including Nancy Maragioglio, Kelly Sonnack, Dave Cella, Chuck Crumly, and Sarah Hajduk. We thank them all.

All chapters in this book were peer reviewed, and the reviewers all provided thorough, thoughtful, and constructive reviews, generally in a timely manner. We thank Giacomo Bernardi, Brian Bowen, John Bruno, Stan Cobb, Rebecca Fisher, Graham Forrester, Michael Graham, Ron Karlson, Phil Levin, Diego Lirman, Milton Love, Jon Lovett–Doust, Jean Marcus, Camilo Mora, Lance Morgan, Helen Murphy, John Olney, Hugh Possingham, Tom Quinn, Robert Vadas, and several others (who preferred to remain anonymous) for their reviews.

Kritzer would like to say a special thank you to his parents, John and Irene, and his sister Becky for all they have done throughout his life, and to Helen Murphy for her friendship and support over many years, especially during production of this book. Sale extends his love and a heartfelt thanks to Donna. She knows why.

Jake Kritzer	Peter Sale	
New York, New York	Windsor, Ontario	March 2006

ABOUT THE EDITORS

JACOB P. KRITZER—Dr. Kritzer completed his PhD in Marine Biology at James Cook University in Queensland, Australia, in association with the Cooperative Research Centre for the Great Barrier Reef World Heritage Area. There, he studied spatial patterns in the demography of an exploited tropical snapper. He then held a postdoctoral fellowship with the coeditor of this volume at the University of Windsor in Ontario, Canada, studying dispersal, recruitment, and connectivity of coral reef fishes along the Mesoamerican Barrier Reef System. Currently, Dr. Kritzer is a marine scientist with Environmental Defense in New York City, where he works in both research and environmental advocacy. His focus has shifted from tropical reefs to the cold-water coastal ecosystems of the northeastern United States, especially diadromous fish populations and their habitats. However, he retains his interest in spatial structure and its implications for population dynamics and natural resource management. *Marine Metapopulations* is his first edited volume.

PETER F. SALE—Following his education at University of Toronto and University of Hawaii, Peter Sale continued moving south and west, and spent 20 years at the University of Sydney, where he witnessed the coming of age of Australia's tropical marine science community. He returned to North America in December 1987, first to the University of New Hampshire and then to Canada and the University of Windsor in 1994. His research on reef fishes has helped make coral reefs relevant to the wider world of ecology, while challenging conventional ideas on community organization, and the importance of larval settlement and recruitment. Working in the Caribbean after many years on the Great Barrier Reef brought home vividly just how massively humanity has impacted the world's coastal oceans, and current research on population connectivity seeks answers that may help enhance sustainable management of reef fisheries using no-take marine reserves. Prior to working on *Marine Metapopulations,* he edited two definitive books on reef fish ecology: *Ecology of Fishes on Coral Reefs* (1991) and *Coral Reef Fishes* (2002). He and his wife, Donna, live in Canada, but he dives where water is warmer and fish more colorful.

CONTRIBUTORS

SUSAN S. BELL, Department of Biology, University of South Florida, Tampa, Florida, USA

LOUIS W. BOTSFORD, Department of Wildlife, Fish, and Conservation Biology, University of California, Davis, California, USA

LARRY B. CROWDER, Nicholas School of the Environment, Duke University Center for Marine Conservation, Beaufort, North Carolina, USA

PATRICK T. DRAKE, Long Marine Lab, University of California, Santa Cruz, California, USA

CALVIN DYTHAM, Department of Biology, University of York, York, UK

WILL F. FIGUEIRA, Department of Environmental Sciences, University of Technology, Sydney, Australia

MICHAEL J. FOGARTY, National Marine Fisheries Service, Woods Hole, Massachusetts, USA

BRIAN GAYLORD, Marine Science Institute, University of California, Santa Barbara, California, USA

DONALD R. GUNDERSON, School of Aquatic & Fisheries Sciences, University of Washington, Seattle, USA

ILKKA HANSKI, Metapopulation Research Group, Department of Biological and Environmental Sciences, University of Helsinki, Helsinki, Finland

ALAN HASTINGS, Department of Environmental Science and Policy, University of California, Davis, California, USA

MICHAEL HELLBERG, Department of Biological Sciences, Louisiana State University, Baton Rouge, Louisiana, USA

M. FORREST HILL, Institute of Theoretical Dynamics, University of California, Davis, California, USA

MARK P. JOHNSON, School of Biology and Biochemistry, The Queen's University of Belfast, Belfast, Northern Ireland, UK

CYNTHIA M. JONES, Department of Ocean, Earth, and Atmospheric Sciences, Old Dominion University, Norfolk, Virginia, USA

RONALD H. KARLSON, Department of Biological Sciences, University of Delaware, Newark, Delaware, USA

BRIAN P. KINLAN, Department of Ecology, Evolution, and Marine Biology, University of California, Santa Barbara, California, USA

JACOB P. KRITZER, Environmental Defense, 257 Park Avenue South, New York, New York, USA

LANCE E. MORGAN, Marine Conservation Biology Institute, Glen Ellen, California, USA

LAUREN MULLINEAUX, Biology Department, Woods Hole Oceanographic Institution, Woods Hole, Massachusetts, USA

PETER J. MUMBY, Marine Spatial Ecology Lab, School of Biological and Chemical Sciences, University of Exeter, Exeter, UK

MICHAEL NEUBERT, Biology Department, Woods Hole Oceanographic Institution, Woods Hole, Massachusetts, USA

PETER T. RAIMONDI, Ecology and Evolutionary Biology, University of California, Santa Cruz, California, USA

DANIEL REED, Marine Science Institute, University of California, Santa Barbara, California, USA

JOAN ROUGHGARDEN, Department of Biological Sciences, Stanford University, Stanford, California, USA

PETER F. SALE, Department of Biological Sciences, University of Windsor, Windsor, Ontario, Canada

SCORESBY A. SHEPHERD, South Australian Research and Development Institute, Henley Beach, South Australia, Australia

RUSSELL D. VETTER, Southwest Fisheries Science Center, National Marine Fisheries Service, La Jolla, California, USA

LIBE WASHBURN, Department of Geography, University of California, Santa Barbara, California, USA

PART I

Introduction

CHAPTER 1

The Merging of Metapopulation Theory and Marine Ecology: Establishing the Historical Context

PETER F. SALE, ILKKA HANSKI, and JACOB P. KRITZER

I. Introduction
II. Differences and Similarities between Marine and Terrestrial Systems
   A. Describing Actual Spatial Population Processes and Structures
   B. A Framework for Asking Research Questions
   C. A Paradigm for Population and Conservation Biology
   D. Applications to Marine Populations
III. History and Effects of Predominant Research Questions in Marine Ecology
IV. Conservation and Fisheries Management Influences on Marine Ecology
   A. Scale of Study
   B. Spatial Resolution
   C. Marine Protected Areas
V. Summary
   References

I. INTRODUCTION

The metapopulation concept, first formalized by Levins (1969, 1970), has roots dating back to the work of Andrewartha and Birch (1954), and MacArthur and Wilson (1967). The origin and subsequent evolution of this concept into an important component of modern ecological theory have been described by Hanski and Simberloff (1997), and Hanski (1999). Previous discussions were

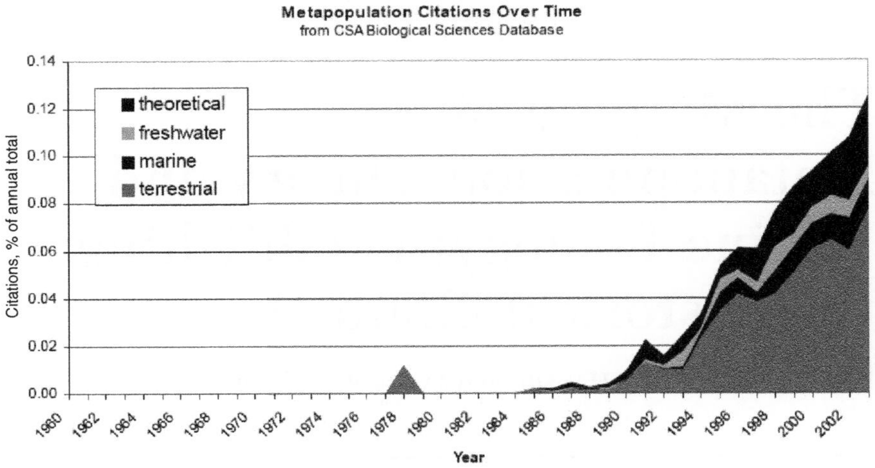

Metapopulation Citations Over Time
from CSA Biological Sciences Database

FIGURE 1-1. The pattern of growth in citations to **metapopulation** in the CSA's Biological Sciences database, by ecosystem studied. "Theoretical" studies are those that discussed metapopulation concepts or theory but could not be assigned to a specific habitat. Terrestrial studies greatly outnumber either freshwater or marine ones, and theoretical studies also exceed the total of aquatic studies. Citation search and figure provided by W. Figueira.

largely focused on theoretical developments and on applications of the metapopulation concept in terrestrial systems. Our goal here is to examine the use of metapopulation concepts and theory in marine systems. In doing so, we will be setting the stage for this volume.

Despite the first appearance of the term *metapopulation* in Levins' papers more than three decades ago, the concept was not widely used until the early 1990s, and it exhibited a steady increase in usage through the mid 1990s (Hanski and Simberloff, 1997), with no apparent plateau in citations into the early years of this century (Hanski and Gaggiotti, 2004). Applications of metapopulation theory to aquatic ecology appeared in the early 1990s, not long after the exponential increase in metapopulation citations generally (Hanski and Gaggiotti, 2004), but these references were to freshwater systems (Figueira, 2002), and use of metapopulation concepts in marine systems was rare prior to the mid 1990s (Fig. 1-1). These have since increased dramatically (Figueira, 2002; Grimm et al., 2003). For example, in the first of two edited books on the ecology of coral reef fishes (Sale, 1991), there is but a single reference to metapopulations, and that is only a parenthetical note (Doherty, 1991, p. 264). In contrast, the second book (Sale, 2002a) contains metapopulation references on 25 pages (approximately 6% of all pages), spanning eight different chapters by different authors on different topics.

Chapter 1 The Merging of Metapopulation Theory and Marine Ecology   5

Citations using metapopulation concepts in marine systems are now about as common as those referencing freshwater systems, although they remain but a minor fraction of terrestrial metapopulation citations (Figueira, 2002). Unlike the case for terrestrial studies, the rate of marine citations of metapopulation concepts does not appear to be increasing greatly. This is despite the fact that more than 70% of the earth's surface is covered by marine habitat and a considerable proportion of global biodiversity dwells in the sea.

In this chapter, we consider three general explanations for the pattern of use of metapopulation ideas in marine ecology. First we examine whether there are fundamental differences between marine and terrestrial systems (apart from the surrounding medium) that render theory developed primarily in one context of limited use for the other. We do this through a critical assessment of how metapopulation theory is used in terrestrial ecology and suggest likely differences in application of metapopulation theory to marine systems.

We then explore whether the slow adoption of a metapopulation paradigm in marine ecology might have been a direct consequence of the pattern of development of ideas and questions in marine ecology. The early absence and eventual integration of metapopulation theory into marine ecology might be driven by a change in the types of questions investigated by marine ecologists, because not all ecological issues fall directly within the domain of metapopulation ecology. An early focus by marine ecologists on topics outside this domain provides a simple explanation for why adoption of metapopulation concepts lagged. We examine this possibility through a historical review of marine ecology during the past 30 years.

Finally, we examine the extent to which marine environmental management and conservation may have helped initiate and now continue to drive the application of metapopulation theory. Scientific inquiry rarely unfolds in a vacuum. Instead, science is often heavily shaped by the rest of human society (Diamond, 1997). Hanski (1999) has described how the development of metapopulation ecology as a whole was spurred by applications in environmental conservation, and we explore this in the context of marine conservation.

Before proceeding, we must have a clear definition of a metapopulation. This clarification is necessary to identify how a given characteristic of marine systems, research topic in marine ecology, or marine management issue fits into our perception of a metapopulation. Recently, Smedbol et al. (2002) and Grimm et al. (2003) have put forth arguments to the marine ecology and fisheries science communities that metapopulations should be defined in the spirit of the classic Levins model, with real likelihood of extinction of local populations a necessary feature. In contrast, we have all put forth arguments for a different perspective (Hanski and Simberloff, 1997; Hanski, 1999; Hanski and Gaggiotti, 2004; Kritzer and Sale, 2004)—one we will adopt here. Metapopulation ecology has grown considerably since Levins' landmark papers, and modern metapopulation research often considers more than simply local presence or absence of organisms. Even without local extinctions, metapopulation structure has implications for abundance of local populations, as well as those populations' age, size, and genetic structures. We define a metapopulation as a system in which (1) local populations inhabit discrete habitat patches and (2) interpatch dispersal is neither so low as to negate significant demographic connectivity nor so high as to eliminate any independence of local population dynamics, including a degree of asynchrony with other local populations.

II. DIFFERENCES AND SIMILARITIES BETWEEN MARINE AND TERRESTRIAL SYSTEMS

How have ecologists used the metapopulation framework in the study of terrestrial populations, and are these approaches applicable to marine systems? Our purpose here is not to imply that the terrestrial comparison is somehow fundamental for marine ecology, and even less that marine ecologists should try to emulate what terrestrial ecologists have already done. But given the 15-year experience of terrestrial metapopulation ecologists, it is worth asking, in this context, about the current standing of the field. We distinguish three ways of using the metapopulation concept, although the boundaries are not necessarily sharp. These are its use to describe spatial processes and structures in actual populations, its use as a framework for asking research questions, and its use as a paradigm for population and conservation biology.

A. Describing Actual Spatial Population Processes and Structures

Many species on land live in highly fragmented landscapes (Hanski, 1999), in which suitable habitat accounts for only a small fraction, often a few percent, of the total landscape area and mostly occurs as small, discrete patches. Such situations are common, especially for invertebrates, many of which are habitat specialists and have small body size, high rate of population increase, and short generation time—all factors that make it likely that the species has a metapopulation structure (Murphy et al., 1990). Small body size means that the number of individuals in even small fragments of habitat may be large enough to constitute a local breeding population, especially because small species tend to have lower migration rates than large ones (excluding very small organisms that disperse passively and very widely; Finlay and Clarke, 1999; Finlay, 2002). A high population growth rate implies that, after population establishment, local populations quickly grow to the local carrying capacity, unless they go extinct. Short generation time means, among other things, that stochastic events are not buffered by great longevity of individuals, which would reduce the risk of population extinction. Finally, habitat specialists with short generation times are likely to be greatly affected by interactions between large-scale weather perturbations and local environmental conditions.

Butterflies have been studied most intensively in the metapopulation context. These satisfy the aforementioned criteria but also possess features that make them especially amenable to field studies (for a review see Thomas and Hanski, 2004). Numerous other terrestrial species have comparable spatial population structures, although studying many of them would be much more difficult than investigating butterflies. For example, thousands of species of insects, fungi, mosses, and other taxa live in dead and decaying tree trunks in forests. Individual decaying logs represent discrete habitat patches, containing local populations of species that necessarily are organized as a metapopulation because of the ephemeral nature of the habitat (Hanski, 1999).

Richard Levins' (1969, 1970) simple metapopulation model captured the bare essentials of the dynamics of species that exist regionally in a balance between local extinctions and recolonizations of currently unoccupied habitat patches. Since then, the theory has developed in many directions, recently reviewed in the edited volume by Hanski and Gaggiotti (2004). The theoretical development that remains closest in spirit to Levins' original concept with local extinctions and recolonizations has been dubbed the *spatially realistic metapopulation theory* (Hanski, 2001; Hanski and Ovaskainen, 2003; Ovaskainen and Hanski, 2004). The key elements of this theory are a finite number of patches (infinite in the Levins model), variation in patch qualities (identical in the Levins model), and variation in patch connectivities resulting from inevitable differences in their spatial locations and distance-dependent migration in most species (all patches equally connected in the Levins model). These features lead to sufficiently realistic models for many real metapopulations to allow statistically rigorous estimation of model parameters (for a review see Etienne et al., 2004), which, in the best cases, allow the construction of predictive models (Wahlberg et al., 1996; Moilanen et al., 1998; Hanski, 2001; Thomas and Hanski, 2004). One limitation that remains is that these models do not include a description of local dynamics, hence the label *(stochastic) patch-occupancy models* (SPOM). Because the emphasis in SPOMs is in the dynamics of assemblages of many small local populations, they cannot profitably be applied to systems consisting of a small number of populations or populations of very large size (Baguette, 2004). Spatially realistic simulation models have been advocated as an efficient tool for the latter situations, and many applications of generic simulation models have appeared in the conservation literature (Akcakaya et al., 2004). Unfortunately, such models tend to be so complex that there are hardly ever sufficient data to estimate the model parameters properly, and the modeling that has been done seldom takes into account uncertainty in parameter values (Hanski, 2004; for an example that does incorporate uncertainty in parameter estimates see Dreschler et al., 2003).

In summary, the defining feature of the terrestrial systems in which spatially realistic metapopulation models have been successful is a highly fragmented landscape structure. The theory has been useful because such highly fragmented habitats are common in terrestrial systems, and they are becoming increasingly common because human land use practices tend to fragment habitats that were more continuous in the past. Apart from the quantitative analysis of real metapopulations referred to earlier, the theory has allowed more qualitative analysis of key phenomena, such as the extinction threshold for long-term persistence (Hanski and Ovaskainen, 2000) and the transient dynamics following perturbations (Ovaskainen and Hanski, 2002; Hanski and Ovaskainen, 2003).

B. A Framework for Asking Research Questions

Over the years, the perceived success of the metapopulation approach has invited applications to systems for which the value of the concept is less clear. Practically all populations have a patchy distribution at some spatial scale, making the metapopulation approach tempting, but patchy populations differ in many ways from each other. Two questions in particular are worthy of closer scrutiny here. First, to what extent is patchiness of populations attributable to patchiness of the physical environment? And second, at what spatial scale does patchiness occur?

Metapopulation theory is primarily a theory for cases when the spatial structure of populations is imposed by habitat patchiness. One of us has even viewed current spatially realistic metapopulation theory as an amalgamation of classic metapopulation theory and landscape ecology (Hanski, 2001; Hanski and Ovaskainen, 2003). If a species has a patchy population structure in the absence of obvious habitat patchiness, there must presumably be some other key factor or process that causes this spatial structure. In this case it would be most unlikely that fundamental understanding of the system could be gained by ignoring such a key process and applying theory that assumes habitat patchiness. Thus, one important message is that rather than focusing on the spatial patterns, one should focus on the processes that underlie those patterns (Hanski, 1999).

That being said, it is nonetheless clear that if the population is patchily distributed, for whatever reason, the patchy population structure has important implications for population dynamics, maintenance of genetic variation, evolutionary response to environmental change, and so forth. Thus, at one level, it is useful to recognize "metapopulation structure" in the sense of spatially structured population, regardless of the causes of that structure (which remains a key research question), because this leads to new questions.

Turning to the issue of spatial scale, metapopulation theory also requires that local populations are at least potentially connected via migration. If the patchiness is at such a large scale that different geographic populations are hardly at all connected to each other, these populations represent practically independent entities, and migration among them makes no real difference for their dynamics. Migration on a large spatial scale may be interesting and important, but it is not addressed by metapopulation theory.

The first questions to ask about metapopulations include what difference it makes that interactions are localized within local populations. Levins (1969, 1970) gave a partial answer for ecology with his model—an example of a so-called island model with an infinite number of identical patches and populations. Other examples of island models include Wright's (1931) population genetic island model and Maynard Smith's (1974) haystack model in evolutionary theory. The island models have allowed researchers to ask questions about the consequences of spatial population structure in the very simplest setting. More recently, researchers have asked questions about the role of dissimilar attributes of local populations in metapopulations (several chapters in Hanski and Gaggiotti, 2004). In ecology, these questions were asked partly in response to criticism of the simplicity of the Levins model (Harrison, 1991, 1994), and this led to the spatially realistic metapopulation theory described earlier. In population genetics, the island model has the unfortunate and unrealistic feature of equalizing the contributions of all local populations (demes) to the next generation, which greatly inflates the effective size of the metapopulation in comparison with what is likely to exist in reality (Whitlock, 2004). The question of effective metapopulation size has been posed but is still without a satisfactory answer. Research on source–sink population structures, when focused on the actual patterns of migration and gene flow among local populations, has opened a new avenue for research that touches, for example, the process of local adaptation (Kawecki, 2004). Observations about the demographic effects of inbreeding in small local populations have stimulated a new round of research on the significance of inbreeding for population persistence (for a review see Gaggiotti and Hanski, 2004). Even this short list of examples demonstrates that the metapopulation concept has stimulated a wide range of important research questions.

C. A Paradigm for Population and Conservation Biology

It is striking how suddenly metapopulation theory became adopted as a new conceptual and theoretical framework (paradigm) for conservation biology. Papers published in the early 1990s made it clear, either explicitly (Hanski, 1989; Merriam, 1991) or implicitly (Wilson, 1992; Noss, 1993), that the metapopulation concept had replaced the dynamic theory of island biogeography as the leading conservation paradigm. Hanski and Simberloff (1997) have examined the reasons for this "paradigm shift." One reason was the perception that island theory is an equilibrium theory whereas metapopulation theory is not. In reality there is no such difference: The core models by MacArthur and Wilson (1967) and Levins (1969) are special cases of a single, more general model (Hanski, 2001).

However, the primary reason why metapopulation theory appeared to provide a better framework for conservation than island theory was the absence of a mainland and mainland populations in the former. In island theory, no species that is present in the mainland pool will go extinct on an island, because there is always the possibility of recolonization from the mainland. Such a vision is not adequate for a fragmented landscape, from which species may and do go permanently extinct. Indeed, only metapopulation theory posed the key question about an extinction threshold, which depends both on the structure of landscapes and on the biological properties of species (Hanski, 1994).

D. Applications to Marine Populations

The three roles of the metapopulation concept in the study of terrestrial systems may all apply to marine systems: Metapopulation theory may provide an effective modeling tool to analyze the actual spatial population dynamics and structures of marine populations, metapopulation theory may be a helpful framework in which to ask new research questions, and metapopulation theory may represent a useful new paradigm.

Application of realistic metapopulation models to actual marine populations is dependent on the physical structure of the habitat, in the same manner as in terrestrial situations. Such applications are therefore restricted to species that have discrete local populations on shores or on the seafloor. Examples include populations of barnacles on rocky shores (Iwasa and Roughgarden, 1986), of crabs in estuaries (Botsford et al., 1994), of various demersal fishes (McQuinn, 1997; Stephenson, 1999), and of animals such as seals that occupy terrestrial breeding colonies but spend their time otherwise at sea. A recent study of the gray seal metapopulation in the Orkney Islands contributed not only to a better understanding of seal biology, but to metapopulation biology in general. Gaggiotti et al. (2002, Gaggiotti, 2004) have developed Bayesian methods to extract information in multilocus genotypes to infer patterns of migration of seals among breeding colonies. This work is especially valuable in showing how effectively to combine ecological (demographic) and genetic information in the same analysis.

Many marine species have highly dispersive juvenile stages, and hence dispersal and gene flow may be widespread (see the following section). In this feature, they appear different from many terrestrial populations. This need not restrict the usefulness of the metapopulation approach, because it can accommodate a broad range of rates of dispersal. However, in cases when dispersal is so extensive as to homogenize the demography of the set of local groups, it has nothing to tell that a conventional, spatially nonstructured, demographic analysis would reveal.

One obvious way in which metapopulation theory may help to guide marine ecological research concerns the question of patchiness. Patchiness is a common feature of the vast majority of both terrestrial and marine populations, although the cause of patchiness is often less clear for the latter. Although it is important to understand the causes of patchiness in marine just as in terrestrial systems, it is both valid and necessary to ask questions about the consequences of observed patchiness. A metapopulation approach provides one way to do this.

Finally, we see a genuine difference between terrestrial and marine environments when it comes to considering the nature of human impacts. The primary threat of human activities to terrestrial biodiversity, although clearly not the only one, is posed by habitat loss and fragmentation. Most habitats in modern landscapes in areas with high human population density are already highly fragmented, and becoming more so. It appears obvious that a theory that is specifically focused on the dynamics of species in such highly fragmented landscapes is not only useful, but essential, for research and management. The situation in marine environments is different. Although patchy loss of habitat, leaving remnant patches untouched, may result from effects of storm-tossed debris on rocky shores, or from crown-of-thorns starfish outbreaks or bleaching on coral reefs, each of these agencies can result in widespread degradation if extreme (Connell et al., 2004), and anthropogenic impacts on marine habitats generally result in degradation and homogenization of habitat across broad areas. Coastal pollution tends to spread out over large areas downstream of the source (Rabalais et al., 2002), and commercial trawling has been documented to simplify and homogenize the predominantly biogenic structure of benthic habitats over extensive areas of continental shelves (Thrush and Dayton, 2002).

Rather than any effects on habitat, the major human impact on marine systems is arguably overfishing (Reynolds et al., 2002). This reduces abundance and truncates age distributions of target and by-catch populations, and its effects vary spatially in correspondence with the spatial variation in fishing pressure. Although fishing pressure is rarely spatially homogeneous, certain types of management action (particularly, the use of protected areas or regional seasonal closures) are spatially explicit and serve to create or strengthen spatial variation in fishing pressure, with the result that populations come to be more patchily distributed, and more variable demographically, than they were before (Polunin, 2002). Thus, although patchiness of populations may be increasing in many marine populations, the reasons are largely different than in terrestrial systems. Therefore, the metapopulation paradigm, if and when it helps guide marine research, most likely will have distinct uses from those that serve in terrestrial systems. This also means that marine ecologists are likely to make their own novel contributions to the development of metapopulation concepts and theory.

III. HISTORY AND EFFECTS OF PREDOMINANT RESEARCH QUESTIONS IN MARINE ECOLOGY

Marine ecology seems always to have been the younger sibling of terrestrial ecology. Although marine ecologists have appreciated the substantial differences between the marine and the terrestrial environment, and their consequences for ecology, terrestrial ecology was always there to offer the tested paradigms. Fisheries science, and chemical and physical oceanography all contribute ideas and approaches, but terrestrial paradigms have prevailed, particularly for those aspects of marine ecology that concern populations and communities. Given this pattern, it was perhaps inevitable that marine ecologists would incorporate metapopulation theory into their work, but do so after its embrace by their terrestrial colleagues. What is interesting in exploring the progress of marine ecology, is to see how and why terrestrial concepts become incorporated into the marine context. Our adoption of a metapopulation paradigm grew out of an enthusiastic rediscovery of the powerful effects on populations of larval dispersal and subsequent recruitment.

The early days of marine ecology, a half century or more ago, were a time for detailed descriptive studies that cataloged what species lived where, their characteristics, their trophic interactions, and the circulation of energy and materials through their ecosystems. Studying marine systems called for innovative methods that usually involved machinery to grab samples and haul them into boats, where the ecologists would examine what was caught and try and make sense of what it might have been doing before it was abruptly interrupted. During this period, ecologists probably developed a greater appreciation for the unique properties of marine systems than was evident later on in what might be called the *hypothesis-testing phase* of marine ecology. This began with hesitant steps early in the 20th century, and really came of age in the 1970s and 1980s.

Doing proper, hypothesis-testing science requires that you have some ability to manipulate the system being studied. To do such science in the marine environment initially required a retreat to its edges—those parts of the marine world that were more or less accessible to the scientist. Thus we find that the science of marine ecology is strongly rooted on the intertidal rocky shore; in the shallow sub- and intertidal mud flats, salt marshes, and estuaries; and on rocky and coral reefs. Some brave souls persisted in studying plankton and neuston of the open ocean, but it was virtually impossible to do experiments out there, and the taxonomy was horrendous. Even today, the gap (as reflected in patterns of citation and conferences attended) between the "blue water" biological oceanographers and the coastal marine ecologists is arguably larger than the gap between the latter and terrestrial ecologists. The development of the science of marine ecology during the last 50 years has been driven from the edges, as is well illustrated by a recent text on marine ecology (Bertness et al., 2001) that, in its eight chapters (of 19 total) dealing with "community types," deals exclusively with inshore and benthic systems.

Two of the most profound features of the ocean are its vastness and its continual motion. Even drifting passively, marine organisms are able to travel long distances during relatively short periods of time. Ignoring for the moment the creatures that occupy the pelagic zone throughout their lives, those demersal, and sessile plants and animals that live on firm substrata are overwhelmingly organisms with pelagic dispersive (usually egg or larval) stages. As a result, local populations, occupying particular shores, bays, estuaries, or reefs, can be extensively interconnected by the dispersal of their propagules. Following pelagic life, larvae settle to suitable substrata and are recruited to populations of juveniles and adults living there. The importance of this fact of life was clearly recognized by ecologists studying marine communities at the mid century (e.g., Scheltema, 1974; Thorson 1950, 1966), but it was largely neglected by those who interpreted observations and experimental results on demersal and often sessile organisms using the concepts of terrestrial ecology. During the '60s and '70s, the prevailing assumption was that larvae were always available to replenish populations.

Intertidal ecology moved from description to experimentation, with the pronounced zonation on shores acting as a major driver. This universal and conspicuous pattern was initially interpreted using tolerances of individual species to emersion and submersion. When that did not suffice (because too many well-zoned species could tolerate conditions far more extreme than any they would experience on a shore), the competitive and predatory interactions among them became the chief focus of attention (Underwood and Denley, 1984). Some of the finest field experimentation anywhere has been that done to explore the interactions of plants and animals of the rocky shore. Yet most of it dealt with the sessile and sedentary organisms easy to find stranded at low tide, as if they existed in nicely contained local communities, and it ignored, until the mid 1980s, the fact that local "populations" of barnacles, mussels, or whelks were the product of large numbers of individual settlement events bringing single larvae to that shore from places outside (at least on the scale of tens of meters at which the ecological studies were done; Morgan, 2001).

A similar pattern of maturation of the science can be seen in coral reef studies (where zonation was also an early fascination), and in studies of subtidal mud flats, sea grass beds, mangroves, and other coastal marine environments. Open-water studies evolved very differently, and never had the impact on the wider science of ecology, perhaps because experimentation was so much more difficult. Deep-sea studies evolved slowly because of the technical difficulties and expense, but included experiments and focused on local biotic interactions in much the same way that intertidal studies did (Etter and Mullineaux, 2001). In the 1960s and '70s, to pose experimentally testable hypotheses, marine ecologists worked with demersal and sessile species, and studied processes that took place on very local scales—often scales of centimeters. The vastness and continual motion of the oceans were fine for sunset watching at the end of the day, but they did not contribute to the science.

Classic field experiments that had profound impacts on the development of community ecology examined the competitive interactions of barnacles of different species (e.g., Connell, 1961a,b, 1978), and the predatory interactions of mussels, whelks, and starfishes (e.g., Paine, 1966; Dayton, 1971). Superficially simple but important concepts, such as escapes in space or in growth, or that of disturbance as an agent to maintain community structure, came from such studies. More complex studies involving urchins, kelp, and sea otter (e.g., Estes and Duggins, 1995), or the creatures that colonize fouling plates (e.g., Sutherland, 1974) gave us alternate stable states, and the related idea that ecological changes could be sudden phase shifts as well as gradual successional processes. Yet it was not until late in the 20th century that marine ecologists remembered that profoundly important feature of virtually all the organisms they studied—pelagic dispersal of larval stages and subsequent settlement to juvenile and adult demersal and sessile populations.

The idea that recruitment of dispersive larvae to populations of sessile or sedentary juveniles and adults was an important ecological process was "discovered" virtually simultaneously by marine ecologists working on rocky shores and on coral reefs. On the rocky shores, Dayton's (1971) early work, which drew attention to the importance of disturbance to maintaining community structure, indirectly recognized that it was settlement and recruitment of larval stages that led the recovery from disturbances. In an important review, Connell (1985) pointed to the likelihood that recruitment processes would be found to be quite variable, and play a role in the development and maintenance of intertidal communities. However, it was not until Caffey's (1985) work that a serious effort to document spatial and temporal patterns of settlement or recruitment of intertidal organisms was undertaken. Note that it has been traditional in marine ecology to measure recruitment very close to the time of settlement from the pelagic stage, rather than later in life, as is more typical of fisheries science, when the interest is in recruitment to the catchable population—typically a size-related event.

Caffey set out to quantify spatial and temporal patterns in settlement and in recruitment about one month after settlement for the barnacle *Tesseropera rosea* on a series of rocky shores extending over 500 km along the east coast of Australia. His nested design demonstrated profound (order of magnitude) spatial

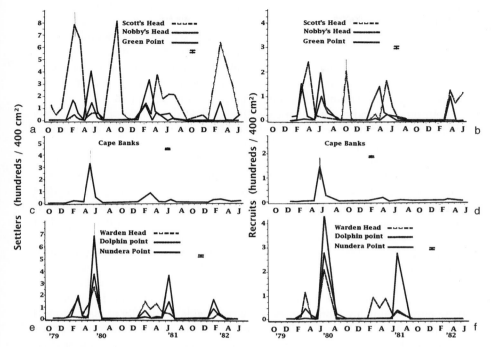

FIGURE 1-2. Geographical and temporal variation in patterns of settlement and recruitment of the barnacle *Tesseropora rosea* on rocky shores of New South Wales, Australia. Settlement was defined as the density of animals 30 days old or less at sampling; recruitment was density of animals 30 to 60 days old at sampling. Graphs represent the mean rates of settlement (over 24 replicates) and recruitment (over 12 replicates) at each shore. (Figure redrawn from Caffey, 1985, and with permission of the Ecological Society of America.)

variations in settlement on scales of less than 3 m, 20 to 50 m, and 10 to 500 km, and very considerable variance in the relationship between settlement and recruitment one month later (Fig. 1-2). His study also demonstrated large differences in the spatial patterns in two successive years. Other intertidal researchers had begun to monitor recruitment and settlement on local scales, and over short time periods, but Caffey's work was the first to show the rich patterns in these processes for an intertidal organism. Nearly simultaneously, and continuing over several more years, Roughgarden et al. (1985, 1988) demonstrated equally complex patterns in barnacle recruitment on the California and Oregon coasts, and they were able to relate some of this variability to regional-scale processes, particularly the effects of wind-induced coastal upwellings.

On coral reefs, Sale (1977), in proposing lottery competition as the mechanism permitting coexistence of territorial damselfishes, had argued for the importance of, but had not attempted to quantify, variability in settlement. However, Williams and Sale (1981) monitored the settlement of a broad range of fish species to coral colonies of the same type that had been set out in four grids on the shallow sandy floor of a lagoon about 2 km in diameter. Grids of corals were about 1 km apart, and daily collections of all fish settling to them were made over two successive summer settlement seasons. The results showed pronounced variation among sites, within and between years, and among species.

Both on reefs and in intertidal areas there followed a period of intense study of settlement and recruitment dynamics using a variety of species of organism. The fact that recruitment was highly variable in time and in space, and seemingly on many scales, was confirmed numerous times, and Hjort (1914) was suddenly being cited with enthusiasm by marine ecologists. Throughout these investigations, there has been a broadly accepted appreciation that the larvae settling to local sites have been supplied at least partially, and perhaps substantially, by populations located elsewhere. Local marine populations are open, and their recruitment is subsidized by neighboring populations.

If documenting the variability in recruitment was easy, developing explanations for that variability has been far more difficult. Roughgarden's (Roughgarden et al., 1988) early success in explaining a considerable portion of recruitment variation of barnacles has been repeated for other organisms and locations, but there have been many more studies that were unable to identify causal factors of any importance (Caley et al., 1996). The reason for this appears to be that the process of larval dispersal and subsequent settlement is a strongly biophysical process, with multiple interacting causes, in all except those species with behaviorally simple larvae that are pelagic for only hours. The movement of water is complex at any but the global scale, particularly close to shores and substrata where the demersal and sessile species live, and it varies with time as well as space. The larvae may well be passively transported when very young, but they undergo rapid development and are frequently highly adapted for their pelagic existence. Fish and crustacea, in particular, may be pelagic for months, and are very capable pelagic organisms by the end of this period. To understand the path followed by a long-lived larva, it is necessary to have detailed knowledge of a complex and variable hydrodynamics, and of a set of sensory and behavioral skills that change as larval life progresses (Sale and Kritzer, 2003). One consequence of this difficulty in specifying causes of recruitment variation has been an inability to specify either the spatial scale on which local populations are interconnected or the extent of the subsidy of recruitment to each local population as a result of dispersal from other sources.

Recognition that openness and larval dispersal play major roles in the ecology of marine populations and communities leads logically to the consideration of metapopulation ideas. Perhaps because marine ecologists have tended to focus their studies on quite local populations and communities, executing experimental treatments on replicate 1-m² plots or replicate small boulders on rocky shores or replicate small patch reefs or coral heads on coral reefs, attention to metapopulation concepts was slow to develop. Such attention began during the 1980s with the rediscovery of recruitment dynamics and a move to larger spatial scales. Roughgarden et al. (1985, 1988) sampled barnacles over approximately 100 km of the central coast of California. Caffey (1985) did the same across 500 km of Australian coasts. Doherty and colleagues (Doherty and Williams, 1988; Fowler et al., 1992) sampled reef fish recruitment at sites on sets of reefs spread across 1000 km of the Great Barrier Reef. Hughes et al. (2002) documented recruitment dynamics and adult population sizes for corals across the same 1000-km range. Each of these studies has concerned recruitment dynamics primarily, and population structure secondarily, if at all. But each provides data consistent with a metapopulation paradigm, and Roughgarden and Iwasa (1986, Iwasa and Roughgarden, 1986) were the first to use a metapopulation model explicitly. Although marine ecologists currently have trouble applying metapopulation theory, chiefly because of the difficulty of quantifying the larval dispersal that provides the connectivity among local populations, the value of the metapopulation paradigm is widely recognized, and the need to develop techniques for measuring connectivity among marine populations is broadly recognized as a high priority for this science (Kritzer and Sale, 2004; Sale, 2004).

IV. CONSERVATION AND FISHERIES MANAGEMENT INFLUENCES ON MARINE ECOLOGY

A. Scale of Study

The preceding section documents shifts in the focal questions of academic marine research from within-patch dynamics that occur in isolation from other populations to within-patch dynamics influenced by replenishment from unidentified external sources, to demographic linkages among distinct populations. Concurrently, the spatial scale of study has increased to often oceanwide scales (e.g., Hughes et al., 2002; Planes, 2002; Mora et al., 2003). In contrast, fisheries science has long adopted very large-scale perspectives. For example, the broad geographic scope of Sinclair's (1988) stimulating essay on population ecology of northern temperate marine fisheries species is comparable to that adopted 20 years earlier by Harden Jones (1968) in his classic text on fish migration.

The historical large-scale approach of fisheries science is likely due in part to the large geographic areas that fisheries management agencies are charged with overseeing, as well as the larger areas from which scientists can obtain data through fishery-dependent sampling. At any rate, the metapopulation concept is fundamentally about ecology at large scales, and the larger geographic scope of fisheries science seems to have created a research climate conducive to adopting metapopulation theory once that theory had found its way into marine science. For example, Shepherd and Brown (1993) called for metapopulation thinking in managing abalone fisheries several years before the documented increase in general marine metapopulation references (Grimm et al., 2003). Also, Botsford and coworkers' studies of commercially harvested crab stocks were among the first to make use of the metapopulation concept in a marine context (Botsford, 1995; Botsford et al., 1994, 1998; Wing et al., 1998; and Chapter 9). Of course, the earliest marine metapopulation studies were not conducted in a fisheries context (Iwasa and Roughgarden, 1986; Roughgarden and Iwasa, 1986), but those are the exceptions rather than the rule. Although fisheries science and academic marine ecology have often operated across an unnecessary divide (Sale, 2002b), the influence of metapopulation approaches in fisheries on broader marine ecological research must be acknowledged.

B. Spatial Resolution

The historically larger focal scale of fisheries science certainly helped facilitate incorporation of a large-scale approach such as metapopulation ecology, but fisheries science had held that broad-scale perspective for many years before metapopulation theory was invoked explicitly. This is due in part to the fact that metapopulation ecology is not simply about ecology at large spatial scales, but also considers in detail the spatial arrangement of and relationships among local populations.

Fisheries science initially tended to downplay consideration of smaller scale structure among local populations. Although Atlantic herring biologists have examined spatial patterning related to migration and the distribution of herring "races" since the mid 1900s (reviewed by McQuinn, 1997), comparable attention to other species has been slower to emerge. However, the 1980s saw a groundswell of studies examining stock structure of a variety of marine fisheries resources, particularly the identification of substocks within a larger management area (e.g., Fairbairn, 1981; Shaklee et al., 1983; Grant, 1985; Oxenford and Hunte, 1986; Cooper and Mangel, 1999). Investigation of stock structure has now become a standard component of fisheries science (Begg et al., 1999). Molecular genetics has been a prominent tool in this area, but morphometric, life history, and other approaches have also played important roles (Begg and Waldman, 1999).

Initially, resolving spatial structure of marine fish stocks was primarily a descriptive exercise in distinguishing between discrete spawning units. Now, there is often greater attention to the dynamics within and interactions among distinct substocks, which are akin to local populations in metapopulation terminology and therefore invite a metapopulation perspective. For example, McQuinn (1997) reviewed a wide range of data on genetics, morphometrics, life history traits, and population dynamics of Atlantic herring. He then developed a metapopulation model that explains how seasonally distinct spawning populations exchange individuals as a consequence of within-cohort variability in development rates, yet retain their characteristic behavioral traits through social transmission of behaviors to immigrants from other spawning populations. The utility of merging dynamic ecological processes with spatial patterns in distribution and migration has invoked calls for increased application of metapopulation theory to fisheries science (Stephenson, 1999; Frank and Brickman, 2000).

The spatial arrangement and interrelationships of local populations in metapopulation ecology (or of substocks in fisheries science) is necessarily interwoven with the structure and distribution of marine habitats that support those populations. In the United States, the amendment of the Magnuson-Stevens Fishery Conservation and Management Act that introduced the need to identify and protect "essential fish habitat" as a management responsibility has spurred an increase in attention to the topic in the latter half of the 1990s. Recognition of essential habitat has implications for regulating human activities that might affect adjacent marine areas (e.g., Dewey, 2000) and for planning restoration efforts (e.g., Coen et al., 1999). This focus on habitat is necessarily spatially explicit and perhaps has helped direct interest to the substock structure within fishery stocks.

The most widespread application of essential fish habitat research is direct protection of that habitat through use of marine protected areas (MPAs), or areas permanently protected from development or extractive activities (e.g., Conover et al., 2000, and associated papers). Attention to essential fish habitat may not have had a direct influence on the integration of metapopulation ecology and marine ecology; however, it has at least had indirect influence through fostering MPA research, which is probably the largest single contribution of marine management science to the integration of metapopulation ecology and marine ecology.

C. Marine Protected Areas

The potential benefits and optimal design of MPAs are arguably the most actively researched and vigorously debated topic in contemporary marine conservation biology and fisheries science (Polunin, 2002; Lubchenco et al., 2003; Willis et al., 2003; Sale et al., 2005). MPAs are not a new concept. Beverton and Holt (1957) discussed area closures in fisheries management in their classic monograph, and MPAs have existed in many locations across the globe well before the explosion in their popularity since 1990. However, MPA literature prior to 1990 was focused largely on the role of MPAs in protecting pockets of biodiversity, isolated from human activity. For example, Polunin et al. (1983) and Minchin (1987) report inventories of organisms found within MPAs in Bali and Ireland respectively, and Salm (1984) offered guidelines of MPA locations that should ensure they contain representative biodiversity. This objective of sequestering biodiversity from human impacts does require attention to spatial ecology, but it utilizes the static equilibrium perspective that informed terrestrial SLOSS debates (Soulé and Simberloff, 1986; Simberloff, 1988) along with a classical view that emphasizes the role of internal biotic interactions rather than dispersal across boundaries in maintaining community structure. It does not call upon the perspective of a dynamic, interconnected patch network that characterizes metapopulation ecology.

As discussed earlier, renewed attention by ecologists in the 1980s to patterns of recruitment variability focused both on consequences (e.g., Iwasa and Roughgarden, 1986; Roughgarden and Iwasa, 1986; Sale, 1988; Victor, 1983), and causes, including larval production, dispersal, and pelagic survival processes (Underwood and Fairweather, 1989; Sale, 1990). This change brought "pure" marine ecology closer to the domain of fisheries science (Doherty, 1991; Rothschild, 1998). The shift also led Fairweather (1991) to consider the implications of larval production, dispersal, and recruitment patterns for management using MPAs, and Roberts and Polunin (1991) to call for increased attention to the services MPAs can provide to unprotected areas outside their borders. Around the same time, Polacheck (1990) and DeMartini (1993) began to model small-scale linkages between MPAs and adjacent fished areas through adult movement. A perspective that views marine populations as grouped into metapopulations has now become well established (Fig. 1-3). Man et al. (1995) provided the first marine fisheries model explicitly couched within the metapopulation framework. Numerous other marine metapopulation modeling studies exploring fisheries management issues soon followed (Supriatna and Possingham, 1998; Cooper and Mangel, 1999; Crowder et al., 2000; Tuck and Possingham, 2000; Lipcius et al., 2001; Sanchirico and Wilen, 2001; Lockwood et al., 2002; Smedbol and Wroblewski, 2002; Wilen et al., 2002). Mathematical models are important stimuli for development of both empirical research and theory, and this growing body of literature addressing metapopulation models to MPAs has no doubt provided important stimuli for metapopulation-oriented marine research.

V. SUMMARY

Marine ecology overwhelmingly concerns those benthic or demersal populations that are found in association with solid substrata. Such populations are frequently distributed patchily, along with the habitats they occupy. Our knowledge of pelagic populations is less extensive, and although they also appear to be

Chapter 1 The Merging of Metapopulation Theory and Marine Ecology

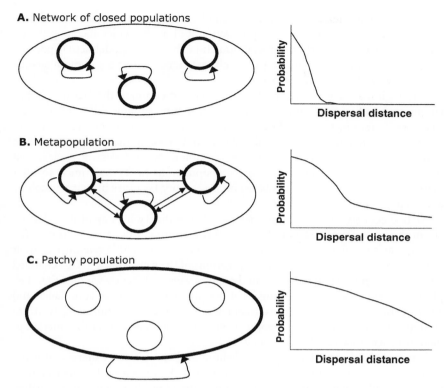

FIGURE 1-3. The way in which the extent of larval dispersal, relative to interpopulation spacing, determines the nature of the linkages among a set of local populations. These variously form a set of independent, closed populations (A); a metapopulation (B); or a single, large, but spatially subdivided population (C). Diagram from Kritzer and Sale (2004) with permission of Blackwell Publishing Ltd.

patchily distributed, it is unclear to what extent habitat factors usually drive this patchiness.

Although it might be logical for marine ecology and fisheries management science to interact closely, there has been a long history of marine ecology borrowing concepts and theories from terrestrial ecology. The recent adoption of metapopulation theory is a case in point.

Metapopulation theory, in its terrestrial context, is highly relevant to the description and exploration of sets of small populations scattered over small patches of suitable habitat—a condition that applies to a broad and growing range of species as human impacts continue to reduce and to subdivide patches of formerly more contiguous habitat. It deals explicitly with the dynamics of small populations that interact through dispersal of individuals among them, and considers dynamics both at the local population and at the wider metapopulation scale. Terrestrial ecologists use metapopulation theory to model actual populations, as a framework for guiding research questions, and as a paradigm against which to view population ecology.

Although marine and terrestrial systems have in common the fact that populations are patchily distributed, it is clear that marine systems differ from terrestrial ones. In marine systems, the causes of the patchy distributions are not always so clearly related to patchiness of habitat, and most human impacts tend to reduce, rather than increase, patchiness of habitat. Nevertheless, it is clear that there are many situations in which metapopulation theory has much to offer marine ecology. Time will tell how useful this theory becomes in marine ecology and management.

Marine ecology came to adopt metapopulation theory by way of its rediscovery of the importance of larval dispersal and recruitment dynamics. Prior to the 1980s, marine ecologists examined local populations with little reference to their interactions with nearby or distant populations of the same species. This changed with the recognition that recruitment dynamics were sufficiently variable that they played a major role in structuring local populations in terms of abundance and age distribution. Given that recruitment was perceived as coming largely, if not entirely, from outside the local population, it was perhaps logical that metapopulation concepts would be embraced.

The worldwide trend toward overfishing and the loss of fisheries, and the growing interest in marine conservation, have led to a search for management tools that will protect marine species, and particularly those with fishery value. The use of spatially explicit management strategies, such as the use of no-take marine fishery reserves and other types of MPA is being widely advocated and enthusiastically endorsed at a time when marine ecologists are becoming interested in the use of metapopulation theory to model the populations with which they work. As a consequence, metapopulation theory is now being applied frequently to marine systems. The following chapters illustrate the current "state of play" and provide glimpses of the likely future.

REFERENCES

Akcakaya, H.R., Burgman, M.A., Kindvall, O., Wood, C.C., Sjögren–Gulve, P., Hatfield, J.S., and McCarthy, M.A., eds. (2004). *Species conservation and management: Case studies.* Oxford: Oxford University Press.

Andrewartha, H.G., and Birch, L.C. (1954). *The distribution and abundance of animals.* Chicago: University of Chicago Press.

Baguette, M. (2004). The classical metapopulation theory and the real, natural world: A critical appraisal. *Basic Appl. Ecol.* 5, 213–224.

Begg, G.A., Friedland, K.D., and Pearce, J.B. (1999). Stock identification and its role in stock assessment and fisheries management: An overview. *Fish. Res.* **43**, 1–8.
Begg, G.A., and Waldman, J.R. (1999). An holistic approach to fish stock identification. *Fish. Res.* **43**, 35–44.
Bertness, M.D., Gaines, S.D., and Hay, M.E. (2001). *Marine community ecology.* Sunderland, MA: Sinauer Associates.
Beverton, R.J.H., and Holt, S.J. (1957). *On the dynamics of exploited fish populations.* London: H.M. Stationery Office.
Botsford, L.W. (1995). Population dynamics of spatially distributed, meroplanktonic, exploited marine invertebrates. *ICES Mar. Sci. Symp.* **199**, 118–128.
Botsford, L.W., Moloney, C.L., Hastings, A., Largier, J.L., Powell, T.M., Higgins, K., and Quinn, J.F. (1994). The influence of spatially and temporally varying oceanographic conditions on meroplanktonic metapopulations. *Deep-sea Res.* **41**, 107–145.
Botsford, L.W., Moloney, C.L., Largier, J.L., and Hastings, A. (1998). Metapopulation dynamics of meroplanktonic invertebrates: The Dungeness crab (*Cancer magister*) as an example. *Can. Spec. Publ. Fish. Aquat. Sci.* **125**, 295–306.
Caffey, H.M. (1985). Spatial and temporal variation in settlement and recruitment of intertidal barnacles. *Ecol. Monogr.* **55**, 313–332.
Caley, M.J., Carr, M.H., Hixon, M.A., Hughes, T.P., Jones, G.P., and Menge, B.A. (1996). Recruitment and the local dynamics of open marine populations. *Annu. Rev. Ecol. Syst.* **27**, 477–500.
Coen, L.D., Luckenbach, M.W., and Breitburg, D.L. (1999). The role of oyster reefs as essential fish habitat: A review of current knowledge and some new perspectives. *Am. Fish. Soc. Symp.* **22**, 438–454.
Connell, J.H. (1961a). The effects of competition, predation by *Thais lapillus*, and other factors on natural populations of the barnacle, *Balanus balanoides*. *Ecol. Monogr.* **31**, 61–104.
Connell, J.H. (1961b). The influence of interspecific competition and other factors on the distribution of the barnacle *Chthamalus stellatus*. *Ecology.* **42**, 710–723.
Connell, J.H. (1978). Diversity in tropical rain forests and coral reefs. *Science.* **199**, 1302–1310.
Connell, J.H. (1985). The consequences of variation in initial settlement vs. postsettlement mortality in rocky intertidal communities. *J. Exp. Mar. Biol. Ecol.* **93**, 11–45.
Connell, J.H., Hughes, T.P., Wallace, C.C., Tanner, J.E., Harms, K.E., and Kerr, A.M. (2004). A long-term study of competition and diversity of corals. *Ecol. Monogr.* **74**, 179–210.
Conover, D.O., Travis, J., and Coleman, F.C. (2000). Essential fish habitat and marine reserves: An introduction to the second Mote Symposium in Fisheries Ecology. *Bull. Mar. Sci.* **66**, 527–534.
Cooper, A.B., and Mangel, M. (1999). The dangers of ignoring metapopulation structure for the conservation of salmonids. *Fish. Bull.* **97**, 213–226.
Crowder, L.B., Lyman, S.J., Figueria, W.F., and Priddy J. (2000). Source–sink population dynamics and the problem of siting marine reserves. *Bull. Mar. Sci.* **66**, 799–820.
Dayton, P.K. (1971). Competition, disturbance and community organization: the provision and subsequent utilization of space in a rocky intertidal community. *Ecol. Monogr.* **41**, 351–389.
DeMartini, E.E. (1993). Modeling the potential of fishery reserves for managing Pacific coral reef fishes. *Fish. Bull.* **91**, 414–427.
Dewey, W.F. (2000). Endangered species act and sustainable fisheries act implications for molluscan shellfish culture management. *J. Shellfish Res.* **19**, 628.
Diamond, J. (1997). *Guns, germs, and steel.* New York: W.W. Norton.
Doherty, P.J. (1991). Spatial and temporal patterns in recruitment. In *The ecology of fishes on coral reefs* (P.F. Sale, ed., pp. 261–293). San Diego: Academic Press.
Doherty, P.J., and Williams, D.McB. (1988). The replenishment of coral reef fish populations. *Oceanogr. Mar. Biol. Annu. Rev.* **26**, 487–551.

Drechsler, M., Frank, K., Hanski, I., O'Hara, R.B., and Wissel, C. (2003). Ranking metapopulation extinction risk: From patterns in data to conservation management decisions. *Ecol. Appl.* **13**, 990–998.

Estes, J.A., and Duggins, D.O. (1995). Sea otters and kelp forests in Alaska: Generality and variation in a community ecological paradigm. *Ecol. Monogr.* **65**, 75–100.

Etienne, R.S., van ter Braaka, C.J.F., and Vos, C.C. (2004). Application of stochastic patch occupancy models to real metapopulations. In *Ecology, genetics, and evolution of metapopulations* (I. Hanski and O.E. Gaggiotti, eds., pp. 105–132). Amsterdam: Elsevier Academic Press.

Etter, R.J., and Mullineaux, L.S. (2001). Deep-sea communities. In *Marine community ecology* (M.D. Bertness, S.D. Gaines, and M.E. Hay, eds., pp. 367–393). Sunderland, MA: Sinauer Associates.

Fairbairn, D.J. (1981). Which witch is which? A study of the stock structure of witch flounder (*Glyptocephalus cynoglossus*) in the Newfoundland region. *Can. J. Fish. Aquat. Sci.* **38**, 782–794.

Fairweather, P.G. (1991). Implications of "supply-side" ecology for environmental assessment and management. *Trends Ecol. Evol.* **6**, 60–63.

Figueira, W.F. (2002). *Metapopulation dynamics of coral reef fish: understanding habitat, demography and connectivity in source-sink systems.* PhD dissertation. Durham, North Carolina: Duke University.

Finlay, B.J. (2002). Global dispersal of free-living microbial eukaryote species. *Science.* **296**, 1061–1063.

Finlay, B.J., and Clarke, K.J. (1999). Ubiquitous dispersal of microbial species. *Nature.* **400**, 828.

Fowler, A.J., Doherty, P.J., and Williams, D.McB. (1992). Multiscale analysis of recruitment of a coral reef fish on the GBR. *Mar. Ecol. Prog. Ser.* **82**, 131–141.

Frank, K.T., and Brickman, D. (2000). Allee effects and compensatory population dynamics within a stock complex. *Can. J. Fish. Aquat. Sci.* **57**, 513–517.

Gaggiotti, O.E. (2004). Multilocus genotype methods for the study of metapopulation processes. In *Ecology, genetics, and evolution of metapopulations* (I. Hanski and O.E. Gaggiotti, eds., pp. 367–386). Amsterdam: Elsevier Academic Press.

Gaggiotti, O., and Hanski, I. (2004). Mechanisms of population extinction. In *Ecology, genetics, and evolution of metapopulations* (I. Hanski and O. Gaggiotti, eds., pp. 337–366). Amsterdam: Elsevier Academic Press.

Gaggiotti, O.E., Jones, F., Lee, W.M., Amos, W., Harwood, J., and Nichols, R.A. (2002). Patterns of colonization in a metapopulation of grey seals. *Nature.* **416**, 424–427.

Grant, W.S. (1985). Biochemical genetic stock structure of the southern African anchovy, *Engraulis capensis* Gilchrist. *J. Fish. Biol.* **27**, 23–29.

Grimm, V., Reise, K., and Strasser, M. (2003). Marine metapopulations: A useful concept? *Helgoland Mar. Res.* **56**, 222–228.

Hanski, I. (1989). Metapopulation dynamics: Does it help to have more of the same? *Trends Ecol. Evol.* **4**, 113–114.

Hanski, I. (1994). A practical model of metapopulation dynamics. *J. Anim. Ecol.* **63**, 151–162.

Hanski, I. (1999). *Metapopulation ecology.* Oxford: Oxford University Press.

Hanski, I. (2001). Spatially realistic theory of metapopulation ecology. *Naturwissenschaften.* **88**, 372–381.

Hanski, I. (2004). Metapopulation theory, its use and misuse. *Basic Appl. Ecol.* **5**, 225–229.

Hanski, I., and Gaggiotti, O.E., eds. (2004). *Ecology, genetics, and evolution of metapopulations.* Amsterdam: Elsevier Academic Press.

Hanski, I., and Ovaskainen, O. (2000). The metapopulation capacity of a fragmented landscape. *Nature.* **404**, 755–758.

Hanski, I., and Ovaskainen, O. (2003). Metapopulation theory for fragmented landscapes. *Theor. Popul. Biol.* **64**, 119–127.

Hanski, I., and Simberloff, D. (1997). The metapopulation approach, its history, conceptual domain and application to conservation. In *Metapopulation biology: Ecology, genetics, and evolution* (I. Hanski and M.E. Gilpin, eds., pp. 5–26). San Diego: Academic Press.

Harden Jones, F.R. (1968). *Fish migration*. London: Edward Arnold.
Harrison, S. (1991). Local extinction in a metapopulation context: An empirical evaluation. *Biol. J. Linn. Soc.* **42**, 73–88.
Harrison, S. (1994). Metapopulations and conservation. In *Large-scale ecology and conservation biology* (P.J. Edwards, R.M. May, and N.R. Webb, eds., pp. 111–128). Oxford: Blackwell Scientific Press.
Hjört, J. (1914). Fluctuation in the great fisheries of northern Europe reviewed in the light of biological research. *Rapp. P.-V. Reun., Cons. Perm. Int. Explor. Mer.* **20**, 1–228.
Hughes, T.P., Baird, A.H., Dinsdale, E.A., Moltschaniwskyj, N.A., Pratchett, M.S., Tanner, J.E., Willis, B.L., and Harriott, V.J. (2002). Detecting regional variation using meta-analysis and large-scale sampling: latitudinal patterns in recruitment. *Ecology.* **83**, 436–451.
Iwasa, H., and Roughgarden, J. (1986). Interspecific competition among metapopulations with space-limited subpopulations. *Theor. Popul. Biol.* **30**, 194–213.
Kawecki, T.J. (2004). Ecological end evolutionary consequences of source–sink population dynamics. In *Ecology, genetics, and evolution of metapopulations* (I. Hanski and O.E. Gaggiotti, eds., pp. 387–414). Amsterdam: Elsevier Academic Press.
Kritzer J.P., and Sale, P.F. (2004). Metapopulation ecology in the sea: From Levins' model to marine ecology and fisheries science. *Fish and Fisheries.* **5**, 131–140.
Levins, R. (1969). Some demographic and genetic consequences of environmental heterogeneity for biological control. *Bull. Entomol. Soc. Am.* **15**, 237–240.
Levins, R. (1970). Extinction. In *Some mathematical problems in biology* (M. Desternhaber, ed., pp. 77–107). Providence, RI: American Mathematical Society.
Lipcius, R.N., Stockhausen, W.T., and Eggleston, D.B. (2001). Marine reserves for Caribbean spiny lobster: Empirical evaluation and theoretical metapopulation recruitment dynamics. *Mar. Freshwat. Res.* **52**, 1589–1598.
Lockwood, D.R., Hastings, A., and Botsford, L.W. (2002). The effects of dispersal patterns on marine reserves: Does the tail wag the dog? *Theor. Popul. Biol.* **61**, 297–309.
Lubchenco, J., Palumbi, S.R., Gaines, S.D., and Andelman, S. (2003). Plugging a hole in the ocean: The emerging science of marine reserves. *Ecol. Appl.* **13**, S3–S7.
MacArthur, R.H., and Wilson, E.O. (1967). *The theory of island biogeography*. Princeton, NJ: Princeton University Press.
Man, A., Law, R., and Polunin, N.V.C. (1995). Role of marine reserves in recruitment to reef fisheries: A metapopulation model. *Biol. Cons.* **71**, 197–204.
Maynard Smith, J. (1974). *Models in ecology*. Cambridge: Cambridge University Press.
McQuinn, I.H. (1997). Metapopulations and the Atlantic herring. *Rev. Fish Biol. Fish.* **7**, 297–329.
Merriam, G. (1991). Corridors and connectivity: Animal populations in heterogenous environments. In *Nature conservation 2: The role of corridors* (D.A. Saunders and R.J. Hobbs, eds., pp. 133–142). Chipping Norton, Australia: Surrey Beatty & Sons.
Minchin, D. (1987). Fishes of the Lough Hyne marine reserve. *J. Fish Biol.* **31**, 343–352.
Moilanen, A., Smith, A.T., and Hanski, I. (1998). Long-term dynamics in a metapopulation of the American pika. *Am. Nat.* **152**, 530–542.
Mora, C., Chittaro, P.M., Sale, P.F., Kritzer J.P., and Ludsin, S.A. (2003). Patterns and processes in reef fish diversity. *Nature.* **421**, 933–936.
Morgan, S.G. (2001). The larval ecology of marine communities. In *Marine community ecology* (M.D. Bertness, S.D. Gaines, and M.E. Hay, eds., pp. 159–181). Sunderland, MA: Sinauer Associates.
Murphy, D.D., Freas, K.E., and Weiss, S.B. (1990). An environment-metapopulation approach to population viability analysis for a threatened invertebrate. *Conserv. Biol.* **4**, 41–51.
Noss, R.F. (1993). Wildlife corridors. In *Ecology of greenways* (D.S. Smith & P.C. Hellmund, eds., pp. 43–68). Minneapolis: University of Minnesota Press.
Ovaskainen, O., and Hanski, I. (2002). Transient dynamics in metapopulation response to perturbation. *Theor. Popul. Biol.* **61**, 285–295.

Ovaskainen, O., and Hanski, I. (2004). Metapopulation dynamics in highly fragmented landscapes. In *Ecology, genetics, and evolution of metapopulations* (I. Hanski & O.E. Gaggiotti, eds., pp. 73–104). Amsterdam: Elsevier Academic Press.

Oxenford, H.A., and Hunte, W. (1986). A preliminary investigation of the stock structure of the dolphin, *Coryphaena hippuru*, in the western central Atlantic. *Fish. Bull.* **84**, 451–460.

Paine, R.T. (1966). Food web complexity and species diversity. *Am. Nat.* **100**, 65–75.

Planes, S. (2002). Biogeography and larval dispersal inferred from population genetic analysis. In *Coral reef fishes: Dynamics and diversity in a complex ecosystem* (P.F. Sale, ed., pp. 201–220). San Diego: Academic Press.

Polacheck, T. (1990). Year around closed areas as a management tool. *Nat. Resour. Model.* **4**, 327–354.

Polunin, N.V.C. (2002). Marine protected areas, fish and fisheries. In *Handbook of fish biology and fisheries* (P.J.B. Hart and J.D. Reynolds, eds., vol. 2, pp. 293–318). Malden, MA: Blackwell Science Ltd.

Polunin, N.V.C., Halim, M.K., and Kvalvagnaes, K. (1983). Bali Barat: An Indonesian marine protected area and its resources. *Biol. Cons.* **25**, 171–191.

Rabalais, N.N., Turner, R.E., and Wiseman, W.J., Jr. (2002). Gulf of Mexico hypoxia, a.k.a. "the dead zone." *Annu. Rev. Ecol. Syst.* **33**, 235–263.

Reynolds, J.D., Dulvy, N.K., and Roberts, C.M. (2002). Exploitation and other threats to fish conservation. In *Handbook of fish biology and fisheries* (P.J.B. Hart and J.D. Reynolds, eds., vol. 2, pp. 319–341). Malden, MA: Blackwell Science Ltd.

Roberts, C.M., and Polunin, N.V.C. (1991). Are marine reserves effective in management of reef fisheries? *Rev. Fish Biol. Fish.* **1**, 65–91.

Rothschild, B.J. (1998). Notes on recruitment in commercially fished stocks. In *ReeFish '95: Recruitment and population dynamics of coral reef fishes* (Jones, G.P., Doherty, P.J., Mapstone, B.M., and Howlett, L., eds., pp. 87–99). Townsville, Australia: CRC Reef Research Centre.

Roughgarden, J., Gaines, S., and Possingham, H. (1988). Recruitment dynamics in complex life cycles. *Science.* **241**, 1460–1466.

Roughgarden, J., and Iwasa, Y. (1986). Dynamics of a metapopulation with space-limited subpopulations. *Theor. Popul. Biol.* **29**, 235–261.

Roughgarden, J., Iwasa, Y., and Baxter, C. (1985). Demographic theory for an open marine population with space-limited recruitment. *Ecology.* **66**, 54–67.

Sale, P.F. (1977). Maintenance of high diversity in coral reef fish communities. *Am. Nat.* **111**, 337–359.

Sale, P.F. (1988). Perception, pattern, chance and the structure of reef fish communities. *Env. Biol. Fish.* **21**, 3–15.

Sale, P.F. (1990). Recruitment of marine species: Is the bandwagon rolling in the right direction? *Trends Ecol. Evol.* **5**, 25–27.

Sale, P.F., ed. (1991). *The ecology of fishes on coral reefs.* San Diego: Academic Press.

Sale, P.F., ed. (2002a). *Coral reef fishes: Dynamics and diversity in a complex ecosystem.* San Diego: Academic Press.

Sale, P.F. (2002b). The science we need to develop for more effective management. In *Coral reef fishes: Dynamics and diversity in a complex ecosystem* (P.F. Sale, ed., pp. 361–376). San Diego: Academic Press.

Sale, P.F. (2004). Connectivity, recruitment variation, and the structure of reef fish communities. *Integr. Comp. Biol.* **44**, 390–399.

Sale, P.F., and J.P. Kritzer (2003). Determining the extent and spatial scale of population connectivity: Decapods and coral reef fishes compared. *Fish. Res.* **65**, 153–172.

Sale, P.F., Cowen, R.K., Danilowicz, B.S., Jones, G.P., Kritzer, J.P., Lindeman, K.C., Planes, S., Polunin, N.V.C., Russ, G.R., Sadovy, Y.J., and Steneck, R.S. (2005). Critical science gaps impede use of no-take fishery reserves. *Trends Ecol. Evol.* **20**, 74–80.

Salm, R.V. (1984). Ecological boundaries for coral-reef reserves: Principles and guidelines. *Environ. Cons.* **11**, 209–215.

Sanchirico, J.N., and Wilen, J.E. (2001). A bioeconomic model of marine reserve creation. *J. Environ. Econ. Manage.* **42**, 257–276.

Scheltema, R.S. (1974). Biological interactions determining larval settlement of marine invertebrates. *Thal. Jugosl.* **10**, 263–296.

Shaklee, J.B., Brill, R.W., and Acerra, R. (1983). Biochemical genetics of Pacific blue marlin, *Makaira nigricans*, from Hawaiian waters. *Fish. Bull.* **81**, 85–90.

Shepherd, S.A., and Brown, L.A. (1993). What is an abalone stock? Implications for the role of refugia in conservation. *Can. J. Fish. Aquat. Sci.* **50**, 2001–2009.

Simberloff, D. (1988). The contribution of population and community biology to conservation science. *Annu. Rev. Ecol. Syst.* **19**, 473–511.

Sinclair, M. (1988). *Marine populations: An essay on population regulation and speciation.* Seattle: University of Washington Press.

Smedbol, R.K., McPherson, A., Hansen, M.M., and Kenchington, E. (2002). Myths and moderation in marine "metapopulations"? *Fish and Fisheries.* **3**, 20–35.

Smedbol, R.K., and Wroblewski, J.S. (2002). Metapopulation theory and northern cod population structure: Interdependency of subpopulations in recovery of a groundfish population. *Fish. Res.* **55**, 161–174.

Soulé, M.E., and Simberloff, D. (1986). What do genetics and ecology tell us about the design of nature reserves? *Biol. Cons.* **35**, 19–40.

Stephenson, R.L. (1999). Stock complexity in fisheries management: A perspective of emerging issues related to population sub-units. *Fish. Res.* **43**, 247–249.

Supriatna, A.K., and Possingham, H.P. (1998). Optimal harvesting for a predator–prey metapopulation. *Bull. Math. Biol.* **60**, 49–65.

Sutherland, J.P. (1974). Multiple stable points in natural communities. *Am. Nat.* **108**, 859–873.

Thomas, C.D., and Hanski, I. (2004). Metapopulation dynamics in changing environments: Butterfly responses to habitat and climate change. In *Ecology, genetics, and evolution of metapopulations* (I. Hanski and O.E. Gaggiotti, eds., pp. 489–514). Amsterdam: Elsevier Academic Press.

Thorson, G. (1950). Reproduction and larval ecology of marine bottom invertebrates. *Biol. Rev.* **25**, 1–45.

Thorson, G. (1966). Some factors influencing the recruitment and establishment of marine benthic communities. *Neth. J. Sea Res.* **3**, 267–293.

Thrush, S.F., and Dayton, P.K. (2002). Disturbance to marine benthic habitats by trawling and dredging: Implications for marine biodiversity. *Annu. Rev. Ecol. Syst.* **33**, 449–473.

Tuck, G.N., and Possingham, H.P. (2000). Marine protected areas for spatially structured exploited stocks. *Mar. Ecol. Prog. Ser.* **192**, 89–101.

Underwood, A.J., and Denley, E.J. (1984). Paradigms, explanations, and generalizations in models for the structure of ecological communities on rocky shores. In *Ecological communities: Conceptual issues and the evidence* (D. Strong, D. Simberloff, L.G. Abele, and A.B. Thistle, eds., pp. 151–180). Princeton, NJ: Princeton University Press.

Underwood, A.J., and Fairweather, P.G. (1989). Supply-side ecology and benthic marine assemblages. *Trends Ecol. Evol.* **4**, 16–20.

Victor, B.C. (1983). Recruitment and population dynamics of a coral reef fish. *Science.* **219**, 419–420.

Wahlberg, N., Moilanen, A., and Hanski, I. (1996). Predicting the occurrence of endangered species in fragmented landscapes. *Science.* **273**, 1536–1538.

Whitlock, M.C. (2004). Selection and drift in metapopulations. In *Ecology, genetics, and evolution of metapopulations* (I. Hanski and O. Gaggiotti, eds., pp. 153–173). Amsterdam: Elsevier Academic Press.

Wilen, J.E., Smith, M.D., Lockwood, D., and Botsford, L.W. (2002). Avoiding surprises: Incorporating fisherman behavior into management models. *Bull. Mar. Sci.* **70**, 553–575.

Williams, D.McB., and Sale, P.F. (1981). Spatial and temporal patterns of recruitment of juvenile coral reef fishes to coral habitats within One Tree Lagoon, Great Barrier Reef. *Mar. Biol.* **65**, 245–253.

Willis, T.J., Millar, R.B., Babcock, R.C., and Tolimieri, N. (2003). Burdens of evidence and the benefits of marine reserves: Putting Descartes before des horse? *Environ. Cons.* **30**, 97–103.

Wilson, E.O. (1992). *The diversity of life*. Cambridge, MA: Harvard University Press.

Wing, S.R., Botsford, L.W., and Quinn, J.F. (1998). The impact of coastal circulation on the spatial distribution of invertebrate recruitment, with implications for management. *Can. Spec. Publ. Fish. Aquat. Sci.* **125**, 285–294.

Wright, S. (1931). Evolution in mendelian populations. *Genetics.* **16**, 97–159.

PART II

Fishes

CHAPTER 2

The Metapopulation Ecology of Coral Reef Fishes

JACOB P. KRITZER and PETER F. SALE

I. Introduction
II. Spatial Structure
   A. Geographic Extent
   B. Spatial Subdivision
   C. Interpatch Space
III. Biology and Ecology of Coral Reef Fishes
   A. Postsettlement Life Stages
   B. Dispersal and Connectivity
   C. Metapopulation Dynamics
IV. Factors Dissolving Metapopulation Structure
   A. Spawning Aggregations
   B. Nursery Habitats
V. Summary
   References

I. INTRODUCTION

The ecology of coral reef fishes seems to invite a metapopulation perspective. Metapopulation structure is determined by the spatial arrangement of local populations, coupled with life history traits that allow metapopulation dynamics to be enacted. At a minimum, distinct local populations must be identifiable and organisms generally need to remain within those subpopulations, but there needs to be some mechanism for interpopulation dispersal. Coral reef fishes seem to meet these criteria, and reef fish ecologists have readily adopted the concept (e.g., Armsworth, 2002; Doherty, 2002; James et al., 2002). But the devil is ultimately in the details, and subpopulations with a dispersal mechanism alone are not enough to create metapopulation structure. Therefore, whether reef fishes actually form metapopulations awaits more detailed consideration. Smedbol et al. (2002) have expressed dissatisfaction with the eager and perhaps uncritical adoption of metapopulation concepts by many marine ecologists. Grimm et al.

(2003) have been more forgiving, allowing that the metapopulation concept can be a useful working hypothesis, but that eventually a more critical judgment is needed.

One difficulty in performing this final assessment is agreement on what exactly constitutes a metapopulation (i.e., What are the details in which the devil hides?), not to mention confidently identifying those features in nature (Hanski and Gagiotti, 2004). Smedbol et al. (2002) and Grimm et al. (2003) agree that a metapopulation must retain some features of Levins' (1969, 1970) original formulation. Specifically, habitat patches must be discrete, movement among them must occur but not be excessive, and the probability of local extinctions must be nonnegligible. We generally agree with the first two criteria (although we modify the former somewhat herein; see Section II.B.), but do not see the value in the third and have outlined our argument against this criterion elsewhere (see Kritzer and Sale, 2004; Sale et al., this volume and references therein). In summary, what we see as unique about metapopulations is the coupling of spatial scales. The dynamics of local populations are determined in large part by local demography and self-recruitment, but these dynamics are modified by replenishment from external populations in ways that preclude viewing the local population in isolation. Ultimately, definitively answering the question of whether coral reef fishes (or any taxonomic group, for that matter) form metapopulations is less important than understanding how the scales and processes associated with the metapopulation concept affect fundamental questions in population ecology (e.g., What determines the persistence or extinction of local populations? What factors drive fluctuations in abundance?).

Is the interaction of local and regional processes that defines a metapopulation characteristic of coral reef fish populations? The answer is currently unclear, but the stage is certainly set for a critical examination. The study of coral reef fish biology and ecology has seen a progressive increase in focal scale since the early 1970s. If we assume that, as the highest ranking scientific journals, publications in *Science* and *Nature* accurately reflect the major directions within a field of study, tracking papers on coral reef fishes in these journals clearly demonstrates a move among reef fish ecologists toward metapopulation-scale research and, indeed, beyond (Fig. 2-1; for related discussion, see Sale et al., this volume). More important, however, the early focus on individual and local population-level research has much to offer metapopulation ecology, and should not be forgotten or discontinued as the needed work at larger scales continues. One of our objectives in this chapter is to bring these earlier research traditions into the domain of metapopulation ecology as applied to coral reef fishes.

We approach the issue of coral reef fish metapopulations from the perspective of Grimm et al. (2003), using metapopulation structure as a working hypothesis. We therefore use the term before answering the question. We first examine the spatial arrangement of coral reef habitat as the structural basis for reef fish

Chapter 2 The Metapopulation Ecology of Coral Reef Fishes 33

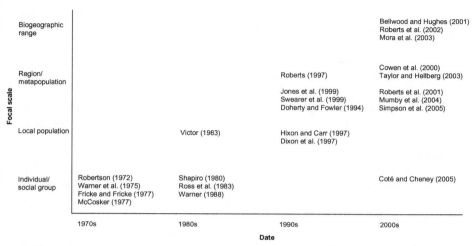

FIGURE 2-1. Historical trends in the focal scale of publications on coral reef fishes appearing in *Science* and *Nature*. We position the studies by Jones et al. (1999), Swearer et al. (1999), and Simpson et al. (2005) between the local population and metapopulation scales because each focuses on a single population, but considers questions about dispersal pathways that are very relevant to metapopulation ecology. Although the study by Simpson et al. (2005) addresses individual behavior and sensory abilities, its real questions of interest deal with larval dispersal and connectivity, which are metapopulation-level issues. We position the studies by Doherty and Fowler (1994) and Mumby et al. (2004) in between the local population and metapopulation scales because each focuses on local population dynamics, but builds comparisons over a larger metapopulation-level scale. We position the study by Roberts et al. (2001) in between these scales, because, although it examines linkages among distinct spatial units, it is unclear whether these units represent local populations or components of a single local population. Excluded is the study by Thresher (1982), which examines an individual life history trait (egg size), but compares it on a global scale, and is therefore difficult to position within this scheme. Also excluded are studies that adopt an ecosystem-level perspective and include fishes as one component.

metapopulations, and find that the potential for metapopulation structure is generally evident. Then, we review the biology and ecology of coral reef fishes to illustrate the processes that potentially allow metapopulation dynamics to play out and to see whether species consequently exhibit metapopulation dynamics on the spatial template provided by their habitat. In the third section, we address two unique attributes of coral reef fish ecology—spawning aggregations and use of nonreef nursery habitats—which have the potential to negate metapopulation structure. Finally, we return in the final section to the bigger question of whether the evidence supports metapopulation structure among coral reef fishes, what specific forms those metapopulations take, and what needs to be done to get answers when they cannot be generated at present. The most important objective, here and in future considerations of this issue, has less to do with producing a yes or no answer and more to do with developing an understanding of the processes that determine the population dynamics of coral reef fishes.

II. SPATIAL STRUCTURE

Fish species seen as characteristic of coral reef systems can be found on both reefs built by corals and inorganic reefs made of rock and other materials (Robertson, 1998). However, the vast majority of coral reef fish populations inhabit reefs of organic origin. Consequently, the spatial structure of their metapopulations is inextricably tied to the spatial structure of the hard coral metapopulations that form the basis of their habitat. As living animals, the corals that provide underlying habitat for reef fishes will themselves experience population dynamics, resulting in shifts in abundance of particular species and therefore changes in coral community structure (see Mumby, 1999; and Mumby and Dytham, this volume, for more on metapopulation ecology of corals). Although within-patch changes in habitat structure will likely affect reef fish populations (Jones, 1991), especially for habitat-specialist species (Jones et al., 2002), in this section we focus on the general arrangement of habitat patches, with some consideration of how the biology of reef fishes interacts with habitat structure. We temporarily ignore changes in the size and composition of those patches, many of which will be gradual, but do not deny their importance.

A. GEOGRAPHIC EXTENT

The maximum possible extent of any metapopulation is set by a species' biogeographic range. Availability and continuity of suitable habitat, coupled with the presence or absence of dispersal barriers, will set limits to all species' ranges within a major biogeographic region such as the Indo-west Pacific, tropical eastern Pacific, or Caribbean Sea. However, species-specific attributes such as evolutionary origin and subsequent dispersal ability (Mora et al., 2003), competitive ability in the face of limited resources (Bellwood and Hughes, 2001), and the range of tolerance to environmental conditions (Mora and Robertson, 2005) will collectively determine a particular species' range within its biogeographic region. Consequently, range sizes show tremendous variability among coral reef fishes. For example, in the Indo-west Pacific, some species are distributed across hundreds of thousands of square kilometers or more, whereas others are endemic to one particular location (Hughes et al., 2002; Mora et al., 2003). Therefore, there is considerable potential for metapopulations of very different size.

Of course, most subpopulations of coral reef fishes will not interact demographically across the full biogeographic range. All metapopulation models assume that the system is open within, but closed as a whole. This complete closure of an interconnected network may never fully occur in nature, but it is still important to know for both scientific and management purposes where the ecological interactions between populations in two geographic areas are effectively irrelevant. Unfortunately, we do not yet know the boundaries for any reef fish metapopulations, beyond endemic species. Genetic data can show major phylogeographic breaks that effectively mark the division between metapopulations (Hellberg, this volume), although we are not aware of any such data for coral reef fishes. Nevertheless, a lack of clear genetic division does not mean continuous demographic connectivity, given the potential for even minimal gene flow to maintain genetic homogeneity (Hellberg, this volume). For example, Shulman and Bermingham (1995) report generally uniform genetic structure of several reef fish species across several thousand kilometers in the Caribbean, but the analysis of microchemical tracers by Swearer et al. (1999) and the dispersal modeling work of Cowen et al. (2000) suggest that demographic exchange of individuals occurs on much smaller scales in the region.

It is still possible, however, that reef fish metapopulations might function across the full extent of species ranges if dispersal follows a steppingstone model rather than an island or larval pool model (Hellberg, this volume; Palumbi, 2003). Steppingstone structure can connect very distant populations by a series of intermediate steps, even if most direct exchange of individuals occurs more locally. There are important questions to be answered regarding the limits of demographic connectivity among reef fish populations. These are being addressed through continued refinements in genetic resolution and dispersal models, development of new behavioral understanding, use of natural and artificial tagging methods for pelagic larvae, and, especially, multidisciplinary integration of these approaches (Mora and Sale, 2002; Sale and Kritzer, 2003; see Section III.B).

B. Spatial Subdivision

The most basic concept of a metapopulation involves a binary spatial structure: There are suitable habitat patches distributed within a field of unsuitable nonpatch space. Each occupied patch is inhabited by a local population of organisms that freely and randomly interbreed within the patch. However, although random mating within an area of contiguous suitable habitat is assumed by metapopulation models and by other population genetic and ecological models, many species will not conform to this. Additional spatial structuring is common within habitat patches, and this can be driven by microhabitat structure within the patch, individual or social behavior, or a combination of these factors. Coral reef fishes are certainly no exception to this multiscale complexity in spatial structure (Sale, 1998).

The most basic and possibly the only clearly definable grouping of reef fishes is the local breeding group. This is the group of individuals that is actually mating with one another, and will involve different numbers of individuals and exist at different spatial scales for different species. For example, a single large coral formation (or *bommy*, in Australian parlance) within a platform reef on the Great Barrier Reef might have several colonies of digitate *Acropora* corals growing along its sides, each housing several gobies of the genus *Gobiodon*. Each of these groups will be restricted to its coral and each will constitute a local breeding group. Atop the bommy, a school of planktivorous *Chromis* damselfishes may swim in the immediate water column and shelter within the coral. These fish are unlikely to stray from the bommy and they constitute a single local breeding group on that bommy to go along with the several goby groups. The bommy and several others nearby may collectively form the territory of a dominant male *Scarus* parrotfish, which, along with the females in his harem, constitute another local breeding group. Past the multiple bommies comprising the parrotfish territory might swim a large *Plectropomus* grouper that, at certain times of the year, will gather at a strategic location with others of its kind living on that reef to spawn (see Section IV.A). All the groupers on the reef make up a single local breeding group.

Therefore, a single contiguous habitat patch can have very different numbers of local breeding groups for different species. For the larger and more motile species, the local breeding group may be equivalent to a local population, in that there is some large degree of recruitment to the group derived from reproductive output produced by the group. But for other species, perhaps including the territorial parrotfishes and certainly including the small-bodied, site-attached, and especially habitat-specialized species, nearly 100% of the recruitment to the local breeding group will originate externally. In these instances, to define ecologically meaningful populations will require pooling multiple local breeding groups. When enough local breeding groups have been pooled, the majority of recruitment to the included local breeding groups will be the result of production from the included local breeding groups. Such a local population comprised of a number of local breeding groups has been termed a *mesopopulation* by Forrester et al. (2002). In the example of an Australian platform reef described earlier, a convenient place to stop this pooling will be at the limits of the reef, so that the contiguous habitat patch now aligns with a single local population comprised of the numerous local breeding groups it contains. Our "population" now consists of individuals that do not freely and randomly interbreed, a more important condition for genetic theories, but which can account for a significant proportion of its own replenishment—a key condition in population ecology definitions.

The small islands and platform reefs of the Great Barrier Reef provide a convenient habitat template on which to assemble local breeding groups as local populations for metapopulation analyses (e.g., James et al., 2002; Fig. 2-2). Other reef systems are less straightforward. Unlike the thousands of individual reefs that

Chapter 2 The Metapopulation Ecology of Coral Reef Fishes 37

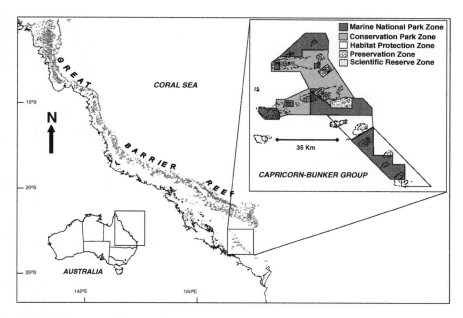

FIGURE 2-2. Map of the Great Barrier Reef illustrating the numerous individual reefs (habitat patches) that collectively comprise the world's largest barrier reef system. The detail of the Capricorn–Bunker Group at the southernmost extent of the Great Barrier Reef shows different types of MPAs, and the different ways in which they can be designated. Some MPAs include an individual reef, some group multiple reefs, and others include only part of a single reef.

comprise the Great Barrier Reef, the world's second largest barrier reef system, the Mesoamerican Barrier Reef System, is comprised of a few very large offshore reef developments and a nearly contiguous network of barrier and fringing reefs along the coasts of Mexico and Belize (Fig. 2-3). The offshore atolls of Belize and Mexico, and the Bay Islands of Honduras can span 50 km in their greatest dimension, and the coastal reef tract is more than 300 km long. To use any of these contiguous habitat features as the basis for defining a single local population is likely to be too coarse and too uninformative to understand population dynamics in a meaningful way, especially given the relative small spatial scales over which variability in demographic rates like growth (e.g., Kritzer, 2004) and recruitment (reviewed by Doherty, 1991) is evident. Therefore, within these contiguous systems, it might be useful to define local populations (again, as collections of local breeding groups) that are not truly discrete but are nonetheless conceptually divided for functional purposes. We will understand a great deal more about population dynamics by imposing functional divisions where real divisions do

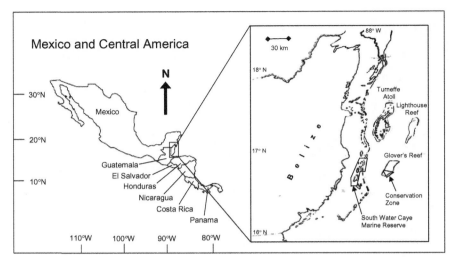

FIGURE 2-3. Map of the Belizean section of the Mesoamerican Barrier Reef System. This system is comprised of two primary habitat types: the Belize Barrier Reef and the three large offshore atolls. The barrier reef is a primarily contiguous yet heterogeneous reef tract that continues from the coastal fringing reef along the Mexican Yucatan. Two of the seven marine protected areas in Belizean waters are illustrated. The Glover's Reef Conservation Zone includes only the southern portion of the otherwise contiguous Glover's Reef System. The South Water Caye Marine Reserve includes a section of the barrier reef.

not necessarily exist, while avoiding the intractability of treating each local breeding group as a separate entity.

How to make this division within an otherwise continuous population is not straightforward, and will not be possible until we better understand larval dispersal, adult movement, and consequent connectivity patterns. Essentially, the process will involve assessing the scale over which local breeding groups will need to be pooled to achieve a degree of closure that characterizes a local population. So, in the example in Figure 2-4, if a threshold of 50% self-recruitment is set to define a local population, local breeding groups will need to be pooled over a scale such that each composite local population contains four groups. Imposition of these subdivisions and treatment of the defined units as separate local populations for research and modeling purposes invites a metapopulation perspective to the study of systems that might not otherwise suggest this approach. We see this as a useful way forward.

Jones (this volume) has drawn a distinction between *naturally evolved* and *fragmentation-induced* metapopulations. The former, she argues, are more typical of marine systems wherein habitat has a naturally patchy distribution and species have evolved to exist within this context. In contrast, many terrestrial

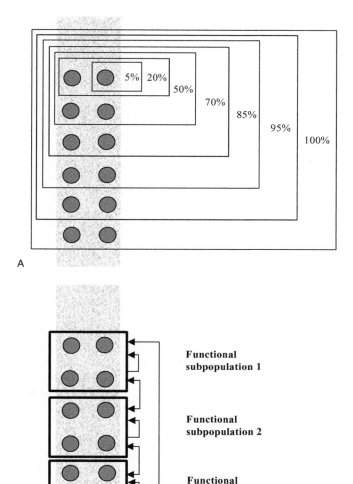

FIGURE 2-4. Adopting a "functional" metapopulation approach in contiguous reef systems. Local breeding groups (gray circles) exist along a contiguous tract of reef (stippled strip). (A) Increasingly greater proportions of overall replenishment can be accounted for by considering increasing greater numbers of local breeding groups. (B) By combining sets of four local breeding groups, local production of offspring accounts for a significant amount of replenishment—here, chosen to be 50%. The system can now be studied in terms of a more manageable number of units, linked by larval dispersal (arrows), allowing evaluation of the importance of processes occurring at local and regional scales. The coastal reef track within the Mesoamerican Barrier Reef System is an example of an area that is amenable to this functional metapopulation approach. Note that real systems will contain many more local breeding groups than this stylized illustration, and there will be many possible ways in which they can be grouped. Often, groupings will be arbitrary unless habitat changes or management units dictate boundaries between functional subpopulations. Otherwise, adjacent breeding groups may be combined into one functional subpopulation or situated on the edge of two different functional subpopulations.

metapopulations once occupied a single, contiguous tract of habitat that has now been carved into numerous, discrete habitat patches by anthropogenic modification. We propose a *functional metapopulation* as an additional metapopulation type to describe systems in which local breeding groups are lumped into local populations for the purposes of research and modeling. It is distinct from Jones' categories in that it exists not in nature, but rather is a structure imposed conceptually upon a system to assist in understanding ecological dynamics. Subpopulations are still discrete, conceptually at least, but interpatch space becomes nonexistent. The similarity of environmental conditions experienced by different individuals or social groups will decrease as a function of distance for all but the largest scale phenomena, which provides justification for treating subdivisions of a larger contiguous system as "separate" local populations within a metapopulation. It is possible that the functional subdivisions will have an ecological basis, given that habitat structure is rarely homogenous along extensive reef systems, so habitat patches can be delineated based on reef type, even if no nonreef space separates them. Perhaps a more effective basis on which to divide functional subpopulations will be where discontinuities in demographic rates occur.

A functional metapopulation perspective is especially relevant when MPAs are designated within a contiguous system. MPAs create a type of anthropogenically induced subdivision different from that created by fragmentation. Rather than physically restructuring habitat, MPAs demographically restructure populations by reducing mortality within their borders relative to that outside (see Gunderson and Vetter, this volume). Therefore, when MPAs are established, divisions are no longer purely functional in terms of tractability, but rather take on an ecological basis. Some MPAs will contain an entire subpopulation, such as the whole reef closures on the Great Barrier Reef (Fig. 2-2). In these cases, the natural metapopulation structure is retained, but differences in local demography are exacerbated. On the other hand, some MPAs protect only part of an otherwise contiguous reef system. For example, the well-studied MPAs at Sumilon and Apo Islands in the Philippines (Alcala and Russ, 1990; Russ and Alcala 1989, 1996) and the partial closure of Glover's Reef in Belize (Fig. 2-3) represent the division of an otherwise contiguous system into protected and unprotected parts.

For both MPAs and the functional metapopulation approach to be effective in contiguous systems, organisms within the prescribed division must be sufficiently sedentary that they can be treated as a distinct subpopulation. That reef fishes within MPAs tend to be larger and more numerous (see the review by Coté et al., 2001; Halpern and Warner, 2002; Crowder and Figueria, this volume) suggests that many species are sufficiently sedentary that real demographic differences are achieved (see Section III.A.3.1). Of course, the lack of hard boundaries at the borders of MPAs that would truly close part of a contiguous system means some blurring of the division between functional subpopulations is inevitable (e.g., Chapman and Kramer, 2000). However, the overall mean demographic differ-

Chapter 2 The Metapopulation Ecology of Coral Reef Fishes   41

ences often seen between MPAs and open areas suggest that the blurring does not negate the usefulness of either the management tool or the conceptual framework. The demographic differences successfully imposed on MPA populations might not be adequate to ensure their persistence if connectivity patterns do not allow adequate self-recruitment (see Botsford and Hastings, this volume). Therefore, continued empirical research on dispersal and connectivity is needed to reveal the scales at which MPAs and functional subpopulations need to be set (see Section III.B).

C. INTERPATCH SPACE

A corollary of the criterion that subpopulations inhabit discrete habitat patches is that the area in between patches is uninhabitable (Hanski, 1999). However, recent surveys of interreef seafloor on the Great Barrier Reef suggest that this binary distinction between patch and nonpatch area might be too simplistic. In between the major reef structures on the Great Barrier Reef can be found large sponges, soft corals, sea fans, and individual hard coral colonies (Poiner et al., 1998). These features are found less as discrete groups forming distinct reefs, and more as scattered and often isolated individual structures, which might provide shelter for small groups of fishes. It is possible that such features can be ignored in the metapopulation context on the assumption that any resident fishes will contribute little to population biomass and replenishment (although their collective area might make them more important than immediately apparent). This has not been studied, but even if true it does not necessarily negate the importance of interreef habitat. Levin (1995) argues that individual plants found in between major local populations in terrestrial systems provide an important bridge along which pollen and seeds can travel, thereby promoting connectivity. It is unlikely that coral reef fishes will require similar bridges for movement of reproductive propagules, given that larval reef fishes are predominantly pelagic and have little interaction with benthic habitat until settlement. However, interreef habitat might provide a conduit for movement of postsettlement individuals. This can add another layer of connectivity to that determined by larval dispersal. Although postsettlement coral reef fishes are typically seen as sedentary, successful movement of even a very few adult fishes can have long-term demographic effects given the extensive longevity of many reef fish species (see Section III.A.2.a; also, see Gunderson and Vetter, this volume).

The presence and potential importance of interreef structure might call for a more complex perspective than the traditional binary patch/nonpatch dichotomy. This perspective will borrow concepts from landscape ecology to consider the composition of both patches and the intervening landscape matrix, and how these affect within-patch dynamics and interpatch dispersal (Wiens, 1997; Murphy and

Lovett–Doust, 2004). Although use of the term *landscape ecology* among marine ecologists is rare, this perspective has essentially been adopted in the study of within-patch dynamics. For example, Sale and Douglas (1984) relate community structure of reef fishes to patch attributes. A landscape perspective is also being adopted in studying dispersal among patches. The marriage of physical oceanography with marine ecology in mapping interreef connectivity (reviewed by Cowen, 2002; Sale and Kritzer, 2003) reflects awareness that the space in between habitat patches is not uniform and that simple linear distance alone will not determine dispersal. Rather, current speed and direction (e.g., Cowen and Schultz, 1994), temperature and salinity gradients (e.g., Pepin, 1991), and planktonic patchiness (e.g., Doherty et al., 1985) will all affect larval growth, survival, and transport. So, the pelagic realm is being seen as a heterogeneous "landscape." In systems like the Great Barrier Reef, it might now be necessary to apply this same perspective on the seafloor to examine whether previously hidden benthic connections exist among the more obvious habitat patches. This will likely be most important for larger and more mobile species, many of which are targets of commercial and recreational fisheries. In contrast, it might less important in the Caribbean, where depths in interpatch space can often be substantial.

III. BIOLOGY AND ECOLOGY OF CORAL REEF FISHES

We turn now to the biology and ecology of coral reef fishes, and how these allow and shape metapopulation structure. We first examine facets of life on the reef and how they affect reproductive output and population structure. Next, we briefly discuss larval dispersal as the basis for both local population closure and interpopulation connections characteristic of metapopulation structure. Finally, we address different types of population dynamics that arise through the interaction between postsettlement and pelagic processes.

A. Postsettlement Life Stages

There has been a clear shift in the ecological study of coral reef fishes from a perspective focused on patterns and processes within individual reef systems to a larger scale perspective that examines patterns and connections among reefs (Fig. 2-1). This increase in scale has been a logical and necessary outgrowth of the scientific questions posed by ecologists studying reef fishes and management needs, and has fostered the adoption of metapopulation concepts (Sale et al., this volume). However, the earlier tradition of research on within-reef ecology should not be lost as the study of reef fishes shifts to a metapopulation framework.

The development of metapopulation ecology since the original papers of Levins (1969, 1970) has involved relaxation of the original model's assumptions and increasing resolution of population dynamics (Kritzer and Sale, 2004). For example, relaxing the assumption of equal size, spacing, and demography of local populations has resulted in concepts such as the source–sink metapopulation, in which inequities in reproductive potential among local populations are a key feature (Pulliam, 1988). These more complex models have considered population size and intrinsic population growth rates (e.g., Foley, 1997; Crowder et al., 2000), and have incorporated the constituent demographic rates that collectively determine population growth, forging a link between individual life histories and population dynamics (e.g., Armsworth, 2002; Kritzer and Davies, 2005). Few metapopulation studies have considered behavior (but see McQuinn, 1997), but this might be the next logical degree of complexity in model resolution. Detailed understanding of the benthic ecology of coral reef fishes can contribute toward understanding how interpopulation inequities are generated, so we consider population size, demography, and behavior in turn.

1. Population Size

The number of fish living on a given reef is the first and possibly the most important metric determining the reproductive potential of a local population. Most information on reef fish abundance is reported as density, with fewer estimates of overall population size. Yet, a dense but very small population is likely to be less important in reproductive output than a less dense but very large population (although demography and behavior can have important interactions with density that can give fitness advantages to populations of both high and low density; see Sections II.A.2 and II.A.3). Population size will clearly be determined in large part by habitat area, and advances in GIS and remote sensing technology (Mumby et al., 1997) will enable density estimates to be readily expanded to population size estimates. The most extensive study to model a reef fish metapopulation to date (James et al., 2002) uses habitat area as an index of local population size in the absence of large-scale abundance data.

Habitat area is not a perfect predictor of population size, however, due in part to the influence of habitat structure on local density (reviewed by Jones, 1991). The utility of habitat area as a predictor of relative population size seems greatest for proximal populations, although the relationship between habitat area and population size can still be quite variable among neighboring reefs. For example, Kritzer (2002) estimated a fourfold difference in population size of *Lutjanus carponotatus* between Pelorus and Orpheus islands in the Palm group on the Great Barrier Reef, but the habitat area of these populations differs approximately sevenfold. Given the uncertainty in the density estimates on which these population size estimates are based, coupled with the potential for temporal variability in population size and the imperfect relationship between population size and reproductive output, the disparity between the relative difference in habitat area and the relative difference in population size might be sufficiently small to allow the former to serve as a useful proxy for the latter. Sensitivity analyses will be important in determining the degree of precision in population size estimates required for metapopulation modeling efforts.

In contrast, on a larger spatial scale, the relationship between habitat area and population size is more strongly decoupled. Kritzer (2002) estimates that the Lizard Island group reef system, some 400 km from the Palm group islands, houses a population of *L. carponotatus* that is approximately equal to that of Pelorus Island and one fourth that of Orpheus Island, despite having nine times the habitat area of the former location and nearly the same area as the latter. However, the Lizard Island group is not only very distant from the Palm Island group, but also in a different continental shelf position. It is a midshelf complex whereas the Palm group is an inshore complex. Greater differences in relative abundance of both corals (Done, 1982) and fishes (Williams and Hatcher, 1983) on the Great Barrier Reef are evident along a cross-shelf axis spanning tens of kilometers than on a latitudinal axis spanning hundreds of kilometers. Newman and Williams (1996) found that catch rates of *L. carponotatus* in fish traps differed ninefold between inshore and midshelf reefs on the Great Barrier Reef, which is similar to the fivefold to ninefold density differences reported between the Lizard Island group and the Palm group islands by Kritzer (2002). Therefore, habitat area can be a useful proxy for population size within but not among continental shelf positions on the Great Barrier Reef, at least within the degree of resolution likely needed for large-scale modeling, and James et al. (2002) accordingly scaled reproductive output by shelf position in their metapopulation modeling study. Large-scale patterns in abundance for regions beyond the Great Barrier Reef are needed to determine whether habitat area is a useful proxy for abundance, and the extent to which this relationship needs to be adjusted for local populations in different areas.

2. Demography

Fecundity among fish and other species that produce large clutches is dependent upon body size, so biomass is a better index of reproductive output than simple numerical abundance. When population size estimates for *L. carponotatus* at the Lizard Island group are converted to biomass estimates, the population is much less different from the Palm group populations than is apparent when comparing numerical population sizes (Kritzer, 2002). This is because the mean growth trajectory at the Lizard group reaches larger body sizes, coupled with a series of strong cohorts at older age classes that shift the population structure toward older age classes. Indeed, reproductive output by a population is determined by the combined effects of recruitment, growth, mortality, maturation, and size-specific fecundity. Therefore, understanding spatial patterns in these traits and their individual effects on reproductive output can illustrate the importance of distinct populations for the persistence of a larger metapopulation.

a. Life History Patterns

The advent of age-based approaches to demographic studies of coral reef fish occurred in the 1990s and has been an important development in understanding evolutionary, population, and fisheries biology (Choat and Robertson, 2002). Not unexpectedly, age-based demographic studies have revealed a tremendous diversity of life history patterns among coral reefs fishes matching the diversity of colors, morphologies, and behaviors that reef fishes exhibit. Very short-lived species have been found, generally within small-bodied families such as the Gobiidae (reviewed by Munday and Jones, 1998). However, some very small species have surprisingly long life spans, with longevities of 20+ years in some damselfishes (Meekan et al., 2001), and some larger species show maximum ages exceeding 30, 40, or even 50 years (Choat and Axe, 1996; Newman et al., 1996a). Interestingly, body size does not correlate well with maximum age among reef fishes, contrary to predictions of life history theory (Roff, 1992).

Longevity is important in the context of population biology for at least two reasons. First, longevity is an important determinant of population fluctuations and extinction risk, both key foci of metapopulation ecology. Specifically, short-lived species show higher population turnover, and populations of short-lived species have less capacity to persist through periods of poor recruitment via storage effects (Warner and Chesson, 1985). In fact, Levins (1969) developed his original metapopulation model with agricultural pest insects as the focal populations. These are short-lived organisms with a high potential for local extinction. The second reason longevity is important is that it typically correlates with other demographic rates. Higher maximum age is typically associated with later age at maturity, slower somatic growth, and lower reproductive rates (Roff, 1992). Therefore, although greater longevity can allow some populations to endure droughts of replenishment, they might have less potential to rebuild and/or recolonize because of their lower reproductive capacity.

Many coral reef fishes are difficult to place within this "live fast and die young" versus "live slow and die old" dichotomy. Some long-lived species mature relatively early in life at 15% or less of their maximum age (see Kritzer, 2004, and references therein; also see Choat and Robertson, 2002). Many of these species also grow very quickly, reaching their asymptotic body size in only a few years, with the onset of maturity often coinciding with the cessation of somatic growth (Day and Taylor, 1997). After maturation, spawning often takes place over a protracted season within the year (Kritzer, 2004). Reef fishes following this life history pattern exhibit early and intensive reproductive investment typical of high-turnover species, but over a long life span. Of course, other life history patterns are also evident. Labroid fishes (wrasses and parrotfishes) exhibit a range of maximum ages from less than 5 years to more than 30, but many species show more continuous somatic growth and later relative ages at maturity than those species with a pronounced asymptote in the growth trajectory (Choat and Robertson, 2002). The relationship between recruitment patterns and life history schedules has not been examined for coral reef fishes, but presents an exciting opportunity for testing and development of life history theory, a body of theory from which coral reef fishes have been largely absent (Caley, 1998).

b. Spatial Variation in Demographic Traits

Demographic rates are important for population dynamics generally, but spatial patterns in demographic rates are additionally important for metapopulation dynamics. Spatial patterns determine the inequities in local reproductive output and population dynamics that can create source–sink dynamics and rescue effects. Also, if a functional metapopulation approach is to be applied, demographic discontinuities may be the best basis for the delineation of functional subpopulations (see Section II.B).

Growth seems to be the most variable demographic trait among coral reef fishes (Jones, 1991), and studies examining spatial patterns in growth nearly always report differences in either growth rate or asymptotic body size among populations at some scale. Some of the comparisons are on sufficiently large spatial scales that they might represent genetically determined differences (Meekan et al., 2001; Kritzer, 2002; Williams et al., 2003), whereas others compare neighboring populations and likely represent responses to local environmental conditions (Choat and Axe, 1996; Gust et al., 2002; Kritzer, 2002; Newman et al., 1996b; lack of genetic basis demonstrated by Dudgeon et al., 2000). Local environmental conditions can also drive differences in mortality rates among neighboring local populations (Newman et al. 1996b; Gust et al., 2002; Kritzer, 2002), or even among distinct groups within a single reef system (Aldenhoven, 1986). However, quite similar mortality rates among neighboring populations have also been reported (Hart and Russ, 1996). Spatial patterns in reproductive traits have received less attention than those in growth and mortality. Kritzer (2004) reports large-scale consistency in age and size at maturity of *L. carponotatus*, but Adams et al. (2000) report large-scale differences in schedules of both maturation and sex change in *P. leopardus*. These studies of spatial patterns in demography of coral reef fish show potentially important differences in several traits, but exhibited over multiple spatial scales (tens of meters to hundreds of kilometers). Until more information is compiled on the spatial scales at which reef fish populations are interconnected, it is difficult to determine which of these patterns will have the greatest effects.

Variation in different demographic traits can have unequal consequences for reproductive output and metapopulation dynamics. Ebert (1985) conducted a sensitivity analysis of a matrix population model of a sea urchin. He found that varying mortality, age at maturity, and growth rate all had important effects on the intrinsic rate of population growth, but asymptotic body size did not. In the metapopulation context, this suggests that spatial patterns in the former traits, but not the latter, might lead to inequities in offspring production, and therefore could be the roots of source–sink status and other interpopulation disparities. Kritzer and Davies (2005) reexamined this question with a spatially structured stochastic simulation model using demographic data for *L. carponotatus* to determine whether Ebert's conclusion about the lack of importance of variation in asymptotic size was contingent upon the deterministic equilibrium assumptions of matrix models. They found that variation in asymptotic size still had little effect in many ecological scenarios modeled, but that a subpopulation with larger asymptotic size played a role in keeping the overall metapopulation size above very small (and, potentially, more extinction-prone) population sizes when cyclical recruitment patterns periodically shifted the population structure toward older age classes. The combined lessons of these studies are that spatial variation in a given demographic trait is not necessarily indicative of significant differences among populations in terms of reproductive potential, but also that the importance of spatial variation in a given demographic trait can be very context specific.

c. Density Dependence

Density dependence is important in the metapopulation context for several reasons. External sources of replenishment may be unimportant if local production of offspring is sufficient to saturate local carrying capacity. If so, a local population might receive, and therefore appear to be reliant upon, a significant supply of new recruits from other local populations, yet would fare just as well in their absence. Or, the addition of recruits from other sources might increase local density to a level where density-dependent demographic changes take effect. At the very least, modeling reproductive output of each local population requires understanding the extent to which maturity, fecundity, growth, mortality, and movement vary as a function of density, and therefore need to be adjusted through time.

The extent to which coral reef fish populations are regulated by density-dependent mechanisms has been a topical issue. Typically, the focus is on early postsettlement life stages. Hixon and Webster (2002) approached this issue from a generally binary perspective: Mortality is defined as either density dependent or density independent. They find that density dependence is more often the case than not. In contrast, Doherty (2002) adopts a more continuous approach and examines the strength of density dependence. He argues that density dependence is certainly evident, but that recruitment rates are often low enough that density-dependent effects are minimal. Armsworth (2002) illustrates the same point using a mathematical model. He highlights the fact that population size can be determined by recruitment fluctuations, but that the magnitude of consequent population fluctuations can be dampened by density dependence without losing the recruitment signal.

In contrast to early postsettlement mortality, larval demography is seen as highly variable, but generally density independent. After the first few weeks or months on the reef, a process of demographic stabilization is often assumed to occur. Density-independent larval demography is almost certainly true, especially given the stochastic nature of oceanic conditions and the low density of reef fish larvae in the pelagic environment (Leis, 1991). Whether demographic stabilization occurs after the early postsettlement period is difficult to assess because there is a lack of data. Most studies of density dependence focus on early postsettlement life stages (Hixon and Webster, 2002). Those that do not, typically use adults of small site-attached species because they are easier to manipulate and monitor. Density-dependent mortality can be observed among adults of these small-bodied species (Forrester and Steele, 2000), but more often mortality is fairly robust to the effects of density whereas growth suffers more (Jones, 1991). Whether density affects demography of larger bodied species is more difficult to assess, and will likely rely on correlations rather than experiments, complicating interpretation of the patterns. For example, Gust et al. (2001, 2002) show inverse relationships between density and mean size, mean age, and maximum age of scarids on the Great Barrier Reef. Whether this truly reflects density dependence is unclear, because the populations occupy very different habitats, so environmental effects could be driving differences in both density and demography. Better understanding of the effects of density on reef fish demography will be important because the potential advantages of larger populations can be compromised if reproductive output is reduced by density. The assumption that density dependence must always occur in all populations is likely unwarranted, and restricts consideration of other factors in population growth.

3. Behavior

The field of sociobiology arose from an awareness of the important effects that population biology has upon behavior and the associated limitations of viewing behavior in isolation from the population context (Wilson, 1975). Similarly, behavior can influence population dynamics, but there is a tendency among population biologists, demographers, and other quantitative ecologists to view organisms as sets of numbers (age, size, fecundity) and binary distinctions (mature or immature, male or female). These numbers are then grouped into frequency distributions that characterize a population. However, the actions of an organism can have profound effects on ecological processes (Wainwright and Bellwood, 2002). Although the focal scale of population biology, particularly the metapopulation perspective, makes it difficult to consider individuals, it is important to ask whether this difficulty must be overcome or whether a coarser resolution ignoring behavior is adequate.

a. Movement

Coral reef fishes have long been viewed as sedentary, with few formal tests of this assumption, likely because experimental ecological research on coral reef fishes has focused on very site-attached species such as damselfishes (Jones, 1991). However, tagging studies repeatedly show that even larger and/or more mobile taxa such as lutjanids, lethrinids, serranids, siganids, scarids, acanthurids, and mullids rarely move distances greater than 500 m (Chapman and Kramer, 2000; Davies, 1995; Holland et al., 1993; Meyer et al., 2000; Zeller, 1997; Zeller and Russ, 1998). Although we cannot define the minimum scale of a local population until we have more information on larval dispersal, it is unlikely to be less than several kilometers. Therefore, most mixing of demersal individuals will occur on a scale that is less than that of a local population. The exceptions to the rule of small home ranges include species that undertake extensive migrations to spawning aggregations. Although long-distance movement among local populations by relatively few individuals will not compromise the identity of local populations or the structure of the metapopulation, mass migration to spawning aggregations has tremendous potential to negate metapopulation structure. This topic is treated separately in Section IV.A.

b. Social and Mating Systems

The earliest studies on coral reef fishes appearing in the leading scientific journals focused on behavioral ecology, especially mating systems (Fig. 2-1). Since those early studies, behavioral ecology has become less studied in coral reef fishes. However, Petersen and Warner (2002) have revisited the importance of behavioral ecology, including its implications for large-scale ecological and management questions. There are at least two main issues in behavioral ecology that can affect population dynamics. The first is social control of the sex ratio in hermaphroditic species, and the second is variability in mating systems and consequent effects on fertilization success.

Ecologists typically model population dynamics in terms of females, assuming that female fecundity limits reproduction, and Hilborn and Walters (1992) have suggested the same approach for modeling fish population dynamics. For gonochoristic species with an equal sex ratio, this can be achieved by simply halving the biomass (although when sex-specific growth patterns cause biomass to be unequally distributed between the sexes [e.g., Davis and West, 1992; Kritzer,

2004], a proportion different from 50% should be considered). Of course, age- or size-specific differences in the relationship between female size and fecundity (Sadovy, 1996) or unequal offspring quality among age or size classes (Berkeley et al., 2004) can mean that aggregate biomass is an inadequate index, but requires accounting for population structure without necessarily considering behavior. On the other hand, the sex ratio of hermaphroditic species is determined by a complex (and not completely understood) interplay between individual life history (e.g., Adams and Williams, 2001) and social setting (Warner, 1988), with possible genetic constraints also having an influence. Most hermaphroditic coral reef fishes are protogynous (sex change from female to male), but protandrous (male to female) and bidirectional sex changers and simultaneous hermaphrodites are also found (reviewed by Petersen and Warner, 2002). Like most fishes and other animals, protogynous species may be generally egg limited. However, the fact that male status is the final stage of development and that sex ratios are usually female biased (but see Adams et al., 2000), coupled with greater vulnerability to harvest of the larger bodied males, allows the possibility of sperm limitation. If so, then male abundance and size structure become the important metrics. The social mechanisms that determine sex structure in hermaphroditic species need to be understood to best predict the reproductive output by a local population.

Coral reef fishes can display intraspecific variation in mating systems within or between reef systems. Specifically, a single species can use several modes of spawning, including pair, small group, or mass spawning, with or without the presence of "sneaker" males joining the spawning event. In terms of understanding population dynamics, these differences are effectively irrelevant if the net result is comparable fertilization success. If so, then population size and structure (in terms of sex, size, and age distributions) need only be accounted for to predict reproductive output. For example, Kiflawi et al. (1998) measured comparable fertilization rates between small group and mass aggregation spawning events by the surgeonfish *Acanthurus nigrofuscus*. In contrast, Petersen (1991) found higher fertilization success in the wrasse *Halichoeres bivittatus* when streaking males joined the spawning pairs. Complicating the issue, Marconato et al. (1997) measured greater fertilization rates of *Thalassoma bifasciatum* spawning in groups compared with pair spawning on one reef in St. Croix, but not on another. Therefore, distinguishing between mating systems might be important for some species at some times or places, but not others.

In reviewing these studies, Petersen and Warner (2002) note that all fertilization rates are generally high and that the differences between them seem to have little effect on zygote production. Hence, unlike social control of sex structure in hermaphroditic species, variation in mating systems might be relatively unimportant in the metapopulation context, especially given the large focal scales (in both space and time) and associated lower local resolution at which metapopulation studies and models operate. Indeed, modeling studies are using, and may be fully justified in using, habitat area as a proxy for population size rather than direct abundance data (e.g., James et al., 2002; see Section III.A.1), and it seems counterintuitive to combine such a fine level of behavioral resolution with such coarse abundance estimates. Still, the fusion of behavioral ecology with metapopulation ecology remains a largely untapped area of study, and heuristic modeling might reveal that certain behavioral effects are greater than intuition and available data suggest.

B. Dispersal and Connectivity

Larval dispersal and consequent interpopulation connectivity are perhaps the most central issues in determining whether metapopulation structure exists among coral reef fishes and, if so, what form it takes. Yet these are perhaps the most poorly understood aspects of coral reef fish ecology. These are also the most actively researched topics in the field at present, and we can expect considerable progress in mapping dispersal pathways in the next decade. A number of reviews have outlined the major issues and approaches, and synthesized the growing information on dispersal, while emphasizing the many as yet unanswered questions (Cowen, 2002; Kinlan and Gaines, 2003; Kritzer and Sale, 2004; Mora and Sale, 2002; Planes, 2002; Sale and Kritzer, 2003; Sale et al., 2005; also see Lubchenco et al., 2003, and associated papers on marine reserve design). We will not repeat these reviews here, but instead will highlight four major points concerning larval dispersal among coral reef fishes. First, behavioral studies are showing that larvae are not inanimate particles, but rather have quite developed behavioral and sensory capabilities (reviewed by Leis and McCormick, 2002). Second, although these capabilities afford larvae the *potential* to resist passive transport, there is as yet no demonstration of how they use their swimming abilities in the field, and therefore it is unclear how likely is recruitment back to the natal population. Third, there is evidence that self-recruitment and short-distance dispersal, consistent with behavioral resistance to advective transport, do occur (e.g., Jones et al., 1999; Planes, 1993; Swearer et al., 1999). But, fourth, we have good estimates of neither relative dispersal rates over varying distances nor of overall mean dispersal distances. Theory suggests that the latter is especially important in metapopulation persistence (Lockwood et al., 2002), and therefore estimating this parameter should be a priority of future research.

C. Metapopulation Dynamics

The ultimate goal of adopting a metapopulation perspective is to better explain the processes governing changes in the abundance of organisms in both space and time—one of, if not the, central objectives of ecology generally. Understanding dispersal, local demographic rates, behavior, and other aspects of a species' ecology is done to understand and predict how population size and structure are generated. Although an integrated understanding of the disparate processes governing metapopulation dynamics has not yet been reached, we do have some insights into what those resultant dynamics look like, and we now turn our attention to those patterns.

1. Extinction–Recolonization Events

Levins' (1969, 1970) original metapopulation analyses were built around a simple presence–absence patch-occupancy model. Although metapopulation ecology has expanded considerably since that time, especially with respect to marine systems (Kritzer and Sale, 2004), extinction–recolonization dynamics are still critical, especially in the conservation context. Although the high fecundity, "r-selected" life history, and broad dispersal potential of many marine organisms mean that extinction–recolonization dynamics are rare, some species may be naturally prone to these changes, and anthropogenic impacts are driving more species in that direction (Roberts et al., 2002).

a. Natural Extinctions

If the extinction–recolonization dynamics of classic metapopulation structure are to be taken as the defining characteristic of metapopulations in general, then many reef fishes should seemingly be excluded from the concept. The most frequently studied reef fish species, such as *Pomacentrus mollucensis* and *P. amboinensis* on the Great Barrier Reef and *T. bifasciatum* in the Caribbean, exhibit tremendously high densities (i.e., one fish per square meter and greater; see literature reviewed by Jones, 1991; Sale, 1991) and are very unlikely to go locally extinct in the absence of severe anthropogenic disturbance. Even some larger predators exist as local populations on the order of 10^5 individuals or more (Kritzer, 2002; Zeller and Russ, 2000; see Section III.A.1) and do not seem at risk of naturally induced local extinctions. However, there is a clear bias against rare species in ecology at large, and this holds for the study of coral reef fishes (Jones et al., 2002). Jones et al. (2002) examined the relationship between rank abundance and population density of butterflyfishes, and found that, in many regions, populations of the most abundant species were many times larger than other species, and that the majority of species exist as relatively small populations. Therefore, the research bias toward a few species that are readily observed, collected, and manipulated might have engendered a false impression that most reef fishes are abundant and unlikely to experience local extinction.

At One Tree Reef in the Capricorn–Bunker group at the southern extent of the Great Barrier Reef, two species of butterflyfishes, *Chaetodon plebius* and *C. rostratus*, exist in appreciable numbers but show no signs of maturation among females (Fowler, 1991). Therefore, these populations must be established initially by colonization from elsewhere. They can persist for many years as a result of the extensive life span of many butterflyfishes, and may therefore be unlikely to go locally extinct before receiving additional replenishment from other sources in the form of a rescue effect (Brown and Kodric–Brown, 1977). However, a connectivity relationship with some external source population is certain even in the absence of observable recolonization following local extinction, because the arrested sexual development of resident fishes precludes self-recruitment. It is unclear whether replenishment comes from nearby reefs in the Capricorn–Bunker group or whether the group as a whole cannot support breeding populations and receives new recruits from northern reefs (Fowler, 1991). In either case, there is a clear source–sink relationship between One Tree Reef and some other external location. It might be argued that a nonreproducing population is effectively irrelevant from a population ecological perspective, and this argument has merit. However, the metapopulation dynamics that establish even nonreproducing populations can affect multispecies interactions, including fisheries, in the outlying areas, because fish will eat and be eaten at their destination reef.

b. Anthropogenic Extinctions

Larger predators in particular are susceptible to human-induced local extinctions because they are less abundant on average than small-bodied species and are typically the targets of coral reef fisheries. For example, a Caribbean-wide survey program conducted by Reef Check suggests local extinctions of Nassau grouper at several sites (Pennisi, 2003). It is possible that these surveys simply missed these animals given that survey methods aimed at large predators that live in naturally low numbers need to involve very large transects to maximize the likelihood of detecting fish (Newman et al., 1997). Still, drastic declines in Nassau grouper populations have been documented through time (Sala et al., 2001), and the Reef Check data certainly argue for populations at best perilously close to local extinction. Therefore, fishing pressure might be imposing extinction and (hopefully) recolonization cycles on populations that do not naturally exhibit these dynamics.

2. Synchrony of Population Fluctuations

The expanded view of metapopulation ecology does not require that we observe or establish the likelihood of local extinction to adopt the term. But the new view does require some degree of asynchrony in fluctuations among local populations, primarily as a consequence of the requisite partial closure, and therefore partial independence, of those populations (Hanski, 1999; Kritzer and Sale, 2004). Unfortunately, there is a lack of long-term time series of population size for most coral reef fish populations, let alone such data for multiple local populations potentially interacting as a metapopulation. We are therefore unable, at present, to assess the degree of asynchrony for any presumed reef fish metapopulation.

Despite the lack of long-term and spatially replicated population trajectories for coral reef fishes with which to assess asynchrony of population dynamics, there are data of this nature for recruitment to reef fish populations (i.e., settlement of postlarval juveniles). The importance of recruitment as a determinant of population fluctuations will vary inversely with longevity. Warner and Hughes (1988) used a mathematical model to show that population fluctuations of a short-lived species closely match recruitment fluctuations. Similarly, Armsworth (2002) showed that, even with density-dependent mortality of early postsettlement juveniles, recruitment strength was reflected in cohort size of long-lived species many years later. This reflects recruitment limitation, whereby recruitment largely determines postsettlement population structure (reviewed by Doherty, 2002), and is the basis of the storage effect, whereby strong cohorts persist and drive replenishment for many years (Warner and Chesson, 1985). Empirical demonstrations of cohort persistence among coral reef fishes include those by Doherty and Fowler (1994) and Russ et al. (1996). However, populations of long-lived species combine many recruitment events within a single population, so the capacity for any one to modify population size is limited (Sale, 2004). Moreover, although differences in cohort size are greatest at settlement, the importance of a cohort to population replenishment will peak at some later age when proportional maturation, somatic growth, survivorship, and offspring quality combine to maximize cumulative annual reproductive success.

Despite the limitations of recruitment data in reflecting overall population fluctuations, recruitment is undoubtedly important in determining population size and represents the best available means of assessing the degree of asynchrony in coral reef fish population fluctuations. Recruitment data are available in two major forms. First, there is a voluminous literature reporting direct surveys of the abundance of new recruits (reviewed by Doherty and Williams, 1988; Doherty, 1991; Sale and Kritzer, 2003). Second, age frequency data derived from analysis of either otolith microincrements (e.g., Pitcher, 1988) or macroincrements (e.g., Meekan et al., 2001) can show in a "snapshot" sample the strength of different age groups within or between years respectively, although the former has been used surprisingly infrequently. Studies directly surveying abundance of recruits typically offer greater spatial replication than analyses of otolith annuli (but see Hart and Russ, 1996; Newman et al., 1996b; Kritzer, 2002; Williams et al., 2003).

Age-based studies, on the other hand, typically offer a longer time series given the extensive longevity of many coral reef fishes. Doherty and Fowler (1994) provide a unique study combining age frequency and direct survey indices of recruitment for several neighboring reefs.

Whichever type of data is used, the message tends to be the same: Recruitment strength of coral reef fishes is rarely synchronous in space. Recruitment events are often timed synchronously, but their magnitude can differ greatly among neighboring reefs. Peaks and valleys in population age structures rarely coincide among local populations. For example, the strong cohort of *Plectropomus leopardus* tracked by Russ et al. (1996) was not evident on neighboring reefs.

Asynchrony in metapopulations is the result of the independent dynamics within local populations, but the existence of asynchronous recruitment fluctuations alone is not sufficient to argue for metapopulation structure among coral reef fishes. Although there is evidence for unique demography and dynamics within local populations of coral reef fishes, recruitment strength in local populations might be driven by extrapopulation factors that are nonuniform in space, regardless of the sources of recruiting larvae. In other words, recruitment to a local population might be of primarily nonlocal origin, but oceanographic, meteorological, and planktonic conditions in the vicinity of a reef might ultimately determine settlement success (e.g., Dixon et al., 1999). Therefore, recruitment data suggest the needed asynchrony of population fluctuations characteristic of metapopulation dynamics, but whether this asynchrony arises from metapopulation structure is uncertain, pending further mapping of connectivity patterns.

IV. FACTORS DISSOLVING METAPOPULATION STRUCTURE

As discussed throughout this chapter and elsewhere, a critical feature of a metapopulation is the presence of identifiable local populations with some degree of population closure. This closure cannot be complete and there must be an element of external influence, which results in the coupling of local and regional spatial scales (Kritzer and Sale, 2004). However, we must be able to define local breeding populations, even if only by subdividing large, contiguous habitat patches into functional subpopulations (see Section II.B). Determination of whether such partially self-reliant local populations exist among reef fishes has focused on the relative magnitude of larval retention at and dispersal away from the natal reef, with high rates of the latter reducing the degree of closure in local population dynamics. However, there are also aspects of postlarval ecology that can negate metapopulation structure. In this section we focus on two such aspects: spawning aggregations and nonreef nursery habitats.

A. Spawning Aggregations

Many reef fish species spawn in small groups at locations regularly spaced throughout a reef (e.g., most damselfishes and other demersal-spawning species; Thresher, 1984). However, a number of species aggregate at a few widely dispersed places at particular times to spawn en masse (e.g., Nassau grouper *Epinephalus striatus*; Sala et al., 2001; see general review by Domeier and Colin, 1998). Although these species appear to be distributed as discrete populations most of the time, their reproduction takes place when they have coalesced temporarily as a single, composite population. Specific recruitment events are likely to occur at local habitat patches, within which most postsettlement demography will be enacted. This will result in patch-specific fluctuations in population size through site- and/or time-specific rates of recruitment, mortality, and growth (see Section III.A.1). However, local fecundity will not determine local replenishment. Rather, aggregationwide fecundity will determine local replenishment, with the local contribution being difficult or impossible to identify.

The existence of spawning aggregations does not guarantee that metapopulation structure does not exist. The critical feature will be the scale of the catchment from which aggregating spawners are drawn relative to the scale of a local habitat patch. For certain species, aggregations might be created by the movement of individual fish within a single contiguous reef system to one especially advantageous location within that reef system (Fig. 2-5A). For example, aggregations of *Acanthurus nigrofuscus* in the Red Sea appear to consist of local animals that have traveled at most 1 or 2 km within a single reef system (Mazeroll and Montgomery, 1995). In these cases, the aggregation still represents a distinct local breeding population, the distribution of animals in which is simply changed from nonspawning periods. If, on the other hand, fish travel great distances and cross expanses of nonreef habitat, thereby leaving discrete habitat patches, to join an aggregation, then metapopulation structure is dissolved (Fig. 2-5B). For example, Zeller (1998) and Bolden (2000) have tracked large serranids traveling from one reef system to another to join spawning aggregations. Whether the majority of individuals of aggregating species leave their resident patch to spawn en masse or whether they simply congregate within their home patch is unclear. When considerable interpatch travel does occur, the apparent distribution of animals among discrete patches during the majority of the year is an illusion of a metapopulation because no local population closure exists.

An interesting contrast to reef fish spawning aggregations is the nonspawning aggregations of coastal sciaenids and anadromous alosines and salmonids (Jones, this volume). These fish intermingle at offshore sites as a composite group when not spawning, but spread themselves among estuaries (sciaenids) and inland rivers (alosines and salmonids) when spawning. Larvae are entrained and juveniles grow within the estuaries and rivers, resulting in distinct populations of

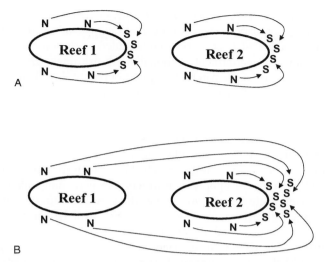

FIGURE 2-5. Schematic diagrams of the redistribution of fish forming spawning aggregations from nonspawning (N) to spawning (S) periods. (A) Fish can move to aggregate within a contiguous habitat patch, preserving the integrity of local populations, and therefore metapopulation structure. (B) Alternatively, fish can leave their home reef and travel greater distances to join an aggregation, dissolving metapopulation structure by eliminating independent local production of offspring.

immature life stages. More important, after maturing and moving to an offshore habitat, a large proportion of fish return to spawn in their natal estuary or river, resulting in a degree of population closure. Therefore, although there can be an element of homogenous demography among populations when they intermix offshore, variability among populations using distinct estuaries is likely to induce an element of asynchrony in population fluctuations. Although those reef fishes that migrate extensively to traditional spawning aggregation sites might have an illusionary metapopulation structure as a result of their generally patchy distribution when not spawning (e.g. Nassau grouper; Sala et al., 2001), among sciaenids, alosines, and salmonids, metapopulation structure is real, although arguably masked by homogenous distribution during part of the life cycle.

B. Nursery Habitats

Mangrove forests and sea grass beds often contain small individuals of common reef fish species (reviewed by Beck et al., 2001; Heck et al., 2003). Typically, these small-bodied fish are assumed to be juveniles that will eventually move to reef habitat at maturation, although this assumption has rarely been tested (but see

Sheaves, 1995; Chittaro et al., 2004). If nonreef-dwelling fish are destined to join the reef-dwelling population, then a potentially complicating (from the metapopulation perspective, at least) extra step is added to the process linking offspring production and recruitment. Although adult fish might grow, reproduce, and die within a single habitat patch, if they do not recruit directly to that habitat patch, the requisite partial closure of local populations becomes blurred or lost. With the blurring or loss of local population identity comes the dissolution of metapopulation structure.

Like the effect of spawning aggregations on metapopulation structure, the effect of nursery areas is largely dependent upon issues of scale. Although nursery areas might be distinct habitats, if they can be linked with specific patches of reef then we can still identify metapopulation structure by combining different habitat types into a composite patch. We might then be able to distinguish self-recruitment and external supply (Fig. 2-6A). This will be most likely when nursery habitats are contained within reef systems, such as sea grass beds or mangrove-fringed islands within reef lagoons, or are located in close proximity. Mumby et al. (2004) have documented strong differences in population size of reef fishes along the Mesoamerican Barrier Reef System between reefs adjacent to mangroves and those with no mangroves nearby. This provides indirect evidence of a strong link between adjacent reef and mangrove habitats, and suggests that they can be combined into a single composite population. The existence of metapopulation structure would then rest upon a high proportion of larvae that settle into the nursery habitat being produced by the same reef-dwelling population that is in turn fed by that nursery habitat.

In contrast, if recruits from mixed sources settle within nursery habitat that is distinct and distant from the adult reef habitat (Fig. 2-6B), then we will be less likely to delineate metapopulation structure. Sheaves (1995) examined lutjanid and serranid populations in coastal estuaries and on offshore reefs on the Great Barrier Reef, and found that fish in estuaries were considerably smaller that those on reefs. This pattern has been reported elsewhere (e.g., Nagelkerken et al., 2001; Nagelkerken and van der Velde, 2002; Mumby et al., 2004), but does not necessarily imply the use of nonreef habitat as a nursery ground. It is possible, for example, that environmental differences cause stunted growth, and the size differences are the result of dwarfing and not the presence of a large juvenile population. This effect has been reported for scarid populations on outer shelf reefs of the Great Barrier Reef (Gust et al., 2001, 2002). However, Sheaves (1995) was able to provide more convincing evidence that estuaries are nurseries. He was not able to document movement of tagged fish from an estuary to a reef, but he was able to demonstrate a fundamental demographic divide between the two habitats. Specifically, there was minimal overlap in age structures in estuaries and on reefs, and all estuarine fish were immature whereas all reef-dwelling fish were mature. Therefore, it had to be the case that fish lived in estuaries while young

Chapter 2 The Metapopulation Ecology of Coral Reef Fishes   59

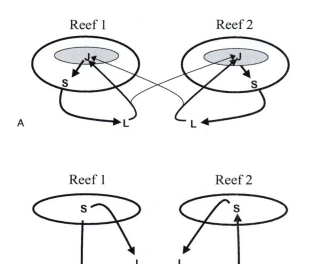

FIGURE 2-6. Schematic diagrams of the possible arrangement of nonreef nursery habitat (stippled) relative to reef habitat. (A) Nursery habitat adjacent to or enclosed within a reef system allows metapopulation structure to exist by preserving a link between local production of larvae (L) by the spawning population (S) and recruitment of juveniles (J) back to that spawning population. This is contingent, of course, upon sufficient self-recruitment to achieve some independence of local populations coupled with some external replenishment. (B) A nursery habitat some distance from the reef habitat that is supplied by multiple local populations and, in turn, replenishes multiple local populations is less likely to allow partial closure of local populations.

and immature before moving to reefs at older, reproductively active age classes.

Notably, the estuaries and reefs examined by Sheaves (1995) were not close to one another, instead being separated by tens of kilometers. This geographic spacing decreases the likelihood of strong connections between individual nursery habitats and individual reefs, instead making it more likely that nursery habitats will both receive fish from and supply fish to multiple adult populations (Fig. 2-6B). This will decouple local production and recruitment of offspring, and dissolve metapopulation structure. Of course, it is possible that fish recruiting to and intermingling within a shared nursery habitat will relocate to their natal reefs at maturation, therefore preserving the identity of local populations and, by extension, metapopulation structure. This will be difficult to discern, but advances in genetic, microchemical, and tagging techniques might reveal these patterns (see Chittaro et al., 2004).

V. SUMMARY

Not surprisingly, there is no single answer to the question of whether coral reef fishes form metapopulations. The answer will depend upon species, geographic setting, and definition. We generally lack the information needed to propose an answer for any species in any place using any but the loosest definitions (i.e., "populations connected by dispersal"). Especially important and most notably lacking are detailed data on dispersal and connectivity patterns that will determine to what degree local populations are closed and to what degree their dynamics are influenced by external sources of replenishment.

The critical knowledge gaps notwithstanding, there is certainly evidence to suggest that coral reef fishes may form metapopulations, at least when metapopulations are defined with reference to spatial structure and connectivity rather than extinction–recolonization dynamics. The latter phenomenon may be evident at the edge of species ranges, such as within the Capricorn–Bunker group of the Great Barrier Reef. Extinction–recolonization events might also occur naturally or might be caused by fishing pressure for large species that occur in small numbers and for other rare species. However, for many of the best-studied and nonexploited species, local extinction is and will likely remain an infrequent phenomenon.

We view metapopulation dynamics as something more than extinction–recolonization dynamics, and see greater potential for the presence of metapopulations among coral reef fishes. Habitat patches can form discrete templates for local populations as on the Great Barrier Reef (Fig. 2-2), or species can be sufficiently sedentary that we can combine sets of local breeding groups into functional subpopulations within otherwise contiguous habitats (Fig. 2-4). Demographic traits and behavior can vary even among very proximal populations, or among adjacent reef zones, reflecting part of the independence of these populations. These differences mean that numbers of fish, age and size structure, and reproductive output will vary spatially as well, which can create interpopulation inequities, making some populations more important than others in terms of sustaining the larger system, at least under certain conditions. The population dynamics that result from dispersal patterns and local demography typically show asynchrony among populations, at least as indicated by recruitment data. However, it is unclear whether this is the result of a high degree of independence of local populations, or whether it results from populations being replenished from a common regional larval pool but with supply to local populations being differentially modified by local oceanographic conditions. Although the connectivity picture is still very much in the initial sketch phase, early evidence suggests the potential for self-recruitment that will create the local independence characteristic of metapopulation dynamics.

To provide better answers, reef fish ecologists need, first and foremost, to continue the burgeoning trend of large-scale, interdisciplinary research on larval dispersal and demographic connectivity. At the same time, we should not accept too blindly the assumption that postsettlement reef fishes are sedentary, and we should be looking to test this assumption in situations where it seems likely to be violated. The longer tradition of demographic and behavioral research focused on local populations should not be lost, and will need to be focused on those populations identified through dispersal and movement studies to be interacting demographically. In addition to these more interesting process-oriented questions, a key goal should be to compile long-term and spatially replicated data sets on trends in abundance of reef fishes, because these are the data that will ultimately test how dispersal, recruitment, local demography, and behavior at multiple scales interact. Whether some or all reef fish species turn out to be structured as true metapopulations, reef fish ecology owes a great deal to metapopulation ecology for helping to encourage larger scale and longer term thinking, and awareness of the importance of processes acting both locally and regionally to determine persistence and changes in abundance.

Finally, reef fish are eminently amenable to field experimentation and long-term studies of local groups. Our suggestion that many sedentary fish that do not migrate to breed, and are distributed over an extensive contiguous habitat nevertheless breed as if they were a set of adjacent but functionally independent populations should be directly testable. Under the right conditions, such sets of functionally independent populations could function as metapopulations. If confirmed, this would further extend the concept of the metapopulation in a way that might be applicable to other organisms that are distributed across contiguous habitats but do not interbreed freely because of limits of movement by individuals. The importance of spatial relationships among individuals and groups may become as important to understanding metapopulation biology as the nesting of spatial scales that is the hallmark of this phenomenon.

REFERENCES

Adams, S., Mapstone, B.D., Russ, G.R., and Davies, C.R. (2000). Geographic variation in the sex ratio, sex specific size, and age structure of *Plectropomus leopardus* (Serranidae) between reefs open and closed to fishing on the Great Barrier Reef. *Can. J. Fish. Aquat. Sci.* **57**, 1448–1458.

Adams, S., and Williams, A.J. (2001). A preliminary test of the transitional growth spurt hypothesis using the protogynous coral trout *Plectropomus maculatus*. *J. Fish. Biol.* **59**, 183–185.

Alcala, A.C., and Russ, G.R. (1990). A direct test of the effects of protective management on abundance and yield of tropical marine resources. *J. Cons. Int. Explor. Mer.* **47**, 40–47.

Aldenhoven, J.M. (1986). Local variation in mortality rates and life-expectancy estimates of a coral reef fish *Centropyge bicolor* (Pisces: Pomacanthidae). *Mar. Biol.* **92**, 237–244.
Armsworth, P.R. (2002). Recruitment limitation, population regulation, and larval connectivity in reef fish metapopulations. *Ecology.* **83**, 1092–1104.
Beck, M.W., Heck, K.L., Able, K.W., Childers, D.L., Eggleston, D.B., Gillanders, B.M., Halpern, B., Hays, C.G., Hoshino, K., Minello, T.J., Orth, R.J., Sheridan, P.F., and Weinstein, M.P. (2001). The identification, conservation, and management of estuarine and marine nurseries for fish and invertebrates. *Bioscience.* **51**, 633–641.
Bellwood, D.R., and Hughes, T.P. (2001). Regional-scale assembly rules and biodiversity of coral reefs. *Science.* **292**, 1532–1534.
Berkeley S.A., Hixon M.A., Larson R.J., and Love M.J. (2004). Fisheries sustainability via protection of age structure and spatial distribution of fish populations. *Fisheries.* **29**, 23–32.
Bolden, S. (2000). Long-distance movement of a Nassau grouper (*Epinephelus striatus*) to a spawning aggregation in the central Bahamas. *Fish. Bull.* **98**, 642–645.
Brown, J.H., and Kodric–Brown, A. (1977). Turnover rates in insular biogeography: Effects of immigration on extinction. *Ecology.* **58**, 445–449.
Caley, M.J. (1998). Age-specific mortality rates of reef fishes: Evidence and implications. *Aust. J. Ecol.* **23**, 241–245.
Chapman, M., and Kramer, D. (2000). Movements of fishes within and among fringing coral reefs in Barbados. *Env. Biol. Fishes.* **57**, 11–24.
Chittaro, P.M., Usseglio, P., and Sale, P.F. (2004). Variation in fish density, assemblage composition and relative rates of predation among mangrove, sea grass and coral reef habitats. *Env. Biol. Fishes.* **72**, 175–187.
Choat, J.H., and Axe, L.M. (1996). Growth and longevity in acanthurid fishes: An analysis of otolith increments. *Mar. Ecol. Prog. Ser.* **134**, 15–26.
Choat, J.H., and Robertson, D.R. (2002). Age-based studies. In *Coral reef fishes: Dynamics and diversity in a complex ecosystem* (P.F. Sale, ed., pp. 57–80). San Diego: Academic Press.
Coté, I.M., and Cheney, K.L. (2005). Animal mimicry: Choosing when to be a cleaner-fish mimic. *Nature.* **433**, 211–212.
Coté, I.M., Mosqueira, I., and Reynolds, J.D. (2001). Effects of marine reserve characteristics on the protection of fish populations: A meta-analysis. *J. Fish Biol.* **59**, 178–189.
Cowen, R.K. (2002). Larval dispersal and retention and consequences for population connectivity. In *Coral reef fishes: Dynamics and diversity in a complex ecosystem* (P.F. Sale, ed., pp. 149–170). San Diego: Academic Press.
Cowen, R.K., and Schultz, E.T. (1994). Recruitment of coral-reef fishes to Bermuda: Local retention or long-distance transport? *Mar. Ecol. Prog. Ser.* **109**, 15–28.
Cowen, R.K., Lwize, K.M.M., Sponaugle, S., Paris, C.B., and Olson, D.B. (2000). Connectivity of marine populations: Open or closed? *Science.* **287**, 857–859.
Crowder, L.B., Lyman, S.L., Figueira, W.F., and Priddy, J. (2000). Source–sink population dynamics and the problem of siting marine reserves. *Bull. Mar. Sci.* **66**, 799–820.
Davies, C.R. (1995). *Patterns of movement of three species of coral reef fish on the Great Barrier Reef.* PhD thesis, Department of Marine Biology. Townsville, Australia: James Cook University.
Davis, T.L.O., and West, G.J. (1992). Growth and mortality of *Lutjanus vittus* (Quoy and Gaimard) from the North West Shelf of Australia. *Fish. Bull.* **90**, 395–405.
Day, T., and Taylor, P.D. (1997). Von Bertalanffy's growth equation should not be used to model age and size at maturity. *Am. Nat.* **149**, 381–393.
Dixon, P.A., Milicich, M.J., and Sugihara, G. (1999). Episodic fluctuations in larval supply. *Science.* **283**, 1528–1530.
Doherty, P.J. (1991). Spatial and temporal patterns in recruitment. In *The ecology of fishes on coral reefs* (P.F. Sale, ed., pp. 261–293). San Diego: Academic Press.

Doherty, P.J. (2002). The impact of variable replenishment upon populations of coral reef fishes. In *Coral reef fishes: Dynamics and diversity in a complex ecosystem* (P.F. Sale, ed., pp. 327–355). San Diego: Academic Press.
Doherty, P., and Fowler, T. (1994). An empirical test of recruitment limitation in a coral reef fish. *Science.* **263**, 935–939.
Doherty, P.J., and Williams, D.M. (1988). The replenishment of coral reef fish populations. *Oceanogr. Mar. Biol. Ann. Rev.* **26**, 487–551.
Doherty, P.J., Williams, D., and Sale, P.F. (1985). The adaptive significance of larval dispersal in coral reef fishes. *Env. Biol. Fishes.* **12**, 81–90.
Domeier, M.L., and Colin, P.L. (1998). Tropical reef fish spawning aggregations: Defined and reviewed. *Bull. Mar. Sci.* **60**, 698–726.
Done, T.J. (1982). Patterns in the distribution of coral communities across the central Great Barrier Reef. *Coral Reefs.* **1**, 95–107.
Dudgeon, C.L., Gust, N., and Blair, D. (2000). No apparent genetic basis to demographic differences in scarid fishes across continental shelf to the Great Barrier Reef. *Mar. Biol.* **137**, 1059–1066.
Ebert, T.A. (1985). Sensitivity of fitness to macroparameter changes: An analysis of survivorship and individual growth in sea urchin life histories. *Oecologia.* **65**, 461–467.
Foley, P. (1997). Extinction models for local populations. In *Metapopulation biology: Ecology, genetics and evolution* (I. Hanski and M.E. Gilpin, eds., pp. 215–246). San Diego: Academic Press.
Forrester, G.E., and Steele, M.A. (2000). Variation in the presence and cause of density-dependent mortality in three species of reef fishes. *Ecology.* **81**, 2416–2427.
Forrester, G.E., Vance, R.R., and Steele, M.A. (2002). Simulating large-scale population dynamics using small-scale data. In *Coral reef fishes: Dynamics and diversity in a complex ecosystem* (P.F. Sale, ed., pp. 275–301). San Diego: Academic Press.
Fowler, A.J. (1991). Reproductive biology of bisexual and all-female populations of chaetodontid fishes from the southern Great Barrier Reef. *Env. Biol. Fishes.* **31**, 261–274.
Fricke, H., and Fricke, S. (1977). Monogamy and sex change by aggressive dominance in coral reef fish. *Nature.* **266**, 830–832.
Grimm, V., Reise, K., and Strasser, M. (2003). Marine metapopulations: A useful concept? *Helgoland Marine Research.* **56**, 222–228.
Gust, N., Choat, J.H., and Ackerman, J.L. (2002). Demographic plasticity in the tropical reef fishes. *Mar. Biol.* **140**, 1039–1051.
Gust, N., Choat, J.H., and McCormick, M.I. (2001). Spatial variability in reef fish distribution, abundance, size and biomass: A multiscale analysis. *Mar. Ecol. Prog. Ser.* **214**, 237–251.
Halpern, B.S., and Warner, R.R. (2002). Marine reserves have rapid and lasting effects. *Ecol. Lett.* **5**, 361–366.
Hanski, I.A. (1999). *Metapopulation ecology.* New York: Oxford University Press.
Hanski, I., and Gaggiotti, O.E. (2004). Metapopulation biology: Past, present, and future. In *Ecology, genetics and evolution of metapopulations* (I. Hanski and O.E. Gaggiotti, eds., pp. 3–22). San Diego: Academic Press.
Hart, A.M., and Russ, G.R. (1996). Response of herbivorous fishes to crown-of-thorns starfish *Acanthaster planci* outbreaks. III. Age, growth, mortality and maturity indices of *Acanthurus nigrofuscus. Mar. Ecol. Prog. Ser.* **136**, 25–35.
Heck, K.L., Hays, G., and Orth, R.J. (2003). Critical evaluation of the nursery role hypothesis for sea grass meadows. *Mar. Ecol. Prog. Ser.* **253**, 123–136.
Hilborn, R., and Walters, C.J. (1992). *Quantitative fisheries stock assessment: Choice dynamics and uncertainty.* New York: Chapman and Hall.
Hixon, M.A., and Carr, M.H. (1997). Synergistic predation, density dependence, and population regulation in marine fish. *Science.* **277**, 946–949.

Hixon, M.A., and Webster, M.S. (2002). Density dependence in reef fish populations. In *Coral reef fishes: Dynamics and diversity in a complex ecosystem* (P.F. Sale, ed., pp. 303–325). San Diego: Academic Press.

Holland, K.N., Peterson, J.D., Lowe, C.G., and Wetherbee, B.M. (1993). Movements, distribution and growth rates of the white goatfish *Mulloides flavolineatus* in a fisheries conservation zone. *Bull. Mar. Sci.* **52**, 982–992.

Hughes, T.P., Bellwood, D.R., and Connolly, S.R. (2002). Biodiversity hotspots, centres of endemicity and the conservation of coral reefs. *Ecol. Lett.* **5**, 775–784.

James, M.K., Armsworth, P.R., Mason, L.B., and Bode, L. (2002). The structure of reef fish metapopulations: Modelling larval dispersal and retention patterns. *Proc. Roy. Soc. Lond. B.* **269**, 2079–2086.

Jones, G.P. (1991). Postrecruitment processes in the ecology of coral reef fish populations: A multifactorial perspective. In *The ecology of fishes on coral reefs* (P.F. Sale, ed., pp. 294–328). San Diego: Academic Press.

Jones, G.P., Caley, M.J., and Munday, P.L. (2002). Rarity in coral reef fish communities. In *Coral reef fishes: Dynamics and diversity in a complex ecosystem* (P.F. Sale, ed., pp. 81–102). San Diego: Academic Press.

Jones, G.P., Milicich, M.J., Emslie, M.J., and Lunow, C. (1999). Self-recruitment in a coral reef fish population. *Nature.* **402**, 802–804.

Kiflawi, M., Mazeroll, A.I., and Goulet, D. (1998). Does mass spawning enhance fertilization in coral reef fish? A case study of the brown surgeonfish. *Mar. Ecol. Prog. Ser.* **172**, 107–114.

Kinlan, B.P., and Gaines, S.D. (2003). Propagule dispersal in marine and terrestrial environments: A community perspective. *Ecology.* **84**, 2007–2020.

Kritzer, J.P. (2002). Variation in the population biology of stripey bass *Lutjanus carponotatus* within and between two island groups on the Great Barrier Reef. *Mar. Ecol. Prog. Ser.* **243**, 191–207.

Kritzer, J.P. (2004). Sex-specific growth and mortality, spawning season, and female maturation of the stripey bass (*Lutjanus carponotatus*) on the Great Barrier Reef. *Fish. Bull.* **102**, 94–107.

Kritzer, J.P., and Davies, C.R. (2005). Demographic variation within spatially structured reef fish populations: When are larger-bodied subpopulations more important? *Ecol. Mod.* **182**, 49–65.

Kritzer, J.P., and Sale, P.F. (2004). Metapopulation ecology in the sea: From Levins' model to marine ecology and fisheries science. *Fish and Fisheries.* **5**, 131–140.

Leis, J.M. (1991). The pelagic stage of reef fishes: The larval biology of coral reef fishes. In *The ecology of fishes on coral reefs* (P.F. Sale, ed., pp. 183–230). San Diego: Academic Press.

Leis, J.M., and McCormick, M.I. (2002). The biology, behavior and ecology of the pelagic, larval stage of coral reef fishes. In *Coral reef fishes: Dynamics and diversity in a complex ecosystem* (P.F. Sale, ed., p. 171–200). San Diego: Academic Press.

Levin, D.A. (1995). Plant outliers—an ecogenetic perspective. *Am. Nat.* **145**, 109–118.

Levins, R. (1969). Some demographic and genetic consequences of environmental heterogeneity for biological control. *Bulletin of the Entomological Society of America.* **15**, 237–240.

Levins, R. (1970). Extinction. In *Some mathematical problems in biology* (M. Desternhaber, ed., pp. 77–107). Providence, RI: American Mathematical Society.

Lockwood, D.L., Hastings, A., and Botsford, L.W. (2002). The effects of dispersal patterns on marine reserves: Does the tail wag the dog? *Theor. Pop. Biol.* **61**, 297–309.

Lubchenco, J., Palumbi, S.R., Gaines, S.D., and Andelman, S. (2003). Plugging a hole in the ocean: The emerging science of marine reserves. *Ecol. Appl.* **13**, S1–S7.

Marconato, A., Shapiro, D.Y., Petersen, C.W., Warner, R.R., and Yoshikawa, T. (1997). Methodological analysis of fertilization rate in the bluehead wrasse *Thalassoma bifasciatum*: Pair versus group spawns. *Mar. Ecol. Prog. Ser.* **161**, 61–70.

Mazeroll, A.I., and Montgomery, W.L. (1995). Structure and organization of local migrations in brown surgeonfish (*Acanthurus nigrofuscus*). *Ethology.* **99**, 89–106.

McCosker, J.E. (1977). Fright posture of the plesiopid fish *Calloplesiops altivelis*: An example of batesian mimicry. *Science.* **194**, 400–401.
McQuinn, I.H. (1997). Metapopulations and the Atlantic herring. *Rev. Fish Biol. Fish.* **7**, 297–329.
Meekan, M.G., Ackerman, J.L., and Wellington, G.M. (2001). Demography and age structures of coral reef damselfishes in the tropical eastern Pacific Ocean. *Mar. Ecol. Prog. Ser.* **212**, 223–232.
Meyer, C.G., Holland, K.N., Wetherbee, B.M., and Lowe, C.G. (2000). Movement patterns, habitat utilization, home range size and site fidelity of whitesaddle goatfish, *Parupeneus porphyreus*, in a marine reserve. *Environ. Biol. Fish.* **59**, 235–242.
Mora, C., and Robertson, D.R. (2005). Factors shaping the range–size frequency distribution of the endemic fish fauna of the tropical eastern Pacific. *J. Biogeog.* **32**, 277–286.
Mora, C., and Sale, P.F. (2002). Are populations of coral reef fish open or closed? *Trends Ecol. Evol.* **9**, 422–428.
Mora, C., Chittaro, P.M., Sale, P.F., Kritzer, J.P., and Ludsin, S.A. (2003). Patterns and processes in reef fish diversity. *Nature.* **421**, 933–936.
Mumby, P.J. (1999). Can Caribbean coral populations be modelled at metapopulation scales? *Mar. Ecol. Prog. Ser.* **180**, 275–288.
Mumby, P.J., Green, E.P., Edwards, A.J., and Clark, C.D. (1997). Coral reef habitat mapping: How much detail can remote sensing provide? *Mar. Biol.* **130**, 193–202.
Mumby, P.J., Edwards, A.J., Arlas–Gonzalez, J.E., Lindeman, K.C., Blackwell, P.G., Gall, A., Gorcyznska, M.I., Harborne, A.R., Pescod, C.L., Renen, H., Wabnitz, C.C.C., Llewellyn, G. (2004). Mangroves enhance the biomass of coral reef fish communities in the Caribbean. *Nature.* **427**, 533–536.
Munday, P.L., and Jones, G.P. (1998). The ecological implications of small body size among coral reef fishes. *Oceanogr. Mar. Biol. Ann. Rev.* **36**, 373–411.
Murphy, H.T., and Lovett–Doust, J. (2004). Context and connectivity in plant metapopulations and landscape mosaics: Does the matrix matter? *Oikos.* **105**, 3–14.
Nagelkerken, I., and van der Velde, G. (2002). Do non-estuarine mangroves harbor higher densities of fish than adjacent shallow-water and coral reef habitats in Curacao (Netherlands Antilles)? *Mar. Ecol. Prog. Ser.* **245**, 191–204.
Nagelkerken, I., Kleijnen, S., Klop, T., van den Brand, R.A.C.J., de la Moriniere, E.C., and van der Velde, G. (2001). Dependence of Caribbean reef fishes on mangroves and sea grass beds as nursery habitats: A comparison of fish faunas between bays with and without mangroves/sea grass beds. *Mar. Ecol. Prog. Ser.* **214**, 225–235.
Newman, S.J., and Williams, D.M. (1996). Variation in reef associated assemblages of the Lutjanidae and Lethrinidae at different distances offshore in the central Great Barrier Reef. *Environ. Biol. Fishes.* **46**, 123–138.
Newman, S.J., Williams, D.M., and Russ, G.R. (1996a). Age validation, growth and mortality rates of the tropical snappers (Pisces: Lutjanidae) *Lutjanus adetii* (Castelnau, 1873) and *L. quinquelineatus* (Bloch, 1790) from the central Great Barrier Reef, Australia. *Mar. Freshwater Res.* **47**, 575–584.
Newman, S.J., Williams, D.M., and Russ, G.R. (1996b). Variability in the population structure of *Lutjanus adetii* (Castelnau, 1873) and *L. quinquelineatus* (Bloch, 1790) among reefs in the central Great Barrier Reef, Australia. *Fish. Bull.* **94**, 313–329.
Newman, S.J., Williams, D.M., and Russ, G.R. (1997). Patterns of zonation of assemblages of the Lutjanidae, Lethrinidae and Serranidae (Epinephelinae) within and among mid-shelf and outer-shelf reefs in the central Great Barrier Reef. *Mar. Freshwater Res.* **48**, 119–128.
Palumbi, S.R. (2003). Population genetics, demographic connectivity, and the design of marine reserves. *Ecol. Appl.* **13**, S146–S158.
Pennisi, E. (2003). Survey confirms coral reefs are in peril. *Science.* **297**, 1622–1623.
Pepin, P. (1991). Effect of temperature and size on development, mortality, and survival rates of the pelagic early life history stages of marine fish. *Can. J. Fish. Aquat. Sci.* **48**, 503–518.

Petersen, C.W. (1991). Variation in fertilization rate in the tropical reef fish, *Halichoeres bivattatus*: Correlates and implications. *Biol. Bull.* **181**, 232–237.
Petersen, C.W., and Warner, R.R. (2002). The ecological context of reproductive behavior. In *Coral reef fishes: Dynamics and diversity in a complex ecosystem* (P.F. Sale, ed., pp. 103–119). San Diego: Academic Press.
Pitcher, C.R. (1988). Validation of a technique for reconstructing daily patterns in the recruitment of coral reef damselfish. *Coral Reefs.* **7**, 105–111.
Planes, S. (1993). Genetic differentiation in relation to restricted larval dispersal of the convict surgeonfish *Acanthurus triostegus* in French Polynesia. *Mar. Ecol. Prog. Ser.* **98**, 237–246.
Planes, S. (2002). Biogeography and larval dispersal inferred from population genetic analysis. In *Coral reef fishes: Dynamics and diversity in a complex ecosystem* (P.F. Sale, ed., pp. 201–220). San Diego: Academic Press.
Poiner, I.R., Glaister, J., Pitcher, C.R., Burridge, C., Wassenberg, T., Gribble, N., Hill, B., Blaber, S.J.M., Milton, D.A., Brewer, D., and Ellis, N. (1998). *The environmental effects of prawn trawling in the far northern section of the Great Barrier Reef Marine Park: 1991–1996*. Final report to GBRMPA and FRDC. CSIRO Division of Marine Research—Queensland Department of Primary Industries Report.
Pulliam, H.R. (1988). Sources, sinks and population regulation. *Am. Nat.* **132**, 652–661.
Roberts, C.M. (1997). Connectivity and management of Caribbean coral reefs. *Science.* **278**, 1454–1455.
Roberts, C.M., Bohnsack, J.A., Gell, F., Hawkins, J.P., and Goodridge, R. (2001). Effects of marine reserves on adjacent fisheries. *Science.* **294**, 1920–1923.
Roberts, C.M., McClean, C.J., Veron, J.E.N., Hawkins, J.P., Allen, G.R., McAllister, D.E., Mittermeier, C.G., Schueler, F.W., Spalding, M., Wells, F., Vynne, C., and Werner, T.B. (2002). Marine biodiversity hotspots and conservation priorities for tropical reefs. *Science.* **295**, 1280–1284.
Robertson, D.R. (1972). Social control of sex reversal in a coral reef fish. *Science.* **177**, 1007–1009.
Robertson, D.R. (1998). Do coral-reef fish faunas have a distinctive taxonomic structure? *Coral Reefs.* **17**, 179–186.
Roff, D.A. (1992). *The evolution of life histories*. New York: Chapman and Hall.
Ross, R.M., Losey, G.S., and Diamond, M. (1983). Sex change in a coral-reef fish: Dependence of stimulation and inhibition on relative size. *Science.* **217**, 574–576.
Russ, G.R., and Alcala, A.C. (1989). Effects of intense fishing pressure on an assemblage of coral reef fishes. *Mar. Ecol. Prog. Ser.* **56**, 13–27.
Russ, G.R., and Alcala, A.C. (1996). Do marine reserves export adult fish biomass? Evidence from Apo Island, central Philippines. *Mar. Ecol. Prog. Ser.* **132**, 1–9.
Russ, G.R., Lou, D.C., and Ferreira, B.P. (1996). Temporal tracking of a strong cohort in the population of a coral reef fish, the coral trout, *Plectropomus leopardus* (Serranidae: Epinephelinae), in the central Great Barrier Reef. *Can. J. Fish. Aquat. Sci.* **53**, 2745–2751.
Sadovy, Y.J. (1996). Reproduction of reef fishery species. In *Reef fisheries* (N.V.C. Polunin and C.M. Roberts, eds., pp. 15–59). New York: Chapman and Hall.
Sala, E., Ballesteros, E., and Starr, R.M. (2001). Rapid decline of Nassau grouper spawning aggregations in Belize: Fishery management and conservation needs. *Fisheries.* **26**, 23–29.
Sale, P.F. (1991). Reef fish communities: open nonequilibrial systems. In *The ecology of fishes on coral reefs* (P.F. Sale, ed., pp. 564–598). San Diego: Academic Press.
Sale, P.F. (1998). Appropriate spatial scales for studies of reef-fish ecology. *Aust. J. Ecol.* **23**, 202–208.
Sale, P.F. (2004). Connectivity, recruitment variation, and the structure of reef fish communities. *Integrative and Comparative Biology.* **44**, 390–399.
Sale, P.F., Cowen, R.K., Danilowicz, B.S., Jones, G.P., Kritzer, J.P., Lindeman, K.C., Planes, S., Polunin, N.V., Russ, G.R., and Sadovy, Y.J. (2005). Critical science gaps impede use of no-take fishery reserves. *Trends Ecol. Evol.* **20**, 74–80.

Sale, P.F., and Douglas, W.A. (1984). Temporal variability in the community structure of fish on coral patch reefs and the relation of community structure to reef structure. *Ecology.* **65**, 409–422.

Sale, P.F., and Kritzer, J.P. (2003). Determining the extent and spatial scale of population connectivity: Decapods and coral reef fishes compared. *Fish. Res.* **65**, 153–172.

Shapiro, D.Y. (1980). Serial female sex changes after simultaneous removal of males from social groups of a coral reef fish. *Science.* **209**, 1136–1137.

Sheaves, M. (1995). Large lutjanid and serranid fishes in tropical estuaries: Are they adults or juveniles? *Mar. Ecol. Prog. Ser.* **129**, 31–40.

Shulman, M.J., and Bermingham, E. (1995). Early life histories, ocean currents, and the population genetics of Caribbean coral reef fishes. *Evolution.* **49**, 897–910.

Simpson, S.D., Meekan, M., Montgomery, J., McCauley, R., and Jeffs, A. (2005). Homeward sound. *Science.* **308**, 221.

Smedbol, R.K., McPherson, A., Hansen, M.M., and Kenchington, E. (2002). Myths and moderation in marine "metapopulations?" *Fish and Fisheries.* **3**, 20–35.

Swearer, S.E., Caselle, J.E., Lea, D.W., and Warner, R.R. (1999). Larval retention and recruitment in an island population of a coral-reef fish. *Nature.* **402**, 799–802.

Taylor, M.S., and Hellberg, M.E. (2003). Genetic evidence for local retention of pelagic larvae in a coral reef fish. *Science.* **299**, 107–109.

Thresher, R.E. (1982). Interoceanic differences in the reproduction of coral-reef fishes. *Science.* **218**, 70–72.

Thresher, R.E. (1984). *Reproduction in reef fishes.* Neptune City, NJ: TFH Publishing.

Victor, B.C. (1983). Recruitment and population dynamics of a coral reef fish. *Science.* **219**, 419–420.

Wainwright, P.C., and Bellwood, D.R. (2002). Ecomorphology of feeding in coral reef fishes. In *Coral reef fishes: Dynamics and diversity in a complex ecosystem* (P.F. Sale, ed., pp. 33–56). San Diego: Academic Press.

Warner, R.R. (1988). Traditionality of mating-site preferences in a coral reef fish. *Nature.* **335**, 719–721.

Warner, R.R., and Chesson, P.L. (1985). Coexistence mediated by recruitment fluctuations: A field guide to the storage effect. *Am. Nat.* **125**, 769–787.

Warner, R.R., and Hughes, T.P. (1988). The population dynamics of reef fishes. *Proc. Int. Coral Reef Symp.* **1**, 149–155.

Warner, R.R., Robertson, D.R., and Leigh, E.G. (1975). Sex change and sexual selection. *Science.* **190**, 633–638.

Wiens, J.A. (1997). Metapopulation dynamics and landscape ecology. In *Metapopulation biology: Ecology, genetics and evolution* (I. Hanski and M.E. Gilpin, eds., pp. 43–67). San Diego: Academic Press.

Williams, A.J., Davies, C.R., Mapstone, B.D., and Russ, G.R. (2003). Scales of spatial variation in demography of a large coral-reef fish: An exception to the typical model? *Fish. Bull.* **101**, 673–683.

Williams, D.M., and Hatcher, A.I. (1983). Structure of fish communities on outer slopes of inshore, mid-shelf and outer shelf reefs of the Great Barrier Reef. *Mar. Ecol. Prog. Ser.* **10**, 239–250.

Wilson, E.O. (1975). *Sociobiology.* Cambridge, MA: Harvard University Press.

Zeller, D.C. (1997). Home range and activity patterns of the coral trout *Plectropomus leopardus* (Serranidae). *Mar. Ecol. Prog. Ser.* **154**, 65–77.

Zeller, D.C. (1998). Spawning aggregations: Patterns of movement of the coral trout *Plectropomus leopardus* (Serranidae) as determined by ultrasonic telemetry. *Mar. Ecol. Prog. Ser.* **162**, 253–263.

Zeller, D.C., and Russ, G.R. (1998). Marine reserves: Patterns of adult movement of the coral trout (*Plectropomus leopardus* [Serranidae]). *Can. J. Fish. Aquat. Sci.* **55**, 917–924.

Zeller, D.C., and Russ, G.R. (2000). Population estimates and size structure of *Plectropomus leopardus* (Pisces: Serranidae) in relation to no-fishing zones: Mark–release–resighting and underwater visual census. *Mar. Freshwater Res.* **51**, 221–228.

CHAPTER 3

Temperate Rocky Reef Fishes

DONALD R. GUNDERSON[1] and RUSSELL D. VETTER[2]

I. Introduction
II. Geological Processes and the Types and Distribution of Temperate Reef Habitat
    A. Subduction, Volcanism, and Faulting
    B. Glaciation
    C. Fluvial and Dynamic Submarine Erosive Processes
III. Temperate Reef Fish Communities
    A. Biogeographic Provinces and Temperate Reef Fish Communities
    B. Depth as a Master Variable in Temperate Reef Fish Communities
    C. Typical Fish Fauna of Temperate Rocky Reef Communities
IV. The Role of Oceanography in Metapopulation Structuring
    A. Major Oceanographic Domains
    B. Dispersal and Retention Mechanisms
V. Climate, Climate Cycles, and Historical Metapopulation Structuring
VI. The Role of Life History in Metapopulation Structuring
    A. Early Life History
    B. Juveniles
    C. Adults
    D. Longevity
VII. Empirical Approaches to Measuring Dispersal and Metapopulation Structure
VIII. Population Genetic Studies in North Pacific Rocky Reef Fishes

[1]University of Washington School of Aquatic & Fishery Sciences, Box 355020, Seattle, WA 98195-5020
[2]Southwest Fisheries Science Center, National Marine Fisheries Service, 8604 La Jolla Shores Dr., La Jolla, CA 92038

Marine Metapopulations, Jacob P. Kritzer and Peter F. Sale
Copyright © 2006 by Elsevier. All rights of reproduction in any form reserved.

IX. Human Impacts
X. Future Directions for Metapopulation Studies of
    Temperate Reef Fishes
    References

I. INTRODUCTION

The application of the metapopulation concept to the study and management of temperate rocky reef fishes will succeed only to the extent that the concept has greater explanatory and predictive power than other approaches, such as island biogeography (MacArthur and Wilson, 1967), regional stocks, or clinal variation. As noted by Hanski and Gilpin (1997), "The metapopulation approach is based on the notion that space is not only discrete but that there is a binary distinction between suitable and unsuitable habitat types." Levins (1969, 1970) originally emphasized the importance of extinction-recolonization of local populations but Kritzer and Sale (2004) have emphasized the "degree of demographic connectivity" whereby populations that are primarily self recruiting but with some significant external replenishment are the essence of the metapopulation concept. North Pacific temperate reef fishes generally fit this notion but sometimes with the added complexity of ontogenetic migration between pelagic larval, benthic juvenile, and benthic adult habitats (Vetter and Lynn, 1997). Furthermore, the combination of extended pelagic phases coupled with unpredictable changes in ocean climate (e.g. El Niño events) allows for demographically significant but temporally rare linkages between local populations. In an era of intense overfishing and ecosystem change (Pauly et al., 1998; Jackson et al., 2001; Dayton et al., 2002; Myers and Worm, 2003), temperate reef communities are becoming increasingly fragmented as inaccessible natural refugia (Yoklavich et al., 2000) and intentionally created (Airamé et al., 2003) marine protected areas (MPAs) become islands within a matrix of heavily exploited habitat. Thus the metapopulation dynamics of the past may not be the metapopulation dynamics of the future. In this chapter we consider two classes of potential metapopulation effects on temperate reef fishes: short-term demographic effects and long-term genetic effects. These can also be considered as ecological and evolutionary effects. Demographic effects are those resulting from differences in the dispersal of propagules and resultant effects on local and regional recruitment and population dynamics. Genetic effects are those that result from the collective, long-term effects of consistent patterns in metapopulation dynamics and the impact of processes such as extinction-recolonization, gene flow, genetic drift, and local selection. In a management context, both demographic and genetic effects of metapopulation dynamics on temperate reef fish communities need to be considered. Fisheries managers need to evaluate the utility of incorporating metapopulation concepts into existing ideas of fish stock dynamics, recovery plans, and regional management strategies. Conservation biologists need to consider the utility of incorporating metapopulation concepts into the provisions of the Endangered Species Act and the Act's concept of a distinct population segment (DPS), or evolutionarily significant unit (ESU). Conservation biologists may also need to incorporate metapopulation concepts such as extinction-recolonization into recovery strategies and the design of MPA networks.

As noted throughout this volume, marine metapopulations differ from terrestrial metapopulations because of the tendency of many marine species to produce large numbers of pelagic eggs and larvae that are capable of dispersing beyond adjacent populations to distant populations in demographically significant numbers. However, this does not mean that all members of marine fish communities behave similarly or that dispersal potential equates to realized dispersal.

Since many temperate reef species have long life-spans (sometimes over 100 or even 200 years, Table 3-1) the effects of chance on favorable recruitment events can be quite long lasting for local populations. A single strong year class can take decades to progress through the population, first as a dominant prey item, and then as a dominant predator. The presence of several age groups tends to stabilize the abundance of adults and the species composition of temperate marine communities (Berkeley et al., 2004), and longevity and persistence of adults may also limit the risk of stochastic extinction of an allele or haplotype. In this chapter we confine our frame of reference to temperate rocky reef fish communities along the continental margins of the Northeast Pacific Ocean. We describe the different types of reef habitat, the geological origins of habitats, the fish communities that inhabit them at different depths and latitudes, the present day oceanographic and climatic forces that drive variations in local demography and the longer term forces such as glaciation and sea level change that may have shaped their genetic legacy. Temperate rocky reef communities are largely absent from the Northwest Atlantic south of Canada but are very common and share similarities in other tectonically or glacially active coastlines such as the Northwest Pacific coast, New Zealand and the west coast of South America. As a point of departure we suggest that the metapopulation concept may best be employed to describe variations in recruitment dynamics between oceanographic domains such as the Alaska Gyre, the California Current and the California Counter Current (Fig. 3-1).

II. GEOLOGICAL PROCESSES AND THE TYPES AND DISTRIBUTION OF TEMPERATE REEF HABITAT

While coral reef fish communities rely on biogenic coral growth to provide critical habitat structure, temperate reef habitats typically occur in geologically active areas. Stable rocky areas quickly become covered in sediments unless kept clear by prevailing currents. In the Northeast Pacific, temperate reef habitats are commonly formed along subduction margins, and in areas where geological thrust

TABLE 3-1. Common members of temperate marine reef fish communities, female age at maturity, and maximum age

		Age at Maturity (females)	Max. Age (years)	Max. Size (cm)
San Diego Province (S. California)				
Shallow (less than 40 m)				
black-and-yellow rockfish	Sebastes chrysomelas	4	30?	39
blue rockfish	S. mystinus	6–11	44	53
brown rockfish	S. auriculatus	4–5	34+	56
gopher rockfish	S. carnatus	4	30?	43
grass rockfish	S. rastrelliger	4	23+	56
greenblotched rockfish	S. rosenblatti	–	50+	48
kelp rockfish	S. atrovirens	4	25+	43
olive rockfish	S. serranoides	5	30+	61
rosy rockfish	S. rosaceus	–	14	36
squarespot rockfish	S. hopkinsi	4–5	19	29
starry rockfish	S. constellatus	6–9	30+	46
treefish	S. serriceps	–	23+	41
California scorpionfish	Scorpaena guttata	2	21	43
California sheephead	Semicossyphus pulcher	3–6	50	91
kelp greenling	Hexagrammos decagrammus	4	16	53
giant sea bass	Stereolepis gigas	11–13	75+	226
kelp bass	Paralabrax clathratus	3–5	34	71
barred sand bass	P. nebulifer	3–5	24	66
cabezon	Scorpaenichthys marmoratus	4–5	20+	99
striped seaperch	Embiotoca lateralis	3	10?	38
black surfperch	E. jacksoni	1–2	9+	39
Deep (50–500 m)				
blackgill rockfish	Sebastes melanostomus	20	87+	61
bocaccio	S. paucispinis	4	50	91
chilipepper	S. goodei	3–4	35	59
cowcod	S. levis	11	55+	94
greenspotted rockfish	S. chlorostictus	6	33+	47
halfbanded rockfish	S. semicinctus	2	15	25
swordspine rockfish	S. ensifer	3+	43+	25
vermilion rockfish	S. miniatus	5	60+	76
Oregonian Province (Central California-eastern Gulf of Alaska)				
Shallow (less than 40 m)				
black rockfish	Sebastes melanops	6–8	50	69
blue rockfish	S. mystinus	6–11	44	53
brown rockfish	S. auriculatus	4–5	34+	56
china rockfish	S. nebulosus	4–5	79+	45
copper rockfish	S. caurinus	6	50	66
Puget Sound rockfish	S. emphaeus	2	22	18
quillback rockfish	S. maliger	7–11	95	61
tiger rockfish	S. nigrocinctus	–	116?	61
lingcod	Ophiodon elongatus	4–5	25?	152
kelp greenling	Hexagrammos decagrammus	4	16	53

TABLE 3-1. *Continued*

		Age at Maturity (females)	Max. Age (years)	Max. Size (cm)
cabezon	*Scorpaenichthys marmoratus*	4–5	20+	99
striped seaperch	*Embiotoca lateralis*	3	10?	38
Deep (50–500 m)				
bank rockfish	*Sebastes rufus*	10+	85+	55
bocaccio	*S. paucispinis*	4	50?	91
canary rockfish	*S. pinniger*	7–9	84+	76
harlequin rockfish	*S. variegatus*	–	47+	38
Pacific ocean perch	*S. alutus*	10	100+	53
pygmy rockfish	*S. wilsoni*	–	26	23
redstripe rockfish	*S. proriger*	7	55+	51
rosethorn rockfish	*S..helvomaculatus*	10	87+	41
rougheye rockfish	*S. aleutianus*	20	205	97
sharpchin rockfish	*S. zacentrus*	6	58+	45
silvergray rockfish	*S. brevispinis*	–	82+	73
tiger rockfish	*S. nigrocinctus*	–	116	61
widow rockfish	*S. entomelas*	5–7	60+	59
yelloweye rockfish	*S. ruberrimus*	19–22	118+	91
yellowmouth rockfish	*S. reedi*	–	99+	58
yellowtail rockfish	*S. flavidus*	10	64+	66
lingcod	*Ophiodon elongatus*	4–5	25?	152
Aleutian Province (Central and Western Gulf of Alaska)				
Shallow (less than 40 m)				
black rockfish	*Sebastes melanops*	10	50	69
copper rockfish	*S. caurinus*	6	50	66
dusky/dark rockfish	*S. variabilis/S. ciliatus*	11	67	53
quillback rockfish	*S. maliger*	11	95	61
lingcod	*Ophiodon elongatus*	4–5	25?	152
Deep (50–500 m)				
northern rockfish	*Sebastes polyspinis*	13	57+	48
Pacific ocean perch	*S. alutus*	10	100+	53
rougheye rockfish	*S. aleutianus*	20	205	97
shortraker rockfish	*S. borealis*	–	157?	120
yelloweye rockfish	*S. ruberrimus*	19–22	118+	91
Atka mackerel	*Pleurogrammus monopterygius*	4	15	54

Principal Sources: Leet et al., 2001, Love et al., 2002.

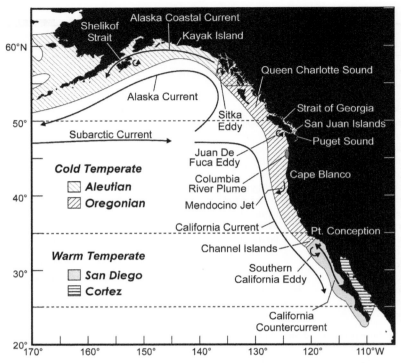

FIGURE 3-1. Biogeographic regions of the Northeast Pacific that contain temperate rocky reef fish communities. Communities in the Cortez Province are thought to be mostly isolated due to warmer waters at the tip of Baja California. Major geologic headlands, bays, and islands that are discussed in the context of metapopulations are shown. Persistent oceanographic features that may affect dispersal, retention, or create barriers to dispersal are also shown.

faults have occurred. Island archipelagoes are common in such areas, and areas such as the San Juan Islands, and the Channel Islands supply a mosaic of rocky banks, and steep walls that are ideal habitats for species such as rockfish and lingcod. The amounts and spatial distributions of temperate reef habitat can vary greatly over glacial-interglacial cycles (Fig. 3-2).

A. Subduction, Volcanism, and Faulting

Subduction along the continental margin results in the formation of steep eroded cliffs and exposed rocky faces (Fig. 3-3a) that typify much of the region's coast, intertidal zone, and narrow continental shelf (Ricketts et al., 1992). Submerged volcanoes and lava fronts with exposed columnar basalt and lava fields provide another type of high-relief habitat required by rockfish on the continental shelf and slope

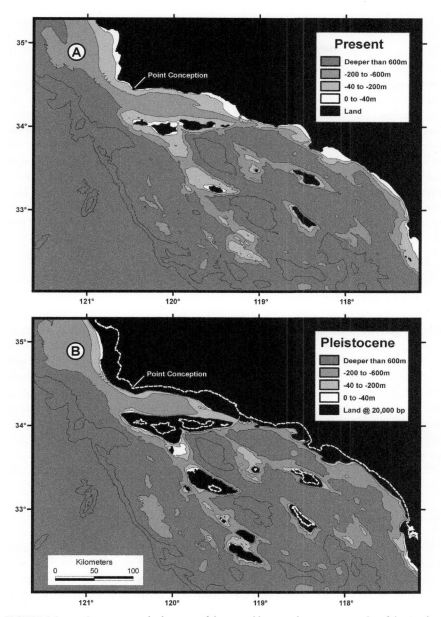

FIGURE 3-2. a. Contemporary bathymetry of the typical basin and range topography of the Southern California Bight. Narrow contours generally suggest high relief and places where rocky reef habitat may be expected but recent studies using ROVs and side scan sonar indicate a complex and often unpredictable mosaic of soft bottom and rocky reef habitats that are highly dependent on current scour and sediment deposition. Note that for nearshore rocky reef fishes inhabiting kelp forests the four northern Channel Islands are separate but for deeper reef species the islands are contiguous habitat. b. Large changes in sea level, habitat availability, and habitat connectivity are driven by glacial cycles and in the future perhaps by anthropogenic climate change. As recently as 20,000BP at the end of the Pleistocene sea level was up to 200m lower, the northern Channel Islands were joined, and several offshore banks were islands. Circulation was limited by the ridge extending down to the now submerged banks at 119° west.

off Alaska (O'Connell and Carlile, 1993), and on numerous seamounts (Dower and Perry, 2001). In the eastern Gulf of Alaska and Southern California Bight other tectonic forces are dominant. Here the Pacific and North American Plates grind past each other and strike-slip faulting within the transform fault boundary results. In the Southern California Bight a series of uplifted east-west trending ridges form the Santa Barbara Basin and northern Channel Islands. A series of northwest-southeast trending ridges form the southern Channel Islands and associated offshore banks of the Southern California Borderland. The offshore banks and steep island faces are composed primarily of volcanic, metamorphic, and differentially eroded sedimentary rock. This results in a patchwork of linear ridges, undercut ledges, and sand channels (Fig. 3-3b). The habitat mosaic shown in Figure 3-3b is the most typical type of temperate rocky reef in the Northeast Pacific. In shallow water the substrate is overgrown with algal turf and large macrophytes such as kelp. In deeper water the substrate is covered by sessile, filter-feeding, invertebrate communities.

B. Glaciation

Glacial activity creates habitat for rocky reef fishes via two processes. Glaciers scour away surface sediment resulting in submerged fjords and extensive areas of bare rock. Ice sheets can deposit extensive boulder fields or moraines that are also ideal habitat for rocky reef fishes. The most recent Cordilleran Ice Sheet extended to just south of the present day Puget Sound, and to near the edge of the continental shelf. In the northern part of the Northeast Pacific such forces have formed extensive areas of rocky habitat throughout Alaska, British Columbia, Washington, and Oregon (O'Connell and Carlile, 1993; Kreiger, 1992; Murie et al., 1994; Richards, 1986; Mathews and Richards, 1991; Nasby-Lucas et al., 2002). Some species and size classes prefer boulder fields to high relief bedrock.

C. Fluvial and Dynamic Submarine Erosive Processes

Fluvial processes that incise the continental shelf during periods of low sea level and subsequent submersion can create features such as submarine canyons. Dynamic submarine erosive processes, such as turbidity currents, debris flows, and other gravity-transported sediment can also form gullies and submarine canyons. High-negative relief areas such as Monterey Canyon or Scripps Canyon off California are known to support high concentrations of temperate reef fishes (Yoklavich et al., 2000; Vetter pers. obs.). Despite the high relief, in many cases the substrate is actually quite soft because the canyon's walls have been either eroded into soft unconsolidated sediments or covered by recent sedimentation.

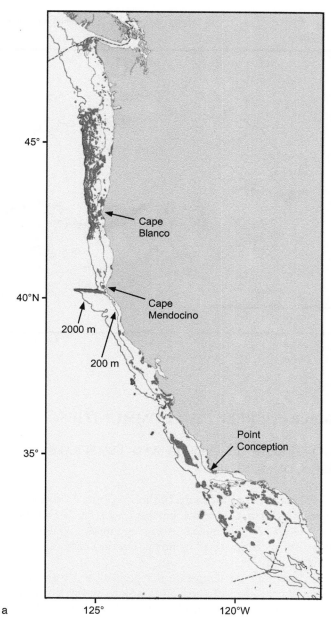

FIGURE 3-3. a) Preliminary map of rocky substrate shallower than 2000 m off the Pacific coast of the United States (after NOAA, 2005). b) Habitat map of the south side of San Miguel Island. This is the westernmost of the four northern Channel Islands shown in Figure 2a. Side-scan imaging, together with visual ROV surveys, complements high resolution multi-beam bathymetry to reveal fine scale details of rock substrate and the attached flora and fauna that create rocky reef habitat. In this simplified image habitat types are grouped into rock (dark grey), sand (light grey), and rock with kelp (black). The linear nature of the sedimentary rock ledges, interspersed with sand channels, can be readily seen. In shallow water the rock ledges are often covered by a dense mat of giant kelp. Under ideal conditions, such as occur at San Miguel Island, the kelp forest adds a vertical component to rocky reef habitat that extends from the bottom to the surface in depths up to 40 m (after Cochrane et al., 2002).

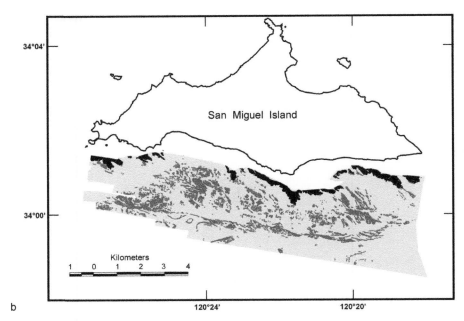

FIGURE 3-3. *Continued*

III. TEMPERATE REEF FISH COMMUNITIES

A. BIOGEOGRAPHIC PROVINCES AND TEMPERATE REEF FISH COMMUNITIES

Briggs (1974) characterized the fauna of the temperate Northeast Pacific, using three provinces: Aleutian, Oregonian, and San Diego (Fig. 3-1). Differences in the species composition of the temperate reef communities conform closely to the biogeographic provinces described by Briggs, but they have the added dimension of structuring by depth (Table 3-1).

The break between cold temperate (Oregonian) and warm temperate (San Diego) provinces near Point Conception is striking in the larger picture but the ranges of many species do not coincide with this biogeographic break and ranges often tend to move north and south with oceanic regime changes (Wares et al., 2001). Near Pt. Conception many families that are common to tropical and subtropical reefs such as the wrasses (Labridae), damselfishes (Pomacentridae) and sea basses and groupers (Serranidae) reach their northern limits and are replaced by cold temperate reef families such as greenlings (Hexagrammidae), sculpins

(Cottidae) and surfperches (Embiotocidae). In general the northern boundary of warm temperate species is more defined than the southern boundaries of cold temperate species (Wares et al., 2001). From a metapopulation perspective perhaps the most interesting families are those that contain species that span biogeographic boundaries and are subject to different temperature regimes, oceanographic processes, and biotic interactions in different parts of their range (see Section IV.A). The Family Scorpaenidae, which includes the rockfishes (*Sebastes*), thornyheads (*Sebastolobus*), and scorpionfishes (*Scorpaena*), is the most prominent example of these types of distributions. For example, a dominant deep reef species such as the bocaccio rockfish must adapt its life history to recruit in the upwelling-dominated systems of the central California coast as well as within the complex basin and range hydrography of the Southern California Bight. Other commercial temperate reef species that span biogeographic boundaries include the lingcod and the cabezon.

B. Depth as a Master Variable in Temperate Reef Fish Communities

Geological forces create the basic habitats for rocky reef fishes and latitude and current boundaries form the basis for biogeographic provinces and intraspecific phylogeographic barriers to dispersal (above). However, for reasons partly known and unknown, depth creates a profound degree of bathymetric structure within temperate rocky reef fish communities. Sunlight, wave energy, and the presence of kelp forests and attached algae create obvious differences between shallow reefs and those only slightly deeper. However, depth also profoundly influences fish communities below 100 m (Vetter et al., 1994; Vetter and Lynn, 1997). Temperate reef species complexes at a given depth are often more similar between Alaska and California than they are a few hundred meters shallower or deeper on the same offshore bank (Table 3-1). Depth preferences often differ with ontogenetic stage (Jacobson and Vetter, 1996; Vetter and Lynn, 1997) yet they remain very predictable. Physiological and biochemical effects of hydrostatic pressure are well known (Somero, 1991), but the bases for depth structuring and ontogenetic migration are complex and probably reflect more than biochemical limitations because the early life stages usually occur in surface waters. The importance of depth as a habitat variable is well manifested even on deep-water oil production platforms where the fish fauna changes dramatically but predictably from the surface to the bottom of the structures (Love et al., 2000). One aspect of this depth stratification that is particularly interesting, in terms of metapopulation structure, is that fish on the same reef with different depth preferences view the world as connected or non-connected depending on depth (Fig. 3-2). Bocaccio and cowcod rockfishes living at 100–300 m view the Channel Islands and offshore banks as more or less connected habitat while rockfishes living in the kelp forests (0–40 m contour) view the available habitat as a series of unconnected patches.

C. Typical Fish Fauna of Temperate Rocky Reef Communities

There are numerous field and photographic guides that illustrate the variety of North Pacific temperate reef fishes (e.g. Humann, 1996; Love et al., 2002). Shallow water communities are dominated by macrophytes such as giant kelp, *Macrocystis pyrifera*, that attach to hard substrates and increase vertical structure. Species such as the olive rockfish (2–3 cm) utilize the kelp canopy for their initial settlement site (Fig. 3-4a). Larger juveniles such as this calico rockfish (10–20 cm), *Sebastes dalli*, settle to the bottom (Fig. 3-4b) and may shelter in kelp holdfasts, sea urchin spines, and natural cavities in the rock (Fig. 3-4b). Below the photic zone, reefs are often covered by invertebrates that provide shelter as well as forage. Although presently overfished, the bocaccio rockfish is one of the most common large rockfish species (50–60 cm), inhabiting deeper reefs (Fig. 3-4c). Once recruited, many temperate reef fishes will remain on a reef throughout their lives. Here a yelloweye rockfish, more common north of Pt. Conception, coexists with a vermilion rockfish on an offshore bank in the Southern California Bight (Fig. 3-4d). Given the size of the yelloweye rockfish (30–40 cm) this fish probably recruited to the area over 50 years ago. Lingcod (70–80 cm) with their huge gape and needle-sharp teeth are a dominant predator from inshore reefs to depths of over 400 m (Fig. 3-4e). Although highly resident, adult lingcod also migrate yearly to shallow water reefs to breed and guard egg masses attached to boulders. Rockfishes give birth to live young. When fully gravid, species such as the cowcod (60–80 cm; Fig. 3-4f) may show diminished movement and feeding activity. As with species that produce attached egg masses, the release of live young rather than free-drifting eggs may decrease the time of passive dispersal and encourage larval retention in rocky reef species.

IV. THE ROLE OF OCEANOGRAPHY IN METAPOPULATION STRUCTURING

Demographic asynchrony among local and regional temperate reef populations is one of the hallmarks of metapopulations. Oceanography is the primary driving factor in temperate reef metapopulation structuring because ocean currents dictate in large measure the survival of recruits and their delivery to settlement habitats. For invertebrate larvae, oceanographic effects have been shown to be important at all spatial and temporal scales from boundary layer phenomena, through mesoscale features, up to persistent major currents (Botsford, 2001).

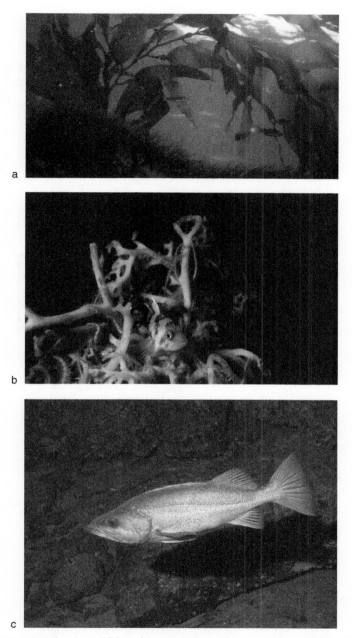

FIGURE 3-4. Typical rocky reef fishes of the Northeast Pacific. a. newly recruited juvenile rockfish sheltering in kelp canopy (J. Hyde). b. benthic juvenile calico rockfish sheltering in a kelp holdfast (J. Hyde). c. bocaccio rockfish, (J. Butler). d. yelloweye rockfish, at rear, and vermilion rockfish, in front, (J. Butler). e. lingcod, (J. Butler). f. gravid cowcod, (J. Butler). (See color plate)

FIGURE 3-4. *Continued* (See color plate)

Early stages of fishes have been more difficult to monitor but a number of novel empirical approaches such as continuous egg pumps (Watson et al., 1999), and the use of standardized settlement traps (Ammann, 2004) are now being used to study local recruitment pulses and regional recruitment variability. Indirect methods such as genetics also hold much promise (Section VIII). Much of this research is driven by the interest in networks of Marine Protected Areas (MPAs, Section X).

A. Major Oceanographic Domains

Large-scale structuring of the northeast Pacific is characterized by the gyres formed by the California and Alaska Currents (Fig. 3-1). Within the California Current system, the influence of warm temperate and cold temperate water differs substantially North and South of Pt. Conception, and the fish communities in these areas are strikingly different (Table 3-1).

Complex oceanography at the convergence of the southward flowing California Current and the Southern California Eddy (Fig. 3-1 and Harms and Winant, 1998) dominates the biology of the Southern California Bight and the associated Channel Islands. Each island has a predictable but different fish fauna composed of a mix between the cold temperate Oregonian Province and warm temperate San Diego Province faunas (Ebeling et al., 1980; Engle, 1993). Strong El Niño currents can transport fish larvae long distances resulting in the northward displacement of southern species. Self-sustaining populations usually do not persist but such displacement phenomena do illustrate the types of long-distance larval transport that can presumably occur within a normal species' range (Lea and Rosenblatt, 2000).

The California Current system, north of Pt. Conception, is characterized by strong, alongshore winds. Mean monthly advection over the continental shelf is often on the order of 10–20 cm/sec (Hickey, 1998). Reversals in alongshore winds result in dramatic shifts from upwelling to downwelling on both the event scale and seasonally (Hickey, 1998). The theoretical net transport for passively drifting eggs and larvae above the thermocline can often exceed 1000 km over a two-month period (Rooper, 2002).

Puget Sound's glacial origin has resulted in the presence of a number of sills and basins, and it interacts with the adjacent ocean much like an estuary. The fish communities within Puget Sound are sufficiently isolated that unique stocks have evolved for Pacific hake (*Merluccius productus*), and several species of flatfish and rockfish. A number of basins, separated from the main sound by sills, as well as rocky island archipelagoes exist throughout Puget Sound and the Strait of Georgia, resulting in a complex current structure with a variety of semi-permanent and tidal eddies (Ebbesmeyer et al., 1991).

The eastern Gulf of Alaska off British Columbia and Southeast Alaska receives substantial freshwater inputs from several major river systems there, and is characterized by the presence of a strong northward-flowing coastal current over the inner 10–25 km of the continental shelf (Royer, 1998). The Alaska Current flows north and westward throughout the year, although current speeds are higher in the winter than during summer. Maximum nontidal current coastal current speeds fluctuate from about 50 cm/sec off the British Columbia coast to more than 180 cm/sec near the apex of the Gulf of Alaska (Royer, 1998). Simulated 90-day drift trajectories (Ingraham et al., 1998) indicate that passively drifting eggs and larvae released near the coast would frequently be transported over 600 km. The Exxon Valdez oil spill occurred on March 24, 1989, near the time when many of the fish in the area were actively spawning, and the leading edge of the spill traveled 760 km in 56 days (Royer, 1998). Westward surface velocities exceeding 50 cm/sec and approaching 100 cm/sec have been observed off Atka Island in the Aleutians (Favorite et al., 1976).

Flow in the Gulf of Alaska is characterized by a number of mesoscale eddies of varied origin (Royer, 1998). In Shelikof Strait, these eddies originate from forcing due to fresh-water inputs and local winds, while they are produced by the interaction of submarine banks and canyons with along-shelf flow in areas such as Chiniak trough and Amatuli Canyon south of Kodiak Island (Royer, 1998). The origins of several large, semi-permanent, eddies off Vancouver Island, Sitka, and Kayak Island are not yet fully understood (Royer, 1998).

B. Dispersal and Retention Mechanisms

Once an egg floats to the surface or is advected offshore into open water, dispersal over wide distances is a simple matter of passive drift given the strong alongshore advection throughout most of the California and Alaska Current systems (Rooper, 2002; Ingraham et al., 1998). In contrast, understanding local retention of eggs and larvae is a far more formidable problem.

A number of mesoscale (10–100 km) processes could act to retain eggs and larvae around local features. These include fronts associated with upwelling or the intersection of different current systems (Bakun, 1996; Shanks et al., 2000), internal waves (Shanks, 1983; Kingsford and Choat, 1986), gyres above banks, seamounts and canyons (Hickey, 1997), or upwelling shadows on the lee-side of headlands such as Cape Blanco and Cape Mendocino (Hickey, 1998). Fronts and gyres can also develop over much smaller scales (1–10 km), and are evident in areas such as the San Juan Islands, where tidal currents are strong, and the bottom topography is complex and rugged. The Columbia River plume is a strong feature off Washington and Oregon, and results in local, eddy-like features (Hickey, 1998). These mechanisms would tend to retain eggs and larvae, even if they were drifting passively, but directed movements by larvae probably act in concert with these processes (Bradbury and Snelgrove, 2001; Sponaugle et al., 2002) to facilitate "self recruitment" over local reef or reef complexes (see Section VI.A). Direct field sampling has shown that retention of rockfish larvae within the Southern California Eddy is influential in determining larval distribution patterns within the Southern California Bight (Taylor, 2004). A large body of recent theoretical and empirical work has investigated recruitment variability in invertebrates along the upwelling-dominated central California coast (see Botsford and Fogarty, this volume). The role of advection-relaxation events as a means of moving larvae offshore to grow and then returning them inshore to settle is compelling. The role of retention zones created by upwelling shadows (Graham and Largier, 1997; Wing et al., 1998) also seems important. Fewer studies have addressed fish but it seems clear that retention zones can operate similarly for *Sebastes* larvae (Wing et al., 1998).

At smaller scales boundary layer phenomena, or "sticky water," may be extremely important in limiting dispersal of shallow-water, temperate-reef species. Close to shore, particularly in kelp beds, advective flows are minimal (Jackson and Winant, 1983) and alongshore water movements are often bi-directional with the tidal cycle (Zeidberg and Hamner, 2002). In this nearshore environment, dispersal may be dominated by diffusive forces (Largier, 2003) and a small amount of directed movement may greatly alter retention probabilities. Adhesive eggs, nest guarding, and production of free-swimming young rather than eggs are all common behaviors in shallow reef species. While decreases in mortality are often suggested as reasons for producing more developed offspring, it is becoming clear that limiting the passive drift stage as eggs or larvae can have important consequences for retention (see Section VI.A).

V. CLIMATE, CLIMATE CYCLES, AND HISTORICAL METAPOPULATION STRUCTURING

Climate cycles impact every marine system but in temperate rocky reefs of the North Pacific the effects are dramatic and occur at several temporal scales. Glacial cycles have occurred at an average period of 100,000 years for the past million years. Aside from temperature and salinity effects they have a large effect on sea level and the relative amounts of kelp, cobble, and sand habitats (Graham et al., 2003).

At the height of the last glacial maximum, roughly 18,000 to 20,000 BP, the Puget Sound-Georgia Basin ecosystem did not exist (Thorson, 1980; Clague, 1989). Coastal reef habitats to the north were generally covered by ice except for small ice-free refugia such as the Haida Gwaii off Queen Charlotte Island (Clague, 1983; Mann and Hamilton, 1995). The past 18,500 years BP have been charac-

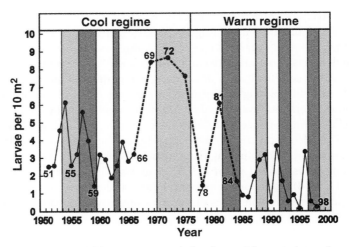

FIGURE 3-5. Bocaccio rockfish. Average annual abundance of bocaccio larvae from CalCOFI ichthyoplankton stations. Cold and warm regimes of the Pacific Decadal Oscillation (PDO) are indicated at the top, and Southern Oscillation (ENSO) events by vertical bars (El Niño, dark grey; La Niña, light grey). The wide bar represents a series of three consecutive La Niña events (after Moser et al., 2000).

terized by isostatic rebound of the land and rapid increases in sea level. This rapid increase in sea level had little effect on the open coast but had profound effects in the Puget Sound-Georgia Basin and the Southern California Bight. In the Southern California Bight the Northern Channel Islands changed from a large and continuous landmass to a series of four islands (Fig. 3-2).

In addition to the long-term pattern of Holocene warming and sea level rise, there is a persistent oscillation termed the Pacific Decadal Oscillation (PDO). Shifts in the PDO occur at roughly 20–30 year intervals, alternating between "warm" and "cool" phases that have significant impacts on temperature and primary production (Brodeur and Ware, 1992; Francis et al., 1998; McGowan et al., 1998) and this may have an effect on larval production and recruitment success (Fig. 3-5, Moser et al., 2000). The great longevity seen in many Northeast Pacific temperate reef species (Table 3-1) could, in part, represent an evolutionary adaptation to long-term climate change (Leaman and Beamish, 1984; Longhurst, 2002; Berkeley et al., 2004).

Wind and current patterns throughout the California-Alaska region can also vary substantially on a shorter temporal scale. El Niño events, which result in the enhanced advection of tropical water northward occur intermittently, resulting in reduced upwelling and increased surface temperatures. The impact of El Niño events is strongest off California, and relatively minor north of British Columbia. The El Niños of 1982 and 1998 have been among the strongest on record, resulting in the northward displacement of many fish species, and reduced recruitment for others. A number of species of rockfish (both exploited and unexploited) showed poor young-of-the-year survival off California during the 1983 and 1992 El Niño events (Ralston and Howard, 1995; Moser et al., 2000; Fig. 3-5).

VI. THE ROLE OF LIFE HISTORY IN METAPOPULATION STRUCTURING

The report "Open vs Closed Marine Populations" (Warner and Cowen, 2002) summarizes empirical and theoretical approaches to the question of "How connected are local populations?". If populations are closed (totally self recruiting) or open (recruitment from a common larval pool) there is no need to apply metapopulation concepts. Where each rocky reef species falls between these extremes will in large part determine the utility of applying the metapopulation concept. Reproductive mode, dispersal during egg and larval stages, and migrations during the juvenile and adult stages all play an important role in determining the linkage between temperate reefs and banks.

A. Early Life History

Many of the species that characterize temperate reefs are either viviparous (all rockfishes and surfperches) or spawn demersal eggs and guard the nests (lingcod, kelp greenling, Atka mackerel, cabezon). However, the dispersal strategies of temperate rocky reef fishes are difficult to generalize. Some, such as the surfperches (family Embiotocidae) produce a few large precocious young that do not have a pelagic dispersal phase. Others, such as the rockfishes (family Scorpaenidae) produce large numbers of larvae that remain pelagic for several months. Most of the rockfishes brood a single batch of young per year and release larvae over a short time window, while others such as the California sheephead, a member of the wrasse family, spawn every few days and produce eggs throughout the year except for a brief period in winter (Moser et al., 1993). Egg and larval surveys usually indicate widespread coherence in abundance over broad spatial scales (Moser et al., 2000), probably the result of the extensive alongshore transport characteristic of the Northeast Pacific. Coherence in recruitment patterns over relatively broad spatial scales has also been reported for pelagic and newly settled juveniles (Ralston and Howard, 1995).

Although relatively few rockfish larvae can be identified to the species level, Moser and Boehlert (1991) have shown that the distribution of species such as the bocaccio have a broad distribution from central Baja California to well north of the CalCOFI sampling area off northern California, while shortbelly rockfish

(*Sebastes jordani*) are more coastally and latitudinally confined. The duration of the larval and pelagic juvenile phase varies from a few months (copper rockfish) to a year (splitnose rockfish, *S. diploproa*), and clearly influences the extent of larval distribution. Despite this, benthic recruitment of juvenile rockfish often shows a predictable seasonal regularity (Larson, 1980a; Larson, 1980b; Love et al., 1991; Buckley, 1997), suggesting that they are able to orient themselves and undertake directed movements. This could be accomplished by taking advantage of the wide variety of localized oceanographic features that have been described previously. Moser and Boehlert (1991) noted that a variety of rockfish larvae are captured in midwater trawls, and explain their existence there by hypothesizing a shoreward return mechanism whereby late pelagic juveniles could migrate downward to a depth of about 200 m, where the geostrophic flow is shoreward. The 50+ years of the CalCOFI ichthyoplankton survey provides tremendous insights into patterns of larval dispersal. It is important to note that most ichthyoplankton surveys of the west coast, including CalCOFI surveys, do not venture close to land and may not present a clear picture of larvae that tend to be retained in shallow water kelp beds. Also, species such as surfperch that produce precocious young that can avoid nets are not well represented. As molecular methods of identification become more routine and methods for fine scale sampling such as continuous underway egg pumps are developed for use over rocky reef habitats (Watson et al., 1999), the degree of resolution of mesoscale variability in larval distribution patterns will undoubtedly improve.

B. Juveniles

In general, juveniles recruit to shallower depths than those occupied by adults. This property, which has been called ontogenetic migration or bathymetric demography, (Vetter et al., 1994; Jacobson and Vetter, 1996; Vetter and Lynn, 1997) has important implications for how populations are sampled and how these communities function since availability of juvenile and adult habitat may differ. Most deep-reef species such as Pacific ocean perch, yellowtail rockfish, sharpchin rockfish, rosethorn rockfish, widow rockfish, vermilion rockfish, and canary rockfish recruit directly to the bottom, and do not require kelp canopy (Carlson and Haight, 1976; Richardson and Laroche, 1979; Laroche and Richardson, 1980; Laroche and Richardson, 1981; Carr, 1991). However, recruitment of several species occupying shallow-water reefs as adults takes place where macrophytes such as kelp and eelgrass exist (Love et al., 1991). Macrophytic habitats may be utilized for only a few months in some cases, but provide habitat during the period immediately following settlement, when extensive mortality due to predation (Hobson et al., 2001), can occur. Initial recruits of copper, kelp, and gopher rockfishes are most abundant in the canopy of kelp forests, but with increased size, recruits move down the plants and then to adjacent habitats dominated by lower-growing macrophytes (Haldorson and Richards, 1987; Carr, 1991). Olive rockfish show a similar pattern of habitat use off southern California, with recently settled recruits aggregating along the shoreward edge of kelp beds, while older juveniles disperse over adjacent beds of low-growing brown algae at night. Large numbers of recruiting juvenile lingcod, black rockfish, bocaccio, and other shallow and deep-water rockfish associate to varying degrees with macrophytic habitats (Leaman, 1976; Love et al., 1991). In the case of lingcod and copper rockfish in Puget Sound, juveniles leave the macrophytic habitats after a few months, and move to deeper water reef or crevice habitats in the fall (Buckley, 1997). Using both mechanical tags and parasite markers, Love (1980) demonstrated that juvenile olive rockfish rarely, if ever, move off shallow (4–20 m) reefs off southern California, even when other suitable habitats are within 2 km. Blue rockfish tagged on breakwaters off central California moved little or not at all after 6 months (Miller and Geibel, 1973).

Seasonal migrations of juveniles from shallow to deep reefs, probably associated with the onset of winter storms have been reported for many shallow-water species including black, blue, yellowtail, vermilion, canary, and olive rockfish (Love et al., 1991). One exception to the general pattern of limited juvenile migrations is found in yellowtail rockfish. Adult yellowtail rockfish are rare in Puget Sound, which acts as a nursery area for larvae of oceanic origin, and juveniles tagged there commonly migrate to adult habitats 144 km away, and probably further (Mathews and Barker, 1983). Juvenile yellowtail rockfish tagged in Alaska moved from 425–1400 km (Stanley et al., 1994).

While fewer observations of juveniles have been made below SCUBA depth, observations from submersibles have shown that many rockfish juveniles use deep-water reef habitat, such as Heceta Bank (Nasby-Lucas et al., 2002). Yelloweye rockfish prefer wall, boulder, and high-relief rock habitats, with fish 20 cm and smaller being most abundant at 20–80 m, and those greater than 20 cm at 60–100 m (Richards, 1986; O'Connell and Carlile, 1993). Juvenile rougheye rockfish averaging seven years old were captured at 100 m in the protected waters of Southeast Alaska (Nelson and Quinn, 1987), while adults were most common in high relief and boulder habitats from 300–500 m deep (Ito, 1999; Soh et al., 2001). Cowcod juveniles recruited to shallow (40–100 m), soft-bottom habitats in Monterey Bay when 100 days old, although it is likely that these habitats were near low-lying patches of sand and cobble or low-relief rock outcrops (Johnson et al., 2001). Adult cowcod (Fig. 3-4f) are found much deeper, inhabiting high-relief habitats from 100–300 m (Love, 1996; Yoklavich et al., 2000).

Recruitment of juveniles to temperate reef adult habitats often appears to be limited by the carrying capacity of these habitats. Field observations and manipulation experiments showed clearly that given an adequate supply of larvae, kelp bass recruitment off southern California was dependent on the quality and quantity of kelp habitat available (Carr, 1994). In Puget Sound, densities of copper rockfish, quillback rockfish, and kelp greenling populations on an isolated reef increased almost three-fold in a single month following experimental removal of 85% of the fish present (Barker, 1979). However, the mean weights of the fish that recruited to the reef were 36–41% (depending on the species) less than those present before the experiment. Similarly, subadult (<6 cm) rockfish colonized a newly created artificial reef near Monterey Bay, CA within a few weeks (Mathews, 1985), attaining higher densities than at natural reefs that were 0.8–1.6 km away. In both Puget Sound and Monterey Bay, it appeared that large numbers of transient juveniles were actively searching for suitable habitat in which to live.

Annual recruitment to adult habitats is highly variable, a feature that is partially attributable to variations in climate and ocean conditions during the larval period (Fig. 3-5). However, while larval and juvenile supply may sometimes limit recruitment to adult habitats, competition with the long-lived adults already present (Carlson, 1986), and predation by other species could also be implicated. Whatever the mechanism, Ralston and Howard (1995) noted a substantial damping of the variation in rockfish year class strength between the pelagic juvenile stage ($CV = 0.96$–2.25) and entry into the fishery ($CV = 0.6$–1.39) for four species of rockfish, and suggested density dependent mortality as the most likely cause.

At present, our inability to determine the original source populations for those juveniles that have successfully established themselves in adult habitats presents the greatest obstacle in understanding the metapopulation dynamics of Northeast Pacific temperate reef communities. We are currently unable to delineate the spatial linkages between spawning areas, juvenile rearing areas, and adult habitats, although recent genetic evidence (Section VIII) indicates that the scales that these processes operate over can be much smaller (tens of kilometers for some species) than had previously been supposed.

C. Adults

The largest, most prominent fish associated with temperate reefs in the northeast Pacific are predominately rockfish (genus *Sebastes*) or greenlings (family Hexagrammidae: greenlings, lingcod, and Atka mackerel). Two Serranids (kelp and barred sand bass), a Polyprionid (giant sea bass), a Cottid (cabezon), and a Labrid (the California sheephead) are common off southern California (south of Pt. Conception).

Adult obligate reef-dwellers typically have very limited home ranges. Conventional and ultrasonic tagging have shown typical copper and quillback rockfish home ranges decrease with habitat quality, being $13\,m^2$ in optimal, high-relief habitat and $400\,m^2$ on low-relief rocky reefs (Mathews et al., 1987), and both species usually returned to within 10 m² of their release site when displaced 400 m. Larson (1980a,b) found that black-and-yellow and gopher rockfishes off California defend territories consisting of a shelter hole and nearby feeding area, and no movement greater than 1.2 km has ever been observed. Olive rockfish tagged with external and parasite tags showed no movement between reefs that were 2 km apart (Love, 1980), and the maximum movement of any tagged fish was 0.8 km. None of the 1,536 adult blue rockfish tagged by Lea et al. (1999) showed any movement away from their tagging location.

Shallow water species other than rockfish also show highly restricted movements. Acoustic telemetry tagging shows that both kelp bass (Lowe et al., 2003) and California sheephead have restricted home ranges. Conventional external tags have shown typical home ranges on the order of 11 km for lingcod (Chatwin, 1956), although more extensive seasonal migrations sometimes occur during the spawning season (Martell et al., 2000). While most (81%) adult lingcod tagged off the Washington coast were recaptured within 8 kilometers of their release site, 7% had moved over 50 km (Jagielo, 1990).

Seasonal movements of copper and quillback rockfishes to deeper water or alternative reef sites are also associated with turbulent winter conditions (Moulton, 1977) or the reduction in the availability of kelp habitat in winter. Shallow-water species that undergo extensive migrations are exceptional, but black rockfish is one of these. While most (69%) black rockfish tagged off the coast of Washington and northern Oregon were recaptured within 19 km of their release site, 12% were recovered more than 93 km away, and one of these had moved 656 km (Culver, 1987).

Many deep-water species of temperate reef rockfish cannot be tagged owing to distension of the swim bladder. While some deep-water species have home ranges that are just as small as those found shallower, there seems to be an increase in home range size with depth. Green-spotted rockfish tagged with ultrasonic tags spent most of their time within an area less than 2 sq. km (Starr et al., 2002). Stanley et al. (1994) found that 75% of adult yellowtail rockfish tagged were recovered within 25 km of their release point, although maximum movements of up to 400 km were observed. Pearcy (1992) tracked yellowtail rockfish using ultrasonic tags, and found that 8 out of 12 fish remained within 1.3 km of their capture location for a month. Displaced yellowtail rockfish have also demonstrated an ability to home, returning to their release site after being displaced up to 22.5 km (Carlson and Haight, 1972). Ten out of 16 adult bocaccio tracked by Starr et al. (2002) using ultrasonic tags spent less than 10% of their time within the approximately 12 km² observation area. However, none of the 86 bocaccio tagged by Lea et al. (1999) showed any significant movement away from their tagging site after 161–545 days at liberty.

Atka mackerel do not have a swim bladder, and have been successfully captured at 100–200 m and tagged. Fish tagged in 2000 were all recovered within

65 km of their tagging location after being at liberty 32–111 days. All Atka mackerel tagged in a single haul (11 out of 554 fish) near Seguam Pass in the Aleutian Islands homed to their capture site when displaced 9 km (McDermott, 2003).

Several of the species in Table 3-1, while commonly observed in rocky reef areas, spend much of their time away from high-relief habitats, and have relatively large home ranges. Widow rockfish are known to move into the water column at night and form dense schools that are targeted by commercial fishermen. Lea et al. (1999) tagged 50 canary rockfish off central California, and these had moved 6–704 km after being at liberty 1,114–1,439 days. Pacific ocean perch are known to make seasonal bathymetric migrations, and are most abundant in areas of low bottom relief (Mathews and Richards, 1991). Nevertheless, indirect evidence such as genetic analyses (Withler et al., 2001), parasite fauna and small-scale response to overfishing indicate that migrations over scales greater than about 56 km are uncommon (Gunderson, 1997). Localized depletion of deep-water species such as Pacific ocean perch has been documented (Gunderson, 1997), suggesting that little adult movement occurs to replenish depleted reefs.

D. Longevity

Adults of many temperate reef species don't mature until 10–20 years of age, and are long-lived (Table 3-1). Copper rockfish females mature at age 6 and live to be at least 50, while yelloweye and rougheye rockfish mature at age 20 years and live to be well over 100. Mean longevity is higher in the Gulf of Alaska than off southern California, but even there the species that dominate the temperate reef community typically have longevities in excess of 20 years. As a result of this high longevity, together with the broad-scale spatial patterns in the distribution of eggs and larvae, the species composition of temperate reef communities within a given biogeographic province and depth range tends to be relatively predictable (Weinberg, 1994; Williams and Ralston, 2002).

Longevity can also result in "demographic smoothing" at a given reef, where adult abundance remains roughly stable, although age composition changes over time as a series of dominant year classes pass through the population. Using SCUBA, Carlson (1986) observed a school of yellowtail rockfish that occupied a sunken ship off southeastern Alaska for 6 years, and noticed that the age structure reflected the progression of a few strong year classes. Recruitment to the school appeared to have taken place over 1–2 years, with little or no additional recruitment over subsequent years. Carlson concluded that this was probably due to exclusion of recruits by adults already present, which are far more adept at competing for space, cover, and highly motile food sources (Hobson et al., 2001).

Longevity also results in linkage between metapopulations over several decades. In fact, the clearest example of metapopulation structuring we are aware of occurs in the case of black and yellowtail rockfish populations in Puget Sound, which rely on recruitment pulses from oceanic populations for their existence. SCUBA surveys indicated the virtual disappearance of yellowtail rockfish between surveys conducted near San Juan Island in 1975 and 1991, and they were absent between 1991 and 2001. However, recent surveys in other areas indicate that yellowtail rockfish populations in other parts of Puget Sound are beginning to rebound (Eisenhardt, 2001).

The factors that have shaped such extreme "K-selected" life history patterns are still being debated. Some authors have suggested that great longevity can be explained by the decreased metabolic rates associated with low temperature or dissolved oxygen concentration (Cailliet et al., 2001), while others suggest that it is an adaptation to reproductive uncertainty (Leaman and Beamish, 1984; Longhurst, 2002). Certainly great longevity is more common in the more northerly portions of the Northeast Pacific (Table 3-1), and in deep water where temperatures and dissolved oxygen concentrations are relatively low. The reduced metabolic rates and low oxidative damage that characterize these environments would certainly be expected to enhance longevity (Cailliet et al., 2001), although this does not appear to be the case within the thornyheads (Pearson and Gunderson, 2003). However, any explanation that relies solely on environmental conditions is at odds with the fact that many other temperate reef species that overlap in their distribution with longer-lived species (Table 3-1) have evolved a far more "r-selected" life history pattern. Even within the genus *Sebastes*, short-lived "forage" species such as Puget Sound rockfish and pygmy rockfish can be found over the same reef complexes as long-lived apex predators. Purely "environmental" explanations also fail to account for the fact that many tropical reef species have also evolved K-selected life history patterns (Coleman et al., 2002; Kritzer and Sale, this volume).

As a result, it seems likely that the high longevities attained in Table 3-1 could also have evolved as the result of some combination of increasing survival, reproductive effectiveness, or ability to compete for resources (space, food, etc.) with age. As such, the oldest members of the population are likely to be the most valuable in terms of reproductive value. Fecundity increases exponentially with size for most species in Table 3-1, and this is true for both long and short-lived species. Survival of gestating embryos (maternal effect) has been shown to be highest in older fish for black rockfish (Bobko and Berkely, 2004), although this is not the case for copper rockfish (Cooper, 2003). Black rockfish larvae from older females survived over twice as long and grew more than three times faster than larvae of young females in laboratory rearing experiments (Berkeley et al., 2004). Current fisheries management practices do not take the differential value of older fish into account, yet the older members of the population are rapidly depleted when fishing begins. This could eventually have demographic and evolutionary consequences that we cannot yet begin to understand (Berkeley et al., 2004).

VII. EMPIRICAL APPROACHES TO MEASURING DISPERSAL AND METAPOPULATION STRUCTURE

The original metapopulation concept depended heavily on the idea of extinction-recolonization (Levins, 1969, 1970). The reliance on extinction-recolonization events might have been partly driven by the fact that recolonization was the only case where dispersal from other areas could be unequivocally demonstrated. Kritzer and Sale (2004) have argued for a broader definition for marine populations that de-emphasizes extinction-recolonization and emphasizes populations that are mostly self-recruiting but are influenced by a non-trivial component of dispersal among local populations. To a great degree concepts of larval dispersal have changed as methods of studying dispersal have developed (reviewed by Swearer et al., 2002). Studies of rocky intertidal invertebrates of the Northeast Pacific have to a large degree provided empirical data and driven theory regarding the connectivity of marine populations (see chapters by Shepherd and Morgan and Botsford and Fogarty, this volume). Cases of synchrony and asynchrony in recruitment and demographics among populations distributed along the Northeast Pacific coastline, and the oceanographic conditions that might drive recruitment give hints as to whether populations are self-seeding or draw upon a common pool of pelagic larvae. However, differences in local survival can give a false impression of metapopulation structure when none exists.

In temperate rocky reef fishes several "overfishing experiments", both intentional and otherwise have served to show the insular nature of rocky reef populations, through the demonstration of localized depletion. The "patchiness" in age composition structure that can result from spatial differences in fishing was first shown by Westrheim et al. (1974), and Gunderson et al. (1977, Fig. 3-6), where

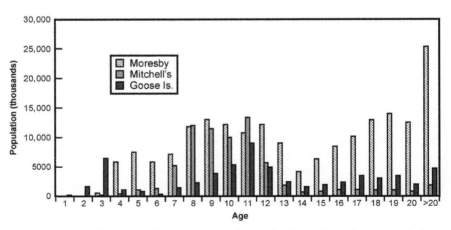

FIGURE 3-6. Pacific ocean perch. Population size, by age, for three gullies within Queen Charlotte Sound (after Gunderson et al., 1977).

Pacific ocean perch populations in two adjacent gullies in Queen Charlotte Sound, separated by no more than about 30 km showed striking differences in age composition, even though the dominant year classes indicated that recruitment patterns were similar in the two areas. A similar situation was reported by Leaman (1991), off northern British Columbia, and Gunderson (1997) off the northern Washington coast. Adult migrations are limited for Pacific ocean perch, so that any linkages between local populations must occur during the larval or juvenile stages. Nevertheless, these linkages were sufficiently weak to allow the persistence of substantial differences in age composition between local populations. These relatively weak linkages are also reflected in the fact that "rebuilding" depleted populations of Pacific Ocean perch typically takes several decades. The existence of localized population structure within Queen Charlotte Sound is also reflected in significant differences in growth (Gunderson, 1972), parasite fauna (Leaman and Kabata, 1987), and microsatellite DNA (Withler et al., 2001). Nevertheless, outright extinctions of local populations appear to be rare in Northeast Pacific temperate reef communities, perhaps once again due to the longevity that typifies many of these species.

As with invertebrate species, metapopulation studies of temperate rocky reef fishes can proceed only as methods of measuring dispersal are improved. Recently, new empirical approaches to the study of larval dispersal and self-recruitment have been developed that are directly applicable to the study of temperate rocky reef fishes. Direct methods include: observation of pelagic stages and their orientation to settlement habitats (e.g., Leis et al., 1996; Leis, 2003), the intentional chemical marking of eggs and larvae (e.g., Jones et al., 1999), the use of naturally occurring chemical markers of natal origin (e.g., Swearer et al., 1999; Miller and Shanks, 2004) and the genetic signatures of larval cohorts (Davies et al., 1999). Indirect methods of estimating dispersal and connectivity generally rely on the study of genetic disequilibria between local populations and a comparison of within and between population measures of genetic variance. If studies indicate limited movements of adults then genetic differences should indicate the degree of exchange during the pelagic larval dispersal stage. Helberg (Helberg et al., 2002; Helberg, this volume) does an excellent job reviewing the approaches and assumptions of genetic assessments of connectivity among marine populations. Similarly Thorrold et al. (2002) review the utility of using natural markers such as otolith microchemistry for estimating retention or dispersal. Both types of data should be interpreted with caution. It is early in the game, but for Northeast Pacific rocky reef fishes we suggest that genetic approaches may overestimate connectivity while natural markers may underestimate connectivity. Genetic methods estimate evolutionarily significant migration rather than demographically significant migration. A few migrants per generation may mix gene pools. Also a rare event such as a once per century current reversal during an El Niño can provide migrants that will mix gene pools even though during most years there is no demographically significant mixing. This makes genetic methods highly vulnerable to type II errors (falsely accepting the proposition that two populations are connected when in fact there is little demographic connectivity (Taylor and Dizon, 1996; Waples, 1998). Natural markers such as otolith microchemistry are highly vulnerable to type I error (falsely accepting that there is isolation between populations when occasional but demographically significant connectivity exists. Otolith microchemistry has no population memory beyond the lifetime of the individual. A study that demonstrates population separation and self-recruitment even for a few years may fail to detect a decadal pulse of outside recruitment that would erase genetic differences. As noted above, occasional outside recruitment events may be important for species with longer life spans than the investigator! We advocate the need for combining approaches to get a more accurate picture of metapopulation dynamics in rocky reef fishes. Otolith microchemisty studies of North Pacific rocky reef fishes are being undertaken (Miller and Shanks, 2004) but to date most of what we know about population structure in these fishes comes from a substantial recent effort in applying indirect molecular genetic approaches.

VIII. POPULATION GENETIC STUDIES IN NORTH PACIFIC ROCKY REEF FISHES

Here we present four larval dispersal models that build on the more abstract models of Kritzer and Sale (2004) and the larval replenishment models of Carr and Reed (1993) but designed specifically for Northeast Pacific rocky reef fishes (Fig. 3-7). While there is not a simple link between early life history characters, oceanography, and metapopulation structure, there are beginning to be enough empirical genetic studies to begin to examine how various reproductive strategies and oceanographic conditions are perhaps reflected in population genetic structure. We emphasize that historical influences such as effects of post-glacial recolonization and sea level change may play heavily into observed present day patterns as would unobserved movements of adults.

Broad Dispersal (Fig. 3-7a) This scenario suggests that larvae are broadly advected away from their natal origin and disperse freely among populations along the coast. The assumption is that local recruitment, while variable, is determined by propagule rain from a common pool with no self-recruitment or metapopulation structure. In this scenario local populations can be managed as a common coastwide stock. In fact this is the starting assumption of most fishery management plans. The short- and longspine thornyheads, *Sebastolobus alascanus* and *S. altivelus*, are deepwater scorpaenid fishes that inhabit soft bottom as well as rocky reefs along the outer continental shelf and slope (Jacobson and Vetter, 1996; Vetter and Lynn, 1997). These two species have extraordinarily long plank-

Chapter 3 Temperate Rocky Reef Fishes 97

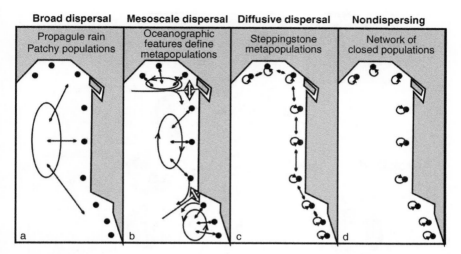

FIGURE 3-7. Propagule dispersal models and population structure: a. broad advective dispersal typical of species with extended early planktonic stages. b. mesoscale dispersal typical of species whose early life stages develop in the plankton but are retained in an oceanographic feature such as the Southern California Eddy. c. diffusive dispersal describes nearshore species whose eggs and larvae remain in nearshore boundary layers subject to diffusive rather than advective flows. d. non-dispersing describes species that produce large precocious young capable of swimming and not subject to passive dispersal in currents.

tonic juvenile phases of 14–15 months for *S. alascanus* and 18–20 months for *S. altivelis* (Moser, 1974), and occur from the Aleutian Islands to Baja California, Mexico. Population genetic studies based on mitochondrial DNA sequences (mtDNA) support the prediction of wide ranging dispersal with little to no population structure (Stepien et al., 2000).

Mesoscale Dispersal (Fig. 3-7b) We believe that this is a common dispersal strategy for many deep reef species with extended pelagic phases. Here larvae and pelagic juveniles are advected away from their natal habitat and remain in the plankton for weeks to months. During the pelagic phase they are entrained in mesoscale oceanographic features such as upwelling fronts (Wing et al., 1998), eddies (Dower and Perry, 2001; Taylor, 2004), upwelling shadows (Graham and Largier, 1997; Wing et al., 1998) and inland basins and channels (Withler et al., 2001; Buonaccorsi et al., 2002). Populations are self-recruiting on a regional rather than local scale with limited dispersal between oceanographic domains. The metapopulation structure is defined by mesoscale oceanographic barriers to dispersal. Many of these potential boundaries in the Northeast Pacific are shown in Figure 3-1 and studies involving the Alaska Gyre, Queen Charlotte Sound,

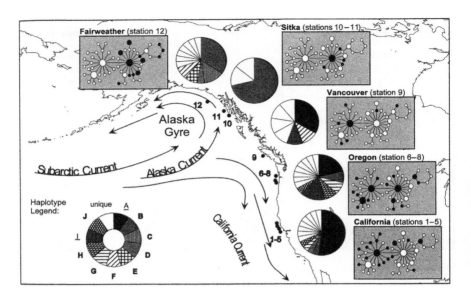

FIGURE 3-8. Rosethorn rockfish. Map of study area showing the sampling stations (1–12). In each putative population, pie charts represent the frequency of 10 haplotype classes (A–J) defined by ignoring autapomorphic sites. Classes A and I (underlined) belong to clade A. Classes B-H and J (bold) belong to group B. Unique haplotypes can be in either one. In each population solid circles in the minimum spanning network indicate the presence of a given haplotype. The major oceanographic features of the Northeast Pacific are also presented (after Rocha-Olivares and Vetter, 1999).

Puget Sound, the Cape Mendocino Jet, and the Southern California Eddy have been conducted. This type of metapopulation structure has added significance because the provisions of the Endangered Species Act consider metapopulations such as the Puget Sound population of copper rockfish as a DPS worthy of separate legal status (Stout et al., 2001).

Sebastes helvomaculatus, the rosethorn rockfish, was one of the first examples of mesoscale population structure (Fig. 3-8, from Rocha-Olivares and Vetter, 1999). In this study there was a clear separation between populations in the northward flowing Alaska Gyre and those occurring within the southward flowing California Current. In this study historical factors such as post glacial recolonization and its effect on observed genetic diversity were acknowledged but not formally modeled regarding contemporary oceanographic barriers to dispersal, gene flow, and the maintenance of metapopulation differences.

The greater Puget Sound Basin and Queen Charlotte Sound have provided an interesting oceanographic context to examine rocky reef metapopulation structure in a variety of species including Pacific ocean perch in Queen Charlotte

Sound (Seeb and Gunderson, 1988; Withler et al., 2001), quillback, brown, and copper rockfish in Puget Sound (Seeb, 1998; Buonaccorsi et al., 2002) and lingcod in Puget Sound (Jagielo et al., 1996). The existence of metapopulation structure is highly convincing but the oceanographic barriers to dispersal are in some cases not fully understood given the enormous complexity of coastal currents. As genetic methods have improved, greater inferences can be made. A recent study of copper rockfish compared populations within Puget Sound and the Canadian Gulf islands to populations on the outer coast (Buonaccorsi et al., 2002). The multiple locus microsatellite DNA results clearly showed that Puget Sound had low diversity indicative of a postglacial founder effect. However, modeling of the decay of F_{ST} or the accumulation of genetic divergence upon isolation of two populations could test the likelihood that the observed results were purely historical or resulted from a continuing barrier to larval dispersal between the outer coast and Puget Sound. The results indicate that the predominantly outward flowing fresher water that dominates surface flow from Puget Sound may be a significant barrier to larval exchange.

Promontories such as Cape Mendocino can also form barriers to dispersal by advecting waters offshore, presumably causing the demise of larvae dependent on benthic settlement. Blue rockfish are abundant from Oregon to the Southern California Bight. Pelagic juveniles are found well offshore and are prominent in midwater juvenile rockfish surveys where they occur at an average depth of 80 m, well below the thermocline (Ross and Larson, 2003). Despite a pelagic phase of roughly four months prior to settlement in nearshore rocky reef habitats, Cope (2004) found a pronounced genetic break north and south of the Cape Mendocino Jet (Fig. 3-1). On the basis of mtDNA differences in haplotype lineages it appears that populations north of this point are derived fairly recently (post-Pleistocene) from southern populations but the two regions of the coast are presently quite isolated and should be treated as separate management units. A remarkably similar result was recently found in the barnacle *Balanus glandula* using similar mtDNA markers (Sotka et al., 2004). Here again modeling gave persuasive proof that the continued existence of the genetic break was dependent on contemporary barriers to larval exchange near Cape Mendocino.

For some intertidal invertebrates the importance of Pt. Conception (Fig. 3-1) as a barrier to larval dispersal and hence its significance as a boundary for metapopulations is at present not well supported and quite controversial (Burton, 1998; Wares et al., 2001). However, it is far too early to generalize since animals spawning at different seasons, depths and locations will interact with vastly different oceanographic forces.

In a recent study of nuclear microsatellite-based genetic structure in the grass rockfish, disequilibria were detected between populations in northern California-southern Oregon and central California and between central California and Southern California (Buonaccorsi et al., 2004). If the results from multidimensional scaling and adjacent sample pooling are correct, larval dispersal is limited across the region generally referred to as Pt. Conception. These regional differences are overlain on even finer scale isolation by distance structure (see below). A recently completed study of the kelp rockfish also indicates differences north and south of Pt. Conception (Taylor, 2004). When enough samples and sample locations are available for fine scale analyses, it appears that the metapopulation structure near Pt. Conception reflects the boundary between the southward flowing California Current and the Southern California Eddy and not Pt. Conception per se (Buonaccorsi et al., 2004; Taylor, 2004). A recent study of bocaccio rockfish also indicates population differences near the Pt. Conception faunal break (Matala et al., 2004).

Diffusive Dispersal (Fig. 3-7c) Nearshore rocky reefs, particularly those with associated kelp forests, function in a zone of "sticky water" where advective processes are limited and along shore flows are dominated by reversing tidal currents (Zeidberg and Hamner, 2002). Kelp forests can increase entrainment and the dominance of diffusive processes such that the residence times of water parcels increase dramatically (Jackson and Winant, 1983). Largier (2003) has pointed out that larvae must get offshore to be advected along shore. It is reasonable that fishes that build and guard adhesive egg nests or produce live young are trying to maximize local retention and remain in the diffusive regime, at least until swimming and directed movements are better developed. For these types of species local populations may be largely self-recruiting and external recruitment is from adjacent habitats rather than a common external larval pool. Here populations can exist in a stepping stone arrangement (Kimura and Weiss, 1964) and genetic distances may increase with geographic distance if only neighboring reefs tend to exchange larvae. Many nearshore rockfish display this type of genetic structure, sometimes in combination with regional structure indicative of the broader scale oceanographic barriers discussed above (see Helberg chapter for invertebrate examples).

The copper rockfish is a common nearshore rockfish that occurs from northern Baja California to Alaska. As discussed above, copper rockfish within Puget Sound do not freely exchange larvae with the outer coast. Populations along the outer coast display a strong isolation by distance structure suggesting that larval dispersal is diffusive rather than advective or eddy driven and that the most likely source of external recruits to a local population is from neighboring populations (Fig. 3-9). Grass rockfish (Buonaccorsi et al., 2004) show isolation by distance and stepping stone dispersal over shorter distances within oceanographic regimes but also show regional differences that suggest oceanographic barriers to dispersal at the larger scale (see above). Within a single oceanographic regime the slopes of the isolation by distance regressions for these two related rockfish species that share habitat and larval characteristics are quite similar. Based on the methods of Rousset (1997) and assuming symmetrical, exponential dispersal along a linear

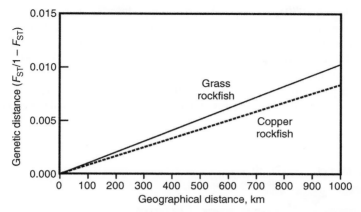

FIGURE 3-9. Comparison of isolation by distance coastal regression slopes for grass and copper rockfishes. For the purpose of comparing slopes the y-intercept was set to zero (Buonaccorsi et al., 2004, used with permission).

coast, mean dispersal distances were estimated for grass and copper rockfish assuming different adult densities from 100 to 10,000 per km of shoreline. This was thought to bracket historical and present day abundances. Mean dispersal distance ranged from 1 to 13 km depending on species and choice of population size. In all scenarios mean dispersal was remarkably small (Buonaccorsi et al., 2004). Kelp rockfish (Taylor, 2004), and brown rockfish (Buonaccorsi et al., 2005) also show a strong isolation by distance signal suggesting that stepping stone dispersal may be a common property of the *Pteropodus* subgeneric group of nearshore rockfishes.

Non-Dispersing (Fig. 3-7d) A few temperate reef species, most notably the surfperches, Embiotocidae, but also rocky reef elasmobranchs such as the horn shark, *Heterodontus francisci*, have eliminated larval and juvenile dispersal stages by producing large, live-born juveniles (surfperch) or producing sessile egg cases that release fully developed juveniles that do not disperse (horn shark). In this scenario all recruitment is local and populations are essentially closed. Bernardi (2000) examined gene flow in the black surfperch, *Embiotoca jacksoni*, along the mainland of central and southern California as well as between the Channel Islands and the mainland of southern California. He found pronounced population structure at all spatial scales and, based on the mtDNA differences, concluded that both deep water channels between islands and mainland, and sandy expanses along the mainland, severely limit gene flow. The four northern Channel Islands were linked prior to recent sea level decline and still share shallow water connections (Fig. 3-2). These islands share haplotype lineages that are separate from those occurring at the more southern islands, which are separated from the north by deep water (Fig. 3-10). Bernardi estimated the separation between populations on the southern Channel Islands and the mainland is on the order of tens of thousands of years.

The study of population genetic structure of temperate rocky reef fishes is in its infancy and the case studies presented almost certainly contain the ghosts of

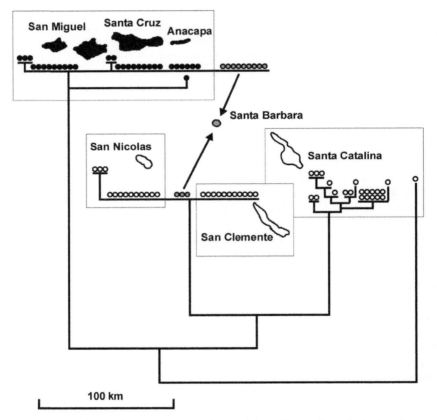

FIGURE 3-10. Black surfperch. Superimposed map of the California Channel Islands and a simplified (polymorphisms were removed) neighbor-joining phylogenetic tree of the corresponding island populations of black surfperch. Each sample is represented by a circle. Black circles represent the northern Channel Islands, open circles represent the southern islands, and gray circles represent Santa Barbara Island. Samples from Santa Barbara Island clustered with northern Channel Islands (nine samples) and with San Nicolas/San Clemente Island (three samples). One sample from Santa Catalina Island clustered with those from the mainland (after Bernardi et al., 2000).

population dynamics past. In addition to population structure, examination of overall and local genetic diversity is likely to shed light on metapopulation structure and extinction/recolonization dynamics. The apparent lack of a relationship between census size (N_c) and effective population size (N_e) in many marine species has long intrigued marine geneticists (Grant and Waples, 2000). Most marine species display a low N_e relative to N_c, which may indicate periodic population crashes or the disproportionate contributions of only a few individuals to the population (Hedgecock, 1994). To date it appears that temperate rocky reef fishes differ markedly from this typical pattern. Mitochondrial DNA haplotype diversities (h) for *Sebastes* are usually above 95% (rosethorn rockfish, 94.5%, Rocha-Olivares and Vetter, 1999; *S. capensis*, 86.7%, Rocha-Olivares et al., 1999a; *S. oculatus*, 95%, Rocha-Olivares et al., 1999a; gopher rockfish, 97.0%, Narum, 2000; black-and-yellow rockfish, 95.4%, Narum, 2000). Nucleotide diversity (Π) is also high. High haplotype and nucleotide diversity is not common in marine fishes (Grant and Bowen, 1998; Grant and Waples, 2000). Great ages for taxon lineages or high mutation rates do not appear to be the explanation in *Sebastes* (Rocha-Olivares et al., 1999b,c). This suggests a low rate of stochastic extinction of haplotypes is the cause of high diversity. This diversity may well be due to a metapopulation structure that includes great longevity, a high degree of overlapping generations, and a lattice of persistent, quasi-independent, metapopulations (Slatkin, 1977; Wright, 1977; Wade and McCauley, 1988; McCauley, 1993). Whether this is true or not it deserves further study.

Recent efforts to estimate dispersal distance from the slope of the isolation by distance relationship (Rousset, 1997 and Buonaccorsi et al., 2004) or by simulation modeling (Palumbi, 2003) are rightfully open to strong scrutiny. Historical allopatry, assortative mating, and limited sampling can give the appearance of isolation by distance when there are really several panmictic metapopulations located along the coastline (e.g. Fig.3-7b). However even a low slope of genetic distance vs geographic distance, if consistent over a long stretch of coastline, is compelling evidence of limited dispersal. Information on mean larval dispersal distance is so important for understanding metapopulation dynamics and the design of MPA networks that it is worth trying to improve methods of estimation. Determining the mean dispersal distances of additional species with different life histories, and examining the symmetry and shape of the dispersal functions (presently assumed to be exponential and symmetrical) are current research priorities.

IX. HUMAN IMPACTS

Habitat. Because of the non-migratory nature of the adults, stocks on individual temperate reefs often function as submerged "islands", which are linked by larval drift and juvenile migrations. The absence of a single stepping stone habitat can have profound effects on metapopulation structure and connectivity for species with limited larval dispersal (Barber et al., 2002). Habitat loss is a real problem for temperate rocky reef species, particularly those reliant on vegetated nearshore habitats. Kelp forests and eelgrass are quite vulnerable to damage through shoreline development, eutrophication, and pollution. Maintaining such habitats is critical if the populations that rely on them are to be sustained and remain connected. Some areas of the Puget Sound Basin such as Hood Canal are seeing the effects of hypoxia but the precise cause is under investigation.

The construction and placement of artificial reefs and other man-made structures such as oil platforms have been championed as means of improving the productivity of temperate marine reef communities, although not without controversy. Unless managers are certain that artificial reefs are actually generating new recruits for the adult population rather than serving solely as a place where adults are concentrated for exploitation or, in the case of oil platforms, where pelagic juveniles are attracted without hope of settling and growing to maturity, it is possible that the net effect of the added stepping stone is detrimental (Grossman et al., 1997; Mathews, 1985).

Climate Change. Presently ocean warming and the relaxation of upwelling during regime shifts or El Niño can all but eliminate kelp habitat in the Southern California Bight with a drastic change from a cold temperate to a warm temperate fish community (Stephens et al., 1994). Kelp forests will be particularly sensitive to global warming as will the inland waters of Puget Sound. Oxygen depletion and "dead zones" are typically associated with agricultural run off, eutrophication, and low circulation coastal environments. However, the oxygen minimum zone, (OMZ) is a naturally occurring, recurrent feature of the Northeast Pacific that impacts temperate reefs along the upper continental slope from 600–1000m depth (Vetter et al., 1994; Vetter and Lynn, 1997). Recently it has been discovered that this water can be advected onto the continental shelf smothering nearshore rocky reef habitats during periods of intense upwelling and offshore advection of surface waters (Grantham et al., 2004). Such processes are predicted to increase under some models of global warming and would result in a periodic dead zone along the Oregon coastline effectively separating areas north and south of it (Service, 2004).

Overfishing. Where temperate reefs are located in shallow waters close to population centers, small boat fishermen can readily overfish them with relatively inexpensive traps and hook-and-line gear. As a result, maintaining temperate reef communities and managing them so as to provide sustainable harvests is a daunting task. In fact many of these communities may have been depleted, or drastically altered before their presence was documented scientifically. This certainly would have been the case if the recent explosive growth of the "live-fish" fishery off California had been allowed to continue unregulated for even a few more years, due to the high prices (over $5.00 per pound) and simple gear (small boats or even kayaks) required to participate in the fishery (Bloeser, 1999). Fortunately, sufficient fishery monitoring data and biological background on the species impacted was available for the state of California to recognize the problem and begin formulating a suitable management plan. Overfishing of keystone species such as the California sheephead that feeds on benthic invertebrates can have profound effects on the kelp ecosystem. Loss of adult sheephead shifts the balance of nearshore reefs from kelp forest to urchin barren if purple urchins (*Strongylocentrotus purpuratus*) are allowed to graze kelp holdfasts in the absence of predation.

The patchy distribution and small home range sizes characteristic of temperate reef ecosystems often make it difficult to apply traditional fisheries management methods. Serial depletion of populations on individual reef complexes is a serious danger, and easily goes undetected when "management areas" encompass geographically broad regions and where landings data are not kept on a species basis. Traditional management methods, i.e. stock assessment, establishment of quotas, and effort control (e.g. fleet reductions), are difficult to implement and enforce effectively over the small spatial scales that are often required. As a result, managers of temperate reef communities in California and the Pacific Northwest are exploring the widespread use of areas where fishing is prohibited (Marine Protected Areas, MPAs), and enlisting stakeholder cooperation and community management in developing effective fishery management plans. The range of stakeholders with an interest in temperate reef ecosystems has been broadening in North America, and now includes environmental groups, recreational SCUBA divers, and the ecotourism industry as well as representatives from the more traditional commercial, recreational, subsistence and aboriginal fisheries. Considerable scientific and public debate has arisen over the development of MPA networks in areas such as the San Juan Islands and the Channel Islands National Marine Sanctuary, but the process has moved forward to the point where MPA networks have been established in both areas (Airamé et al., 2003; Palsson, 2002). Nevertheless, serious questions still remain as to metapopulation structure in both areas, and whether or not the existing networks will function as sustainable entities.

X. FUTURE DIRECTIONS FOR METAPOPULATION STUDIES OF TEMPERATE REEF FISHES

Temperate reef communities as study systems. In addition to the need to understand metapopulation structure for effective management, temperate rocky reef systems provide scientific opportunities that are clearly different from tropical reef systems. While tropical systems are among the most environmentally constant habitats, with favorable habitat distributed longitudinally, temperate reef systems are subject to strong latitudinal gradients in temperature. The Northeast Pacific rocky reef systems that we are most familiar with provide opportunities to study the effects of extirpation and recolonization during glacial cycles. Events such as the opening of the Georgia Basin-Puget Sound system are recent and the times since recolonization are known with some precision (Buonaccorsi et al., 2002). An equally interesting area is the Sea of Cortez. This vast inland sea is contiguous with the outer coast of Baja and other temperate reef habitats during cool periods but presently contains a trapped temperate fauna in its upper reaches. Variations in the vagility of adults and the dispersal properties of offspring affect the degree of connectivity between populations living on the outer coast of Baja and within the upper Sea of Cortez. Recent studies have been particularly illuminating in terms of metapopulation structure in temperate reef fishes (Terry et al., 2000; Bernardi et al., 2003; Rocha-Olivares et al., 2003). The Channel Islands also provide interesting island-mainland comparisons that are just beginning to be explored (Bernardi, 2000; Taylor, 2004).

Metapopulations and the conservation of genetic diversity. In the genetic or evolutionary context, it has been argued that metapopulations may increase or decrease overall genetic diversity (Slatkin, 1977; McCauley, 1991). The few empirical studies that specifically address the topic (Hale et al., 2001; Giles and Goudet, 1997) indicate that results can be complex. Many temperate rocky reef fishes seem to be selected for high longevity and maintain high levels of genetic diversity. In an evolutionary context, longevity allows bridging of climate oscillations that affect recruitment. If more adults leave some progeny there will be a higher equilibrium level of genetic diversity compared to those species with sweepstakes recruitment and short life spans (Grant and Bowen, 1998; Hedgecock, 1994). Fishing makes populations smaller but it also makes populations younger and mean life spans shorter, so that adults may not outlive a climate oscillation to leave progeny. This will cause a loss of genetic variation due to greater stochastic extinction of genetic lineages, with unknown consequences. The length of time a local population remains low prior to rebuilding affects the retention of genetic variability. Fishing effects on genetic diversity have been explored in small pelagic fishes (Gaggiotti and Vetter, 1999) and the New Zealand snapper, *Pagrus auratus* (Hauser et al., 2002). The biological consequences of this lost diversity are not known but could affect responses to external selective forces such as climate change.

Metapopulations and Local Adaptation. Neutral genetic markers can be used to infer metapopulation structure and connectivity. However, genes under selection are the frontier of temperate reef metapopulation studies. While total isolation of populations may lead to endemism and extinction when the environment changes, metapopulation structure may allow local adaptations to accumulate while maintaining a degree of genetic connectivity across latitudinal gradients. The high fecundities and high mortalities associated with most marine fishes sets the stage for rapid local selection (Conover and Munch, 2002). Recent studies in cod, *Gadus morhua*, clearly show important selective differences in antifreeze levels and genetic markers between inshore and offshore populations of over-wintering cod (Ruzzante et al., 1996). Similarly, Pogson and Fevolden (2003) have found evidence for selection of pantophysin genes among coastal and oceanic populations of cod off northern Norway. To our knowledge no such studies have been carried out on temperate reef fishes but it would be surprising if a species such as copper rockfish did not display heritable adaptive differences between Puget Sound and the upwelling driven systems of the outer coast (Buonaccorsi et al., 2002) or along a thermal gradient from Alaska to Baja California. Demonstrating local adaptive differences will be enormously important in understanding the consequences of local depletion, the success of migrants verses local recruits, and the design of effective rebuilding schemes.

Designing MPA networks. The movement towards establishing MPA networks along the Northwest Pacific coast and within the Channel Islands and the Georgia Basin is driving many new theoretical and empirical studies with an emphasis on metapopulation processes (Warner and Cowen, 2002; Lubchenco et al., 2003). Attempts are being made to combine field studies of recruitment with genetics and otolith microchemistry to better determine sources of larvae, connectivity, and spillover to fished areas. While the demographic equivalence of reserves vs conventional fisheries management has been much discussed, the genetic consequences of a network of MPAs and intervening fished areas have not been closely studied. Recently it has been shown that some nearshore reef fishes may have mean dispersal distances on the order of tens of kilometers (see above). If the distance between MPAs is greater than the dispersal distance of larvae, then connectivity must be maintained by a multi-generational, stepping-stone pattern of gene flow across fished habitats. If the intentional or de facto result of management is to create high densities of adult spawners in MPAs, with mostly subadults in the fished areas, the N_e (effective population size) of the fished area will be close to zero, and the probability of a dispersed larvae growing to maturity and producing larvae to disperse further becomes very small indeed. The MPAs will become disconnected local populations with no coastwide gene flow (Buonaccorsi et al., 2004, 2005) despite equivalence in yield of MPAs and traditional fishery management (Hastings and Botsford, 1999; Botsford et al., 2003). This point deserves immediate attention.

Spatial management. Whether we are dealing with conventional fisheries management, establishment of MPA networks, artificial reef placement, or the development of hatchery programs, a clear understanding of metapopulation structure and specific requirements for both juvenile and adult habitat is imperative if we are to be successful. The challenges of delineating linkages between local populations of temperate reef species must be met if these valuable resources are to be managed effectively. Our ability to preserve, and even restore the overall productivity of temperate marine reef communities creates a unique opportunity in the Northeast Pacific, and demands an increased focus on the metapopulation dynamics that drive these systems. Appropriate spatial management should be the goal of every modern fishery management plan, and metapopulation-based regional landings data and fishery independent indices of abundance are prerequisites for such an approach. It will be difficult to find new resources, but the effective management of many temperate rocky reef species depends on it.

REFERENCES

Airamé, S., Dugan, J.E., Lafferty, K.D., Leslie, H., McArdle, D.A., and Warner, R.R. (2003). Applying ecological criteria to marine reserve designs: A case study from the California Channel Islands. In "The Science of Marine Reserves", *Ecol. Appl.* **13 (Suppl.)**, S170–S184.

Ammann, A.J. (2004) SMURFs: standard monitoring units for the recruitment of temperate reef fishes. *J. Exp. Mar. Biol. and Ecol.* **299**, 135–154.

Bakun, A. (1996). "Patterns in the Ocean: Ocean Processes and Marine Population Dynamics." California Sea Grant College System, University of California, La Jolla, CA.

Barber, P.H., Palumbi, S.R., Erdmann, M.V., and Moosa, M.K. (2002). Sharp genetic breaks among populations of *Haptosquilla pulchella* (Stomatopoda) indicate limits to larval transport: Patterns, causes, and consequences. *Mol. Ecol.* **11**, 659–674.

Barker, M.W. (1979). Population and Fishery Dynamics of Recreationally Exploited Marine Bottomfish of Northern Puget Sound. Ph.D. Dissertation, University of Washington, Seattle, WA.

Berkeley, S.A., Chapman, C., and Sogard, S.M. (2004). Maternal age as a determinant of larval growth and survival in a marine fish, *Sebastes melanops*. *Ecology* **85**, 1258–1264.

Bernardi, G. (2000). Barriers to gene flow in *Embiotoca jacksoni*, a marine fish lacking a pelagic larval stage. *Evolution* **54**, 226–237.

Bernardi, G., Findley, L., and Rocha-Olivares, A. (2003). Vicariance and dispersal across Baja California in disjunct marine fish populations. *Evolution* **57**, 1599–1609.

Bloeser, J.A. (1999). "Diminishing Returns: The Status of West Coast Rockfish." Pacific Marine Conservation Council, Astoria, OR.

Bobko, S.J., and Berkeley, S. (2004). Maturity, ovarian cycle, fecundity, and age-specific parturiturition of black rockfish (Sebastes melanops). *Fish. Bull.* **102**, 418–429.

Botsford, L.W. (2001). Physical influences on recruitment to California Current invertebrate populations on multiple scales. *ICES J. Mar. Sci.* **58**, 1081–1091.

Botsford and Fogarty, this volume.

Botsford, L.W., Micheli, F., and Hastings, A. (2003). Principles for the design of marine reserves. *Appl. Ecol.* **13 Suppl.**, S25–S31.

Bradbury, I.R., and Snelgrove, P.V.R. (2001). Contrasting larval transport in demersal fish and benthic invertebrates: The roles of behavior and advective processes in determining spatial pattern. *Can. J. Fish. Aquat. Sci.* **58**, 811–823.

Briggs, J.C. (1974). "Marine Zoogeography." McGraw-Hill, New York.
Brodeur, R.D., and Ware, D.M. (1992). Interannual and interdecadal changes in zooplankton biomass in the subarctic Pacific Ocean. *Fish. Oceanogr.* **1**, 32–38.
Buckley, R.M. (1997). Substrate Associated Recruitment of Juvenile *Sebastes* in Artificial Reef and Natural Habitats in Puget Sound and the San Juan Archipelago, Washington. Ph.D. Dissertation, University of Washington, Seattle, WA.
Buonaccorsi, V.P., Kimbrell, C.A., Lynn, E.A., and Vetter, R.D. (2002). Population structure of copper rockfish (*Sebastes caurinus*) reflects postglacial colonization and contemporary patterns of larval dispersal. *Can. J. Fish. Aquat. Sci.* **59**, 1374–1384.
Buonaccorsi, V.P., Kimbrell, C.A., Lynn, E.A., and Vetter, R.D. (2005). Limited realized dispersal and introgressive hybridization influence genetic structure and conservation strategies for brown rockfish, *Sebastes auriculatus. Conserv. Genet.* **6**, 697–713.
Buonaccorsi, V.P., Westerman, M., Stannard, J., Kimbrell, C., Lynn, E., and Vetter, R. (2004). Stepping stone larval dispersal in grass rockfish, *Sebastes rastrelliger. Mar. Biol.* **145**, 779–788.
Burton, R.S. (1998). Intraspecific phylgeography across the Point Conception biogeographic boundary. *Evolution* **52**, 734–745.
Cailliet, G.M., Andrews, A.H., Burton, E.J., Watters, D.L., Kline, D.E., and Ferry-Graham, L.A. (2001). Age determination and validation studies of marine fishes: Do deep-dwellers live longer? *Exp. Gerontology* **36**, 739–764.
Carlson, H.R. (1986). Restricted year-class structure and recruitment lag within a discrete school of yellowtail rockfish. *Trans. Am. Fish. Soc.* **115**, 490–492.
Carlson, H.R., and Haight, R.E. (1972). Evidence for a home site and homing of adult yellowtail rockfish, *Sebastes flavidus. J. Fish. Res. Board Can.* **29**, 1011–1014.
Carlson, H.R., and Haight, R.E. (1976). Juvenile life of Pacific ocean perch, *Sebastes alutus*, in coastal fjords of southeastern Alaska: Their environment, growth, food habits, and schooling behavior. *Trans. Am. Fish. Soc.* **105**, 191–201.
Carr, M.H. (1991). Habitat selection and recruitment of an assemblage of temperate zone reef fishes. *J. Exp. Mar. Biol. Ecol.* **146**, 113–137.
Carr, M.H. (1994). Effects of macroalgal dynamics on recruitment of a temperate reef fish. *Ecology* **75**, 1320–1333.
Carr, M.H., and Reed, D.C. (1993). Conceptual issues relevant to marine harvest refuges—examples from temperate reef fishes. *Can. J. Fish. Aquat. Sci.* **50**, 2019–2028.
Chatwin, B.M. (1956). Further results from tagging experiments on lingcod. *Fish. Res. Board Can., Pacific Prog. Rept.* **107**, 19–21.
Clague, J.J. (1983). Glacioisostatic effects of the Cordilleran ice sheet, British Columbia, Canada. In "Shorelines and Isostasy" (D.E. Smith and A.G. Dawson, Eds.), Institute of British Geographers Special Publication **16**, pp. 321–343.
Clague, J.J. (1989). Quaternary Geology of the Canadian Cordillera, Ch. 1. In "Quaternary Geology of Canada and Greenland. Geological Survey of Canada" (R.J. Fulton, Ed.), Geology of Canada No.1.
Cochrane, G.R., Vetter, R.D., Nasby, N., Taylor, C., and Cosgrove, R. (2002). Egg and larval production from marine ecological reserves. Part 2: Benthic habitat in four marine reserve locations surrounding the Santa Barbara Basin. In "Marine Ecological Reserves Research Program, Research Results 1996–2001," California Sea Grant College Program, La Jolla, CA.
Coleman, F.C., Koenig, C.C., Huntsman, G.R., Musick, J.A., Eklund, A.M., McGovern, J.C., Chapman, R.W., Sedberry, G.R., and Grimes, C.B. (2002). Long-lived reef fishes: The snapper-grouper complex. *Fisheries* **25**, 14–21.
Conover, D., and Munch, S.B. (2002). Sustaining fisheries yields over evolutionary time scales. *Science* **297**, 94–96.
Cooper, D. (2003). Possible Differences in Copper Rockfish (*Sebastes caurinus*) Fecundity and Parturition with Maternal Size and Age. M.S. Thesis, University of Washington, Seattle, WA.

Cope, J.M. (2004). Population genetics and phylogeography of the blue rockfish (*Sebastes mystinus*) from Washington to California. *Can. J. Fish. Aquat. Sci.* **61**, 332–342.

Culver, B.N. (1987). Results from tagging black rockfish (*Sebastes melanops*) off the Washington and northern Oregon coast. In "Proceedings of the International Rockfish Symposium." Alaska Sea Grant Rept. 87-2, pp. 231–239.

Davies, N., Villablanca, F.X., and Roderick, G.K. (1999). Determining the source of individuals: Multilocus genotyping in nonequilibrium population genetics. *Trends Ecol. Evol.* **14**, 17–21.

Dayton, P.K., Thrush, S., and Coleman, F.C. (2002). "Ecological effects of fishing in marine ecosystems of the United States." Pew Oceans Commission, Arlington, VA.

Dower, J.F., and Perry, R.I. (2001). High abundance of larval rockfish over Cobb Seamount, an isolated seamount in the Northeast Pacific. *Fish. Oceanog.* **10**, 268–274.

Ebbesmeyer, C.C., Coomes, C.A., Cox, J.M., and Salem, B.L. (1991). Eddy induced beaching of floatable materials in the eastern Strait of Juan de Fuca, In: Puget Sound Reearch 1991 Proceedings. Vol. 1, 1991b: 86–98.

Ebeling, A.W., Larson, R.J., and Alevison, W.S. (1980). Habitat groups and island-mainland distribution of kelp-bed fishes off Santa Barbara, California. In "The California Islands: Proceedings of a Multidisciplinary Symposium" (D.M. Power, Ed.), pp. 401–431. Santa Barbara Museum of Natural History, Santa Barbara, CA.

Eisenhardt, E. (2001). Effect of the San Juan Islands Marine Preserves on Demographic Patterns of Nearshore Rocky Reef Fish. M.S. Thesis, University of Washington, Seattle, WA.

Engle, J. (1993). Distribution patterns of rocky subtidal fishes around the California Islands. In "Third California Islands Symposium: Recent Advances in Research on the California Islands" (F.G. Hochberg, Ed.), pp. 475–484. Santa Barbara Museum of Natural History, Santa Barbara, CA.

Favorite, F., Dodimead, A.J., and Nasu, K. (1976). "Oceanography of the Subarctic Pacific Region, 1960–1971." International North Pacific Fisheries Commission Bulletin Number 33. Vancouver, Canada.

Francis, R.C., Hare, S.R., Hollowed, A.B., and Wooster, W.S. (1998). Effects of interdecadal climate variability on the oceanic ecosystems of the NE Pacific. *Fish. Oceanogr.* **7**, 1–21.

Gaggiotti, O.E., and Vetter, R.D. (1999). Effect of life history strategy, environmental variability, and overexploitation on the genetic diversity of pelagic fish populations. *Can. J. Fish. Aquat. Sci.* **56**, 1376–1388.

Giles, B.E., and Goudet, J. (1997). A case study of genetic structure in a plant metapopulation. In "Metapopulation Biology: Ecology, Genetics, and Evolution" (I. Hanski and M.E. Gilpin, Eds.), pp. 429–454. Academic Press, San Diego.

Graham, W.M., and Largier, J.L. (1997). Upwelling shadows as nearshore retention sites: the example of northern Monterey Bay. *Cont. Shelf Res.* **17**, 509–532.

Graham, M.H., Dayton, P.K., and Erlandson, J.M. (2003). Ice ages and ecological transitions on temperate coasts. *Trends Ecol. Evol.* **18**, 33–40.

Grant, W.S., and Bowen, B.W. (1998). Shallow population histories in deep evolutionary lineages of marine fishes: Insights from sardines and anchovies and lessons for conservation. *J. Hered.* **89**, 415–426.

Grant, W.S., and Waples, R.S. (2000). Spatial and temporal scales of genetic variability in marine and anadromous species: Implications for fisheries oceanography. In "Fisheries oceanography: a science for the new millennium" (P.J. Harrison and T. Parsons, Eds.), pp. 61–93. Blackwell Science, Oxford, UK.

Grantham, B.A., Chan, F., Nielsen, K.J., Fox, D.S., Barth, J.A., Huyer, A., Lubchenco, J., and Menge, B.A. (2004) Upwelling driven nearshore hypoxia signals ecosystem and oceanographic changes in the northeast Pacific. *Nature* **429**, 749–754.

Grossman, G.D., Jones, G.P., and Seaman, Jr., W.J. (1997). Do artificial reefs increase regional fish production? A review of existing data. *Fisheries* **22**, 17–23.

Gunderson, D.R. (1972). Evidence that Pacific ocean perch (*Sebastes alutus*) in Queen Charlotte Sound form aggregations that have different biological characteristics. *J. Fish. Res. Board Can.* **29**, 1061–1070.

Gunderson, D.R. (1997). Spatial patterns in the dynamics of slope rockfish stocks and their implications for management. *Fish. Bull.* **95**, 219–230.

Gunderson, D.R., Westrheim, S.J., Demory, R.L., and Fraidenburg, M.E. (1977). The status of Pacific Ocean perch (*Sebastes alutus*) stocks off British Columbia, Washington, and Oregon in 1974. *Can. Fish. Mar. Serv. Tech. Rep.* 690, 63pp.

Haldorson, L., and Richards, L.J. (1987). Post-larval copper rockfish in the Strait of Georgia: Habitat use, feeding and growth in the first year. In "Proceedings of the International Rockfish Symposium," Alaska Sea Grant Rept. 87-2, pp. 129–142.

Hale, M.L., Lurz, P.W.W., Shirley, M.D.F., Rushton, S., Fuller, R.M., and Wolff, K. (2001). Impact of landscape management on the genetic structure of red squirrel populations. *Science* **293**, 2246–2248.

Hanski, I., and Gilpin, M.E. (1997). "Metapopulation Biology: Ecology, Genetics, and Evolution." Academic Press, San Diego.

Harms, S., and Winant, C.D. (1998). Characteristic patterns of the circulation in the Santa Barbara Channel. *J. Geophys. Res.* **103 (C2)**, 3041–3065.

Hastings, A., and Botsford, L.W. (1999). Equivalence in yield from marine reserves and traditional fisheries management. *Science* **284**, 1537–1538.

Hauser, L., Adcock, G.J., Smith, P.J., Bernal Ramirez, J.H., and Carvalho, G.R. (2002). Loss of microsatellite diversity and low effective population size in an overexploited population of New Zealand snapper (*Pagrus auratus*). *Proc. Nat. Acad. Sci.* **99**, 11742–11747.

Hedgecock, D. (1994). Does variance in reproductive success limit effective population sizes of marine organisms? In "Genetics and Evolution of Aquatic Organisms" (A.R. Beaumont, Ed.), pp. 122–134. Chapman & Hall, London.

Helberg, M.E., Burton, R.S., Neigel, J.E., and Palumbi, S.R. (2002). Genetic assessment of connectivity among marine populations. *Bull. Mar. Sci.* **70 (Suppl. 1)**, 273–290.

Hickey, B.M. (1997). Response of a narrow submarine canyon to strong wind forcing. *J. Phys. Oceanogr.* **27**, 697–726.

Hickey, B.M. (1998). Coastal oceanography of western North America from the tip of Baja California to Vancouver Island. *In.* "The Sea. The Global Coastal Ocean. Regional Studies and Syntheses" (A. Robinson and K.H. Brink, Eds.), pp. 345–393. John Wiley, New York.

Hobson, E.S., Chess, J.R., and Howard, D.F. (2001). Interannual variation in predation on first-year *Sebastes* spp. by three northern California predators. *Fish. Bull.* **99**, 292–302.

Humann, P. (1996). "Coastal Fish Identification: California to Alaska." New World Publications, Inc., Jacksonville, FL.

Ingraham, W.J., Jr., Ebbesmeyer, C.C., and Hinrichsen, R.A. (1998). Imminent climate and circulation shift in Northeast Pacific Ocean could have major impact on marine resources. *Eos. Trans. Am. Geophys. Union.* **79** (16).

Ito, D. (1999). Assessing Shortraker and Rougheye Rockfishes in the Gulf of Alaska: Addressing a Problem of Habitat Specificity and Sampling Capability. Ph.D. Dissertation, University of Washington, Seattle, WA.

Jackson, G.A., and Winant, C.D. (1983). Effect of a kelp forest on coastal currents. *Cont. Shelf Res.* **2**, 75–80.

Jackson, J.B.C., Kirby, M.X., Berger, W.H., Bjorndal, K.A., Botsford, L.W., Bourque, B.J., Bradbury, R.H., Cooke, R., Erlandson, J., Estes, J.A., Hughes, T.P., Kidwell, S., Lange, C.B., Lenihan, H.S., Pandolfi, J.M., Peterson, C.H., Steneck, R.S., Tegner, M.J., and Warner, R.R. (2001). Historical overfishing and the recent collapse of coastal ecosystems. *Science* **293**, 629–638.

Jacobsen, L.D., and Vetter, R.D. (1996). Bathymetric demography and niche separation of thornyhead rockfish: *Sebastolobus alascanus* and *Sebastolobus altivelis*. *Can J. Fish. Aquat. Sci.* **53**, 600–609.

Jagielo, T.H. (1990). Movement of tagged lingcod *Ophiodon elongatus* at Neah Bay, Washington. *Fish. Bull.* **88**, 815–820.

Jagielo, T.H., LeClair, L.L., and Vorderstrasse, B.A. (1996). Genetic variation and population structure of lingcod. *Trans. Am. Fish. Soc.* **125**, 372–386.

Johnson, K.A., Yoklavich, M.M., and Cailliet, G.M. (2001). Recruitment of three species of juvenile rockfish (*Sebastes* spp.) on soft benthic habitat in Monterey Bay, California. *CCOFI Rep.* **42**, 153–166.

Jones, G.P., Milicich, M.J., Emslie, M.J., and Lunow, C. (1999). Self-recruitment in a coral reef fish population. *Nature* **402**, 802–804.

Kimura, M., and Weiss, G.H. (1964). The stepping stone model of population structure and the decrease of genetic correlation with distance. *Genetics* **49**, 561–576.

Kingsford, M.J., and Choat, J.H. (1986). Influence of surface slicks on the distribution and onshore movements of small fish. *Mar. Biol.* **91**, 161–171.

Kreiger, K. (1992). Shortraker rockfish, *Sebastes borealis*, observed from a manned submersible. *Mar. Fish. Rev.* **54**, 34–37.

Kritzer, J.P., and Sale, P.F. (2004). Metapopulation ecology in the sea: From Levins' model to marine ecology and fisheries science. *Fish Fisheries* **4**, 1–10.

Largier, J.L. (2003). Considerations in estimating larval dispersal distances from oceanographic data. In "The Science of Marine Reserves," *Ecol. Appl.* **13 (Suppl.)**, S71–S89.

Laroche, W.A., and Richardson, S.L. (1980). Development and occurrence of larvae and juveniles of the rockfishes *Sebastes flavidus* and *Sebastes melanops* (Scorpaenidae) off Oregon. *Fish. Bull.* **77**, 901–924.

Laroche, W.A., and Richardson, S.L. (1981). Development of larvae and juveniles of the rockfishes *Sebastes entomelas* and *S. zacentrus* (family Scorpaenidae) and occurrence off Oregon, with notes on head spines of *S. mystinus*, *S. flavidus*, and *S. melanops*. *Fish. Bull.* **79**, 231–257.

Larson, R.J. (1980a). Competition, habitat selection, and the bathymetric segregation of two rockfish species. *Ecol. Monogr.* **50**, 221–239.

Larson, R.J. (1980b). Territorial behavior of the black and yellow rockfish and gopher rockfish (Scorpaenidae, *Sebastes*). *Mar. Biol.* **58**, 111–122.

Lea, R.N., McAllister, R.D., and VenTresca, D.A. (1999). Biological aspects of nearshore rockfishes of the genus *Sebastes* from central California. *CDFG Fish Bull.* **177**.

Lea, R.N., and Rosenblatt, R.H. (2000). Observations on fishes associated with the 1997–1998 El Niño off California. *CCOFI Rep.* **41**, 117–129.

Leaman, B.M. (1976). The Association Between the Black Rockfish (*Sebastes melanops* Girard) and Beds of the Giant Kelp (*Macrocystis integrifolia* Bory) in Barkley Sound, British Columbia. M.Sc. Thesis, University of British Columbia, Vancouver, Canada.

Leaman, B.M. (1991). Reproductive styles and life history variables relative to exploitation and management of *Sebastes* stocks. *Environ. Biol. Fishes* **30**, 253–271.

Leaman, B.M., and Beamish, R.J. (1984). Ecological and management implications of longevity in some northeast Pacific groundfishes. *Int. North Pac. Fish. Comm. Bull.* **42**, 85–97.

Leaman, B.M., and Kabata, Z. (1987). *Neobrachiella robusta* (Wilson, 1912) (Copepoda: Lernaeopopidae) as a tag for identification of stocks of its host, *Sebastes alutus* (Gilbert, 1890) (Pisces: Teleostei). *Can. J. Zool.* **65**, 2579–2582.

Leet, W.S., Dewees, C.M., Klingbeil, R., and Larson, E.J. (2001). "California's Living Marine Resources: A Status Report." California Dept. Fish and Game.

Leis, J.M., Sweatman, H.P.A., and Reader, S.E. (1996) What the pelagic stages of coral reef fishes are doing out in blue water: Daytime field observations of larval behaviour. *Mar. Freshw. Res.* **47**, 401–411.

Leis, J.M. (2003). What does larval fish biology tell us about the design and efficacy of marine protected areas? In "Aquatic Protected Areas: What works best and how do we know?" (J.P. Beumer, A. Grant, and D.C. Smith, Eds.), pp. 170–180. Proceedings of the World Congress on Aquatic Protected Areas, Cairns, August 2002. Australian Society for Fish Biology, North Beach, WA.

Levins, R. (1969). Some demographic and genetic consequences of environmental heterogeneity for biological control. *Bull. Entomol. Soc. Am.* **15**, 237–240.

Levins, R. (1970). Extinction. In "Some Mathematical Problems in Biology" (M. Gerstenhaber, Ed.), pp. 75–107. American Mathematical Society, Providence, RI.

Longhurst, A. (2002). Murphy's Law revisited: Longevity as a factor in recruitment to fish populations. *Fisheries. Res.* **56**, 125–131.

Love, M.S. (1980). Isolation of olive rockfish, *Sebastes serranoides*, populations off southern California. *Fish. Bull.* **77**, 975–983.

Love, M. (1996). "Probably More Than You Want to Know About the Fishes of the Pacific Coast." Really Big Press, Santa Barbara, CA.

Love, M.S., Carr, M.H., and Haldorson, L.J. (1991). The ecology of substrate-associated juveniles of the genus *Sebastes*. *Environ. Biol. Fishes* **30**, 225–243.

Love, M.S., Caselle J.E., and Snook, L. (2000). Fish assemblages around seven oil platforms in the Santa Barbara Channel. *Fish. Bull.* **98**, 96–117.

Love, M.S., Yoklavich, M., and Thorsteinson, L. (2002). "The Rockfishes of the Northeast Pacific." University of California Press, Berkeley, CA.

Lowe, C.G., Topping, D.T., Cartamil, D.P., and Papastamatiou, Y.P. (2003). Movement patterns, home range, and habitat utilization of adult kelp bass *Paralabrax clathratus* in a temperate no-take marine reserve. *Mar. Ecol. Prog. Ser.* **256**, 205–216.

Lubchenco, J., Palumbi, S.R., Gaines, S.D., and Andelman, S. (2003). Plugging a hole in the ocean: the emerging science of marine reserves. *Ecol. Appl.* **13 (Suppl.)**, S3–S7.

MacArthur, R., and Wilson, E.O. (1967). "The Theory of Island Biogeography." Princeton University Press, Princeton, NJ.

Mann, D.H., and Hamilton, T.D. (1995). Late Pleistocene and Holocene paleoenvironments of the north Pacific coast. *Quatern. Sci. Rev.* **14**, 449–471.

Martell, S.J.D., Walters, C.J., and Wallace, S.S. (2000). The use of marine protected areas for the conservation of lingcod (*Ophiodon elongatus*). *Bull. Mar. Sci.* **66**, 729–743.

Matala, A.P., Gray, A.K., Gharrett A.J., and Love, M.S. (2004). Microsatellite variation indicates population genetic structure of bocaccio. *N. Am. J. Fish. Manage.* **24**, 1189–1202.

Mathews, K.R. (1985). Species similarity and movement of fishes on natural and artificial reefs in Monterey Bay, California. *Bull. Mar. Sci.* **37**, 252–270.

Mathews, K.R., and Richards, L.J. (1991). Rockfish (Scorpaenidae) assemblages of trawlable and untrawlable habitats off Vancouver Island, British Columbia. *N. Am. J. Fish. Manage.* **11**, 312–318.

Mathews, K.R., Miller, B.S., and Quinn, T.P. (1987). Movement studies of nearshore demersal rockfishes in Puget Sound, Washington. In "Proceedings of the International Rockfish Symposium," Alaska Sea Grant Rept. 87-2, pp. 63–72.

Mathews, S.B., and Barker, M.W. (1983). Movements of rockfish (*Sebastes*) tagged in northern Puget Sound. *Washington. Fish. Bull.* **82**, 916–922.

McCauley, D.E. (1991). Genetic consequences of local population extinction and recolonization. *Trends Ecol. Evol.* **6**, 5–8.

McCauley, D.E. (1993). Evolution in metapopulations with frequent local extinction and recolonization. *Oxford Surv. Evol. Biol.* **10**, 109–134.

McDermott, S.F. (2003). Improving Abundance Estimation of a Patchily Distributed Fish, Atka Mackerel (*Pleurogrammus monopterygius*). Ph.D. Dissertation, University of Washington, Seattle, WA.

McGowan, J.A., Cayan, D.R., and Dorman, L.N. (1998). Climate-ocean variability and ecosystem response in the northeast Pacific. *Science* **281**, 281–217.

Miller, D.J., and Geibel, J.J. (1973). Summary of blue rockfish and lingcod life histories; a reef ecology study; and giant kelp, *Macrocystis pyrifera*, experiments in Monterey Bay, California. *CDFG Fish Bull.* **158**.

Miller, J.A., and Shanks, A.L. (2004). Evidence for limited larval dispersal in black rockfish (*Sebastes melanops*): implications for population structure and marine reserve design. *Can. Jour. Fish. Aquat. Sci.* **61**, 1723–1735.

Moser, H.G. (1974). Development and distribution of larvae and juveniles of Sebastolobus (Pisces: family Scorpaenidae). *Fish. Bull.* **72**, 865–884.

Moser, H.G., and Boehlert, G.W. (1991). Ecology of pelagic larvae and juveniles of the genus *Sebastes*. *Environ. Biol. Fishes* **30**, 203–224.

Moser, H.G., Charter, R.L., Smith, P.E., Ambrose, D.A., Charter, S.R., Meyer, C.A., Sandknop, E.M., and Watson, W. (1993). Distributional atlas of fish larvae and eggs in the California Current region: Taxa with 1000 or more total larvae, 1951 through 1984. CalCOFI Atlas 31.

Moser, H.G., Charter, R.L., Watson, W., Ambrose, D.A., Butler, J.L., Charter, S.R., and Sandknop, E.M. (2000). Abundance and distribution of rockfish (*Sebastes*) larvae in the Southern California Bight in relation to environmental conditions and fishery exploitation. *CCOFI Rep.* **41**, 132–147.

Moulton, L.L. (1977). An Ecological Analysis of Fishes Inhabiting the Rocky Nearshore Regions of Northern Puget Sound, Washington. Ph.D. Dissertation, University of Washington, Seattle, WA.

Murie, D.J., Parkyn, D.C., Clapp, B.G., and Krause, G.C. (1994). Observations on the distribution and activities of rockfish, *Sebastes* spp., in Saanich Inlet, British Columbia, from the Pisces IV submersible. *Fish. Bull.* **92**, 313–323.

Myers, R.A., and Worm, B. (2003). Rapid depletion of predatory fish communities. *Nature* **423**, 280–283.

Narum, S. (2000). Assortative Mating and Genetic Structure of Gopher Rockfish (*S. carnatus*) and Black-and-yellow Rockfish (*S. chrysomelas*): A Case of Incipient Speciation. M.Sc. Thesis, University of San Diego, San Diego, CA.

Nasby-Lucas, M., Embley, B.W., Hixon, M.A., Merle, S.G., Tissot, B.N., and Wright, D.J. (2002). Integration of submersible transect data and high-resolution multibeam sonar imagery for a habitat-based groundfish assessment of Heceta Bank, Oregon. *Fish. Bull.* **100**, 739–751.

Nelson, B., and Quinn, III, T.J. (1987). Population parameters for rougheye rockfish (*Sebastes aleutianus*). In "Proceedings of the International Rockfish Symposium," Alaska Sea Grant Rep. 87-2, pp. 209–228.

NOAA (2005). Pacific Coast Groundfish Fishery Management Plan, Essential Fish Habitat Designation and Minimization of Adverse Impacts. Draft Environmental Impact Statement. February 2005.

O'Connell, V.M., and Carlile, D.W. (1993). Habitat-specific density of adult yelloweye rockfish *Sebastes ruberrimus* in the eastern Gulf of Alaska. *Fish. Bull.* **91**, 304–309.

Palsson, W.A. (2002). The development of criteria for establishing and monitoring no-take refuges for rockfishes and other rocky-habitat fishes in Puget Sound. In "Proceedings of the 2001 Puget Sound Research Conference" (T. Droscher, Ed.). Puget Sound Water Quality Action Team, Olympia.

Palumbi, S.R. (2003). Population genetics, demographic connectivity, and the design of marine reserves. In "The Science of Marine Reserves," *Ecol. Appl.* **13 (Suppl.)**, S146–S158.

Pauly, D., Christensen, V., Dalsgaard, J., Froese, R., and Torres, Jr., F. (1998). Fishing down marine food webs. *Science* **279**, 860–863.

Pearcy, W.G. (1992). Movements of acoustically-tagged yellowtail rockfish *Sebastes flavidus* on Heceta Bank, Oregon. *Fish. Bull.* **90**, 726–735.

Pearson, K.E., and Gunderson, D.R. (2003). Reproductive biology and ecology of shortspine thornyhead rockfish, *Sebastolobus alascanus*, and longspine thornyhead rockfish, *S. altivelis*, from the northeastern Pacific Ocean. *Environ. Biol. Fishes* **67**, 117–136.

Pogson, G.H., and Fevolden, S.E. (2003). Natural selection and the genetic differentiation of coastal and Arctic populations of the Atlantic cod in northern Norway: A test involving nucleotide sequence variation at the pantophysin (PanI) locus. *Mol. Ecol.* **12**, 63–74.
Ralston, S., and Howard, D.F. (1995). On the development of year-class strength and cohort variability in two northern California rockfishes. *Fish. Bull.* **93**, 710–720.
Richards, L.J. (1986). Depth and habitat distributions of three species of rockfish (*Sebastes*) in British Columbia: Observations from the submersible PISCES IV. *Environ. Biol. Fishes* **17**, 13–21.
Richardson, S.L., and Laroche, W.E. (1979). Development and occurrence of larvae and juveniles of the rockfishes *Sebastes crameri, Sebastes pinniger, and Sebastes helvomaculatus* (Family Scorpaenidae) off Oregon. *Fish Bull.* **77**, 1–46.
Ricketts, E.F., Calvin, J., Hedgpeth, J.W., and Phillips, D.W. (1992). "Between Pacific Tides," 5th ed. Stanford University Press, Palo Alto, CA.
Rocha-Olivares, A., and Vetter, R.D. (1999). Effects of oceanographic circulation on the gene flow, genetic structure, and phylogeography of the rosethorn rockfish (*Sebastes helvomaculatus*). *Can. J. Fish. Aquat. Sci.* **56**, 803–813.
Rocha-Olivares, A., Rosenblatt, R.H., and Vetter, R.D. (1999a). Cryptic species of rockfishes (*Sebastes*: Scorpaenidae) in the southern hermisphere inferred from mitochondrial lineages. *J. Hered.* **90**, 404–411.
Rocha-Olivares, A., Kimbrell, C.A., Eitner, B.J., and Vetter, R.D. (1999b). Evolution of a mitochondrial cytochrome *b* gene sequence in the species-rich genus *Sebastes* (Teleostei, Scorpaenidae) and its utility in testing the monphyly of the subgenus *Sebastomus*. *Mol. Phylogenetics Evol.* **11**, 426–440.
Rocha-Olivares, A., Rosenblatt, R.H., and Vetter, R.D. (1999c). Molecular evolution, systematics, and zoogeography of the rockfish subgenus *Sebastomus* (*Sebastes*, Scorpaenidae) based on mitochondrial cytochrome *b* and control region sequences. *Mol. Phylogenetics Evol.* **11**, 441–458.
Rocha-Olivares, A., Leal-Navarro, R.A., Kimbrell, C., Lynn, E.A., and Vetter, R.D. (2003). Microsatellite variation in the Mexican rockfish *Sebastes macdonaldi*. *Sci. Mar.* **67**, 451–460.
Rooper, C.N. (2002). "English Sole Transport During Pelagic Stages on the Pacific Northwest Coast, and Habitat Use by Juvenile Flatfish in Oregon and Washington Estuaries." Ph.D. Dissertation, University of Washington, Seattle, WA.
Ross, J.R.M., and Larson, R.J. (2003). Influence of water column stratification on the depth distributions of pelagic juvenile rockfishes off central California. *CCOFI Rep.* **44**, 65–75.
Rousset, F. (1997). Genetic differentiation and estimation of gene flow from F-statistics under isolation by distance. *Genetics* **145**, 1219–1228.
Royer, T.C. (1998). Coastal processes in the northern North Pacific. Coastal segment (9, P). In "The Sea. The Global Coastal Ocean. Regional Studies and Syntheses" (A. Robinson and K.H. Brink, Eds.), pp. 395–414. John Wiley, New York.
Ruzzante, D.E., Taggart, C.T., Cook, D., and Goddard, S. (1996). Genetic differentiation between inshore and offshore Atlantic cod (*Gadus morhua*) off Newfoundland: Microsatellite DNA variation and antifreeze level. *Can. J. Fish. Aquat. Sci.* **53**, 634–645.
Seeb, L. (1998). Gene flow and introgression among three species of rockfishes, *Sebastes auriculatus, S. caurinus*, and *S. maliger*. *J. Hered.* **89**, 393–403.
Seeb, L.W., and Gunderson, D.R. (1988). Genetic variation and population structure of Pacific ocean perch (*Sebastes alutus*). *Can. J. Fish. Aquat. Sci.* **45**, 78–88.
Service, R.F. (2004) New dead zone off Oregon coast hints at sea change in currents. *Science* **305**, 1099.
Shanks, A.L. (1983). Surface slicks associated with tidally forced internal waves may transport pelagic larvae of benthic invertebrates and fishes shoreward. *Mar. Ecol. Prog. Ser.* **13**, 311–315.
Shanks, A.L., Largier, J., Brink, L., Brubaker J., and Hoof, R. (2000). Demonstration of the onshore transport of larval invertebrates by the shoreward movement of an upwelling front. *Limnol. Oceanogr.* **45**, 230–236.

Slatkin, M. (1977). Gene flow and genetic drift in a species subject to frequent local extinctions. *Theor. Popul. Biol.* **12**, 253–262.

Soh, S.K., Gunderson, D.R., and Ito, D.H. (2001). The potential role of marine reserves in the management of shortraker rockfish (*Sebastes borealis*) and rougheye rockfish (*S. aleutianus*) in the Gulf of Alaska. *Fish. Bull.* **99**, 168–179.

Somero, G.N. (1991). Hydrostatic pressure and adaptations to the deep sea. In "Comparative Animal Physiology" (C.L. Prosser, Ed.), pp. 167–204. John Wiley and Sons, New York.

Sotka, E.E., Wares, J.P., Barth, J.A., Grosberg, R.K., and Palumbi, S.R. (2004) Strong genetic clines and geographic variation in gene flow in the rocky intertidal barnacle *Balanus glandula. Mol. Ecol.* **13**, 2143–2156.

Sponaugle, S., Cowen, R.K., Shanks, A., Morgan, S.G., Leis, J.M., Pineda, J., Boehlert, G.W., Kingsford, M.J., Lindeman, K.C., Grimes, C., and Munro, J.L. (2002). Predicting self-recruitment in marine populations: Biophysical correlates and mechanisms. *Bull. Mar. Sci.* **70, Suppl. 1**, 341–375.

Stanley, R.D., Leaman, B.M., Haldorson, L., and O'Connell, V.M. (1994). Movements of tagged adult yellowtail rockfish *Sebastes flavidus*, off the west coast of North America. *Fish. Bull.* **92**, 655–663.

Starr, R.M., Heine, J.N., Felton, J.M., and Cailliet, G.M. (2002). Movements of bocaccio (*Sebastes paucispinis*) and greenspotted (*S. chlorostictus*) rockfishes in a Monterey submarine canyon: Implications for the design of marine reserves. *Fish. Bull.* **100**, 324–337.

Stephens, J.S. Jr., Morris, P.A., Pondella, D.J., Koonce, T.A., and Jordan, G.A. (1994). Overview of the dynamics of an urban artificial reef fish assemblage at King Harbor, California, USA. 1974–1991: A recruitment driven system. *Bull. Mar. Sci.* **55**, 1224–1239.

Stepien, C.A., Dillon, A.K., and Patterson, A.K. (2000). Population genetics, phylogeography, and systematics of the thornyhead rockfishes (Sebastolobus) along the deep continental slopes of the North Pacific Ocean. *Can. J. Fish Aquat. Sci.* **57**, 1701–1717.

Stout, H.A., McCain, B.B., Vetter, R.D., Builder, T.L., Lenarz, W.H., Johnson, L.L., and Methot, R.D. (2001) Status review of copper rockfish (Sebastes caurinus), quillback rockfish (S. maliger), and brown rockfish (S. auriculatus) in Puget Sound, Washington. *NOAA Tech. Memo. NMFS-NWFSC*, **46**, 1–158.

Swearer, S.E., Casselle, J.E., Lea, D.W., and Warner, R.R. (1999). Larval retention and recruitment in an island population of a coral-reef fish. *Nature* **402**, 799–802.

Swearer, S.E., Shima, J.S., Hellberg, M.E., Thorrold, S.R., Jones, G.P., Robertson, D.R., Morgan, S.G., Selkoe, K.A., Ruiz, G.M., and Warner, R.R. (2002). Evidence of self-recruitment in demersal marine populations. *Bull. Mar. Sci.* **70 Suppl. S.**, 251–271.

Taylor, B.L., and Dizon, A.E. (1996). The need to estimate power to link genetics and demography for conservation. *Conserv. Biol.* **10**, 661–664.

Taylor, C.A. (2004) Patterns of early-stage pelagic dispersal and gene flow in rockfish species from the Southern California Bight. Ph.D. dissertation Univ. of Ca. San Diego, San Diego, CA, USA.

Terry, A., Bucciarelli, G., and Bernardi, G. (2000) Restricted gene flow and incipient speciation in disjunct Pacific Ocean and Sea of Cortez populations of a reef fish species, *Girella nigricans. Evolution* **54**, 652–659.

Thorrold, S.R., Jones, G.P., Hellberg, M.E., Burton, R.S., Swearer, S.E., Neigel, J.E., Morgan, S.G., and Warner, R.R. (2002). Quantifying larval retention and connectivity in marine populations with artificial and natural markers. *Bull. Mar. Sci.* **70, Suppl. 1**, 291–308.

Thorson, R.M. (1980). Ice-sheet glaciation of the Puget lowland, Washington, during the Vashon Stade (Late Pleistocene). *Quatern. Res.* **13**, 303–321.

Vetter, R.D., and Lynn, E.A. (1997). Bathymetric demography, enzyme activity patterns, and bioenergetics of deep-living scorpaenid fishes (genera *Sebastes* and *Sebastolobus*): Paradigms revisited. *Mar. Ecol. Prog. Ser.* **155**, 173–188.

Vetter, R.D., Lynn, E.A., Garza, M., and Costa, A.S. (1994). Depth zonation and metabolic adaptation in Dover sole, *Microstomus pacificus*, and other deepliving flatfishes: Factors that affect the sole. *Mar. Biol.* **120**, 145–149.

Wade, M.J., and McCauley, D.E. (1988). Extinction and recolonization: their effects on the genetic differentiation of local populations. *Evolution (Lawrence, Kans.)* **42**, 995–1005.

Waples, R.S. (1998). Separating the wheat from the chaff: Patterns of genetic differentiation in high gene flow species. *J. Hered.* **89**, 438–450.

Wares, J.P., Gaines S.D., and Cunningham, C.W. (2001). A comparative study of asymmetric migration events across a marine biogeographic boundary. *Evolution* **55**, 295–306.

Warner, R.R., and Cowen, R.K. (2002). Open vs Closed Marine Populations: Synthesis and Analysis of the Evidence. *Bull. Mar. Sci.* **70**, **Suppl. 1**, 245–250.

Watson, W., Charter, R.L., Moser, H.G., Vetter, R.D., Ambrose, D.A., Charter, S.R., Robertson, L.L., Sandknop, E.M., Lynn, E.A., and Stannard, J. (1999). Fine-scale distributions of planktonic fish eggs in the vicinities of Big Sycamore Canyon and Vandenberg Ecological Reserves, and Anacapa and San Miguel Islands, California. *CCOFI Rep.* **40**, 128–153.

Weinberg, K.L. (1994). Rockfish assemblages of the middle shelf and upper slope off Oregon and Washington. *Fish. Bull.* **92**, 620–632.

Westrheim, S.J., Harling, W.R., Davenport, D., and Smith, M.S. (1974). *G.B. Reed* groundfish cruise no. 74-4, 4-25 September 1974. *Can. Fish. Mar. Serv. Tech. Rep.* **497**, 37p.

Williams, E.H., and Ralston, S. (2002). Distribution and co-occurrence of rockfishes (family: Sebastidae) over trawlable shelf and slope habitats of California and southern Oregon. *Fish. Bull.* **100**, 836–855.

Wing, S.R., Botsford, L.W., Ralston, S.V., and Largier, J.L. (1998). Meroplanktonic distribution and circulation in a coastal retention zone of the California upwelling system. *Limnol. Oceanogr.* **43**, 1710–1721.

Withler, R.E., Beacham, T.D., Schulze, A.D., Richards, L.J., and Miller, K.M. (2001). Co-existing populations of Pacific ocean perch, *Sebastes alutus*, in Queen Charlotte Sound, British Columbia. *Mar. Biol.* **139**, 1–12.

Wright, S. (1977). "Evolution and the Genetics of Populations, Vol. 3." University of Chicago Press, Chicago, IL.

Yoklavich, M.M., Greene, H.G., Cailliet, G.M., Sullivan, D.E., Lea, R.N., and Love, M.S. (2000). Habitat associations of deep-water rockfishes in a submarine canyon: An example of a natural refuge. *Fish. Bull.* **98**, 625–641.

Zeidberg, L.D., and Hamner, W.M. (2002) Distribution of squid paralarvae, *Loligo opalescens* (Cephalopoda: Myopsida), in the southern California Bight in the three years following the 1997–1998 El Niño. *Mar. Biol.* **14**, 111–122.

CHAPTER 4

Estuarine and Diadromous Fish Metapopulations

CYNTHIA M. JONES

I. Introduction to Metapopulation Concepts in Estuarine and Diadromous Fish
II. Mechanisms that Form Distinct Populations
III. Tools to Quantify Migration in Diadromous Fish
   A. Genetics
   B. Artificial Tags
   C. Natural Tags with Emphasis on Otolith–Geochemical Tags
IV. Diadromous Fish Exemplify Metapopulation Theory
   A. Salmonids
   B. Alosines
   C. Comparing Salmonids and Alosine Herrings
   D. Sciaenids
   E. Atherinids
V. Value of the Metapopulation Concept in Understanding and Managing Diadromous Fisheries
   A. Conservation of Local Populations
   B. Historical Management of Local Populations in a Fisheries Context
   C. Mixed-Stock Analysis
   D. Effect of Demography on Metapopulation Management
   E. Marine Protected Areas as a Spatial Management Tool
   F. Can Fisheries Be Managed as Metapopulations?
VI. Acknowledgments
   References

I. INTRODUCTION TO METAPOPULATION CONCEPTS IN ESTUARINE AND DIADROMOUS FISH

Diadromous fish regularly migrate between the sea and freshwater. These fish include the anadromous species that spawn in fresh water but spend much of their lives in the sea, the catadromous species that spawn at sea but spend most of their lives in freshwater, and those species that live in estuaries for much of their lives and may spawn in the sea or estuaries but are estuarine dependent during important life stages. Some diadromous species are easily seen as metapopulations, such as the alosine herrings and salmonids (Hollis, 1948; Ricker, 1972; Taylor, 1991; Gall et al., 1992; Quinn, 1993; Rieman and Dunham, 2000; King et al., 2001). Others, such as the estuarine-dependent and marine species, have been largely ignored in formulating metapopulation concepts until recently (Thorrold et al., 2001; Beacham et al., 2002; Swearer et al., 2002; Warner and Cowen, 2002). In contrast, it is difficult to envision metapopulation structure in catadromous species, although there is evidence that some may exist for galaxiids (Barker and Lambert, 1988; Waters et al., 2000; but on the contrary see Berra et al., 1996). In part, the application of the metapopulation concept in diadromous fishes requires an understanding of how their life histories affect population structure, and begs a clear understanding of the concept itself (Smedbol and Wroblewski, 2002).

Levins (1970) coined the term *metapopulation* as a "population of populations," as discussed in the introduction to this book. Unfortunately, the term *metapopulation* has been used to describe two different entities without explicitly recognizing the differences. These are the *naturally evolved* metapopulations and the *fragmentation-induced* metapopulations. By fragmentation induced, I mean a population that occurred naturally as a population distributed across a heterogenous landscape with considerable interbreeding along its entire range, which has subsequently been disrupted and thrust into isolated patches through fragmentation of its habitat. In contrast, a *naturally evolved* metapopulation has developed from its inception as clusters of spatially isolated units, where there is just enough interbreeding with other local populations to restrain speciation.

The dichotomy of naturally evolved versus fragmentation-induced metapopulations emphasizes the role that evolutionary–genetic adaptation will have in how the metapopulation responds to change and stress. In a naturally evolved metapopulation at equilibrium, the population will have surmounted the problems of genetic load and inbreeding depression, and, potentially, small population size. This is evidenced by the long-term viability of the smaller, local populations. Through the millennia, these local populations persist, and this is evidence that they have eliminated the effects of deleterious genes that would have surfaced as small populations became inbred. In a population that has undergone recent fragmentation, the newly isolated local populations will not have surmounted these difficulties. These isolated populations become more inbred, which results in an increase in homozygosity and consequent unmasking of deleterious genes (Doak, 1995). In the terrestrial literature we find frequent reports of population demise as a result of habitat fragmentation—for example, in the Iberian lynx (Goana et al., 1998). Waples (2002) points out that ". . . favorable alleles can be lost and deleterious alleles can become fixed by chance alone. (p. 148)" Based on the time to reach equilibrium, we can anticipate that these two types of metapopulation will respond differently to environmental challenges. I show that there are many excellent examples of diadromous fishes as naturally evolved metapopulations later in this chapter. Through weight of evidence, I show that diadromous, estuarine, and marine fish populations form naturally evolved metapopulations under conducive conditions.

Under natural conditions, life histories that result in the segregation of the species into local reproductive groups will lead to retention of local adaptations if there is philopatry and if the environment is stable in regard to its selection pressures. Taken to its extreme, populations that are fully separate for sufficient time will form incipient species. Taken to a lesser extreme, local populations form with distinct genetic variation and with a probability of local extinction (sensu Levins, 1969, 1970). It was Levins who first formalized mathematically the concept of a species broken into separate, local populations that form a larger metapopulation (Levins, 1970). Parenthetically, the origins of the concept of local populations were recognized a century earlier in the "races" among marine fish, especially for herring in the North Sea (Rozwadowski, 2002) and in American shad (Milner, 1876). Close reading of Levins' papers (1969, 1970) indicate that his model assumes a naturally evolved metapopulation when he relies on equilibrium conditions. Similarly, the concepts of source—sink populations and metaphors of populations with mainland—island dynamics rely on the natural evolution of these patterns as the base concept.

Levins' (1969) work was developed in the context of eradicating pest species by clarifying the effect of extirpating a few local populations on the persistence of the species as a whole. He developed this mathematical model in a historical context of ideas in which scientists were uncovering the importance of spatial heterogeneity for the persistence of coupled predator and prey populations. Gause (1935) tried to experimentally test new theories that predator and prey interactions resulted in natural cycles, as articulated most clearly by Elton (1942). However, Gause's coupled populations did not persist in his test tube experiments. When Huffaker (1958) repeated these experiments with mites, he found that predator and prey populations only persisted when the environment was spatially complex and where refuges existed for prey. Subsequently, in 1967, MacArthur and Wilson published their book, *The Theory of Island Biogeography*, in which the concepts of species extinction and colonization were presented in the context of island–mainland spatial separation. Their theory stated that the number of species would reach equilibrium, even though there was an ongoing process of species extinction and colonization. It was a simple but creative step for Levins to apply these concepts to the dynamics within a single species (Hanski and Gilpin, 1991; Tilman and Kareiva, 1997). Levins' model was the first formalization of species colonization and extinction applied at the population level,

Fragmentation-induced metapopulation   Naturally evolved metapopulation

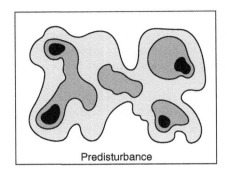
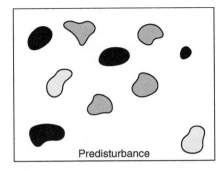

Predisturbance   Predisturbance

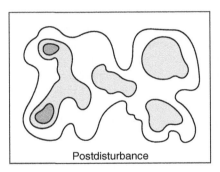
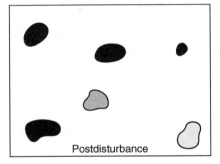

Postdisturbance   Postdisturbance

FIGURE 4-1. Color density indicates abundance. In the fragmentation-induced metapopulation, habitat is lost and, with it, parts of the population. The remaining population is forced into non-contiguous pockets that are newly separated. In the naturally evolved metapopulation, when habitat is lost, whole local populations are lost and remaining populations are spaced farther apart.

and it encouraged the development of metapopulation theory in terrestrial and conservation ecology, and its recent application in marine species as described in Chapter 1.

The application of metapopulation theory differs fundamentally between marine and terrestrial systems (Fig. 4-1). The revitalization of the theory in ecology was in the context of habitat fragmentation of species that were wide ranging, in efforts to stop their decline to extinction (Hanski and Thomas, 1994; Goana et al., 1998; Root, 1998). For example, the endangered Iberian lynx (*Lynx pardinus*) was once widely distributed and free roaming in southwestern areas of the Iberian peninsula (Goana et al., 1998). Human encroachment and land use change has fragmented that habitat over the past millennium and forced this species into isolated pockets. Clearly, the Iberian lynx has lost habitat, but it must now also contend with the genetic consequences of isolation to which it has not adapted over its evolutionary lifetime. Similarly, much of the terrestrial metapopulation literature deals with fragmentation-induced metapopulations such as this.

In contrast, anadromous species such as salmonids and alosines have evolved through geological time to exist as isolated reproductive groups and to contend with the genetic consequences of such isolation, and are naturally evolved metapopulations. For a species that has recently experienced habitat loss and fragmentation (a fragmentation-induced metapopulation), straying and tolerance to isolation may not be as intrinsic to the life history as they are to salmon, for example. Although both types of species exist in isolated populations, their evolutionary histories may dictate different responses to local extirpation. A salmonid model is conceptually closer to Levins' original formulation than a terrestrial model in which habitat has been lost to human activity. In Levins' model, the local population that has been lost is one that was adapted, however well, to isolation. After all, the local population had to be sufficiently isolated to face extinction, and extinction is guaranteed in the classical model. This is a subtle but important difference. Implicit in Levins' model is an assumption of a species that has an inherent straying rate and tolerance to isolation that is intrinsic in its life history as modified through selection and environmental change. Certainly, metapopulation analysis has transcended Levins' initial perspective. Nonetheless, it is instructive to understand the ramifications of habitat loss, particularly how this loss occurs, because it can have a profoundly different impact on population persistence in marine versus terrestrial systems.

The validity of applying the metapopulation concept to diadromous fish lies in the difficulty in defining "population" and "metapopulation" for diadromous fish (Smedbol et al., 2002). Although some anadromous fish exemplify discrete local populations, even in these species genetic distinctions are not clear at fine scale (Ricker, 1972). Moreover, for estuarine-spawning species, the extent of local populations and connectivity are not well understood. Often it appears that estuarine-spawning species exhibit less philopatry (Thorrold et al., 2001) and greater movement at age than anadromous species, and thus lie on a continuum from a nonuniformly distributed population to discrete local populations that form a metapopulation (e.g., Smedbol et al., 2002).

Marine fisheries scientists were among the first to show that marine fishes were composed of distinct, self-sustaining populations (Heincke, 1898; Hjort, 1914), which are called *stocks* in this discipline (Berst and Simon, 1981) or *local populations* in the current ecology literature. By the early 1900s, Heincke had recognized that there were "races" of herring that had different vital rates (birth and death) that resulted in asynchronous abundant cohorts (dominant year classes) across the species range, along with different patterns of migration. Heinke reformulated the variability of catches at different locations and times "into questions about the nature of species and whether species comprised a number of self-sustaining populations" (Rozwadowski, 2002, p. 81). Thus, the concept of local, semi-isolated populations within the larger population existed early in the development of fisheries science and preceded the formulation of the metapopulation concept in ecology by almost a century. This rich tradition of population-level research and management of "races" and "stocks" has been overlooked by fish ecologists, perhaps because it exists in language and literature that is not part of the ecologist's lexicon and libraries. The early identification of local populations resulted from the fisheries scientist's clear advantage of being able to determine age by evaluating the daily and annual bands in bones, scales, and otoliths in fish. Thus, fisheries scientists were able to estimate age-specific vital rates (birth, B and death, D)[1] and connectivity (immigration, I; and emigration, E). The relative magnitude of the birth and death rates in comparison with the magnitude of connectivity effectively determines whether fish exist as a patchily distributed population or distinct local populations that form part of a metapopulation.

This distinction is crucial to evaluating diadromous and estuarine fish as metapopulations and is made clear by Berryman (2002). Berryman (2002, p. 441) defines a population as "a group of individuals of the same species that live together in an area of sufficient size to permit normal dispersal and/or migration behavior and in which numerical changes are largely determined by birth and death processes." Hence, Berryman defines a local population as one in which its abundance is determined almost exclusively by the difference in its own birth and death rates. Although there may be sufficient migration or "straying" to render the population genetically panmictic, nonetheless the population is sufficiently separate that local adaptations largely determine its persistence. Further, Camus and Lima (2002) argue that when there is no clear definition of "population" then "local population" and "metapopulation" become ill-defined and ambiguous terms. The importance of spatial heterogeneity on population structure harken back to Andrewartha and Birch (1954). Clearly, when the births and deaths within the population determine its persistence, then we are addressing a population (sensu Levins, 1969, 1970; Berryman, 2002), whereas in contrast when migration between populations determines their persistence, then we are addressing a patchily distributed population (sensu Camus and Lima, 2002). Levins (1969, 1970) implies that migration is inconsequential in his model. Migration must be sufficient such that empty habitats can be recolonized subsequent to extinction, but Levins' model does not address demography within a population. In the ecological literature, it is understood that the movement between groups is maximum in a patchily distributed population, grading to nonexistent in populations that are so separate that they have begun to form species. However, in the few marine and estuarine fish that have been studied to determine connectivity from a metapopulation perspective (McQuinn, 1997; Ruzzante et al., 2001; Thorrold et

---

[1]Note, however, that the vast majority of study in fisheries concerns estimating and understanding the effect of birth (recruitment) and death (largely by harvest) on population abundance and persistence. For example, even now textbooks in fisheries population dynamics devote little space to migration (38 of 517 pages in Quinn and Deriso, 1999).

al., 2001; Smedbol and Wroblewski, 2002), there is both a degree of philopatry in younger ages but also sufficient migration or gene flow between groups to render the "population" genetically indistinct (for example, contrast Cordes and Graves, 2003, with Thorrold et al., 2001). The problem that lies before us is to quantify the transition between locally distinct populations and the patchily distributed population. At what point does the ratio of increase in population size that comes from birth and death or migration (f[B − D]/f[I − E]) become great enough that it is valid to apply the metapopulation concept? As yet, there is no answer to this (although see Malécot, 1950).

Marine and diadromous populations are typically open and not at equilibrium (see Caley et al., 1996, for marine species). The term *open* is often used without being defined. To some statisticians, an open population is one wherein abundance changes as a result of births and deaths during the period of census, whereas a closed population is one in which there are no appreciable births or deaths to change abundance (White et al., 1982; p. 5). In contrast, population ecologists often use the term *open* to indicate that immigration or emigration occurs, and changes population structure and persistence. For marine populations, the most common source of openness occurs from young recruits that come from off site. (Alexander and Roughgarden, 1996, illustrate these ideas for marine invertebrates.) When applying this concept to diadromous fishes, openness can occur through recruits, as larvae transported between local populations, or through adults that spawn in nonnatal populations. In such open populations, the persistence of a population is decoupled from its own production of young (Alexander and Roughgarten, 1996; Botsford et al., 1998, 2001). In contrast, a closed population is solely dependent on its own production of young to replenish itself.

II. MECHANISMS THAT FORM DISTINCT POPULATIONS

Diadromous fish populations are subject to two competing processes: dispersal and homing (Quinn, 1984; McDowall, 2003). As McDowall (2003; p. 11) points out, "Diadromy suppresses evolutionary divergence and increase in species diversity through facilitation of gene flow among populations that are otherwise isolated . . . ." Dispersal provides opportunity to shift to richer marine habitats but also suppresses adaptations for subsequent specific freshwater habitat upon return. In dispersing forms, natal homing is the primary mechanism to maintain distinct, reproductively isolated, locally adapted populations. The concepts of natal homing in fish originated as the parent–stream hypothesis (Davidson, 1937) to explain the means of separating populations that mix during their adult stage. The first application of this concept was to Pacific salmon populations because of their clear differences geographically. Once this concept of natal homing was accepted, the behavioral mechanisms that reinforced it were sought. Several mechanisms, such as orienting clues, were seen to diminish straying (Leggett, 1977). Among these cues in diadromous fish were sun orientation, polarized light, geomagnetism, temperature, ocean fronts, currents, and olfaction. Although these cues have been well studied in salmonids, little work has been done on marine fishes until recently (but see Atema et al., 2002, and Kingsford et al., 2002, for application to reef fish). Some open ocean marine larvae may "smell" their natal waters, but there has been almost no research to explain the mechanisms by which these species identify natal waters, and olfaction is simply assumed. Thus, the mechanisms that cue natal homing in many species of fish remain to be studied.

Ultimately, though, the metapopulation concept relies on straying and movement between populations to allow for recolonization of habitats in which local populations have gone extinct. To a geneticist this means ". . . movement from one spawning population to another *and* successful contribution of genetic material to the recipient population by the individual who moved" (Gharrett and Zhivotovsky, 2003, p. 141). To a fisheries scientist this means that fish can be (1) identified to their natal habitat and (2) seen spawning on nonnatal grounds in association with other spawners there. The fisheries scientist assumes that spawning results in successful gene flow, although this may not be strictly true. Following Gharrett and Zhivotovsky (2003), migration that results in gene flow alters the gene frequencies in the population that ". . . reflects a balance between its own local dynamics (i.e., selection) and introduced genes. (p. 141)"

Immigration and emigration can have effects on the vital rates of a population through density dependence even in the absence of gene flow. Immigrants will compete with natives for available resources. If these resources are limited, then vital rates will be affected, typically directly through decreased births and increased mortality or through indirect effects such as decreased growth that results in decreased fecundity. Similarly, emigration may free up limited resources to the source population. These effects can occur even if the migrants do not breed. When migrants breed, there may be decreased viability of progeny, which is called *outbreeding depression* and may diminish the degree of apparent population mixing.

Thus, the effects of migration are not straightforward. For Pacific salmon, migration and gene flow depend on their life history patterns (Quinn, 1984). Pink salmon (*Oncorhynchus keta*) have short juvenile residency in natal streams compared with coho (*O. kisutch*), sockeye (*O. nerka*), or chinook (*O. tshawytscha*), which have long juvenile residence in their natal streams. The length of juvenile residence coincides with a tendency to stray; longer residence results in less straying (see also Gharrett and Zhivotovsky, 2003, p. 162). Stream stability is another factor that may affect straying rates. Quinn (1984) hypothesized that unstable streams would have greater straying rates compared with stable streams.

III. TOOLS TO QUANTIFY MIGRATION IN DIADROMOUS FISH

Among the rates of population vitality (B, birth; D, death; I, immigration; E, emigration), fish scientists have developed a plethora of methods to estimate B and D (see for example Ricker, 1975, drawing on work in wildlife biology of Williams et al., 2002). These are the easier rates to measure in fish populations. In contrast, fewer methods exist to estimate I and E, and these rates have remained the most difficult to measure accurately.

The amount of migration and success of gene introgression determine the population structure of a species and the probability of local extinction. The amount and success of migrating individuals in breeding can be measured in diadromous fish using genetic, alternatively or in combination with, artificial and natural tags (see, for example, Tallman and Healey, 1994). Each of these tools has advantages and disadvantages, but note that they do not answer the question of migration and gene introgression in the same way and can give disparate results. Thus, it is important to understand how the tools are used and what limitations they have. The tools that are used also have different levels of specificity. Genetics and natural tags, such as parasites or otolith chemistry, have been used historically to specify population-level migration, whereas many applied tags, such as archival and coded wire, can track individuals.

A. GENETICS

Genetic analysis stands as the ultimate test of population divergence. Molecular genetics studies "have the particular strength of targeting variation that is explicitly genetic, although their power is limited in practice, because even the most comprehensive approaches are based on sampling only a very small part of the genome" (Youngson et al., 2003, p. 195). Several major genetic approaches have been used to determine population differentiation and movement: allozymes, mtDNA, and nuclear or microsatellite DNA. In the past, allozymes were widely used to determine the frequency of codominant alleles in populations and to test for violations of Hardy–Weinberg equilibrium. Recently developed techniques have improved the geneticist's ability to determine population structure and these include nuclear markers such as mtDNA, amplified fragment length polymorphisms, microsatellites, and, most recently, single nucleotide polymorphisms (see Sunnucks, 2000, for a review of these methods). For example, Beacham and his colleagues (2003a) used 13 microsatellite loci to separate Fraser River chinook salmon into seven groups, with 95% of the variation within a population.

Genetic techniques have traditionally been used to determine separation at the aggregate scale of populations and have not been able to assess individual contributions to fitness. However, "Individual contributions to spawning can now be documented by following the transmission of hypervariable mini- or microsatellite alleles between the parental generation and the resulting progeny" (Youngson et al., 2003, p. 198).

A commonly quoted rule when measuring gene flow is that anywhere from 1 to 10 migrants per generation will maintain homozygosity over 90% among populations, which is, in effect, random mating (Lewontin, 1974; Roughgarden, 1979; Allendorf, 1983; Mills and Allendorf, 1996; Hartl and Clark, 1997). Assuming that this rule is correct, minimal straying will yield population homogeneity and lack of population structure. This migration rule also assumes that migrants (or *strays*, as geneticists call them) breed successfully, and this is not supported when tagging studies are compared with genetic differentiation rates (Tallman and Healey, 1994; Hendry et al., 2000). Tagging studies show higher rates of movement than are shown in gene introgression, which indicate that migrants have lower reproductive success than local fish.

Waples (1998) recognized that genetic methods alone could not address the issues of population structure and connectivity in marine species with high gene flow. There are even several examples in which discrete local populations could not be identified through genetic methods alone (Allendorf et al., 1987; Utter et al., 1993). The failure of genetics to identify recently developed population structure leads scientists to turn to other methods of estimating philopatry and migration, such as artificial and natural tags (as discussed next).

B. Artificial Tags

Techniques are well developed for measuring population vital rates and migration with tags that are physically applied by researchers (Wydoski and Emery, 1983; Hilborn et al., 1990; Parker et al., 1990; Schwarz and Arnason, 1990, among others). Physically applied or artificial tags have been the mainstay of measuring migration rates, population mixing, and population structure. Hilborn et al. (1990) state that in some species almost all our knowledge comes from external tags. Artificial tags include those that are external or those placed internally by the investigator. External tags include mass marking such as fin clips, and individual marks such as Peterson, spaghetti, and dart tags. Internal tags include coded wire tags, archival tags, and pigments or dyes inserted under the skin. Otoliths have been mass marked by temperature (Brothers, 1990) and through chemicals. McFarlane et al. (1990) provide at comprehensive review of these tags. Tagging has been used since 1873 in the United States (Rounsefell and Kask, 1943) and much earlier in Europe.

The procedures to estimate movement and migration from tag recovery data are well developed in the statistical literature (Brownie et al., 1985; Burnham et al., 1987; Schwarz and Arnason, 1990). Schwarz and Arnason (1990) present a multiyear, spatially stratified model that could easily be expanded to encompass correct estimation of movement and migration for diadromous species. It is particularly important to note that estimation is facilitated by the fact that a known number of marks are introduced throughout the population's range. Having a known number of marked animals throughout the species' range permits estimation of movement and migration rates, along with spatially explicit survival rates.

C. Natural Tags with Emphasis on Otolith–Geochemical Tags

In contrast to the well-studied nature of physically applied tags, natural tags have been used less frequently, and the underlying statistical theory for estimating migration rates and natal fidelity is almost completely undeveloped (Jones, manuscript in preparation). Historically, the use of natural tags depended on the presence of parasites, unusual marks on bones or otoliths, and differences in meristics and morphometrics that occurred infrequently, in restricted spatial areas, and had to be used opportunistically (Quinn et al., 1987). The inconsistent natural occurrence of these types of tags meant that they were difficult to use in a statistically designed field study. For such reasons, natural tags have been problematic and used less frequently in studies of fish movement and philopatry. However, recent technical development of otolith–geochemical tags to mark fish movement will increase the use of these natural tags in mark–recovery studies.

The value of otolith–geochemical tags lies in their universality. All fish incorporate the chemistry of their environmental waters into their bones, albeit to greater or lesser extent under physiological influence (Campana, 1999). When this chemistry is sufficiently unique in different habitats and when it does not have a biological overprint (uptake is not modified by the fish's physiology), it serves as a mark of an individual's environmental and nursery residence (see, for example, Campana et al., 1994; Thorrold et al., 2001; and Wells et al., 2003). These chemical tags have been used to measure the movement of anadromous fish (Limburg, 1995, 1998; Secor et al., 2001), to differentiate contingents of fish that have separate movement trajectories (Secor, 1999; Secor et al., 2001), and to track their lifetime movements (Rieman et al., 1994; Secor et al., 1995, 1998). This is an area of burgeoning study and, as a nascent field, its future value to the study of fish movement cannot be fully evaluated.

Because the use of natural tags has been infrequent and opportunistic until the recent use of geochemical tags, the development of estimation procedures for these tags has been ignored by statisticians. The use of natural tags to assess movement is problematic in comparison with those that are physically applied by the scientist in known number. Although estimation of movement based on physically applied tags provides true estimates of straying and fidelity, this estimate based on natural tags is compromised by difference in capture probabilities and survival (Jones, manuscript in preparation). Clearly, estimation procedures must be developed to use natural tags knowledgeably and this is an open area for future study.

When we compare the rates of migration and putative gene flow obtained with tags with those obtained from genetic markers, we must realize that we are measuring different things. For example, when correctly estimated, otolith geochemistry measures migration into the receiving population, but does not take account of subsequent successful gene intrusion. In contrast, as Hellberg and his colleagues (2002) point out, genetic markers measure both migration and effective population size. They note that migrants may not breed successfully or their propagules may not survive.

IV. DIADROMOUS FISH EXEMPLIFY METAPOPULATION THEORY

Diadromous fish species provide a broad spectrum of philopatry and exchange, from the clearly distinct local populations of salmonids, to the barely distinct groups of estuarine spawners that may or may not form metapopulations, to catadromous eels that exemplify a single population distributed patchily over broad geographic areas. Ricker (1972) characterized salmonids as having largely closed demography, with many small, reproductively isolated populations. Anadromous fish such as salmonids are antithetical to estuarine-dependent species such as spot (*Leiostomus xanthrus*) and catadromous species such as American eels (*Anguilla anguilla*) that have virtually no reproductive isolation, and where any local adaptation obtained on nursery and adult habitats is subsequently lost in panmictic breeding. Between these two extremes are species that have intermediate degrees of philopatry and straying. "Stepping stone" models can be used to represent spatially structured populations in which immigrants come from the neighboring populations and in which there is some limit to exchange of individuals, hence gene flow, between the populations (Gharrett and Zhivotovsky, 2003). This is a good basic model for estuaries along a linear coast, such as the East Coast of the United States. Where boundaries are diffuse and movement greater, then a model, such as isolation-by-distance, might better represent coastline populations. These lend themselves best to patchily distributed populations (Wright, 1943). Let us now look at specific families to determine whether metapopulation theory can be used to model their dynamics.

A. Salmonids

Salmonids exhibit wide variability in life events and life patterns, such as size at age, age at first maturity, and rate of straying (Ricker, 1972; Youngson et al., 2003). However, for those populations that migrate, all exhibit a strong degree of philopatry, marking these fish as representatives of a Levins type metapopulation structure having local populations that are sufficiently isolated to have a measurable probability of local extinction. This essential feature of their life history means that they are more likely to retain local adaptation to their specific environmental conditions caused by selection. Local adaptation is expressed as alteration in life history traits, behavior, and morphology and meristics (Quinn and Adams, 1996). The most profound adaptations are those alterations in life history that affect reproduction and survival, hence fitness. Moreover, when populations are small, they are also more vulnerable to extinction, and their habitat is subject to subsequent recolonization by strays from other groups.

Pacific salmon (*Oncorhynchus* spp.) are characterized by small populations, reproductive isolation, and habitats that are vulnerable to degradation. These characteristics make local populations prone to extinction. When populations are small and isolated, chance events can drive the population levels so low that they can no longer recover in a process of demographic stochasticity. Their vulnerability is exacerbated by stocking non-locals, a practice that degrades selection for local adaptations that enhance survival (Utter et al., 1993). Although hatchery managers are aware of unintentional selection in captive breeding programs, other traits that are maladaptive in the wild, such as the increase in small egg size in captive breeders, can be introduced when populations are supplemented (Heath et al., 2003).

The mechanism of reproductive isolation of salmonid populations occurs by temporal and geographic means (see, for example, Utter et al., 1989). The degree of philopatry among salmonids in space and time is generally sufficient to form classic naturally evolved metapopulations, but is not unquestioned. In his review paper, Quinn (1993) provides clear evidence of high fidelity to natal streams. However, the degree and motivation for straying among salmon species needs more study using design-based experiments. His review presents further evidence that intrapopulation straying is related to fish age. He shows in salmon, and we see in other fish families, that bigger, older fish stray considerably more than younger, smaller fish and, thus, they are the agents of gene flow between populations. Quinn (1993) also questions whether broad generalizations, such as the idea that pink salmon (*O. gorbuscha*) stray more than other species, may hold up on further experimentally based study.

Pink salmon (*O. gorbuscha*) in the Auke Creek, Alaska, are a good example of a diadromous fish with a high degree of philopatry and with temporal separation among populations. Pink salmon there have even and odd year populations that remain separate and possess different adaptations. When they are crossbred, they exhibit hybrid depression because ". . . each line apparently has a different genetic solution to virtually the same problem of local adaptation" (Gharrett and Smoker, 1993, p. 51). Even though "morphs" within the parental population of these fish experienced the same habitat, they found different solutions to adapting to local conditions. Even and odd year separation between pink salmon has resulted in genetic isolation between year classes (Utter, 1981). To have connectivity between such temporally isolated populations, straying must exist between years—for example, by having premature young or delayed maturity.

Many examples can be found for the spatial influence in the formation of distinct local populations in salmonids (Utter et al., 1989; Gall et al., 1992). A recent example is provided by Beacham et al. (2003a,b), who undertook a comprehensive study of population structure in Fraser River chinook salmon (*O. tshawytscha*) using 13 microsatellite loci to differentiate local populations. Spawning groups of chinook are found in 65 tributaries of the Fraser River. These populations exhibit substantial difference in life history, such as length of freshwater residence, timing of emigration, and size and age at maturity—traits directly linked to fitness. Chinook from the upper and mid Fraser River tend to emigrate in the spring, whereas those from lower river populations leave mid season and those of the Thompson area leave in the fall. Genetic similarity was greatest within a tributary, indicating substantial gene flow between spawning groups. The genetic similarity decreased with regional aggregation, being progressively dissimilar regionally, thus indicating that an isolation-by-distance model may best represent population structure for this species. However, sufficient reproductive isolation exists to form genetically separate regional populations for the purposes of metapopulation analysis.

In a study done a decade earlier, Gall et al. (1992) analyzed the protein products of 78 isozyme loci to determine that chinook salmon from Oregon to California could be separated into five distinct genetic groups with straying rates of two fish per generation between populations. These conclusions are further confirmed by newer technologies in the application of otolith isotope chemistry to mark natal stream origin and to measure subsequent connectivity (Kennedy et al., 2000). These results provide overwhelming evidence that Pacific salmon form discrete populations, either by tributary or over a contiguous region.

Surprisingly for such well-studied fish, very few observations exist of local population extinction and subsequent repopulation—a fundamental process for creating and maintaining classic metapopulation structure. One such example can be found in the Takotna River in Alaska (Mr. Doug Molyneaux, Alaska Department of Fish and Game, personal communication; Stokes, 1985). Over time, the river habitat has remained largely intact. It has, however, experienced bouts of overfishing, and one such incidence extirpated chinook in 1921. Evidence from native peoples indicates that chinook have only recently repopulated this habitat, a lapse of some 60 years for recolonization to occur. This example provides strong evidence that local populations do go extinct and that subsequent colonization will not be immediate. The demise of summer-run Upper Adams sockeye in the Fraser River, British Columbia, is a similar story except that fish were extirpated by a rock slide that severely restricted fish passage at Hell's Gate during railroad construction in combination with overfishing (Williams, 1987). The International Pacific Salmon Fisheries Commission began transplanting eggs and young after 1949 to try to restore a population there. Even with significant human intervention over the years, only small spawning populations have become established. The small resurgence of Upper Adams sockeye is on the same time frame as that of natural restoration. On a geological timescale, habitats have been lost to glaciation and subsequently recolonized as glaciers retreated. Dolly Varden (*Salvelinus malma*), coho (*O. kisutch*), and sockeye (*O. nerka*) are rapid colonizers (Milner et al., 2000). Clearly, these examples are not just patchily distributed populations, but rather provide a powerful demonstration of true metapopulation structure where local populations become extinct and habitats become vacant until subsequent recolonization.

In all, Pacific salmon are one of the best-studied groups and stand as the best example of naturally evolved metapopulation structure in diadromous fish (Fig. 4-2). Over millennia they have colonized streams and tributaries to form relatively isolated local populations that go extinct and are subsequently recolonized.

Other salmonids provide similar results concerning the common formation of discrete, reproductively isolated local populations, but have not shown the radiation and speciation found in the Pacific salmonids. In contrast to the dramatic radiation of Pacific salmon, Atlantic salmon (*Salmo salar*) have evolved less. Local populations of Atlantic salmon form a continuous single species, divided into local populations, but have not diverged to form separate species, as have Pacific salmon (Montgomery, 2000). In part, this may be the result of a more active tectonic regime in the Pacific that has resulted in greater physiographical change. Hence, we can view Atlantic salmon as a model fish for single species that have retained their integrity, even though they have formed discrete, isolated, reproductive populations (King et al., 2001). Similarly Brown trout (*Salmo trutta*), a congener to Atlantic salmon, has spread down to the Mediterranean as a single species, or perhaps two. "Still, there is less species diversity in the Atlantic than in the Pacific" (T.P. Quinn, personal communication).

In the western Atlantic, the story of native Atlantic salmon is one of dramatic decline, loss of population structure, and extirpation. Within the United States, native, large-river populations have been extirpated. Attempts to restore extinguished populations have relied on the reintroduction of Penobscot River fish to other rivers. Despite extensive efforts to restock with nonlocal fish, populations in other large rivers on the East Coast of the United States have failed to rebound. Thus, analysis of metapopulation structure in U.S. waters is thwarted by the

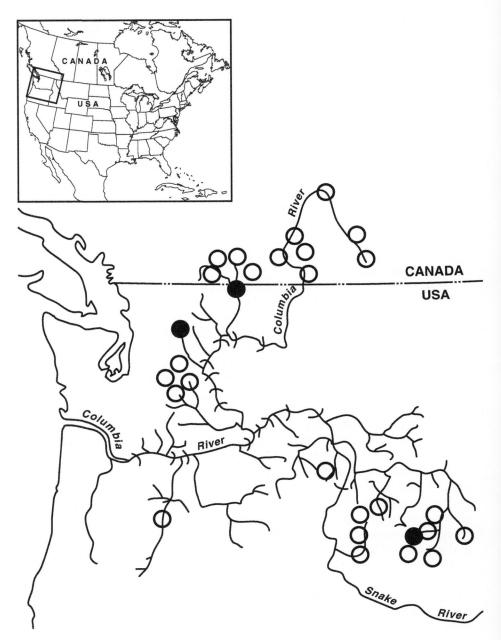

FIGURE 4-2. Structure of local populations of sockeye salmon on the Snake and Columbia rivers. Dark circles indicate extant populations and hollow circles indicate extinct populations. Dams have caused the extirpation of many populations on these rivers. The dendritic nature of these river systems promotes metapopulation formation. Redrawn from Mundy et al., 1995.

distortions of stocking and barely viable restocked populations. Any effort to estimate philopatry and movement is done in a very distorted situation and will not reflect original movements of native populations.

When limiting review to the remaining native salmon populations, we see that life history patterns differ along the western Atlantic range. Most U.S. populations of Atlantic salmon spend two winters at sea and are highly migratory, whereas Canadian populations in the Bay of Fundy characteristically spend only one winter at sea and are nonmigratory in the inner Bay (King et al., 2001). In terms of metapopulation structure, southern populations apparently were adapted to more potential gene introgression and potential straying. The northern and Bay of Fundy populations are more isolated and more closely mimic the true metapopulation structure observed in Pacific salmon. In summary, Atlantic salmon can also be considered a naturally evolved metapopulation. In contrast to Pacific salmon, they have not speciated, although they do show some of the same inclinations in life history patterns. Arguably, Atlantic salmon have also been subject to greater habitat loss or fragmentation throughout their range. In this way they demonstrate that even fish that have evolved to withstand isolation can be severely damaged when extensive habitat is lost.

B. ALOSINES

Alosines in the Atlantic Ocean have been studied almost as extensively as salmon in the Pacific Ocean. Although not as speciose, they do show population structure that ensures local adaptation. Among the most studied are American shad (*Alosa sapidissima*) in the western Atlantic, and allis (*A. alosa*) and twaite (*A. fallax fallax*) in the eastern Atlantic.

American shad has long been considered a diadromous species that exemplifies local populations characterized by important adaptations. This single species has a habitat range of almost 20° latitude—from the St. Johns River, Florida, USA, to the St. Lawrence River in Canada (Brown et al., 1999). Catches have declined from nearly 23,000 metric tons in 1896 to only hundreds of metric tons today (Limburg et al., 2003). Local populations show clinal variation in spawning season, fecundity, juvenile growth rate, and river emigration (Leggett and Carscadden, 1978; Glebe and Leggett, 1981a,b). More than 130 rivers once supported shad populations, but now fewer than 70 support shad populations as a result of a loss of more than 4000 km of spawning and nursery habitat (Limburg et al., 2003).

Local adaptations in reproductive characteristics are the most potent indications of metapopulation structure. In American shad, this structure is evidenced in semelparity in southern rivers and increasing iteroparity in northern rivers (Walburg and Nichols, 1967; Leggett and Carscadden, 1978). Similarly, several studies have shown that fecundity at age is greater in southern populations compared with northern ones (see Figure 5 in Limburg et al., 2003). Note, however, that many studies prior to the mid 1980s did not use modern methods of counting oocytes to determine fecundity. This is particularly problematic when determining the fecundity of batch spawners, such as shad. Based on counts of hydrated oocytes, Olney and McBride (2003) found no differences in batch size between semelparous and iteroparous populations when adjusted for body weight. However without knowledge of annual fecundity (batch size × annual number of batches), the relationship between latitude and fecundity remains in question, although the change to iteroparity from south to north remains an important change in life history pattern.

Besides these differences in important reproductive characteristics, local shad populations also evidence differences in secondary traits, such as size at age. Leggett and Carscadden (1978) showed that southern fish are smaller at age than northern fish. It is interesting to note that one would then expect fecundity to be lower too, but it is not, at least for batch fecundity (Olney and McBride, 2003). However, Limburg et al. (2003) show that size at age is less predictable when more populations are included as a result of more data collection. They show that fish size in the mid-Atlantic rivers is smaller than anticipated from earlier studies. These fish are smaller than in the southernmost rivers. These relationships are further complicated because the role of fishing pressure on size at age is unknown, but Hudson River data suggest an effect (Limburg et al., 2003).

Similar to salmonids, American shad populations mix at sea. Young of the year migrate from their natal rivers as soon as two months of age, or up to the end of their first growing season. Shad migrate long distances to summer feeding and overwintering grounds that serve many local populations (Limburg et al., 2003). Dadswell et al. (1987) demonstrated that fish caught in the mid Atlantic in winter came from rivers as far south as Georgia to as far north as Quebec. Nonetheless, prior to spawning in the spring, these fish return to their natal rivers (Dodson and Leggett, 1973; Melvin et al., 1986; but see Limburg, 1998). Limburg (1998) has shown that the behavior of juvenile migrants is more complex than first thought, and some juveniles may overwinter in the estuaries.

Alosines in the eastern Atlantic Ocean have faced extensive loss of populations and declining abundance similar to American shad. Two species worth mentioning are the twaite (*A. fallax fallax*) and allis shad (*A. alosa*) of Europe. Twaite shad had formed populations from Iceland to Morocco, but now have been reduced or extirpated from many of their former rivers as a result of placement of dams (Aprahamian et al., 2003). Twaite shad populations are now severely reduced to be rare or endangered where they have not been extirpated. In rivers where they have been cut off from their upriver spawning habitats, they have hybridized with allis shad from the forced overlap in current habitats (Baglinière et al., 2003). Evidence for local population structure has relied largely on differences in dynamics, morphometrics, and other metrics, but studies of this species have been hampered by low sampling effort. In contrast, much more is known about allis shad, because they have been harvested extensively (Baglinière et al., 2003). They are similar to American shad in that they have been extirpated from major portions of their original range through habitat loss from dams, habitat degradation, and overharvest. However, little research has been done to confirm genetic isolation or formation of local populations. The genetics that has been done has shown low levels of polymorphism, and hence difficulty in assigning separate populations.

C. COMPARING SALMONIDS AND ALOSINE HERRINGS

Quinn and Adams (1996) have contrasted the temporal scope of adaptation in salmonids and American shad in the Columbia River. They hypothesize that salmonid migratory behavior should evolve more slowly in response to environmental change than would shad because salmonids undergo longer migrations, spawn much later after their arrival to their natal river, and have longer larval development. In contrast, shad spawn in the mainstem of their natal rivers, spawn sooner, and larvae develop more quickly. This comparison was made between native salmonids and shad transplanted to the West Coast more than a century ago. If this hypothesis held true for natives and if greater plasticity were also true for other traits, then we would expect alosines to form discrete populations more readily than salmonids and thus to form more apparent metapopulations. However, this is not the case and other factors, such as greater genetic variability in salmonids perhaps associated with greater heterogeneity of their spawning grounds, may be more important in the formation of locally adapted populations.

Some species experience more mixing than the "model" salmonid and alosine species that I have already discussed. For example, anadromous striped bass (*Morone saxatalis*) have complex population-level migration behaviors in which some contingents within the local group may migrate more than others (Secor, 1999), thereby confounding metapopulation structure and modeling. Nonetheless, if scientists analyze the life history, philopatry, migration patterns, and genetic structure using the framework of a metapopulation as the null hypothesis, they may learn much from the departure of these species from the null model. This is seen in recent efforts to model estuarine-dependent and marine species in the following discussion.

D. SCIAENIDS

Only a few studies have attempted to assess metapopulation structure in estuarine-dependent and marine populations (Ruzzante et al., 2000, 2001; Thorrold et al., 2001). In an estuarine example, Thorrold et al. (2001) measured philopatry in weakfish (*Cynoscion regalis*). Weakfish are distributed along the North American East Coast from Florida to Nova Scotia and are characterized by seasonal migration: inshore to spawn in the spring and summer, and offshore in fall and winter. Their offshore migration offers the opportunity for population mixing, whereas their inshore migration to spawn provides the opportunity for reproductive isolation if natal homing is pronounced. Estuaries along their distributional range are sufficiently separated to allow reproductive isolation. Results using genetic analyses based on mtDNA showed no population structure in weakfish (Cordes and Graves, 2003), which indicated that philopatry was insufficiently strong to result in historical isolation, or that local populations had not been isolated long enough for genetic differences to be apparent. Estuaries along the East Coast of the United States are geologically young, having formed only since the last glacial retreat, compared with their older river systems. mtDNA is maternally inherited and it is less sensitive to recent isolation than nuclear genes. In contrast, geochemical markers of otoliths showed that philopatry varied between 60% and 81% for 1- and 2-year-old weakfish. Sciaenids often remain nearby their natal habitats for the first years of life and migrate extensively thereafter (Gold and Richardson, 1998). This migration of older fish may account for the lack of population structure for these fish despite the high philopatry of young fish.

A congener of weakfish, spotted sea trout (*C. nebulosus*), shows greater population structure along its range. Although it is closely related to weakfish and achieves similar size and longevity, spotted sea trout show a distinct local population structure best modeled as a series of overlapping local populations (Gold et al., 2003; Wiley and Chapman, 2003). Perhaps they have been able to form distinct genetic groups when their congener has not because their geographical range is more southerly and less subject to the effects of glaciation. Artificial tagging studies have shown that interestuarine migrations are rare, except at the northern end of its range in Chesapeake Bay, where winter temperatures force this fish to seek warmer offshore water. Even so, the East Coast of the United States shows greater population structure compared with the Gulf Coast. The lack of movement during juvenile estuarine residency, as told in otolith—geochemical markers (Dorval, 2004), attests to habitat fidelity in this species. However, sufficient gene flow does exist in spotted sea trout to rend it less amenable to metapopulation modeling than salmonids, albeit more so than weakfish.

E. ATHERINIDS

The Atlantic silverside (*Menidia menidia*) is an estuarine species that inhabits tidal marshes along the East Coast of the United States. This fish has served as an excellent candidate for study of biological measures in local populations. David Conover and his colleagues have studied silversides in the field and laboratory in a series of common garden and transplant experiments that have demonstrated that there are latitudinal differences in growth that have a genetic basis. Conover and Present (1990) showed that silversides in the north had higher growth rates compared with those in the south, even though the growing season was shorter and temperatures colder in the north. Thus, by the onset of winter, juveniles have reached a similar size (Conover, 1992). To accomplish this faster growth, northern silversides had to consume twice as much food (Billerbeck et al., 2000). In an experiment that used second and third generation laboratory-reared fish, these scientists determined that this countergradient variation in somatic growth was genetic and, although modified by environmental conditions, northern fish grew consistently faster under the same conditions (Billerbeck et al., 2000).

Unlike other fish presented earlier, Atlantic silversides are a small, annual species that are amenable to experimental study that is impossible with a fish such as shad. Silversides show latitudinal differences in egg production (Klahre, 1997) and lipid storage (Schultz and Conover, 1997). Along the coast, these fish exist as a latitudinal gradient in an isolation-by-distance or steppingstone model. Gene exchange may occur in adults in their offshore winter migration that occurs at an unknown scale. Conover and his colleagues continue to research these relationships (see also Conover, 1998). Although they are not sufficiently distinct to form a true metapopulation, their use in experimental studies provides insight into how genetic selection acts to promote local adaptation that is unavailable in other species.

V. VALUE OF THE METAPOPULATION CONCEPT IN UNDERSTANDING AND MANAGING DIADROMOUS FISHERIES

Ryman and his colleagues (1995, p. 421) state: "Aquatic organisms are the only remaining food resource of mankind that is primarily supplied through direct harvest from the wild populations." Many important exploited species are diadromous, and these species are especially vulnerable to overharvest, loss of nursery and adult habitat, and disruption of prey resources resulting from anthropogenic impacts (Beck et al., 2003). We have lost genetically unique local populations as a result (Ferguson et al., 1995). Arguably, many of these species evolved with metapopulation structure as part of their population dynamics (naturally evolved metapopulations), whereas others have been subjected to habitat fragmentation and loss (fragmentation-induced metapopulations). This is clearly the case for salmonids and shads, and there is growing evidence for spatially linked local populations among other families too. Managing these species with an understanding of the role of spatial structure and connectivity will be an increasingly important tool in maintaining their persistence. Notwithstanding the importance of the metapopulation concept to vulnerable diadromous fish species, the schism between ecology theorists and fisheries scientists, and between marine and freshwater fisheries ecologists has delayed the use of these concepts in fisheries management and resulted in their misapplication (Frank and Leggett, 1994; Smedbol et al., 2002).

A. Conservation of Local Populations

If we return to the Levins model as the basis of describing metapopulation persistence, we observe that he used this model to sketch management's response to a pest species. This early model showed that the best way to eliminate a pest was to apply treatment to all populations simultaneously. Since the publication of this model, others have refined this prediction to show that species that are classic metapopulations can be destroyed if too many of the local populations are extirpated, even with the benefit of a rescue and other effects (Brown and Kodric–Brown, 1977; Hanski, 1982; Gotelli, 1991; Hastings and Harrison, 1994).

Managing fisheries to preserve local populations within a naturally evolved metapopulation has a genetic basis. The local populations have responded to the selective pressures within their specific habitats and have retained those alleles that result in better fitness. Thus, the richness of the genetic pool comes from the presence of each of the local populations and their individual response to regional habitat change over millennia. In contrast, in the fragmentation-induced metapopulation, which previously existed as a single interbreeding population (perhaps as a nonuniformly distributed one), the genetic mix depends more on a random draw of alleles from fragments of the population that became isolated as habitat was lost. The fragments may or may not have the same allelic mix, depending on the degree to which the "founder effect" is present in the isolate and the resulting culling due to small population size (stochastic effects). The founder effect occurs when a small fraction of the original population becomes isolated and, as such, a fraction of the original population's alleles are lost (Ridley, 1996). The isolated fragment has yet to evolve in concert with local conditions and to fine-tune its genetic mix. Fragmentation of this sort may be regarded as a new, uncontrolled form of anthropogenic selection.

B. Historical Management of Local Populations in a Fisheries Context

The value of the metapopulation concept for managing fish and shellfish was understood by a few ecologists and fisheries scientists before its recent popularity (Pulliam, 1988; Frank and Leggett, 1994; Tuck and Possingham, 1994;

Possingham, 1996; Ye and Beddington, 1996; Fogarty, 1998). As early as 1957, Beverton and Holt recognized the importance of spatial structure for fish population dynamics, although Levins' model had yet to be developed. At that time, Beverton and Holt did not have the computing power or the analytical advances that now permit spatial analysis of exploited populations. Since the 1950s, fisheries scientists have concentrated on the temporal dynamics of individual local populations and the effect of harvest on their ability to persist, given that harvest rates were sufficiently high to threaten their persistence. The emphasis on the dynamics of the local population was exemplified by the definition of a stock developed by the United Nations Scientific Workshop for the International Council for the Northwest Atlantic Fisheries/International Council for the Exploration of the Sea/Food and Agriculture Organization of the United States (FAO) in 1957. Under this official definition, a unit stock was a "relatively homogeneous and self-contained population, whose losses by emigration and accessions by immigration, if any, are negligible in relation to the rates of growth and mortality" (as stated in Ong, 1991, p. 2). This definition reflects the emphasis that fisheries scientists put on the discrete population and in estimating the vital rates of birth and death, and their neglect in accessing the importance of connectivity. This shifting of perspective is reminiscent of the psychological perspective drawing that looks either like a vase or two conjoined profiles. Similarly, to manage diadromous species, both ideas of discreteness and connectedness must be held simultaneously. An important point to note is that different dynamics in space can substitute for time, as can time for space in simulating the persistence of populations under exploitation, albeit not perfectly.

Not surprisingly, the impetus to apply metapopulation theory to exploited marine and diadromous fish has come from scientists studying salmonids (see, for example, Nielsen et al., 1999; but see Rieman and Dunham, 2000, for freshwater salmonids). The first use of the formal concept for a fully marine fish was for Atlantic herring, *Clupea harengus*, (McQuinn, 1997), followed shortly after by northern cod, *Gadus morhua* (Ruzzante et al., 2000, 2001; Neilsen et al., 2001), although Heincke (1898) and Iles and Sinclair (1982) clearly identified the same issues much earlier for Atlantic herring without direct use of current theory.

In conservation biology, the focus has been on preserving local populations in an attempt to maintain genetic diversity. This focus on preserving local populations is of similar importance however the metapopulation was formed. Conservation has become an increasing focus of fisheries science, especially in salmonid fisheries, where land and water use has resulted in a loss of spawning habitat throughout much of the species range. Note, however, that anadromous and marine fishes are more fecund than terrestrial organisms and have batch fecundities on the order of thousands to millions of eggs. Only a few individuals survive from the enormous numbers spawned, and this leads to a "sweepstakes-type" selection, with the result that effective population abundance can be considerably less than actual population abundance resulting from the severe selection of genotypes (Grant et al., 1999). Grant and Waples (2000) point out that when population abundance varies greatly, the effective population size is reflected by the harmonic mean of abundance, not the arithmetic mean, and is closer to the smallest abundance in the time series rather than the largest. In the face of exploitation, genetic diversity is further diminished. Hence, fisheries managers may need to maintain larger local populations of fish, compared with those needed for sustaining terrestrial populations maintenance, to preserve sufficient genetic variability of fish metapopulations.

In the late 1980s, a shift in emphasis began as fisheries scientists grappled with the possible extirpation of local populations, and the need to replenish diminished and lost populations. For example, Pulliam (1988) recognized the value of source populations in sustaining sink populations among harvested species. Similarly, Fogarty (1998) attributed the persistence of overexploited local populations of northern lobster to an offshore refuge population that acted as a source of new recruits. If this refuge population were itself to be exploited, then the species would become vulnerable to extirpation and the entire fishery subject to collapse. Frank (1992) and Frank and Leggett (1994) pointed out that ". . . strong year-classes in one population lead to dispersal to the second, thereby inflating the latter's apparent abundance while dampening the realized population increase in the former. (1994, p. 412)" Although these ideas came from the application of island biogeography to offshore bank cod populations, such ideas may also be applicable to metapopulations. Following the Frank and Leggett example, if density-driven connectivity is not identified, then scientists will attribute synchronous recruitment to both, when in fact one population is a source and the other a sink. Without clarity of the underlying spatial dynamics, overexploitation of the source population can result in extirpation of both. This may be one causal agent in the dramatic loss of northern cod population in the mid 1990s.

C. Mixed-Stock Analysis

Traditionally in fisheries science, the focus has been on how to manage fish when their migration brings them together on feeding grounds in mixed-stock (mixed local populations) fisheries. When mixed local populations are fished together, an average fishing mortality is applied across all local populations regardless of the ability of any one group to withstand this level of killing. In mixed stocks, the fishing mortality is not matched to each local population's viability. To conserve all local populations, the fishing mortality that is matched to the most susceptible population should be applied to all in the mixture, a very precautionary approach. This will result in underharvest of populations that can withstand greater fishing mortality. One tool for quantifying the abundance contribution of local populations is mixed-stock analysis (e.g., Pella and Milner, 1987). Mixed-stock analysis has been applied to a variety of anadromous species (American shad [Epifanio et al., 1995], chinook salmon [Utter et al., 1987], sockeye salmon [Grant et al., 1980], and striped bass [Waldman et al., 1988]).

D. Effect of Demography on Metapopulation Management

Metapopulation models have included little or no intrapopulation demography aside from migration rates to measure the probability of recolonization (but also see Hansson, 1991). Rather than model internal demographic processes, Levins only dealt with the resultant spillover of burgeoning populations in modeling emigration. However, when fisheries scientists began to apply metapopulation theory to fisheries, they did consider the effects of demography on local or "stock" dynamics. This is not surprising, given the extensive use of age-based models in this discipline. For example, Frank and Leggett (1994) recognized that they could use the concept of the ideal free distribution (Fretwell and Lucas, 1970; see also Doncaster, 2000, for application to metapopulations) to explain age-specific emigration of adults from one local population to another. In response to increased density, adult migration may also increase (e.g., American lobster; Fogarty, 1998). In recent models, ecologists have begun to incorporate demographic details such as genetics, population density, and age structure (Jollivet et al., 1999; Mumby, 1999; Lopez and Pfister, 2001; Fahrig, 2002; Hixon et al., 2002).

Demographic and genetic structure can influence the persistence of local populations and affect the outcome of management practices. Population density may affect the amount of migration, and hence connectivity. Fish populations have long demonstrated both ontogenetic and age- or size-based migrations. Long-distance dispersal of larval stages keeps many species panmictic, although there is growing evidence that mortality (Cowen et al., 2000) and other factors (Hellberg, this volume) result in more local retention than thought previously. Less widely recognized is the importance of age structure and its influence on connectivity. Many diadromous and marine fish populations show greater migration rates of older, larger fish (Quinn and Brodeur, 1991; Secor and Piccoli, 1996). Large, old adults may ensure panmixia. Older fish are often larger, and thus, better swimmers and often better foragers. When harvested populations are overexploited, the older individuals are typically lost first in a process called *juvenescence*. Juvenescence occurs when populations are harvested; over time, the older fish are lost and mean age declines as abundance declines and as young fish comprise a larger proportion of the population. Hence, the result of harvest is to diminish connectivity between local populations, thereby effecting a form of population isolation similar to that caused by habitat fragmentation in terrestrial systems. With the loss of these older fish that wander afar, local populations are more isolated.

E. Marine Protected Areas as a Spatial Management Tool

The principles of metapopulation theory can complicate the use of MPAs as a tool for fisheries management. Advocates believe that MPAs provide a refuge for populations, that the individuals become larger and more productive in the refuge, and that this productivity spills over into unprotected areas (Roberts et al., 2001; Lubchenco et al., 2003). Area closures have been a historical tool of fisheries managers, and MPAs can extend these closures permanently. Such closures have proved effective for marine benthic species and species with limited movement (Clark, 1996; Collie et al., 1997; Auster and Langston, 1999). However, they are also recommended for highly mobile species (Roberts et al., 2001) for which they are of questionable value. Closures have recently been advocated as protection from destructive fishing gear, such as bottom trawls (Collie et al., 2000). Although broadly advocated, MPAs are less effective for marine pelagic and diadromous fish species whose life history is based on seasonal migration. No specific area provides year-round habitat for these fish as they move about during the year. The difficulty in using MPAs to protect these fish can be seen clearly in the calculation that 50% or so of their habitat must be set aside for mobile species (Polacheck, 1990; Man et al., 1995; Sala et al., 2002). For these fish, rolling closures or quotas may be more effective. Alternatively, closures could be enacted to protect habitats that protect vulnerable life stages. MPAs appear to be a valuable tool for more sedentary species and species in which a portion of the population is contained within the reserves throughout life (Hastings and Botsford, 1999). The value of MPAs to diadromous fish with metapopulation dynamics is complicated by the ontogenic changes in habitat use and by their long-ranging migrations, as well as how they are set up, managed, and enforced (Coleman et al., 2004).

F. Can Fisheries Be Managed as Metapopulations?

The best-studied species among diadromous fish are the Pacific salmon. Arguably, we can expect these species to be the best candidates for management using the concept of metapopulation dynamics, for they form true metapopulations. As yet, they have not been managed this way. Several barriers exist, the most important of which is the lack of knowledge on key factors that determine local population size. Recent reports (National Research Council, 1996a,b) decry the problems in identifying in-season stock composition of the catch and the complication that this brings in limiting harvest on vulnerable stocks in mixed-stock fisheries. Additionally, mortality during the marine life phase, apart from direct harvest, has been recognized as an important source of population decline. However, the mechanisms causing the mortality have yet to be identified. Even with the best-studied species, there is a lot more to understand before we can manage fully with metapopulation concepts.

Managing exploited fish in a metapopulation context requires extensive assessment of population size and demography, connectivity, genetics, and estimation of mortality in pivotal life stages. Terrestrial studies of fragmentation-induced metapopulations dominate much of the literature. Scientists have conducted field studies and developed theoretical and simulation models to test metapopulation concepts for terrestrial organisms. In contrast, there are many fewer studies in diadromous, estuarine, and marine species, possibly because these species are seen as having more gene flow. However, many important species have naturally evolved as metapopulations (e.g., salmonids and alosines), whereas other species have significantly more spatial structure than originally recognized (e.g., Ames, 2004). Additionally, Ryman et al. (1995) point out that even though local populations of anadromous fish are more genetically similar to each other than are freshwater fish or terrestrial organisms, alleles can be disproportionately lost from these populations when abundances decline or when they are extirpated.

Fisheries scientists have long recognized spatial effects in their efforts to manage "stocks" or local populations. They have understood that the spatial heterogeneity of fish population structure provides biocomplexity that protects the species from extinction (Hillborn et al., 2003). As climate and environmental regimes change, different local populations will reach ascendancy and become more productive. This gain and loss of productive dominance will provide a buffer to ensure that some populations survive even under the most severe natural changes. This resiliency can be lost, however, when harvest or climate change occur at too great a rate on vulnerable populations. The value of the metapopulation concept to fisheries management is that it reinforces finer scale regulation and harvest oversight. For example, we know now that northern cod were managed on a too large and aggregated scale. Depletion of this once great population serves as a cautionary tale. As Sinclair (1988) and Wilson and his colleagues (1999) point out, when fish are managed on scales that aggregate local populations, then managers risk "the erosion of spatial scale and the depletion of the resource," thus "scale misperceptions might lead to a different form of over-fishing than that normally hypothesized" (Wilson et al., 1999, p. 244).

In summary, we are led back to the importance of delineating the degree of connectivity between local populations. If local populations do not go extinct, do we need to manage at such fine scale or will migrations replenish depleted local groups? If so, should we manage these groups as parts of a metapopulation or are aggregated management units sufficient? Can harvest rates be sufficiently high that they exceed the metapopulation's ability to replenish extirpated local populations by recolonization? Even in those species that have been studied extensively, scientists do not have the answers to these questions. We must seek the answers to these questions for they are critical to stewarding wild fish populations.

VI. ACKNOWLEDGMENTS

The author acknowledges the support of the National Science Foundation (OCE9876565 and OPP9985884) and Virginia Sea Grant (VGMSC R/CF-42) during the preparation of this manuscript. Many insights about fish metapopulation dynamics were obtained during this grant research. Early drafts of this manuscript were reviewed by T. Miller and K. Limburg. The submitted draft was reviewed by T.J. Quinn and J. Olney. Suggestions from these reviewers improved and broadened the manuscript and the author thanks them.

REFERENCES

Alexander, S.E., and Roughgarden, J. (1996). Larval transport and population dynamics of intertidal barnacles: A coupled benthic/oceanic model. *Ecol. Monogr.* **66**, 259–275.

Allendorf, F.W. (1983). Isolation, gene flow, and genetic differentiation among populations. In *Genetics and conservation: A reference for managing wild animal and plant populations* (C. Schonewald–Cox, S.M. Chambers, B. MacBryde, and W.L. Thomas, eds., pp. 51–65). Menlo Park, CA: Benjamin Cummings.

Allendorf, F., Ryman, N., and Utter, F. (1987). Genetics and fishery management: Past, present, future. In *Population genetics and fishery management* (N. Ryman, and F. Utter, eds., pp. 1–20). Seattle, WA: University of Washington Press.

Ames, E.P. (2004). Atlantic cod stock structure in the Gulf of Maine. *Fisheries* **29**, 10–28.

Andrewartha, H.G., and Birch, L.C. 1954. *The distribution and abundance of animals.* Chicago: University of Chicago Press.

Aprahamian, M.W., Baglinière, J.-L., Sabatié, A.P., Thiel, R., and Aprahamian, C.D. (2003). Biology, status, and conservation of the anadromous Atlantic twaite shad *Alsoa fallax fallax*. In *Biodiversity, status, and conservation of the world's shads* (K.E. Limburg and J.R. Waldman, eds., pp. 103–124). Symposium 35. Bethesda, MD: American Fisheries Society.

Atema, J., Kingsford, M.J., and Gerlach, G. (2002). Larval reef fish could use odour for detection, retention, and orientation to reefs. *Mar. Ecol. Prog. Ser.* **241**, 151–160.

Auster, P.J., and Langton, R.W. (1999). The effects of fishing on fish habitat. *Am. Fish. Soc. Symp.* **22**, 150–187.

Baglinière, J.-L., Sabatié, M.R., Rochard, E., Alexandrino, P., and Aprahamian, M.W. (2003). The allis shad *Alosa alosa*: Biology, ecology, range, and status of populations. In *Biodiversity, status, and conservation of the world's shads* (K.E. Limburg and J.R. Waldman, eds., pp. 85–102). Symposium 35. Bethesda, MD: American Fisheries Society.

Barker, J.R., and Lambert, D.M. (1988). A genetic analysis of populations of *Galaxias maculates* from the Bay of Plenty: Implications for natal river return. *New Zealand Journal of Marine and Freshwater Research.* **22**, 321–326.

Beacham, T.D., Brattey, J., Miller, K.M., Le, K.D., and Withler, R.E. (2002). Multiple stock structure of Atlantic cod (*Gadus morhua*) off Newfoundland and Labrador determined from genetic variation. *ICES J. Mar. Sci.* **59**, 650–665.

Beacham, T.D., Candy, J.R., Supernault, K.J., Wetklo, M., Deagle, B., Labaree, K., Irvine, J.R., Miller, K.M., Nelson, R.J., and Withler, R.E. (2003a). Evaluation and application of microsatellites for population identification in Fraser River chinook salmon (*Onchorhynchus tshawytscha*). *Fish. Bull.* **101**, 243–259.

Beacham, T.D., Supernault, K.J., Wetklo, M., Deagle, B., Labaree, K., Irvine, J.R., Candy, J.R., Miller, K.M., Nelson, R.J., and Withler, R.E. (2003b). The geographic basis for population structure in Fraser River chinook salmon (*Onchorhynchus tshawytscha*). *Fish. Bull.* **101**, 229–242.

Beck, M.W., Heck, K.L., Jr., Able, K.W., Childers, D.L., Eggleston, D.B., Gillanders, B.M., Halpern, B.S., Hays, C.G., Hoshino, K., Minello, T.J., Orth, R.J., Sheridan, P.F., and Weinstein, M.P. (2003). The role of nearshore ecosystems as fish and shellfish nurseries. *Issues in Ecology.* **11**, 2–12.

Berra, T.M., Crowley, L.E.L.M., Ivantsoff, W., and Fuerst, P.A. (1996). *Galaxias maculates*: An explanation of its biogeography. *Marine & Freshwater Research.* **47**, 845–849.

Berryman, A.A. (2002). Population: A central concept for ecology? *Oikos.* **97**, 439–442.

Berst, A.H., and Simon, R. (1981). Introduction to the proceedings of the 1980 Stock Concept International Symposium (STOCS). *Can. J. Fish. Aquat. Sci.* **38**, 1457–1458.

Beverton, R.J.H., and Holt, S.J. (1957). *On the dynamics of exploited fish populations.* U.K. Ministry of Agriculture and Fisheries, Fisheries Investigations (ser. 2). **19**, 1–533.

Billerbeck, J.M., Schultz, E.T., and Conover, D.O. (2000). Adaptive variation in energy acquisition and allocation among latitudinal populations of the Atlantic silverside. *Oecologia.* **122**, 210–219.

Botsford, L.W., Hastings, A., and Gaines, S.D. (2001). Dependence of sustainability on the configuration of marine reserves and larval dispersal distance. *Ecology Letters.* **4**, 144–150.

Botsford, L.W., Moloney, C.L., Largier, J.L., and Hastings, A. (1998). Metapopulation dynamics of meroplanktonic invertebrates: The Dungeness crab (*Cancer magister*) as an example. In *Proceedings of the North Pacific Symposium on Invertebrate Stock Assessment and Management* (G.S. Jamieson and A. Campbell, eds., pp. 295–306). Ottawa: Canadian Special Publication Fisheries and Aquatic Science 125.

Brothers, E. (1990). Otolith Marking. *Amer. Fish. Soc. Symp.* **7**, 183–202.

Brown, B.L., Smouse, P.E., Epifanio, J.M., and Kobak, C.J. (1999). Mitochondrial DNA mixed-stock analysis of American shad: Coastal harvests are dynamic and variable. *Trans. Amer. Fish. Soc.* **128**, 977–994.

Brown, J.H., and Kodric-Brown, A. (1977). Turnover rates in insular biogeography: Effect of immigration on extinction. *Ecology.* **58**, 445–449.

Brownie, C., Anderson, D.R., Burnham, K.P., and Robson, D.S. (1985). *Statistical inference from band-recovery data: A handbook* (2nd ed.). U.S. Fish and Wildlife Service Resource Publication no. 156. Washington, D.C.: U.S. Dept. of the Interior.

Burnham, K.P., Anderson, D.R., White, G.C., Brownie, C., and Pollock, K.H. (1987). *Design and analysis methods for fish survival experiments based on release-recapture.* American Fisheries Society monograph 5. Bethesda, MD: American Fisheries Society.

Caley, M.J., Carr, M.H., Hixon, M.A. Hughes, T.P., Jones, G.P., and Menge, B.A. (1996). Recruitment and local dynamics of open marine populations. *Annual Rev. Ecol. Syst.* **27**, 477–500.

Campana, S.E. (1999). Chemistry and composition of fish otoliths: pathways, mechanisms and application. *Mar. Ecol. Prog. Ser.* **188**, 263–297.

Campana, S.E., Fowler, A.J., and Jones, C.M. (1994). Otolith elemental fingerprinting for stock identification of Atlantic cod (*Gadus morhua*) using laser ablation ICPMS. *Canadian Journal of Fisheries and Aquatic Sciences.* **51**, 1942–1950.

Camus, P.A., and Lima, M. (2002). Populations, metapopulations, and the open–closed dilemma: The conflict between operational and natural population concepts. *Oikos.* **97**, 433–438.

Clark, C.W. (1996). Marine reserves and the precautionary management of fisheries. *Ecol. Appl.* **6**, 369–370.

Coleman, F.C., Baker, P.B., and Koenig, C.C. (2004). A review of Gulf of Mexico marine protected areas: Successes, failures, and lessons learned. *Fisheries.* **29**, 10–21.

Collie, J.S., Escanero, G.A., and Valentine, P.C. (1997). Effects of bottom fishing on the benthic megafauna of Georges bank. *Mar. Ecol. Progress Ser.* **155**, 159–172.

Collie, J.S., Hall, S.J., Kaiser, M.J., and Poiner, I.R. (2000). A quantitative analysis of fishing impacts on shelf-sea benthos. *J. Animal Ecology.* **69**, 785–798.

Conover, D.O. (1998). Local adaptation in marine fishes: Evidence and implications for stock enhancement. *Bull. Mar. Sci.* **62**, 305–311.

Conover, D.O. (1992). Seasonality and the scheduling of life history at different latitudes. *J. Fish. Biol.* **41**, 161–178.

Conover, D.O., and Present, T.M.C. (1990). Countergradient variations in growth rate: Compensation for length of the growing season among Atlantic silversides from different latitudes. *Oecologia.* **83**, 316–324.

Cordes, J.F., and Graves, J.E. (2003). Investigation of congeneric hybridization in and stock structure of weakfish (*Cynoscion regalis*) inferred from analyses of nuclear and mitochondrial DNA loci. *Fish. Bull.* **101**, 443–450.

Cowen, R.K., Lwiza, K.M.M., Sponaugle, S., Paris, C.B., and Olson, D.B. (2000). Connectivity of marine populations: open or closed? *Science.* **287**, 857–859.

Dadswell, M.J., Melvin, G.D., Williams, P.J., and Themelis, D.E. (1987). Influences of origin, life history, and chance on the Atlantic coast migration of American shad. In *Common strategies of anadromous and catadromous fishes* (M.J. Dadswell, R.J. Klauda, C.M. Moffitt, R.L. Saunders, R.A. Rulifson, and J.E. Cooper, eds., pp. 313–330). American Fisheries Society Symposium 1. Bethesda, MD: American Fisheries Society.

Davidson, F.A. (1937). Migration and homing in Pacific salmon. *Science.* **86**, 1–4.

Doak, D. (1995). Why metapopulations are so difficult to manage. *Inner Voice.* **7**, 14–15.

Dodson, J.J., and Leggett, W.C. (1973). Behavior of American shad (*Alosa sapidissima*) homing to the Connecticut River from Long Island Sound. *J. Fish. Res. Bd. Can.* **30**, 1847–1860.

Doncaster, C.P. (2000). Extension of ideal free resource use to breeding populations and metapopulations. *Oikos.* **89**, 24–36.

Dorval, E. 2004. *Determination of essential fish habitat for juvenile spotted sea trout in Chesapeake Bay using trace-element chemistry in surface waters and otoliths.* PhD thesis. Norfolk, VA: Old Dominion University.

Elton, C.S. (1942). *Voles, mice and lemmings.* Oxford: Clarendon Press.

Epifanio, J.M., Smouse, P.E., Kobak, C.J., and Brown, B.L. (1995). Mitochondrial DNA divergence among populations of American shad (*Alosa sapidissima*): How much variation is enough for mixed-stock analysis? *Canadian Journal of Fisheries and Aquatic Sciences.* **52**, 1688–1702.

Fahrig, L. (2002). Effect of habitat fragmentation on the extinction threshold: A synthesis. *Ecological Applications.* **12**, 346–353.

Ferguson, A., Taggart, J.B., Prodöhl, P.A., McMeel, O., Thompson, C., Stone, C., McGinnity, P., and Hynes, R.A. (1995). The application of molecular markers to the study and conservation of fish populations, with special reference to *Salmo. J. Fish Biol.* **47 (suppl. A)**, 103–126.

Fogarty, M.J. (1998). Implications of migration and larval interchange in American lobster (*Homarus americanus*) stocks: Spatial structure and resilience. In *Proceedings of the North Pacific Symposium on Invertebrate Stock Assessment and Management* (G.S. Jamieson and A. Campbell, eds.). *Can Spec. Publ. Fish Aquat. Sci.* **125**, 273–283.

Frank, K.T. (1992). Demographic consequences of age-specific dispersal in marine fish populations. *Can. J. Fish. Aquat. Sci.* **49**, 2222–2231.

Frank, K.T., and Leggett, W.C. (1994). Fisheries ecology in the context of ecological and evolutionary theory. *Annu. Rev. Ecol. Syst.* **25**, 401–422.

Fretwell, S.D., and Lucas, H.L. (1970). On territorial behavior and other factors influencing habitat distribution in birds. I. Theoretical development. *Acta. Biotheor.* **14**, 16–36.

Gall, G.A.E., Bartley, D., Bentley, B., Brodziak, J., Gomulkiewicz, R., and Mangel, M. (1992). Geographic variation in population genetic structure of Chinook salmon from California and Oregon. *Fish. Bull.* **90**, 77–100.

Gause, G.F. (1935). Experimental demonstration of Volterra's periodic oscillation in the numbers of animals. *J. Exp. Biology.* **12**, 44–48.

Gharrett, A.J., and Smoker, W.W. (1993). A perspective on the adaptive importance of genetic infrastructure in salmon populations to ocean ranching in Alaska. *Fisheries Research.* **18**, 45–58.

Gharrett, A.J., and Zhivotovsky, L.A. (2003). Migration. In *Population genetics: Principles and applications for fisheries scientists* (E.M. Hallerman, ed., pp. 141–174. Bethesda, MD: American Fisheries Society.

Glebe, B.D., and Legget, W.C. (1981a). Latitudinal differences in energy allocation and use during the freshwater migrations of American shad (*Alosa sapidissima*) and their life history consequences. *Can. J. Fish. Aquat. Sci.* **38**, 806–820.

Glebe, B.D., and Leggett, W.C. (1981b). Temporal, intra-population differences in energy allocation and use by American shad (*Alosa sapidissima*) during the spawning migration. *Can. J. Fish. Aquat. Sci.* **38**, 795–805.

Goana, P., Ferras, P., and Delibes, M. (1998). Dynamics and viability of a metapopulation of the endangered Iberian lynx (*Lynx pardinus*). *Ecological Monographs.* **68**, 349–370.

Gold, J.R., and Richardson, L.R. (1998). Mitochondrial DNA diversification and population structure in fishes from the Gulf of Mexico and Western Atlantic. *J. of Heredity.* **89**, 404–414.

Gold, J.R., Stewart, L.B., and Ward, R. (2003). Population structure of spotted sea trout (*Cynoscion nebulosus*) along the Texas Gulf Coast, as revealed by genetic analysis. In *Biology of spotted sea trout* (S.A. Bortone, ed., pp. 17–29). Washington, DC: CRC Press.

Gotteli, N.J. (1991). Metapopulation models: The rescue effect, the propagule rain, and the core-satellite hypothesis. *Amer. Nat.* **138**, 768–776.

Grant, W.S., García-Marín, J.L., and Utter, F.M. (1999). Defining population boundaries for fisheries management. In *Genetics in sustainable fisheries management* (S. Mustafa, ed., pp. 27–72). Oxford: Fishing News Books.

Grant, W.S., Milner, G.B., Krasnowski, P., and Utter, F.M. (1980). Use of biochemical genetic variants for identification of sockeye salmon (*Oncorhynchus nerka*) stocks in Cook Inlet, Alaska. *Can. J. Fish. Aquat. Sci.* **37**, 1236–1247.

Grant, W.S., and Waples, R.S. (2000). Spatial and temporal scales of genetic variability in marine and anadromous species: Implications for fisheries oceanography. In *Fisheries oceanography: An integrative approach to fisheries ecology and management* (P.J. Harrison and T.R. Parsons, eds., pp. 61–93). Oxford: Blackwell Science.

Hanski, I. (1982). Dynamics of regional distribution: The core and satellite species hypothesis. *Oikos.* **38**, 210–221.

Hanski, I., and Gilpin, M. (1991). Metapopulation dynamics: Brief history and conceptual domain. *Biol. J. Linnean Soc.* **42**, 3–16.

Hanski, I., and Thomas, C.D. (1994). Metapopulation dynamics and conservation: A spatially explicit model applied to butterflies. *Biol. Cons.* **68**, 167–180.

Hansson, L. (1991). Dispersal and connectivity in metapopulations. *Biol. J. Linnean Soc.* **42**, 89–103.

Hartl, D.L., and Clark, A.G. (1997). *Principles of population genetics* (3rd ed.). Sunderland, MA: Sinauer Associates.

Hastings, A., and Botsford, L.W. (1999). Equivalence in yield from marine reserves and traditional fisheries management. *Science.* **284**, 1537–1538.

Hastings, A., and Harrison, S. (1994). Metapopulation dynamics and genetics. *Annu. Rev. Ecol. Syst.* **25**, 168–188.

Heath, D.D., Heath, J.W., Bryden, C.A., Johnson, R.M., and Fox, C.W. (2003). Rapid evolution of egg size in captive salmon. *Science.* **299**, 1738–1740.

Heincke, F. (1898). *Naturgeschichte des herings. Teil I. Die localformen und die wanderungen des herings in den europäischen meeren.* Berlin: Verlag von Otto Salle.

Hellberg, M.E., Burton, R.S., Neigel, J.E., and Palumbi, S.R. (2002). Genetic assessment of connectivity among marine populations. *Bull. Mar. Sci.* **70 (suppl.)**, 273–290.

Hendry, A.P., Wenburg, J.K., Bentzen, P., Volk, E.C., and Quinn, T.P. (2000). Rapid evolution of reproductive isolation in the wild: Evidence from introduced salmon. *Science.* **290**, 516–518.

Hilborn, R., Quinn, T.P., Schindler, D.E., and Rodgers, D.E. (2003). Biocomplexity and fisheries sustainability. *Proceedings of the National Academy of Science.* **100**, 6564–6568.

Hilborn, R., Walters, C.J., and Jester, D.B., Jr. (1990). Value of fish marking in fisheries management. *Amer. Fish. Soc. Symp.* **7**, 5–7.

Hixon, M.A., Pacala, S.W., and Sandin, S.A. (2002). Population regulation: Historical context and contemporary challenges of open vs. closed systems. *Ecology.* **83**, 1490–1508.

Hjort, J. (1914). Fluctuations in the great fisheries of northern Europe. *Rapports et Procès-verbaux des Réunions, Conseil international pour l'Exploration de la Mer.* **20**, 1–228.

Hollis, E.H. (1948). The homing tendency of shad. *Science.* **108**, 332–333.

Huffaker, C.B. (1958). Experimental studies on predation: Dispersion factors and predator–prey oscillations. *Hilgardia.* **27**, 795–834.

Iles, T.D., and Sinclair, M. (1982). Atlantic herring: Stock discreteness and abundance. *Science.* **215**, 627–633.

Jollivet, D., Chevaldonné, P., and Planque, B. (1999). Hydrothermal-vent alvinellid polychaete dispersal in the eastern Pacific. 2. A metapopulation model based on habitat shifts. *Evolution.* **53**, 1128–1142.

Kennedy, B.D., Folt, C.L., Blum, J.D., and Chamberlain, C.P. (2000). Using natural strontium isotopic signatures as fish markers: Methodology and application. *Can. J. Fish. Aqua. Sci.* **57**, 2280–2292.

King, T.L., Kalinowski, S.T., Schill, W.B., Spidle, A.P., and Lubinski, B.A. (2001). Population structure of Atlantic salmon (*Salmo salar* L.): A range-wide perspective from microsatellite DNA variation. *Molecular Ecology.* **10**, 807–821.

Kingsford, M.J., Leis, J.M., Shanks, A., Lindeman, K.C., Morgan, S.G., and Pineda, J. (2002). Sensory environments, larval abilities and local self-recruitment. *Bull. Mar. Sci.* **70 (suppl.)**, 309–340.

Klahre, L.E. (1997). *Countergradient variation in egg production rate of the Atlantic silverside* Menidia menidia. MS thesis. Stony Brook, NY: State University of New York at Stony Brook.

Leggett, W.C. (1977). The ecology of fish migrations. *Ann. Rev. Ecol. Syst.* **8**, 285–308.

Leggett, W.C., and Carscadden, J.E. (1978). Latitudinal variation in reproductive characteristics of American shad (*Alosa sapidissima*): Evidence for population specific life history strategies in fish. *J. Fish Res. Board of Canada.* **35**, 1469–1478.

Levins, R. (1969). Some demographic and genetic consequences of environmental heterogeneity for biological control. *Bull. Entomological Soc. Amer.* **15**, 237–240.

Levins, R. (1970). Extinction. In *Some mathematical questions in biology: Lectures on mathematics in the life sciences* (M. Gerstenhaber, ed., vol. 2, pp. 75–107). Providence, RI: The Mathematical Society.

Lewontin, R.C. (1974). *The genetic basis of evolutionary change.* New York: Columbia University Press.

Limburg, K.E. (1995). Otolith strontium traces environmental history of subyearling American shad *Alosa sapidissima. Mar. Ecol. Progr. Ser.* **119**, 25–35.

Limburg, K.E. (1998). Anomalous migrations of anadromous herrings revealed with natural chemical tracers. *Can. J. Fish. Aquat. Sci.* **55**, 431–437.

Limburg, K.E., Hattala, K.A., and Kahnle, A. (2003). American shad in its native range. *American Fish. Soc. Symp.* **35**, 125–140.

Lopez, J.E., and Pfister, C.A. (2001). Local population dynamics in metapopulation models: Implications for conservation. *Conservation Biology.* **15**, 1700–1709.

Lubchenco, J., Palumbi, S.R., Gaines, S.D., and Andelman, S. (2003). Plugging a hole in the ocean: The emerging science of marine reserves. *Ecological Applications.* **13 (suppl.)**, S3–S7.

Malécot, G. (1950). Quelques schemas probabilistes sur la variabilité des population naturelles. *Annales de l'Université de Lyon Sciences, Section A.* **13**, 37–60.

Man, A., Law, R., and Polunin, N.V.C. (1995). Role of marine reserves in recruitment to reef fisheries: A metapopulation model. *Biol. Conserv.* **71**, 196–204.

MacArthur, R., and Wilson, E.O. (1967). "The Theory of Island Biogeography." Princeton University Press, Princeton, NJ.

McDowall, R.M. 2003. Shads and Diadromy: Implications for Ecology, Evolution, and Biogeography. *American Fisheries Society Symposium* **35**, 11–23. Bethesda, MD: American Fisheries Society.

McFarlane, G.A., Wydoski, R.S., and Prince, E.D. (1990). Historical review of the development of external tags and marks. *Amer. Fish. Soc. Symp.* **7**, 9–29.

McQuinn, I.H. (1997). Metapopulations and the Atlantic herring. *Reviews in Fish Biology and Fisheries.* **7**, 297–329.

Melvin, G.D., Dadswell, M.J., and Martin, J.D. (1986). Fidelity of American shad, *Alosa sapidissma* (Clupeidae), to its river of previous spawning. *Can. J. Fish. Aquat. Sci.* **43**, 640–646.

Mills, S.L., and Allendorf, F.W. (1996). The one-migrant-per-generation rule in conservation and management. *Cons. Biol.* **10**, 1509–1518.

Milner, J.W. (1876). The propagation and distribution of the shad. A: Operations in the distribution of shad in 1874. In *U.S. Commission of Fish and Fisheries, Report of the Commissioner for 1873–4 and 1874–5*, Part III, Vol 2 (S.F. Baird, ed., pp. 323–326).

Milner, A.M., Knudsen, E.E., Soiseth, C., Robertson, A.L., Schell, D., Phillips, I.T., and Magnusson, K. (2000). Colonization and development of stream communities across a 200-year gradient in Glacier Bay National Park, Alaska, USA. *Can. J. Fish. Aquat. Sci.* **57**, 2319–2335.

Montgomery, D.R. (2000). Coevolution of the Pacific salmon and Pacific Rim topography. *Geology*, **28**, 1107–1110.

Mumby, P.J. (1999). Can Caribbean coral populations be modeled at metapopulation scales? *Mar. Ecol. Progress Ser.* **180**, 275–288.

National Research Council. (1996a). *Upstream: Salmon and society in the Pacific Northwest.* Washington, DC: National Academy of Science Press.

National Research Council. (1996b). *The Bering Sea ecosystem.* Washington, DC: National Academy of Science Press.

Neilson, E.E., Hansen, M.M., and Loescheke, V. (1999). Genetic variation in time and space: Microsatellite analysis of extinct and extant populations of Atlantic salmon. *Evolution.* **53**, 261–268.

Neilson, E.E., Hansen, M.M., Schmidt, C., Meldrup, D., and Grønkjær, P. (2001). Population origin of Atlantic cod. *Nature.* **413**, 272.

Olney, J.E., and McBride, R.S. (2003). Intraspecific variation in batch fecundity of American shad: Revisiting the paradigm of reciprocal latitudinal trends in reproductive traits. *American Fisheries Society Symposium.* **35**, 185–192.

Ong, T.L. (1991). *A pattern recognition approach to stock identification of Atlantic salmon, Salmo salar L., based on derived morphometric features of scales.* PhD dissertation. Providence: University of Rhode Island.

Parker, N.C., Giorgi, A.E., Heidinger, R.C., Jester, D.B., Jr., Prince, E.D., and Winans, G.A., eds. (1990). *Fish-marking techniques.* American Fisheries Society Symposium **7**. Bethesda: American Fisheries Society.

Pella, J.J., and Milner, G.B. (1987). Use of genetic marks in stock composition analysis. In *Population genetics in fishery management* (N. Ryman and F. Utter, eds., pp. 247–276. Seattle, WA: University of Washington Press.

Polachek, T. (1990). Year round closed areas as a management tool. *Nat. Resource Modeling.* **4**, 327–354.

Possingham, H.P. (1996). Decision theory and biodiversity management: How to manage a metapopulation. In *Frontiers of population ecology* (R.B. Floyd, A.W. Sheppard, and P.J. de Barro, eds., pp. 391–398). Melbourne: CSIRO Publishing.

Pulliam, H.R. (1988). Sources, sinks, and population dynamics. *American Naturalist.* **132**, 652–661.

Quinn, T.J., II, and Deriso, R.B. (1999). *Quantitative fish dynamics.* New York: Oxford University Press.

Quinn, T.P. (1984). Homing and straying in Pacific salmon. In *Mechanism of migration in fishes* (J.D. McCleave, G.P. Arnold, J.J. Dodson, and W.H. Neill, eds., pp. 357–362). New York: Plenum Press.

Quinn, T.P. (1993). A review of homing and straying of wild and hatchery-produced salmon. *Fisheries Research.* **18**, 29–44.

Quinn, T.P., and Adams, D.J. (1996). Environmental changes affecting the migratory timing of American shad and sockeye salmon. *Ecology.* **77**, 1151–1162.

Quinn, T.P., and Brodeur, R.D. (1991). Intra-specific variations in the movement patterns of marine animals. *Amer. Zool.* **31**, 231–241.

Ricker, W. (1972). Hereditary and environmental factors affecting certain salmonid populations. In *The stock concept in Pacific salmon* (R. Simon and P. Larkin, eds., pp. 19–160). Vancouver, BC: H.R. MacMillan Lectures in Fisheries, University of British Columbia.

Ricker, W.E. (1975). *Computation and interpretation of biological statistics of fish populations.* Bulletin of the Fisheries Research Board of Canada no. 191. Ottawa: Fisheries Research Board of Canada.

Rieman, B.E., and Dunham, J.B. (2000). Metapopulations and salmonids: A synthesis of life history patterns and empirical observations. *Ecol. Freshwater Fish.* **9**, 51–64.

Rieman, B.E., Myers, D.L., and Nielson, R.L. (1994). Use of otolith microchemistry to discriminate *Oncorhynchus nerka* of resident and anadromous origin. *Can. J. Fish Aquat. Sci.* **51**, 68–77.

Ridley, M. (1996). *Evolution* (2nd ed). Cambridge: Blackwell Science.

Roberts, C.M., Bohnsack, J.A., Gell, F., Hawkins, J.P., and Goodridge, R. (2001). Effects of marine reserves on adjacent fisheries. *Science.* **294**, 1920–1923.

Root, K.V. (1998). Evaluating the effects of habitat quality, connectivity, and catastrophes on a threatened species. *Ecol. Appl.* **8**, 854–865.

Roughgarden, J. (1979). *Theory of population genetics and evolutionary ecology: An introduction.* New York: Macmillan Publishing.

Rounsefell, G.A., and Kask, J.L. (1945). How to mark fish. *Trans. Amer. Fish. Soc.* **73**, 320–363.

Rozwadowski, H.M. (2002). *The sea knows no boundaries.* Seattle, WA: International Council for the Exploration of the Sea, University of Washington Press.

Ruzzante, D.E., Taggart, C.T., Doyle, R.W., and Cook, D. (2001). Stability in the historical pattern of genetic structure of Newfoundland cod (*Gadus morhua*) despite the catastrophic decline in population size from 1964 to 1994. *Conservation Genetics.* **2**, 257–269.

Ruzzante, D.E., Wroblewski, J.S., Taggart, C.T., Smedbol, R.K., Cook, D., and Goddard, S.V. (2000). Bay-scale population structure in coastal Atlantic cod in Labrador and Newfoundland, Canada. *J. Fish. Biol.* **56**, 431–447.

Ryman, N., Utter, F., and Laikre, L. (1995). Protection of intraspecific biodiversity of exploited fishes. *Reviews in Fish Biology and Fisheries.* **5**, 417–446.

Sala, E., Aburto-Oropeza, O., Paredes, G., Parra, I., Barrera, J.C., and Dayton, P.K. (2002). A general model for designing networks of marine reserves. *Science.* **298**, 1991–1993.

Schultz, E.T., and Conover, D.O. (1997). Latitudinal differences in somatic energy storage: adaptive responses to seasonality in an estuarine fish (Atherinidae: *Menidia menidia*). *Oecologia.* **109**, 516–529.

Schwarz, C.J., and Arnason, A.N. (1990). Use of tag-recovery information in migration and movement studies. *Amer. Fish. Soc. Symp.* **7**, 588–603.

Secor, D. (1999). Specifying divergent migrations in the concept of stock: The contingent hypothesis. *Fisheries Research.* **43**, 13–34.

Secor, D.H., Henderson-Arzapalo, A., and Piccoli, P.M. (1995). Can otolith microchemistry chart patterns of migration and habitat utilization in anadromous fishes? *J. Exp. Mar. Biol.* **192**, 15–33.

Secor, D.H., Ota, T., and Tanaka, M. (1998). Use of otolith microanalysis to determine estuarine migrations of Ariake Sea Japanese sea bass *Lateolabrax japonicus*. *Fish. Sci.* **64**, 740–743.

Secor, D.H., and Piccoli, P.M. (1996). Age- and sex-dependent migrations of striped bass in the Hudson River as determined by chemical microanalysis of otoliths. *Estuaries.* **19**, 778–793.

Secor, D.H., Rooker, J.R., Zlokovitz, E., and Zdanowicz, V.S. (2001). Identification of riverine, estuarine, and coastal contingents of Hudson River striped bass based on otolith elemental fingerprints. *Mar. Ecol. Prog. Ser.* **211**, 245–253.

Sinclair, M. (1988). *Marine populations: An essay on population regulation and speciation.* Seattle, WA: University of Washington Press.

Smedbol, R.K., McPherson, A., Hansen, M.M., and Kenchington, E. (2002). Myths and moderation in marine "metapopulations." *Fish and Fisheries.* **3**, 20–35.

Smedbol, R.K., and Wroblewski, J.S. (2002). Metapopulation theory and northern cod population structure: Interdependency of subpopulations in recovery of groundfish population. *Fisheries Research.* **55**, 161–174.

Stokes, J. (1985). *Natural resource utilization of four Upper Kuskokwim communities.* Technical paper no. 86. Juneau: Alaska Department of Fish and Game, Division of Subsistence.

Sunnucks, P. (2000). Efficient markers for population biology. *Trends in Ecology and Evolution.* **15**, 199–203.

Swearer, S.E., Shima, J.S., Hellberg, M.E., Thorrold, S.R., Jones, G.P., Robertson, D.R., Morgan, S.G., Selkoe, K.A., Ruiz, G.M., and Warner, R.R. (2002). Evidence of self-recruitment in demersal marine populations. *Bull. Mar. Sci.* **70 (suppl.)**, 251–271.

Tallman, R.F., and Healey, M.C. (1994). Homing, straying, and gene flow among seasonally separated populations of chum salmon (*Oncorhyncus keta*). *Canadian Journal of Fisheries and Aquatic Science.* **51**, 577–588.

Taylor, E.B. (1991). A review of local adaptation in Salmonidae, with particular reference to Pacific and Atlantic salmon. *Aquaculture.* **98**, 185–207.

Thorrold, S.R., Latkcozy, C., Swart, P.K., and Jones, C.M. (2001). Natal homing in a marine fish metapopulation. *Science.* **291**, 297–299.

Tilman, D., and Kareiva, P. (1997). Preface. In *Spatial ecology: The role of space in population dynamics and interspecific interactions* (D. Tilman and P. Kareiva, eds., pp. vii–xi). Monographs in population biology 30. Princeton: Princeton University Press.

Tuck, G.N., and Possingham, H.P. (1994). Optimal harvesting strategies for a metapopulation. *Bulletin of Mathematical Biology.* **56**, 107–127.

Utter, F. (1981). Biological criteria for definition of species and distinct intraspecific populations of salmonids under the U.S. Endangered Species Act of 1973. *Can J. fish Aquat Sci.* **38**, 1626–1635.

Utter, F., Milner, G., Staahl, G., and Teel, D. (1989). Genetic population structure of Chinook salmon, *Oncorhynchus tshawytscha*, in the Pacific northwest. *Fish. Bull.* **87**, 239–264.

Utter, F., Teel, D., Milner, G., and McIsaac, D. (1987). Genetic estimates of stock compositions of 1983 chinook salmon, *Oncorhynchus tshawytscha*, harvests off the Washington coast and the Columbia River. *Fishery Bulletin.* **85**, 13–24.

Utter, F.M., Seeb, J.E., and Seeb, L.W. (1993). Complementary uses of ecological and biochemical genetic data in identifying and conserving salmon populations. *Fisheries Research.* **18**, 59–76.

Walburg, C.H., and Nichols, P.R. (1967). *Biology and management of the American shad and status of fisheries, Atlantic coast of the United States, 1960.* Washington, DC: U.S. Fish and Wildlife Service Special Scientific Report-Fisheries 550.

Waldman, J.R., Grossfield, J., and Wirgin, I. (1988). Review of stock discrimination techniques for striped bass. *North American Journal of Fisheries Management.* **8**, 410–425.

Waples, R.S. (1998). Separating the wheat from the chaff: Patterns of genetic differentiation in high gene flow species. *J. Heredity.* **89**, 438–450.

Waples, R.S. (2002). Definition and estimation of effective population size in the conservation of endangered species. In *Population viability analysis* (S.R. Beissinger and D.R. McCullough, eds., pp. 147–168). Chicago: The University of Chicago Press.

Warner, R.R., and Cowen, R.K. (2002). Local retention of production in marine populations: Evidence, mechanisms, and consequences. *Bull. Mar. Sci.* **70 (suppl.)**, 245–249.

Waters, M.J., Dijkstra, H.L., and Wallis, P.G. (2000). Biogeography of a southern hemisphere freshwater fish: How important is marine dispersal? *Molecular Ecology.* **9**, 1815–1821.

Wells, B.K., Reiman, B.E., Clayton, J.L., Horan, D.L., and Jones, C.M. (2003). Variation of water, otolith, and scale chemistries within a river basin. *Transactions of the American Fisheries Society.* **132**, 409–424.

White, G.C., Anderson, D.R., Burnham, K.P., and Otis, D.L. (1982). *Capture–recapture and removal methods for sampling closed populations.* Los Alamos, NM: Los Alamos National Laboratory, LA-8787-NERP, UC-11.

Wiley, B.A., and Chapman, R.W. (2003). Population structure of spotted sea trout, *Cynoscion nebulosus*, along the Atlantic Coast of the U.S. In *Biology of spotted sea trout* (S.A. Bortone, ed., pp. 31–40). Washington, DC: CRC Press.

Williams, B.K., Nichols, J.D., and Conroy, M.J. (2002). *Analysis and management of animal populations.* San Diego, CA: Academic Press.

Williams, I.V. (1987). Attempts to re-establish sockeye salmon (*Oncorhynchus nerka*) populations in the Upper Adams River, British Columbia, 1949–84. In *Sockeye salmon population biology and future management* (H.D. Smith, L. Margolis, and C.C. Wood, eds., pp. 235–242). Canadian Special Publication Fisheries and Aquatic Science 96. Ottawa: NRC Press.

Wilson, J., Low, B., Costanza, R., and Ostrom, E. (1999). Scale misperceptions and the spatial dynamics of a social–ecological system. *Ecological Economics.* **31**, 243–257.

Wright, S. (1943). Isolation by distance. *Genetics.* **28**, 114–138.

Wydoski, R., and Emery, L. (1983). Tagging and marking. In *Fisheries techniques* (L.A. Neilsen and D.L. Johnson, eds., pp. 215–237). Bethesda, MD: American Fisheries Society.

Ye, Y., and Beddington, J. (1996). Modeling interactions between inshore and offshore fisheries: The case of the East China hairtail (*Trichiurus haumela*) fishery. *Fisheries Research.* **27**, 153–177.

Youngson, A.F., Jordan, W.C., Vespoor, E., McGinnity, P., Cross, T., and Ferguson, A. (2003). Management of salmonid fisheries in the British Isles: Toward a practical approach based on population genetics. *Fisheries Research.* **62**, 193–209.

PART **III**

Invertebrates

PART III

Invertebrates

CHAPTER 5

Metapopulation Dynamics of Hard Corals

PETER J. MUMBY and CALVIN DYTHAM

I. Introduction
II. Structure of This Chapter
III. Summary of Model Structure and Parameterization
IV. Existing Models of Dynamics on Coral Reefs
V. Development of a Prototype Model of Spatially Structured Coral Reef Communities
   A. Scales
   B. Reproduction
   C. Connectivity
   D. Recruitment
   E. Growth
   F. Mortality
   G. Competition (and Modeling the Dynamics of Competitors)
   H. Herbivory
VI. Testing the Model: Phase Shifts in Community Structure
VII. Sensitivity of Model to Initial Conditions
VIII. Exploration of Model Behavior
   A. Interactions between Initial Coral Cover, Algal Overgrowth Rate, Coral Growth Rate, and Herbivory
   B. Recruitment Scenario and Overfishing of Herbivores
   C. Impact of Hurricane Frequency on Local Dynamics
   D. Effect of Reduced Hurricane Frequency on a Reserve Network
IX. Summary
X. Acknowledgment
References

I. INTRODUCTION

The growth of hard corals leads to some of the largest and most complex naturally occurring structures on earth: coral reefs. The evolutionary success of hard corals may at least in part arise from the existence of a productive mutualistic relationship with dinoflagellate algae (zooxanthellae), which occupy coral tissues and, through photosynthesis, facilitate the rapid deposition of carbonate reef (Veron, 1995). Given the enormous ecological, economic, and social importance of coral reefs worldwide (Done et al., 1996), the imminent threat to coral reefs from climate change (Hoegh–Guldberg, 1999) and other disturbances such as overfishing (Knowlton, 2001) is of grave concern. Not surprisingly, the dynamics of hard corals are of great interest and importance.

Hard corals certainly form spatially structured populations, but whether they form true metapopulations is less clear. A classic metapopulation, as first characterized by Levins (1968, 1969), envisages a very large array of habitat patches that are either occupied by a population or not. Occupied patches become unoccupied through local extinctions, and unoccupied patches become occupied by colonization from occupied patches. Levins was envisaging a system of agricultural fields occupied by a pest insect with colonizations happening frequently, and extinctions partly driven by pesticide use. Clearly, coral populations differ from this type of model system in several respects. The patches that we are considering are much larger areas of reef, with patches perhaps representing the reef of an entire island. Second, local extinctions of coral species are likely to be rare at reef scales, particularly during ecological time (an exception may involve acute outbreaks of coral disease and the impact of mass coral bleaching on monospecific reefs; Aronson et al., 2000). The process of extinction is quite easy to grasp: Chance, predation, disease, or disturbance removes all the individuals of a species in a patch, which either die or emigrate. Colonization, on the other hand, even in the simplest model, describes a multistage process combining production of offspring, dispersal, arrival, settlement, and establishment. Because the original Levins model was spatially implicit, dispersal was equal to all patches. This is analogous to an unlikely "soup of plankton" idea in which all sites are equally likely destinations for settlement, and recruitment is unaffected by oceanography, currents, and tides. The Levins model does make some interesting predictions about patch occupancy. For example, it predicts that, providing the local extinction rate is greater than zero, there will always be some empty habitat patches in a landscape, even though the patches are perfectly suitable for occupation. If hard corals are acting as true metapopulations, we would therefore expect some reefs to lack some species.

Since the rediscovery of the metapopulation paradigm in the 1980s there have been many augmentations and refinements to the model. Hanski (1999) summarizes and defines the characteristics of metapopulations and shows that the spatially implicit model of equal patches is insufficient for most species. Extinction rate, as in the MacArthur and Wilson (1967) model of island biogeography, should be determined by the size and quality of the habitat patch. Colonization should not be independent of isolation from potential sources and patch size, and the probability of a patch being empty or occupied (its *incidence*) should be determined by using a spatially explicit array of patches. The incidence approach predicts that small, isolated patches of habitat will be empty more of the time than large, well-connected ones.

It may be that a relatively small proportion of species operates as true metapopulations. Local extinctions may be very rare, and subpopulations are not easily divided into discrete patches of available habitat. However, it is certain that space is a vital factor in all species, even if they behave as spatially structured populations rather than true metapopulations. Following Kritzer and Sale (2004), we move the focus away from patch-occupancy rates and consider instead coral age and stage structure. We use the terms *metapopulation* and *metacommunity* to describe a spatially structured world where many spatial scales—coral head, reef, island, archipelago—are considered simultaneously. Consideration of space in population and community ecology has raised some difficult questions. How far can individuals disperse from their natal site? How connected are isolated populations? How does the space between habitat patches, the *matrix* in the terminology of metapopulation ecology (Weins, 1997), affect movement between patches? There is also an increasing realization that local management of populations will have effects that are far from local. An augmentation or reduction of a single population will have effects on any patches connected to that population by dispersal. These effects are not instantaneous; indeed, they may take many generations to manifest themselves (see Hengeveld and Hemerik, 2002, for an overview of the link between local and biogeographical scales). Populations can never be properly understood if they are only ever considered in isolation.

A final insight from metapopulation ecology highly relevant to hard coral populations is the *extinction debt*. Tilman et al. (1994, 1997) showed that when habitat patches are lost or degraded, the effects of that loss take a long time to occur. Given the rate of habitat destruction and degradation of both terrestrial and marine habitats across the world, we should expect many species to be already committed to extinction. Species conserved at a local level will be vulnerable to extinction, even when local conservation is fully effective (Dytham, 2000). However, times to extinction may be long, especially in species with low local extinction rates; and, while a species is still extant, there is a potential for reprieve.

Metapopulation models, by definition, consider single species. No species exists in isolation and its fate is the product of both intrinsic processes and interactions with other species: predators, parasites, competitors, and so on. Any model considering more than two metapopulations in the same habitat can be described as a metacommunity model. Modeling an entire community is both difficult to achieve and difficult to interpret, so a better approach is to consider a focal species or functional group and stylize the interactions with other species or groups.

Perhaps the Holy Grail of reef modeling is the creation of spatially realistic community models capable of representing actual reef structures and their hydrological connectivity. Unfortunately, we lack the empirical understanding needed to create these predictive models (Mumby, 1999b), and one objective of this chapter is to identify critical research gaps toward this objective. Such models would provide incredibly powerful management tools with which to implement spatial decision making. For example, larval sources and sinks could be identified and incorporated into networks of MPAs. Similarly, the impact of a proposed coastal development could be predicted by degrading the health of affected reefs and examining the effect on adjacent, or more distant, reefs.

II. STRUCTURE OF THIS CHAPTER

Our aim here is to describe a prototypic, spatially structured model for hard corals and their main competitor: macroalgae. Like any model, we make a number of simplifying assumptions and the parameterization is by no means appropriate for all reefs or habitats. We begin by summarizing the structure and parameterization of the model in a table. Most parameters were derived directly from the literature, and readers familiar with the field should skip the detailed explanation that follows. However, the parameterization for herbivory involved a number of novel analyses that will not be found in the literature. The prototypical model is then refined and partly tested against Hughes' (1994) well-known account of the deterioration of reefs in Jamaica. We close with a brief inspection of model predictions, although a more detailed examination will be published separately. Overall, our aim is to provide a modeling framework that can be adapted to other reef systems and improved upon with new empirical information. Although the connectivity aspects of the models are not spatially realistic, they are at least spatially explicit, and we can gain great insight into reef dynamics by conducting appropriate experiments. For example, under what connectivity scenarios do MPAs have a significant positive benefit on hard corals? It is then the realm of hydrologists and oceanographers to determine where, if anywhere, such scenarios are found.

III. SUMMARY OF MODEL STRUCTURE AND PARAMETERIZATION

A probabilistic cellular automaton was used to model Caribbean forereefs (Fig. 5-1). Full parameterization of the model is given in Table 5-1. To compare

Chapter 5 Metapopulation Dynamics of Hard Corals 161

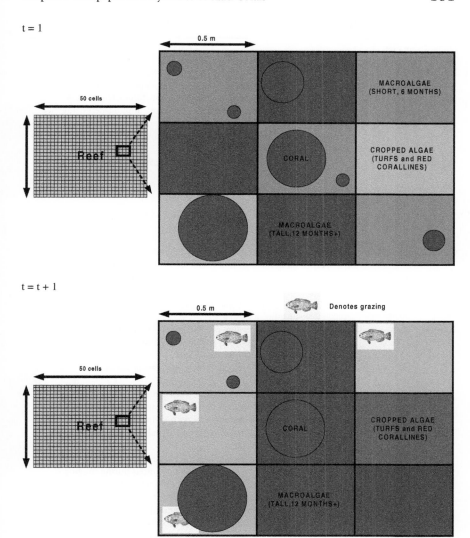

FIGURE 5-1. Diagram of probabilistic cellular automaton over one time interval. Grazing converts macroalgae of either category to the cropped algae class. Two corals recruit into the upper left-hand cell. Two established coral colonies experience partial-colony mortality, and two juvenile corals are overgrown by microalgae (center and bottom right). The center cell is not grazed and advances to a 6-month+ category of cropped algae.

TABLE 5-1.   Parameterization of Simulation Model

Model Parameter	Details
Model structure	
Habitat type	Caribbean forereef at approximately 7–20 m in depth, which may have high vertical relief (rugosity ~1.5–2.0) and be dominated by massive, foliose and encrusting life forms of coral. Most unpublished parameters were acquired from Glover's Atoll, Belize. Assumes low concentration of suspended sediments but dissolved inorganic nutrients levels are typical of atolls.
Size of single population/reef	Lattice of 50 × 50 cells, each representing ~0.25 m^2, arranged on a torus
Metapopulation structure	4 reefs arranged in a loop
Time interval for discrete time model	Updates every 6 months, run for years to decades
Cell content:	Each cell can contain a mix of the following taxa
Ungrazable substrata	Most sparisomid herbivores rarely feed from substrata such as sand, gorgonians, or sponge. A proportion of cells are set aside that are neither grazed nor colonized. These cells do, however, serve to increase the intensity of grazing in surrounding areas and can retard the vegetative spread of macroalgae.
Coral species type 1	An amalgam of coral types that grows with an encrusting to massive morphology, loosely based on species like *Porites astreoides*, *Siderastraea siderea*, and *Diploria strigosa*. Up to 3 colonies (irrespective of species) can occupy a cell. This is a reasonably realistic density of colonies.
Coral species type 2	Slow-growing massive species such as *Montastraea annularis* (results for species type 2 are not reported here because we are seeking to understand the system's behavior of with just type 1 species first)
Algal turf 6-month state	Algal turf represents all cropped substrata including coralline red algae. Corals can settle on algal turf.
Algal turf 12-month state	The age of turf algal patches is separated into 2 classes, implicitly representing turf height. If algal turf is not grazed during the first 6 months, it moves to the 12-month state. If grazed, it remains in the 6-month state. Patches in the 12-month state drop to the 6-month state if grazed, or become macroalgae if not grazed.
Macroalgae 6-month state	Macroalgae (loosely based on the species *Lobophora variegata* and *Dictyota* spp.). Corals cannot settle on either macroalgal class.
Macroalgae 12-month+ state	Macroalgae that is at least 12 months old with potentially greater vertical height and potential to overgrow corals. This represents the final successional stage of ungrazed algae.
Processes	Processes are applied to individual cells in a random sequence
Coral reproduction	Size of puberty, 60 cm^2 (cross-sectional area); size of full maturity, 250 cm^2. Larvae are released twice per year and larval output is scaled to coral size using a hemispherical model for coral growth in which the base is fully attached (i.e., $2*\pi r^2$) and a fecundity of 2 eggs per gonad, 6 gonads per polyp, 18 polyps/cm^2 (i.e., 216 eggs/cm^2). Adolescent corals have only 25% the fecundity of mature colonies.

Chapter 5 Metapopulation Dynamics of Hard Corals

TABLE 5-1. *Continued*

Model Parameter	Details
Dispersal of larvae	Each colony's larval output is summed to give the total larval output of the reef. A proportion of these larvae are then retained by the reef (arbitrarily 5–25%) and a proportion disperse to the nearest reef downstream (arbitrarily 2–10%). The remaining larvae either die or are lost. These values allow 5-fold differences to be simulated in larval retention or dispersal.
Coral recruitment	Total larval input to the reef = number of larvae dispersing from upstream population + number of larvae retained. Three recruitment paradigms were created: 1. Stock–recruitment relationships such that settlement would be maximal on well-connected, healthy reefs, but settlement would decline if either connectivity or adult coral populations decreased. It was assumed that a healthy reef would have at least 30% coral cover where all age classes have an equal chance of being represented. Under these conditions, settlement is maximum, providing that retention and dispersal coefficients are maximal. Conversely, a 10-fold drop in coral cover to 3% was assumed to be unhealthy and that larval output would be inadequate to sustain settlement (i.e., settlement = 0). The mean larval output was calculated for each of 5 equally sized increments of coral cover between 3% and 30%, where the mean was derived from 100 simulations of a coral population of given cover. Larval outputs were then converted to maximal larval inputs assuming high retention and dispersal (i.e., 35% of larval output) giving 6 critical values: $Linput_{coral} = 30\%$ ... $Linput_{coral} = 3\%$. A stock–recruitment relationship was created by assigning a probability of settlement to each critical value of larval input. Thus, the probability of settlement to a cell was 1 if actual larval input $\geq Linput_{coral} = 30\%$ and zero if actual larval input $\leq Linput_{coral} = 3\%$. Intermediate probabilities were assigned at intervals of 0.2 (i.e., 0, 0.2, 0.4, 0.6, 0.8, 1). 2. Stochastic recruitment: the settlement probability category P_s is chosen at random for every time interval (e.g., P_s for 5 consecutive time intervals = 1, 0.2, 0.4, 1, 0). 3. Full recruitment: the population is fully open and saturated with settlers, $P_s = 1$. Probability of settlement dictates the likelihood that turf algae within a given cell will receive a coral recruit. A maximum of 1 coral can settle per cell per time interval (based on observations from Belize). Implicitly, this restriction accounts for early postsettlement mortality in the pre-1-cm size classes.
Coral growth	Coral size is quantified as the cross-sectional, basal area of a colony and only converted to a 3-dimensional hemisphere when calculating colony fecundity. Coral species type 1 grows with a lateral extension rate of 0.8 cm/year, which is an upper level of observations for *P. astreoides*. Corals do not grow beyond the boundaries of a cell, although their growth is indeterminate up to that point. Growth ceases if the combined sizes of corals reaches the area of the cell (2500 cm^2). However, growth can continue if the cell is no longer full of coral.

TABLE 5-1. *Continued*

Model Parameter	Details
Colonization of turf algae	Turf algae arises (1) when macroalgae is grazed and (2) after all coral mortality events except those resulting from macroalgal overgrowth (see coral–algal competition later).
Colonization of macroalgae	Macroalgae become established when algal turf is not grazed for 12 mo.
Macroalgal growth	Macroalgae spread vegetatively across the seabed from cell to cell. The probability that macroalgae will encroach onto the algal turf within a cell $P_{\text{overgrowth}}$ depends on the von Newmann (4-cell) neighborhood. $P_{\text{overgrowth}}$ = proportional cover of macroalgae in the local neighborhood. Note that coral competition can reduce this growth rate and that macroalgae can also overgrow corals (see coral–algal competition).
Competition between corals	Corals reach a competitive "stand-off" if in direct contact with one another. Overgrowth does not occur.
Competition between corals and algal turf	Algal turfs have no impact on coral neighbors or recruit survivorship.
Competition between corals and macroalgae 1: effect of macroalgae on corals	A second, von Newmann neighborhood is examined and the percentage cover of tall, 12-month+ macroalgae calculated (M_{12+}). Only these larger macroalgae contribute to the overgrowth of corals. The probability of coral overgrowth is the product of two subprobabilities: (1) P_{size}, the size of coral, and (2) P_{macro}, extent of tall macroalgae in the immediate neighborhood. (1) is modeled as a negative exponential relationship with increasing coral size: $P_{\text{size}} = 0.83 \exp(-0.0012\chi)$ where x is colony base size in cm^2 (2) $P_{\text{macro}} = 1$ if $M_{12+} \geq 75\%$; $P_{\text{macro}} = 0.75$ if $75\% > M_{12+} \geq 50\%$; $P_{\text{macro}} = 0.50$ if $50\% > M_{12+} \geq 25\%$; $P_{\text{macro}} = 0$ if $M_{12+} < 25\%$. Thus, corals will not be overgrown until the surrounding cover of tall macroalgae exceeds 24%.
Competition between corals and macroalgae 2: effect of corals on macroalgae	The vegetative growth rate of macroalgae is reduced by 25% when at least 50% of the local neighborhood includes coral. A third von Newmann neighborhood is examined and the percent cover of coral calculated and added to that of the focal cell, giving C. If $C \geq 50\%$, $P_{\text{overgrowth}} = 0.75*$proportional cover of macroalgae in the local neighborhood.
Herbivorous fish/fishing	Herbivory is spatially constrained. An unfished community of parrotfishes can graze a maximum of 30% of the seabed per 6-month time interval. Grazers do not discriminate between algal turfs and macroalgae. During a given time interval, cells are visited in a random order. All turf and macroalgae are consumed (and converted to turf 6-month state) until the constraint is reached (e.g., 30% of the total reef area, 1,875,000 cm^2). No further algae are grazed during the time step. Grazing does not eliminate coral recruits. Herbivory is constrained to 10% in overfished systems and a value of 40% may represent an upper ceiling to herbivore biomass.
Diadema antillarum	Urchin grazing is implemented in the same fashion as herbivorous fish. Up to 40% of the seabed can be grazed under normal conditions, and urchins and fish can graze the same areas.

TABLE 5-1. *Continued*

Model Parameter	Details
Partial-colony mortality of corals	Used to represent chronic mortality processes such as parrotfish predation, small patches of disease, etc. Only applied to mature colonies at an incidence of 5% per time interval. Partial mortality reduces the coral's base area size by 15 cm^2.
Whole-colony mortality of corals	Represents total colony mortality during chronic disturbance at an incidence of 2% per time interval (~4% per annum) for adolescent colonies and 1% for mature colonies. These levels of mortality occur in addition to macroalgal overgrowth.
Hurricane impact	The whole-colony mortality of larger corals is represented using a quadratic function. Small colonies avoid dislodgment as a result of their low drag, intermediate-size corals have greater drag and are light enough to be dislodged, whereas large colonies are heavy enough to prevent dislodgment. However, 80% of recruits are removed by scouring of sediment. For nonrecruits, the probability of mortality, P_{hur}, is a minimum of 5% and maximum of 50% and is described by: $$P_{hur} = (0.0000003x^2) + (0.0007x) + 0.05$$ 90% of mature coral colonies experience severe partial mortality. The extent of partial mortality, M_{hur} on each adult coral is modeled using a Gaussian distribution with mean of 0.46 and an SD of 0.20. Each value of M_{hur} represents the percentage of original colony tissue that is lost due to the hurricane. If $M_{hur} \leq 0$, there is no mortality. If $M_{hur} \geq 1$, the entire colony is lost (although this is a rare event). Hurricanes reduce the cover of macroalgae to 10% of its prehurricane level. (So 90% of macroalgal cells are converted to turf 6-month?)
Hurricane frequency	Typical hurricane frequency is stochastic but with a long-term frequency of 1 event per decade per reef or 0.2 per reef per 6-month period. A typical metapopulation scenario allows each reef to be disturbed on independent, stochastic trajectories.
Initial reef structure	
Proportions of taxa	10% of the reef comprises ungrazable substrata (sand, sponges, etc.). Initial coral cover varies from 10–50%. Macroalgal cover is 20% (all set to short category), and turf algae (6-month class) occupy the remainder. Ungrazable substrata, corals, and macroalgae are assigned to cells in a random sequence.
Size distribution of corals	Coral populations are initialized in three different arrangements: (1) each size category has an equal chance of being included (giving a mixed-population structure), (2) the size–frequency distribution is dominated by smaller corals such that only the lowest 75% of possible size classes are included, or (3) where the initial population is dominated by old corals in the top 15% of size categories. Most realizations of the model use arrangement (1).
Alternative scenarios	
Source–sink populations	Alternate sources and sinks where sources have high dispersal (0.1) and sinks have low dispersal (0.02). All populations have high retention (0.25).
Marine reserve network	Alternate populations have different levels of fishing pressure, yielding various levels of fish herbivory. Reserves tend to have high biomasses of herbivores (herbivory, 0.3) whereas nonreserves are either partially overfished (herbivory, 0.2) or fully overfished (herbivory, 0.1).

differences in model design, one author (P.J.M.) used a discrete time (6-month) approach whereas the second author used an event-based approach more akin to continuous time.

IV. EXISTING MODELS OF DYNAMICS ON CORAL REEFS

Surprisingly few studies have attempted to model aspects of coral population or community dynamics. Competitive interactions between other benthic clonal organisms have been investigated using cellular automata (e.g., Karlson and Buss, 1984) and energetic-based models have studied herbivory (McClanahan, 1992) and trophic cascades (Grigg et al., 1984). Population dynamics of corals have been modeled using Leslie matrix models (Hughes, 1984; Done, 1987, 1988; Ruesink, 1997; Hughes and Tanner, 2000). Stone (1995) adapted the spatial patch models of Hastings (1980) and Tilman et al. (1994) to examine the effects of habitat loss on species extinction within a reef community of the Red Sea. We do not review the aforementioned papers in detail, because they do not consider spatially structured populations in which dispersal occurs between populations.

Preece and Johnson (1993) used cellular automata to represent arrays of reefs at various densities and under a gradient of larval retention. Individual cells represented deep water or reefs in various states of coral cover. Both brooding and spawning corals were represented such that the larvae of spawners were dispersed and brooded larvae settled almost immediately. Postsettlement mortality of juvenile corals was varied to represent predation from the crown-of-thorns starfish (*Acanthaster planci*). After an initial disturbance, the recovery of reefs was tracked and found to depend on the overall connectivity scenario and among-reef variance in coral cover. Processes interacted strongly. For example, recovery was less dependent on reef density when retention of larvae was high and reef density was more important for spawners than brooders. Furthermore, brooding corals achieve dominance if reefs are isolated, have a low level of larval retention, and experience high levels of disturbance. This model had the advantage of representing many connectivity regimes, but at the expense of resolving many of the internal ecological processes within reefs.

More recently, Wolanski et al. (2004) modeled ecological processes on the Great Barrier Reef with realistic levels of connectivity. A suite of hydrological models was used to represent larval dispersal and the effluent of river plumes. Local dynamics were represented for four categories of reef biota: adult and juvenile corals, algae, and herbivorous fish. The system was perturbed by river plumes and cyclones, and a fair fit was found between observed and predicted algal cover on 20 reefs, although a number of outliers existed that had been impacted by the crown-of-thorns starfish.

V. DEVELOPMENT OF A PROTOTYPE MODEL OF SPATIALLY STRUCTURED CORAL REEF COMMUNITIES

A. SCALES

The range of possible modeling questions for coral reef communities is bewildering. First, we decided to focus on Caribbean reef ecosystems rather than those of the Indo-Pacific. This decision largely reflects the bias of authors, because P.J.M. has empirical data from reefs in Belize with which to parameterize models, but the decision also simplifies the modeling process because there are fewer species to consider. Second, we elected to model a single type of reef community rather than attempt to represent a full reef seascape encompassing multiple habitats and community types. Although this decision simplifies the model, it also reflects the paucity of information on interactions among adjacent reef habitats (i.e., to what extent do the dynamics of one community affect that of a neighboring community?). Several reef communities could have been modeled, but we focus here on forereef communities at a depth of approximately 7 to 20 m that are often characterized with high rugosity and conspicuous colonies of *Montastraea annularis*. Thus, we ignore shallow reefs of the branching coral, *Acropora palmata*, which have been widely denuded by white band disease (Aronson and Precht, 2001).

Each reef system was represented on a toroidal, square lattice of 2500 cells. Each cell putatively represents approximately $0.25\,m^2$ of reef, such that the maximum recruitment rate of corals is one per cell, which is representative of reefs at this depth in Belize (Mumby, 1999a). At this cell size, each lattice represents a minimum area of $25 \times 25\,m$, which is clearly at odds with the size of individual reefs, which may span kilometers or tens of kilometers. The gulf in scale was partly bridged in two ways. First, by using a toroidal lattice that has no boundary, and second, by rescaling several modeled processes so that they represented the mean behavior at reef scales. For example, hurricane damage may be very patchy at a scale of $25 \times 25\,m$, but less so at the scale of an entire island (Bythell et al., 2000). Reefs in the direct path of a hurricane may experience 50 to 100% whole-colony mortality (Bythell et al., 1993), whereas mortality at the seascape scale may be nearer to 30% (Bythell et al., 1993).

The implications of applying parameters derived from one scale (e.g., the seascape) to another scale (e.g., a reef) warrant consideration. For example, implementing the seascape-averaged impact of a hurricane to a lattice adds a level of mean field behavior to what is intended to be a spatially explicit model. Admittedly, the fate of individual corals will vary stochastically across the reef, but the overall impact is predetermined. An alternative approach would involve stratification of the lattice and varying the hurricane severity in each stratum according to an observed or theoretical distribution of hurricane impact at sublattice scales. The overall seascape-level impact of a hurricane would then be an emergent property of the model. However, it is difficult to conceive an empirical or theoretical justification for such small-scale modeling. Furthermore, we have little information on the responses of other model parameters to such heterogeneity in impact. The biomass of herbivorous fish, for example, is often positively correlated to the structural complexity of reef habitats (van Rooij et al., 1996b; Mumby and Wabnitz, 2002), but does their biomass and grazing respond to small-scale patchiness in rugosity within habitats? In other words, would the grazing rate decline in patches of reef that are razed almost flat by a hurricane or is such patchiness small compared with the territory size of some species? Attempting to build such small-scale sophistication into a model would probably reduce its usefulness (by creating a more uncertain output) and its credibility.

Any model is a compromise between reality and simplicity. If the intention of the model was to *predict* the impact of hurricanes on specific reefs, then a more sophisticated, probabilistic solution to the model would be desirable. Similarly, although a truly predictive model would have uncertainty attached to most, if not all, model parameters, such complexity can obscure the generic insight offered by a simpler model. In this case, the model is not envisaged as a predictive tool, but as a means of understanding the sensitivities and interactions between processes acting on reefs. An example would be quantifying the interaction of algal growth rate and herbivory on the survivorship of corals. To reduce variability in model behavior, which might otherwise obscure the processes of interest, a number of parameters are entered as constants (e.g., coral growth rate). The impact of adjusting these parameters by their expected range is examined during sensitivity analyses, but it is not necessarily important to incorporate, say, phenotypic plasticity in fecundity while investigating the relationship between algal growth rate and herbivory in a system that is not recruitment limited.

There is no fixed size of reef but each is isolated from other reefs by uninhabitable space (usually deep water). Each reef (or lattice in the model) therefore represents a single island or atoll. Four reef systems were represented in the metacommunity, largely because of computational limitations that precluded rapid sensitivity analysis for larger communities. Metacommunity dynamics were investigated on ecological timescales of decades, although these models might also be appropriate for metapopulation questions acting on geological timescales such as the faunal turnover events of the Miocene and early Pleistocene periods (Jackson et al., 1996; Aronson and Precht, 2000). Extending the model to geological time would almost certainly involve a reparameterization, because many processes, such as the intensity of herbivory, may well have acted differently in the past.

B. REPRODUCTION

The puberty size of corals is size dependent (Soong, 1993) and usually larger in species that broadcast rather than brood their larvae (Harrison and Wallace, 1990). Puberty sizes and fecundities in Caribbean corals are well established (Szmant, 1991; Soong and Lang, 1992; Soong, 1993). The corals modeled here were based arbitrarily on a brooding species, *Porites astreoides*. There are approximately two eggs per gonad, six gonads per polyp, and 18 polyps/cm^2 (Szmant, 1986). These parameters were used to generate a linear relationship between coral size and fecundity, but we emphasize that the choice was arbitrary and unimportant. Any scaling relationship could have been chosen, providing that the link between fecundity and recruitment was calculated using the same method. However, the fecundity–recruitment scaling relationship will be affected by the size at which colonies reach puberty and full maturation. Although recent work on the Great Barrier Reef has revealed more complex scaling relationships between colony size and fecundity (Hall and Hughes, 1996), we use a linear scaling relationship here for simplicity. Both *P. astreoides* and the spawner *Siderastrea siderea* reach puberty at a cross-sectional area of around 60 cm^2 and full maturity at around 250 cm^2 (Soong, 1993). We followed the data of Soong (1993) such that adolescent colonies had only 25% of the fecundity of fully mature colonies (where puberty size < adolescent size < maturation size). With a 6-month time step, the model allows reproduction and recruitment to occur twice a year. This is an appropriate frequency for some corals such as *Mycetophyllia* spp. (Szmant, 1986), but twice the frequency of reproduction for spawners and substantially less frequent than some brooding species that may release planulae nine times per year (Szmant, 1986). Despite the difficulty of reconciling such scales, the rate of colonization specified by the model for a healthy reef is consistent with densities found in the field (i.e., one new recruit per 0.25 m^2 of available settlement space; Mumby, unpublished data). This density implicitly incorporates early post-settlement mortality at the previsible stages that are not modeled explicitly.

In its current formulation, the model does not modify fecundity between reefs, yet large-scale patterns of fecundity have been found along the Great Barrier Reef that have a strong predictive relationship with recruitment (Hughes et al., 2000). The difficulty is that we have few data on the effects of environmental and ecological processes on colony fecundity. Nutrification and sedimentation both appear to suppress fertility (Kojis and Quinn, 1984; Tomascik and Sander, 1987), but have not been extensively studied. Competitive interactions also affect fecundity. Tanner (1995) found that the fecundity of *Acropora palifera* on the Great Barrier Reef was halved when in contact with macroalgae. Szmant (1991), working on *Montastraea annularis*, found that zones of polyps a few centimeters wide were infertile if in contact with other macroorganisms as a result of the high defense and growth demands of these polyps. The effects of coral bleaching on fecundity have not been reported and are vital in understanding the full suite of bleaching impacts.

C. CONNECTIVITY

Corals disperse using one of two sexual reproductive modes: broadcast spawning of gametes, which usually occurs on an annual basis; or the brooding of larvae, which are usually released several times a year. Many corals, particularly those with a branching morphology, also disperse locally through fragmentation and scattering of the colony (Highsmith, 1982). Broadcast planulae develop for a minimum of 4 to 6 days prior to becoming competent to settle and metamorphose (see review by Harrison and Wallace, 1990), and settlement must take place within 3 to 4 weeks or else the planulae will have insufficient energy reserves to metamorphose (Richmond and Hunter, 1990). In contrast, brooded planulae may settle within 10 minutes of release from the adult (e.g., *Favia fragum*; see Carlon and Olson, 1993). Other brooding species, such as *Agaricia* spp., may delay metamorphosis by a week while they search for suitable habitat (Carlon and Olson, 1993). However, although larval competency periods have been quantified, little is known about larval mortality rates other than mortality may increase while in the vicinity of reefs, mainly by planktivorous reef fish (Hamner et al., 1988; Westneat and Resing, 1988), and that the larvae of brooding Caribbean corals (e.g., *Agaricia agaricites*, *Porites astreoides*, and *Siderastrea radians*) are highly palatable to damselfish and wrasse (Lindquist and Hay, 1996).

The swimming speeds of coral planulae are orders of magnitude less than those of oceanic currents (Hodgson, 1985; Carlon and Olson, 1993), so assumptions of passive dispersal by oceanic currents may be valid in general (Harrison and Wallace, 1990). However, variations in wind velocity will greatly affect dispersal (Willis and Oliver, 1988) and the buoyancy of larvae tends to decrease with time (Babcock and Mundy, 1996), requiring that vertical mixing in the water column be accounted for. Although models of oceanic circulation are usually vertically stratified (e.g., MICOM; Cowen et al., 2000), the resolution of such models at the reef-ocean boundary is only beginning to be understood. Oceanic models might predict mesoscale circulation reasonably, but it is difficult to estimate the proportion of larvae that ever become entrained in such features. Largier (2003) describes how larvae are often strongly retained by two processes. First, cross-shelf diffusion is often slow such that it typically takes hours to reach even 1 km offshore. Second, frictional forces retard alongshore current velocities near the coast, limiting the degree of advection. A larva of 30-day duration may only move 10 km downstream, because it spends little time in faster alongshore currents. High-resolution numerical simulations of larvae in Australian reef systems suggested high retention, especially when larvae are negatively buoyant and retained by frictional forces near the seabed (Black et al., 1991). More recently, passive advection–diffusion–mortality models for fish larvae released from Caribbean reefs predict very low levels of connectivity between islands (Cowen et al., 2000). In Australia, the Helix experiment of coral settlement around Helix reef found most recruitment within 300 m of the reef (Sammarco and Andrews, 1988). Furthermore, Ayre and Hughes (2000) examined the genotypic diversity and gene flow of both brooding and spawning coral species along the Great Barrier Reef and concluded that the majority of recruitment by corals is local, particularly in brooding corals. No equivalent analysis is available for the Caribbean, but the weight of evidence to date predicts that the dispersal distance of coral larvae is unlikely to be great.

Although empirical data on larval mortality rates and transport between reefs are inadequate, it seems likely that dispersal between populations is limited, at least on ecological timescales. We therefore decided to restrict dispersal so that larvae from one population may reach the nearest reef downstream, but not subsequent reefs. Processes of larval retention and dispersal were represented using a coefficient for each process. Although retention and dispersal were represented separately, they are dependent at extremes. For example, the sum of coefficients cannot exceed unity. Dispersal was lower, with a maximum value of 0.1 (10% of larvae reach the downstream population). Arbitrarily, the maximum retention coefficient was set at 0.25 (i.e., 25% of larval output is retained by the population). A scenario of high retention and high dispersal might represent two reefs in close proximity. Local eddies retain many larvae, but those that become entrained in ocean currents are quickly transported to the downstream reef. Note that these coefficients are not intended to represent realistic levels of reef connectivity because these are often unknown (for an exception see Wolanski et al., 2004). Instead, retention and dispersal coefficients can be used to investigate how the degree of connectivity interacts with other reef processes, such as herbivory. Local dispersal by colony fragmentation was excluded from the model, because it was assumed that appropriate species (*Acropora cervicornis*) are scarce and that the majority of reef species at a depth of 10 m exhibit encrusting or massive morphologies (i.e., ignoring branching species such as *Porites porites* and *Madracis mirabilis*).

D. Recruitment

The recruitment of hard corals remains one of the most enigmatic and challenging questions in reef ecology. Recruitment is the outcome of four factors: larval supply (itself determined by a range of processes from fecundity, oceanography, predation, and so forth), settlement behavior of the planula, the availability of settlement space, and postsettlement mortality. The interaction of all four factors has never been investigated in a single study.

The settlement behavior of planulae has been studied intensively, but many observational studies are inconclusive because of the difficulties in inferring active substratum selection when stochastic larval supply and patchy postsettlement mortality are not recorded. In general, however, larvae appear to prefer rough surfaces (Lewis, 1974; Carleton and Sammarco, 1987; Smith, 1997), which may enhance attachment (see review in Harrison and Wallace, 1990). Morse et al. (1988) found that chemical cues induce the larvae of *Agaricia* spp. to settle on encrusting red algae, although the algal species were not identified (Morse et al., 1988); other coral species (e.g., *Favia fragum*) prefer to settle near conspecifics (Lewis, 1974) whereas others are inhibited by secondary metabolites from other taxa such as soft corals (Maida et al., 1995). Settlement characteristics may be highly species specific (Morse et al., 1988; Carlon and Olson, 1993) and the suitability of specific microhabitats may vary between species and with growth (Babcock and Mundy, 1996).

Sedimentation appears to be particularly important in determining early mortality at both microhabitat scales (Maida et al., 1994; Gleason, 1996) and reef scales (Hunte and Wittenberg, 1992). Sediments can affect larval settlement directly by preventing secure attachment and indirectly by altering larval settlement behavior. Increased water turbidity could make more cryptic habitats too dark, either increasing mortality rates in these refugia (due to decreased coral growth rate) or forcing larvae to settle on more exposed upper surfaces, where sediment levels or grazing intensity may be higher, possibly increasing the mortality rates of juveniles (Maida et al., 1994).

The effects of nutrification (increased ambient nutrient concentrations) on coral settlement are only beginning to emerge. Ward and Harrison (1997) manipulated nitrogen and phosphorus concentrations on adults and larvae of *Acropora longicyathus* and found that increases in both nutrients reduced the settlement of corals and that their effects were synergistic. The mechanisms by which elevated nutrients prevent settlement are not fully understood but may involve blooms of cyanobacteria that release toxins to prevent larval attachment. Conversely, Ward and Harrison (1997) found that larvae released from adults that had been incubated in a nitrogen-rich environment showed high levels of settlement when released in ambient seawater. Corals are nitrogen limited, so the settlement success of larvae developing in nitrogen-rich conditions might be attributable to a release from this limitation, enabling greater amino acid production, which confers greater protection from ultraviolet radiation. One potential implication of this study is that larvae released from eutrophic reefs may have greater survivorship than those from oligotrophic reefs, but that larval settlement on eutrophic reefs may be severely reduced. Whether these patterns are true or extend to other species remains to be seen.

In general, recruitment of corals is space limited. This has been demonstrated in both Jamaica (Hughes, 1985) and the Great Barrier Reef (Connell et al., 1997). Recruitment also shows great temporal variability (Gleason, 1996; Connell et al., 1997) and spatial variability on a wide range of scales, including continental shelves (Sammarco, 1991), between reefs (Fisk and Harriott, 1990), and within sites (Babcock, 1988; Smith, 1992). A multiscale study by Hughes et al. (1999) across the Great Barrier Reef found that recruitment patterns differed between brooders and spawners. The variance in recruitment of brooders was more strongly explained by smaller reef (10–15 km) and site (0.5–3 km) scales than that of spawners, which showed strong variation (52–55% of all variance) at the scale of hundreds of kilometers. The difference in scale of recruitment was attributed to the shorter larval durations of brooders limiting dispersal. Interestingly, there was no concordance in scale between the variation of recruitment and adult abundance for either brooders or spawners. However, patterns of fecundity (the proportion of colonies on each reef containing ripe eggs) strongly predicted the variation in recruitment observed for spawners (Hughes et al., 2000). More important, the relationship between fecundity and recruitment was found to be nonlinear such that a given rise in fecundity may create a disproportionately large increase in recruitment.

Available empirical evidence suggests that postsettlement mortality in corals is high (Smith, 1997), but the density dependence and scale dependence of such processes are poorly understood and difficult to infer from observational studies (Caley et al., 1996). This difficulty arises throughout benthic ecology, because observations of persistent and variable age class strength in a population, which are usually inferred to indicate density-independent mortality (e.g., Victor, 1986), may occur in the presence of density-dependent mortality (Holm, 1990). Further, variable postrecruitment mortality may obscure the relationship between recruitment and year class strength even in the absence of density-dependent mortality (Warner and Hughes, 1988).

Empirical data from the Caribbean are unequivocal in the numerical dominance of recruits from brooding species and the paucity of recruits from species that broadcast their gametes (Bak and Engel, 1979; Rogers et al., 1984; Hughes, 1985; Smith, 1992; Mumby, 1999a). The converse pattern is found in many Indo-Pacific reefs in that broadcasting species may dominate recruitment (Hughes et al., 1999). Whether the contrast between Caribbean and Indo-Pacific systems is attributable to fundamental differences in ecological processes between regions or their respective evolutionary pasts is unknown (Aronson and Precht, 2000). Broadcasting species in the Caribbean tend to be long lived (tens to hundreds of years) and may store their age advantage, only recruiting occasionally when conditions are favorable (Edmunds, 2002). Nevertheless, it is difficult to represent processes of sexual recruitment in these species meaningfully (i.e., does it occur every 10 years, 30 years, or 100 years?). Although it would be possible to distinguish two life history strategies in the model, a separate parameterization for recruitment of spawners would have little credibility.

The settlement of corals was modeled using a simple probabilistic framework with three scenarios: stock recruitment relationships, stochastic settlement, and maximal settlement where the supply of larvae is not limited (Table 5-1). Impacts of sediments and nutrients on recruitment have not been parameterized at this stage.

E. GROWTH

The growth rates of most major reef building corals (e.g., *Montastraea annularis*) decrease with depth (Highsmith et al., 1983; Huston, 1985), although the relationship may be nonlinear. Bosscher and Meesters (1993) found growth to be light saturated to a depth of 15 m and then sharply light limited to 30 m. Juvenile growth rates are high (Moorsel, 1988) and, although growth can be indeterminate (Sebens, 1987), relative growth rate decreases with increasing coral size and age partly as a result of increased rates of partial colony mortality and possibly by greater energetic investment in sexual reproduction (Hughes and Connell, 1987). Bleaching events can halt coral growth even in large corals such as *Montastraea* spp. (Goreau and Macfarlane, 1990), but the anticipated deleterious effects of nutrification and sedimentation on growth rate are less clear.

Tomascik and Sander (1985) found that eutrophication generally decreased the growth rate of *M. annularis* in Barbados, but that suspended particulate matter (SPM) could have variable effects on growth, depending on its concentration. At intermediate SPM concentrations, corals may derive nutrients from particulate matter whereas high SPM concentrations may be deleterious by smothering polyps and reducing light availability to zooxanthellae (Tomascik and Sander, 1985). Increased nutrient concentrations can have indirect effects on coral growth rate by (1) promoting the cover of space competitors including macroalgae (Littler et al., 1993) and sponges (Wilkinson, 1987; Aerts and Van Soest, 1997), which potentially increases the incidence of competitive interactions and reduces growth rate (Tanner, 1995); and (2) by promoting phytoplankton blooms, which reduce the transmission of light to the benthos, thus reducing rates of photosynthesis (D'Elia and Wiebe, 1990). Direct effects of nutrients on coral growth rates are difficult to summarize because of the variation in nutrient concentration, application frequency, and duration of manipulative experiments. It would appear, however, that increased nitrogen can decrease the translocation of photosynthates from zooxanthellae to host, thus reducing calcification rates (McGuire and Szmant, 1997; Steven and Broadbent, 1997), and that phosphorus may interfere with calcification by acting as a crystal poison (Hoegh–Guldberg et al., 1997).

Coral growth could be modeled in several ways. One option would be to make the cells of the lattice small relative to the size of corals and then let corals grow laterally from cell to cell. This method is ideal for examining multispecies competitive interactions (Johnson and Seinen, 2002), but for our objectives it requires undesirable complexity in terms of representing realistic shapes of corals. Not only is this approach difficult to parameterize, but the detailed definition of species boundaries would place heavy emphasis on processes that occur at species boundaries (e.g., competition, overgrowth) that, as we show later, are difficult to parameterize at present. We therefore used a simple approach in which corals were represented as hemispheres with a linear extension rate. Reviewing published data of skeletal extension rates for the species *Porites astreoides*, *P. porites*, *Siderastrea siderea*, *Montastraea annularis*, *Colpophyllia natans*, and *Agaricia agaricites* (Huston, 1985; Chornesky and Peters, 1987; Van Moorsel, 1988), we used a median growth rate of 8 mm/year with a range of 6 to 10 mm/year.

F. MORTALITY

The feeding activities of herbivores can have positive and/or negative effects on coral recruitment. Dense aggregations of *Diadema* (Sammarco, 1980; Rylaarsdam, 1983) and intense fish grazing (Bak and Engel, 1979; Sammarco, 1991) may overgraze the substratum, preventing successful coral recruitment. Conversely, Birkeland (1977) found that intermediate grazing pressure on Caribbean reefs promoted coral settlement by removing fouling turf and macroalgal canopies, and Morse et al. (1988) suggested that grazers may facilitate establishment of specific coralline red algae, which in turn promotes settlement of *Agaricia humilis* through chemical cues. Given the equivocal status of the literature on the effects of herbivores on the mortality of coral recruits, the model ignored most direct interactions between all herbivores and corals (i.e., herbivores do not consume coral recruits, although some predation on adult corals is implied by the rates of partial colony mortality). However, the model does allow herbivores to have indirect impacts on coral colonies through mediation of coral–algal interactions (see Section V.G).

Mortality rates of adult corals are principally size dependent (Hughes, 1984; Hughes and Jackson, 1985; Meesters et al., 1997), and to some extent age dependent (Hughes, 1984; Bak and Meesters, 1998). Whole-colony mortality is greatest for small corals whereas the incidence of partial-colony mortality from various biological (e.g., parrotfish grazing, disease) and physical (e.g., abrasion) sources increases with colony size (Hughes, 1985; Bak and Meesters, 1998). Bythell et al. (1993) studied the mortality of corals in St. Croix for a 26-month period during which reefs were disturbed by Hurricane Hugo, the largest storm to affect the site for 60 years. Deriving data for a 6-month period (one time step in the discrete time model) with no hurricane, the proportion of adult colonies experiencing partial-colony mortality lay between approximately 0 to 3% for *Diploria strigosa*, 1.5 to 9% for *Porites astreoides*, and 3 to 10% for *Montastraea annularis*. An intermediate incidence of partial-colony mortality of 5% was applied to mature colonies in each time interval. Under these levels of chronic disturbance (as opposed to that resulting from hurricanes), individual partial-colony mortality events caused a reduction in colony size of 15 cm^2 in cross-sectional area (Mumby, personal observation).

The data of Bythell et al. (1993) were also used to estimate chronic whole-colony mortality rates. For adolescent colonies, we used the mean, chronic whole-colony mortality rate of *P. astreoides* colonies in the size range 50 to 200 cm^2, rescaled to 6 months (approximately 2%). Chronic whole-colony mortality rates were reduced in larger, mature colonies to 1% (approximately 2% per annum) based on an observed frequency of 1.5% for *P. astreoides* (Bythell et al., 1993). Note that additional whole-colony mortality occurs during hurricanes and as a result of algal overgrowth (discussed later).

At larger spatial scales, storm and hurricane damage are highly patchy (Woodley et al., 1981; Edmunds and Witman, 1991; Rogers, 1993; Bythell et al., 2000). For example, the effects of Cyclone Ivor on the Great Barrier Reef were patchy at scales of tens to hundreds of meters within 50 km of the "eye" of the storm (Done, 1992). Massel and Done (1993) modeled the shear, compression, and tension forces generated by waves acting on massive corals at different depths and in different hydrodynamic settings. They concluded that large size provides a refuge from hurricane dislodgment when corals are weakly or unattached to the substratum and where reefs are exposed to swells but not short waves. This scenario allows small colonies to resist dislodgment because of their low profile, whereas large colonies are sufficiently heavy to resist dislodgment. Although this prediction agrees with observations of small- and large-size refugia in massive colonies during cyclones (Done and Potts, 1992), large size does not provide a refuge from dislodgment if reefs are exposed to short waves, such as might be expected on steep forereefs facing the prevailing storm directly. Given that structurally complex forereef communities tend not to be found in highly exposed locations (Geister, 1977), we used the former model in which corals experience size-based refugia at small and large sizes. The maximum whole-colony mortality rate for intermediate-size corals was set at 50%, after the observed impact of Hurricane Hugo on *P. astreoides* on a heavily impacted forereef in St. Croix (77% reduction in tissue surface area of which approximately 75% was attributable to whole-colony mortality; Bythell et al., 1993). A minimal whole-colony rate of 5% for larger colonies (>2000 cm^2) was estimated as half the reported overall mortality rate of *P. astreoides* colonies in the size class more than 200 cm^2 and assuming that mortality rate continued to decline with increasing size. Coral recruits suffer intense scouring and dislodgment during hurricanes, and a mortality rate of 80% was recorded for coral recruits (2–20-mm diameter) after Hurricane Mitch, which struck Belize in 1998 (Mumby, 1999a).

Hurricane-induced incidences of partial-colony mortality were derived from the impact of Hurricane Mitch on mature colonies of *M. annularis* in Belize (Mumby, unpublished data). At least 90% of entire colonies experienced some kind of partial-colony mortality after the hurricane (n = 90). The mean reduction in living coral tissue, measured 6 months (one time step) after the hurricane, was 46% per colony (SD, 33%; n = 13) at one site and 50% per colony (SD, 23%; n = 17) at a second site, separated by a distance of 3 km. The lower value was inserted into the model.

Models of the effects of storms on reefs (Hughes, 1984; Andres and Rodenhouse, 1993) and empirical studies (e.g., Hughes, 1989; Done, 1992; Connell et al., 1997; Lirman and Fong, 1997; Rogers et al., 1997) point out the importance of considering the historical regime of disturbances when attempting to predict or interpret the effect of any individual disturbance. A potential development of the model would be to include variable susceptibility among corals to hurricane disturbance (i.e., embracing the notion of "weeding out" the less strongly attached colonies so that rapidly occurring hurricanes have progressively minor impacts on the community).

The frequency of hurricanes was modeled stochastically using the analysis of 500 years of hurricane paths in the Lesser Antilles (Treml et al., 1997). The median frequency of direct hurricane disturbance in the region is once per 10 years. In a metacommunity modeling context, hurricanes occurred at a probability of 0.3 per time step, with the reef chosen at random. Thus, the long-term mean hurricane frequency is once per decade per reef, but the incidence of hurricanes on a single reef over ecological time are far less predictable.

Two other major sources of mortality are mass coral bleaching (Hoegh–Guldberg, 1999) and disease (Richardson, 1998). Incidences of mass coral bleaching occur at regional and global scales, and are triggered by anomalously warm temperatures at the hottest time of year. It is widely suspected that other factors such as solar radiation, bathymetry, wind strength, vertical stratification of the water column (W. Skirving, personal communication), and salinity combine with temperature to create a mosaic of environmental stress that may, in the future, be predictable (West and Salm, 2003). The literature offers many accounts of mortality from coral bleaching, but the general patterns appear to be (1) long-lived massive species such as *P. lutea* and *M. annularis* rarely suffer whole-colony mortality (but see Mumby et al., 2001), (2) branching species appear to be highly susceptible to whole-colony mortality (e.g., *Acropora* spp. *Pocillopora* spp.), (3) brooding species in the Caribbean are highly susceptible to whole-colony mortality (Aronson et al., 2000), and (4) coral recruits on Caribbean reefs appear to be resistant to mortality during bleaching events (Mumby, 1999a). Recent studies are attempting to identify thermal thresholds for the initiation of bleaching responses in corals (Berkelmans, 2002) and the probability that prolonged thermal stress will result in mass coral mortality (Berkelmans, personal communication). As such thresholds gain widespread validation, they will increasingly be used to predict the impact of rising sea temperature on the dynamics of corals (e.g., Wooldridge and Done, 2004). However, because the current model is not being used to predict reef futures and it currently only includes one coral species, a detailed exploration of the impact of mass coral bleaching is not undertaken here.

Coral diseases are particularly prevalent in the Caribbean and the incidence of disease appears to be positively correlated to temperature (Harvell et al., 2002). In contrast to mass coral bleaching, long-lived massive coral species appear to be particularly susceptible to diseases. Diseases could be modeled as severe outbreaks in individual reefs rather than phenomena affecting many reefs simultaneously. A number of regional programs are now monitoring the incidence of coral disease (e.g., Caribbean Coastal Marine Productivity or CARICOMP), and should help illuminate the scale at which disease events occur. Disease is not incorporated in the model described here.

G. Competition (and Modeling the Dynamics of Competitors)

Competitive interactions between clonal organisms have been studied and modeled extensively (Buss and Jackson, 1979; Karlson and Jackson, 1981; Karlson and Buss, 1984; Chornesky, 1989; Connolly and Muko, 2003). Coral–coral interactions often form intransitive networks in which competitive dominance reverses between species. However, the importance of coral–coral interactions in structuring reef communities has been questioned (Van Woesik, 2002). The importance of such interactions is particularly questionable on Caribbean reefs, because coral cover has declined markedly with a commensurate increase in macroalgal cover (Gardner et al., 2003). For example, Hughes and Jackson (1985) noted that coral–coral interactions accounted for less than 10% of all coral mortality events observed in Jamaica. Because their observations were made between 1977 and 1980, when macroalgae were relatively scarce on Jamaican reefs, it seems likely that coral–coral interactions are even less common today.

At present, the model does not distinguish coral species, and adjacent corals reach a "standoff" rather than overgrow one another. The functional implication of this parameterization is that large colonies cannot increase in size once corals fill a grid cell. Although competitive standoffs are often observed between corals in the field, we cannot discount the importance of intransitive competitive networks between corals a priori. Although we have "economized" on model complexity by not representing complex competitive networks between corals that rarely meet, emergent properties of competition should be investigated in the future.

The widespread catastrophic mortality of the long-spined urchin, *Diadema antillarum* in 1983/1984 (Lessios et al., 1984) has probably driven regionwide increases in the cover of macroalgae in the Caribbean (de Ruyter van Steveninck and Bak, 1986; Hughes et al., 1987; Lessios, 1988; Levitan, 1988; Carpenter, 1990; Bak and Nieuwland, 1995). Macroalgae occupy space on the reef and therefore preempt available settlement space for coral recruits. Although these algae rarely exceed heights of a few centimeters, they appear truly to compete with corals (Jompa and McCook, 2002). Studying the effects of *Lobophora* on *P. cylindrica* in the Great Barrier Reef, Jompa and McCook (2002) found that coral contact reduced algal growth rate by 0 to 25%. However, despite a reduced growth rate, *Lobophora* was the superior competitor and, in the absence of herbivory, was able to overgrow and kill the coral completely within 12 months. Similarly, de Ruyter van Steveninck et al. (1988) found that growth rates of *L. variegata* were reduced by 35% when in contact with small Caribbean corals, but that algal–coral overgrowth continued. Other studies have found that macroalgae reduce the growth rate of corals, although not necessarily leading to coral mortality. In Australia, Tanner (1995) found that macroalgal competition reduced growth rates of semiencrusting acroporid corals but not bushy (erect) pocilloporids. Unlike macroalgae, algal turfs (mixed-species assemblages usually <1 cm high) are inferior competitors with corals, although coral recruits may be smothered if turfs become heavily laden with sediment. Both coral–algal and algal–coral competitive interactions were entered into the model (Table 5-1).

Modeling the competition between corals and macroalgae is particularly challenging because there is so little empirical information. For example, there are no published data on the overgrowth rate and effects of macroalgae on coral recruits. How long does a recruit need to be smothered before it dies? How does the size of the coral affect its vulnerability to algal overgrowth? Furthermore, there are virtually no data on the patch dynamics of macroalgae. One exception is a study in Curaçao, in which de Ruyter van Steveninck and Breeman (1987) studied the cover of *L. variegata* at a depth of 25 m. They found no significant changes in cover at the reef scale but cover in individual quadrats, measuring 11 × 16 cm, varied from 10 to 90% during the same period (1 year).

The recruitment of macroalgae through dispersal of fragments and germlings appears to be highly constrained in space over short timescales. Working on *Turbinaria* and *Sargassum* in Tahiti, Stiger and Payri (1999) found that the dispersal distance of almost all germlings was limited to within 90 cm of the parent colony. In Curaçao, de Ruyter van Steveninck and Breeman (1987) observed very little colonization of *Lobophora* in plots located just outside the main *Lobophora* stand (zero cover after 10 months, 3% cover after 12 months). In contrast, colonization of space was rapid when plots were located within the mosaic of *Lobophora* patches. When plots (25 × 30 cm) were cleared in the mosaic, they were completely regrown within 6 months, regardless of season. We therefore modeled two forms of macroalgal growth. First, if turf algae is not grazed for a period of 1 year, it becomes macroalgae, because germlings and fragments of macroalgae form part of the turf and only become established if left ungrazed for periods of several months or more, depending on the depth, exposure, and nutrient regime (Lewis, 1986). Second, established macroalgae spread vegetatively. The probability of neighboring cells becoming overgrown by macroalgae depends on (1) the local density of macroalgae, (2) the height of the macroalgae (only taller, patches >12 months can overtop corals), and (3) the nature of the substratum. Algal turfs can be overgrown directly whereas the probability of overgrowing coral will depend on the size of the colony. Unfortunately, there are few empirical data to parameterize this aspect of the model, so several probability density functions were created (Fig. 5-2). Macroalgae such as *L. variegata* and *Dictyota* spp. are able to overgrow recruits (Bak and Engel, 1979; de Ruyter van Steveninck and Bak, 1986), and small corals several years old (R. Bak, personal communication).

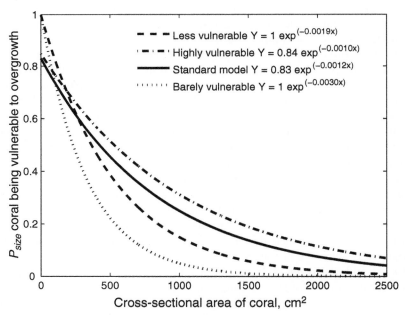

FIGURE 5-2. Postulated probabilities of algal overgrowth for corals of various size. The intermediate scenario (bold) was used routinely in model simulations.

Whether fleshy macroalgae cause rapid whole-colony mortality of larger colonies is unclear, although they can cause extensive partial-colony mortality (Hughes and Tanner, 2000) that may then be followed by whole-colony mortality arising from further algal overgrowth (which may then be more likely to occur once algae become established) or other chronic stresses such as disease.

In the absence of data on patch extinction rates in fleshy macroalgae, it was assumed that any losses of algae resulting from senescence or wave exposure were rapidly replaced vegetatively. An exception was made for hurricanes such that extensive scouring reduces macroalgal cover to 10% of its prehurricane level (Mumby, unpublished data).

H. HERBIVORY

On Caribbean reefs, herbivory is dominated by the long-spined urchin *D. antillarum* and a variety of fish groups, of which parrotfish (Scaridae) are usually the most significant (Carpenter, 1986). Despite the extensive body of literature on herbivory (Sammarco et al., 1974; Hatcher and Larkum, 1983; Carpenter, 1986; Lewis, 1986; Steneck, 1988; Bruggemann et al., 1994b; McClanahan et al., 1996), it is surprisingly difficult to model the influence of herbivory on macroalgae. First of all, the patch dynamics of macroalgal phaeophytes are virtually unknown, although there is some information on calcareous chlorophytes such as *Halimeda* spp., which have been investigated in the context of sediment dynamics (e.g., Neumann and Land, 1975). Second, our understanding of herbivory is much greater for urchins than fish. The two principal groups of herbivores (urchins and fishes) are thought to compete for food (Hay, 1984), but the former group was greatly reduced in 1983/1984 during the catastrophic die-off of *D. antillarum*. Although the feeding ecology of *D. antillarum* had been extensively studied before this event, most of our quantitative insight into the behavior of parrotfish is more recent, thus occurring in the absence of *D. antillarum* (Bruggemann et al., 1994b; Van Rooij et al., 1996a,b; Mumby and Wabnitz, 2002). Whether these recent studies of parrotfish are biased by the absence of parrotfishes' main competitor is difficult to determine, but conceivable. Although there are scattered reports of recovery of *D. antillarum* (Edmunds and Carpenter, 2001; Miller et al., 2003), a key question for reef science is whether herbivorous fish now perform the role vacated by *D. antillarum*.

An elegant study of parrotfish herbivory was carried out on low-rugosity fore-reefs at Ambergris Caye in Belize. Williams et al. (2001) artificially increased coral cover in 5 × 5-m plots and observed a corresponding decrease in macroalgae, although cropped substrata were maintained at an equilibrial level of 50% by the end of the 5 month treatment. They inferred that herbivorous fish were able to graze a maximum of 50% of the substratum. An important implication of this result is that fish herbivory is limited and unlikely to compensate for a large increase in settlement space for macroalgae.

The reefs studied by Williams et al. (2001) have a lower rugosity than that implied by our model, so to adapt their results, and partially to test the generality of their conclusions, we performed a simple comparison. First we compared the rugosity between their site (Ambergris Caye, rugosity = 1.2) and that of Long Cay, Glover's Reef, which better represents the reefs being modeled (rugosity = 1.98). We next compared the total biomass of acanthurid and scarid fishes between sites and found the value from Ambergris Caye (approximately 400 g/ 25 m^2) to lie centrally within the range of biomass found at Long Caye between 1998 and 2003 (225–700 g/25 m^2). The biomass of parrotfish at Ambergris is unusually high for a rugosity of 1.2 (Mumby and Wabnitz, 2002), but this may reflect high proximity to more rugose reefs in deeper water at Ambergris Caye. We then adjusted the surface areas of each reef based on their rugosity and assumed that parrotfish biomasses were directly comparable. Applying the same herbivory to a more topographically complex reef at Long Cay (1.98 vs. 1.20), we conclude that the total cover of cropped substratum would fall from 50% to 30%. We then compared this estimate with the observed cover of cropped substratum at Long Cay in the summer of 1998. At this point, the reef was in extremely good health (42% coral cover) because it had not been severely disturbed since 1978 or earlier (i.e., as close to equilibrial conditions as is likely to be found in Belize). The cover of cropped substrata (algal turfs, filamentous microalgae, and coralline red algae) was 31% in July 1998. This fits the predicted cover of cropped substrata surprisingly well and we therefore decided to constrain herbivory to a maximum of 30% of the substratum.

Constraining grazing to a maximum of 30% of the reef surface area makes the implicit assumption that parrotfish grazing is able to compensate for some decreases in coral cover. In other words, for parrotfish to maintain 30% of the reef in a grazed state, their grazing intensity must increase to compensate for the increased algal colonization that occurs after a coral mortality event. Increases in scarid grazing rate have been reported in response to increased algal production (Carpenter, 1985; Russ and McCook, 1999; Diaz–Pulido and McCook, 2003), but the limits of such upregulation are not well understood. However, we present a coarse preliminary exploration of the generality of a 30% ceiling for grazing efficacy using data from 16 forereefs distributed across 200 km in Belize and sampled in 2002. These reefs were disturbed 4 years previously by mass coral bleaching and Hurricane Mitch, thus increasing the surface area available for algal colonization. The dominant scarid in Belize (*Sparisoma viride*) grows quickly (Choat et al., 2003), and a numerical response to increased food availability is feasible within 4 years (individuals can grow to approximately 250 mm in this time).

Predicted macroalgal cover (M_{pred}) =

$$M_{pred} = \begin{cases} Herbivory \leq 100 - Substratum_{ungrazable}; \\ 100 - Substratum_{ungrazable} - Herbivory \\ Herbivory > 100 - Substratum_{ungrazable}; 0 \end{cases} \quad (1)$$

assuming that the former condition is true, M_{pred} is calculated from Equation 2

$$M_{pred} = 100 - (C + S + Sp + G) - \left[\left(\frac{B_S}{B_{LC}}\right)\left(\frac{R_{LC}}{R_S}\right) \cdot 30\right] \quad (2)$$

where $Substratum_{ungrazable}$ includes the cover of living coral (C), sand (S), sponges (Sp), and gorgonians (G), and the $Herbivory$ constraint (30%) is adjusted for the ratio of scarid biomass at the site (B_S) to that at Long Cay (B_{LC}) and the ratio of rugosity at the site (R_S) to that at Long Cay (R_{LC}).

The mean difference between observed and predicted macroalgal cover on these reefs was 3% (SE, 3%). Not surprisingly, there was a fair degree of spread in the predictions (Fig. 5-3), but the important point is that the greatest dispar-

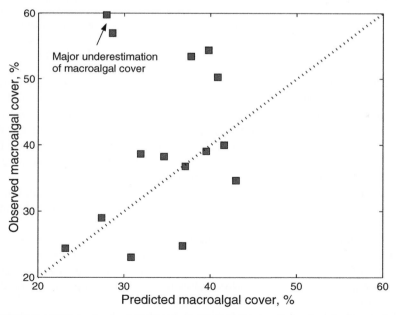

FIGURE 5-3. Efficacy of a simple herbivory model to predict the observed cover of macroalgae on forereefs in Belize. A diagonal line marks positions for perfect agreement.

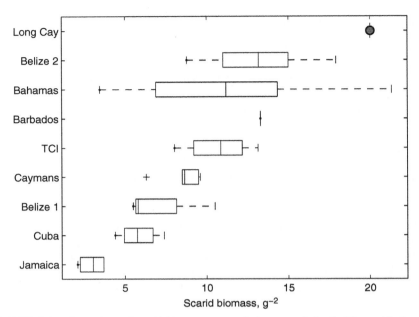

FIGURE 5-4. Comparison of scarid biomass at several sites around the Caribbean. All sites at around 8 to 12 m and ignore individuals (<12 cm standard length). Data for Jamaica, Cuba, Cayman Islands, Barbados, Belize 1 (relatively low-relief reefs; derived from Williams and Polunin, 2000). Data for Belize 2, Bahamas, the Turks and Caicos Islands (TCI), and Long Cay derived from Mumby (unpublished data).

ities occurred when macroalgal cover was *higher* than expected. In other words, there is no clear evidence that parrotfish graze down more than 30 to 40% of the available substratum on these reefs. Indeed, if the spatial constraint to herbivory is increased from 30% to 40% of the substratum, the model overestimates the impact of herbivory in all but one reef. It should also be pointed out that the biomass of scarids at the reference site, Long Cay (Belize), was ordered second greatest in a comparison of 61 Caribbean reefs, including sites from Jamaica, Barbados, the Bahamas, Belize, Cuba, Cayman Islands, and the Turks and Caicos Islands (Fig. 5-4). Therefore, the reference biomass of scarids used in this simple analysis lay at the upper level of observed scarid densities in the region.

When compared with empirical data, the model will tend to underestimate the amount of macroalgae on a reef and the predictions may therefore be conservative with respect to increases in coral cover. The causes of "exceptionally" high macroalgal cover will be explored elsewhere in more sophisticated models of herbivory.

Modeling fish herbivory in this fashion makes the implicit assumption that the dynamics of herbivores are independent from those of algae or corals. Although this is unlikely to be the case at extremes of algal cover, there are insufficient empirical data to create an alternative scenario at this stage. A second assumption is that herbivorous fish (mainly parrotfish) do not discriminate between turf algae and macroalgae. Although parrotfish will avoid very large macroalgal phaeophytes such as *Turbinaria* spp. and *Sargassum* spp., these are uncommon on the forereef. Mumby (unpublished data) recorded several thousands of bites by adult parrotfish and found a ratio of 1:1.05 for bites on macroalgae (chiefly *Dictyota* spp.) to turf algae respectively. Several workers have speculated that avoidance of chemically defended macroalgae by parrotfish may help reinforce a phase change from coral-dominated to algal-dominated community states (e.g., Knowlton, 1992). It is certainly true that relatively few bites are taken from such algae. For example, only 3% of parrotfish bites at Long Cay were taken from *L. variegata*, another important macroalga. However, it does not necessarily follow that herbivorous fish do not influence the dynamics of such algae. Experiments in Australia have documented grazing impacts on both adult and juvenile *Lobophora* (Jompa and McCook, 2002; Diaz–Pulido and McCook, 2003).

Parrotfish may limit the cover of macroalgae by direct consumption (albeit rarely for some algal species) and by preventing the escape of macroalgae from algal turfs. Unfortunately, it is difficult to reconcile the importance of the combined effects of individual grazing processes when the model is simulating the net impact of herbivory at a scale of months, and parrotfish bites occur on a scale of seconds and differ among species at microhabitat scales (Bruggemann et al., 1994a). Finally, the model does not explicitly consider the activities of territorial damselfish, which defend patches of microalgae and macroalgae (Robertson, 1996). Whether the overall framework of herbivory and algal dynamics implicitly captures this behavior is unclear, but there is certainly scope to extend this aspect of the model in future.

The herbivory of *D. antillarum* was modeled using data from the U.S. Virgin Islands. After the die-off of *Diadema*, Carpenter (1988) found that the biomass of algae removed dropped by more than half from 3.74 g dry weight/m^2/day to 1.58 g dry weight/m^2/day. If we let the herbivory of fish be 0.3 (as discussed earlier), then the additional herbivory attributable to urchins is approximately 0.4, based on ([(3.74/1.58) × 0.3] − 0.3). After the die-off of *D. antillarum*, the herbivory attributable to urchins was set to zero.

VI. TESTING THE MODEL: PHASE SHIFTS IN COMMUNITY STRUCTURE

Unfortunately there are no metacommunity-scale empirical data sets against which to test the model. However, there are several long-term accounts of reef health in the Caribbean region at the scale of a single reef or island. Perhaps the best known of these is the change in community structure around Jamaica since the 1970s (Hughes, 1994). In brief, reefs were overfished for a long period but unaffected by hurricanes for an unusually long period of 36 years before Hurricane Allen struck in 1980 (Woodley, 1992). Overfishing seriously depleted the biomass of herbivorous fish such as parrotfish. Hurricane Allen caused patchy, but severe, damage to reefs, but this was followed by a short period of rapid recruitment and recovery. However, the urchin *D. antillarum* disappeared in 1983, causing a massive decrease in total herbivory. Reefs began to decline and were struck by another hurricane (Gilbert) in 1988. Because we do not know the coral cover in 1944, we could not attempt to represent the growth of the reef up to the 1970s. However, we recreated the scenario described by Hughes (1994) to compare model predictions with his empirical data from Discovery Bay at a depth of 10 m. Note that none of the parameters used was derived from case studies in Jamaica. The only parameters entered from Jamaica were (1) initial coral cover, (2) the timing of hurricanes Allen and Gilbert, (3) the time of *D. antillarum* disappearance, and (4) a value for fish herbivory. The latter value was calculated using recent biomass estimates for acanthurids and scarids in Jamaica (Williams and Polunin, 2000). The biomass of herbivorous fish in Jamaica was found to be approximately one third that of Long Cay (Belize), so a spatial constraint of 10% (0.3/3) was entered into the model.

Model predictions were made for each of the algal overgrowth scenarios shown in Figure 5-2 and with two alternative size–frequency distributions of corals (Fig. 5-5). All scenarios captured the qualitative pattern of reef decline observed in Jamaica with the exception of the "barely vulnerable" scenario in which larger corals (>1000 cm^2) were virtually invulnerable to algal overgrowth. The main departure between the model and empirical data occurs shortly after the die-off of urchins. The model predicts a rapid increase in macroalgae but Hughes (1994) observed a slower rate of increase (Fig. 5-5). No explanation was offered for this slow increase in algal cover (Hughes, 1994), and whether it constitutes a local effect or serious misparameterization of the model is unclear. In contrast, a number of other studies reported rapid increases in macroalgal cover after the loss of *Diadema* (Carpenter, 1985; Liddell and Ohlhorst, 1986; Levitan, 1988). Second, the brief decrease in macroalgal cover brought about by Hurricane Gilbert is either inappropriate or it occurred but was not detected because the sampling frequency was too low to detect it.

The most compelling algal overgrowth scenario, denoted "standard model," was used for subsequent simulations.

VII. SENSITIVITY OF MODEL TO INITIAL CONDITIONS

Each time a model is initialized, the spatial arrangement of corals, algae, and ungrazable substratum change. To investigate whether variation in arrangement has an impact on model predictions, a reef was initialized with 30% coral cover,

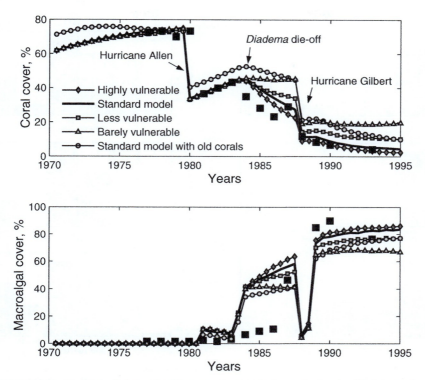

FIGURE 5-5. Model predictions of coral cover and macroalgal cover in Jamaica (line) versus data from Hughes (1994) at Discovery Bay. ■ = 10 m.

10% ungrazable substratum (i.e., other than coral), 20% macroalgal cover, and 40% turf algal cover, and simulated for 15 years in the absence of hurricanes. After 100 simulations, mean percent coral cover at the end of the run was 50.3% and the standard error was extremely small at 0.06%. Similar standard errors were obtained from alternative initial conditions, so we conclude that the small-scale spatial arrangement of cells has no significant impact on model behavior. Although changing the amount of ungrazable substratum will impact the model, this is always held constant at 10%, which is the mean observed cover of noncoral, ungrazable substratum from *Montastraea* reefs in Belize (Mumby, unpublished data).

The impact of coral size–frequency distribution on model behavior was investigated under conditions of no hurricane disturbance, overfishing, or recruitment limitation. An initial coral cover of 30% was specified for the three size–frequency distributions listed in Table 5-1. The trajectory of coral cover varied markedly

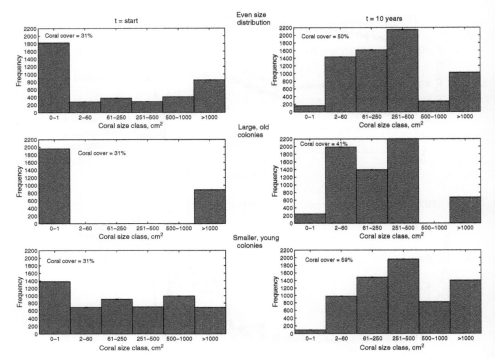

FIGURE 5-6. Impact of the size–frequency distribution of corals on the behavior of the model in the absence of hurricane disturbances, recruitment limitation, and overfishing. Note that reefs were initialized before the first coral settlement event so total coral cover is 1% greater than initialized.

among initial size–frequency distributions (Fig. 5-6). Reefs dominated by large, old colonies showed a modest increase in coral cover of only 11% after 10 years. There was minor mortality of the old corals, but a progressive increase in smaller size classes arising from sexual recruitment (Fig. 5-6). Growth of corals is not indeterminate in the model and capped at 2500 cm^2. Although this may be inappropriate for some long-lived broadcasting species such as *M. annularis*, few Caribbean coral species are regularly observed with sizes greater than 2500 cm^2. Placing a ceiling on coral size does restrict the potential rate of increase in coral cover but only after a considerable number of colonies reach an age of approximately 35 years. The contrasting coral population, comprised of small size classes, increased markedly in 10 years, with a final coral cover of 59%. Rapid increases in coral cover occurred because few corals reached their maximum size. Finally, an even distribution of all possible sizes led to intermediate rises in coral cover (50%). Note that an even distribution of possible sizes does not result in identical histograms when class widths vary between columns!

The latter, even size–frequency distribution was used in all further model simulations. However, the potential rate of increase in coral cover will depend on the disturbance regime. Once corals start accumulating in the largest size category, rates of coral increase will slow and eventually cease.

There is always a danger with modeling of the kind presented here that the results are somehow influenced more by the formulation and construction of the model than by the biology they are hoping to represent. To explore that possibility we generated a parallel modeling approach with the same goals. The alternative approach taken was to use an event-based model, with the same landscape structure of an array of grids connected by dispersal. In an event-based model there is no synchronous updating of the content of cells. Events happen singly and in sequence. This is achieved by picking a cell at random, determining the possible events that could occur in that cell and their relative probabilities and then applying just one event. In each time step there will be, on average, the same number of events as in the synchronous updating model. So, mechanistically, the model is very different because there will be many cells in which no events take place and, conversely, many cells that host a number of events in each time step. One of the advantages of an event-based model is that it can mimic the continuous time models commonly used in community ecology, because the time between events is often infinitesimally small. The results presented in this chapter were derived from the discrete time model, but the results of the event-based model were sufficiently similar to convince us that the effects we report here are not idiosyncrasies of the modeling approach.

VIII. EXPLORATION OF MODEL BEHAVIOR

A. Interactions between Initial Coral Cover, Algal Overgrowth Rate, Coral Growth Rate, and Herbivory

Several scenarios were created to investigate factors influencing the survival of coral recruits. Initial coral cover was set at either 10%, 30%, or 50% and coral growth rate ranged from 0.6 to 1 cm/year. Herbivorous fish were either severely overfished (herbivory, 0.1), partially overfished (0.2), or unfished (0.3). Algae were allowed to compete with corals as in Table 5-1, but two more aggressive scenarios were created. In the first, established macroalgae grow more quickly and are able to overgrow corals after 6 months instead of 12. The second scenario builds on the first and doubles the rate of vegetative growth in macroalgae, representing both an increase in algal fecundity and growth rate. Coral settlement rates were set to their maximum values so connectivity was not investigated per se.

Recruit survival was always low when initial coral cover was low and when reefs were overfished (Fig. 5-7). Increases in initial coral cover and herbivory both increase the survival of recruits, and these processes appear to be more important than coral growth rate or the aggressiveness of algae (Fig. 5-7). Recruit survival was modeled using a binomial, generalized linear model with logit link. The advantage of this method over analysis of variance (ANOVA) is that data are treated as numerical variables (rather than categories) and that the large number of zeros are not problematic. Not surprisingly, most interactions and main effects were found to be significant (Table 5-2).

B. Recruitment Scenario and Overfishing of Herbivores

Four reefs were arranged in a connected ring and were subjected to three recruitment scenarios: stock recruitment, stochastic recruitment, and maximum recruitment in an open population (Table 5-1). All reefs began with a "healthy" coral cover of 30% and were simulated 30 times for a period of 50 years, during which hurricanes occurred stochastically at an average rate of once per decade per reef. It was assumed that larval retention and dispersal coefficients were at their maximal values. In one experiment, all reefs were heavily overfished so that herbivory was confined to 10% of the seabed. In the other, herbivores were not fished and herbivory was set at 30%.

During a 50-year period, coral cover tended to decline in all scenarios, possibly indicating that the incidence of hurricanes was too high, given the somewhat acute impacts being modeled (Fig. 5-8). For this reason, hurricane frequency is revisited in the next section. The most striking result is that overfished reefs declined rapidly, with corals becoming extinct within 30 to 40 years. The response of unfished reefs was far more variable, with periods of recovery and persistence between hurricanes (Fig. 5-8). However, although the contrast between overfished and unfished reefs was obvious, the impact of recruitment scenario was difficult to discern. Indeed, a three-way ANOVA with fishing level and recruitment scenario as fixed factors, and reef number as a random factor showed that coral cover after 50 years was significantly influenced by fishing level and reef, but not by recruitment scenario or any interaction involving recruitment (Table 5-3).

The implication of these results is that connectivity scenarios are unimportant in heavily overfished reefs. Even when recruitment was not limited by larval supply, the processes of mortality dominated the dynamics. Therefore, even though the connectivity scenarios are not realistic, the model predicts that any real connectivity regime will have limited impact once the majority of herbivorous fish (and urchins) are unavailable. Connectivity is only likely to be

Chapter 5 Metapopulation Dynamics of Hard Corals

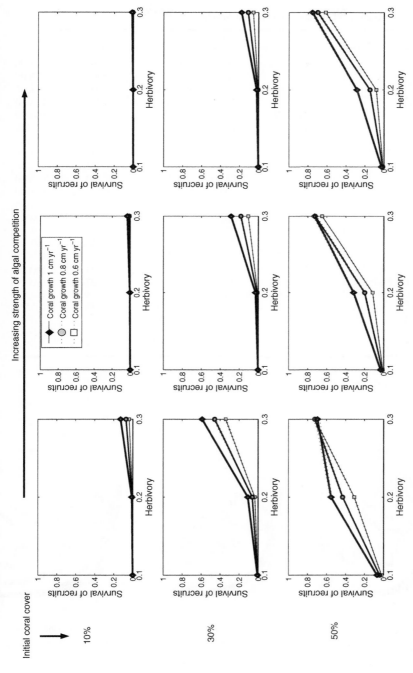

FIGURE 5-7. Effects of initial coral cover, parrotfish herbivory, coral growth rate, and the strength of algal competition on the survival of recruits to adolescence.

TABLE 5-2. Analysis of Deviance for Recruit Survival

Variable	df	Deviance	Resid. df	Resid. dev	F value	Pr(F)
Null			764	112152.1		0.0000
$(ICC)^2$	1	3005.38	763	82146.7	8406.45	0.0000
$(H)^2$	1	52747.99	762	28398.7	15058.29	0.0000
$(CG)^2$	1	7471.10	761	20937.6	2093.14	0.0000
ICC	1	1246.76	760	19680.9	349.30	0.0000
H	1	1514.82	759	18166.0	424.40	0.0000
CG	1	22.24	758	18143.8	6.23	0.0127
AG	2	12308.44	756	5835.4	1724.20	0.0000
$(ICC)^2:(H)^2$	1	111.50	755	5723.9	31.24	0.0000
$(ICC)^2:(CG)^2$	1	161.55	754	5563.3	45.36	0.0000
$(H)^2:(CG)^2$	1	1382.78	753	4179.5	387.41	0.0000
ICC:H	1	17.18	752	4162.4	4.81	0.0285
ICC:CG	1	51.35	751	4111.0	14.39	0.0001
H:CG	1	18.93	750	4092.1	5.30	0.0215
ICC:AG	2	1059.33	748	3032.7	148.39	0.0000
H:AG	2	598.75	746	2434.0	83.87	0.0000
CG:AG	2	367.08	744	2066.9	51.42	0.0000
$(ICC)^2:(H)^2:(CG)^2$	1	134.08	743	1932.9	37.72	0.0000
ICC:H:CG	1	15.85	742	1916.4	4.44	0.0354
ICC:H:AG	2	97.90	740	1818.5	13.71	0.0000
ICC:CG:AG	2	19.10	738	1799.4	2.68	0.0695
H:CG:AG	2	57.75	736	1741.6	8.09	0.0003
ICC:H:CG:AG	2	32.76	734	1708.9	4.59	0.0104

Terms added sequentially from first to last. AG, algal growth scenario; CG, coral growth; H, herbivory; ICC, initial coral cover. Quadratic terms are indicated using ( )².

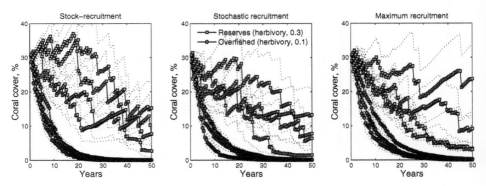

FIGURE 5-8. Comparison of the impacts of recruitment scenario on overfished and unfished coral reefs throughout a 50-year period. Note that networks of reserves and nonreserves were treated independently and were *not* connected to one another within metapopulations.

TABLE 5-3. Crossed ANOVA of Coral Cover at t = 50 Years under 2 Levels of Fishing Pressure, 3 Recruitment Scenarios, and for 4 Reefs

Factors	df	MS	F	P
Fishing level	1	0.467	29.2	<0.001
Recruitment scenario	2	0.008	0.3	0.763
Reef	3	0.210	4.4	0.005
Fishing level × recruitment scenario	2	0.008	0.3	0.772
Error	231	0.016		

Data were arcsin transformed.

important once reefs have a sufficient number of herbivores and/or sufficiently infrequent hurricanes that periods of recovery become feasible.

C. Impact of Hurricane Frequency on Local Dynamics

To investigate the impact of hurricane frequency further, a single population was simulated 30 times for a 50-year period. Recruitment was assumed to be non-limiting, herbivores were not fished, and hurricane frequency decreased from once per decade to less then once every 200 years (Fig. 5-9). It appears that coral cover is maintained at an approximately equilibrial level (approximately 30%) at a frequency of one hurricane per 20 years and that cover will increase when hurricanes are even less frequent. Interestingly, the response of reefs to hurricane frequency is nonlinear. The difference in response between once per decade and once every 20 years is considerable (around 30% in final coral cover) whereas subsequent drops in hurricane frequency have progressively less impact (Fig. 5-9). For example, reducing hurricane frequency from once every 40 years to every 60 years has an impact of around 15% coral cover after 50 years. Therefore, a critical hurricane frequency may exist between once every 10 to 20 years, which dictates whether a reef will experience long-term decline.

On a cautionary note, the model represents the impact of hurricanes when the reef is either directly within the storm's track or when the storm is further away but of sufficiently large size and intensity that its impact is similar (e.g., Hurricane Mitch caused severe coral mortality in Belize even though the eye of the storm missed Belize by some 60 km). Therefore, the hurricane frequencies described here would tend to ignore the occurrence of less destructive hurricanes that do not strike the reef directly.

Although many reefs can be characterized as disturbance-driven systems (Connell, 1978; Done, 1992), it might seem difficult to reconcile the predicted decline of coral cover at observed hurricane incidences of once per decade.

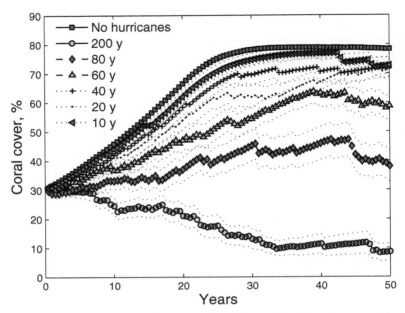

FIGURE 5-9. Impact of hurricane frequency on the dynamics of healthy reef systems with a full complement of parrotfishes.

However, it must be borne in mind that levels of herbivory have declined significantly during the last few hundred years and most notably within the last 20 years (Pandolfi et al., 2003). With unusually high recent hurricane activity in the Caribbean region (Goldenberg et al., 2001) and a possible increase in hurricane frequency being forecast under a warming climate (Knutson et al., 2001), the prognosis for reef health is a matter of grave concern for the Caribbean region.

D. Effect of Reduced Hurricane Frequency on a Reserve Network

Coral cover declined in reserves when hurricanes occurred at decadal intervals (Fig. 5-8). Once hurricane frequencies were halved to a 20-year period, coral cover appeared to be sustainable, albeit with considerable fluctuations (Fig. 5-10). However, even under this relatively low disturbance regime, overfished reefs were unable to maintain corals.

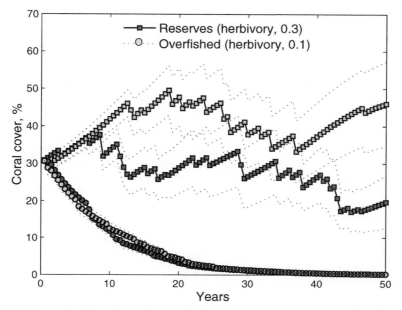

FIGURE 5-10. The effect of a simple network of no-take marine reserves on coral cover throughout a 50-year period, during which hurricanes occur in 20-year periods.

IX. SUMMARY

We present a modeling framework for Caribbean reefs that appears to have some explanatory power at local ecological scales. The connectivity scenarios are unrealistic but suggest that the importance of larval connectivity in coral population dynamics is highly sensitive to the degree of overfishing of herbivores in local populations. Overfished reefs decline even when larvae are not limiting whereas unfished reefs exhibit a variety of dynamics and are influenced by larval supply. Future analyses will explore the relationship between fishing and the postsettlement mortality of corals. Similarly, the implications of these results for the design of no-take marine reserves need further consideration. The model can be improved in several ways, including the use of realistic connectivity matrices between reefs (e.g., James et al., 2002), the addition of outbreaks of coral disease, and the expected impact and frequency of mass bleaching events.

X. ACKNOWLEDGMENT

P.J.M. thanks John Hedley, Laurence McCook, Terry Done, Diego Lirman, Terry Hughes, Al Harborne, Kamila Zychaluk, and Nicola Foster for their discussions on the model; and the Natural

Environment Research Council and Royal Society for funding. C.D. thanks Stephen Simpson and Julie Hawkins for discussions on coral reef ecology.

REFERENCES

Aerts, L.A.M., and Van Soest, R.W.M. (1997). Quantification of sponge/coral interactions in a physically stressed reef community, NE Colombia. *Mar. Ecol. Prog. Ser.* **148**, 125–134.

Andres, N.G., and Rodenhouse, N.L. (1993). Resilience of corals to hurricanes: A simulation model. *Coral Reefs.* **12**, 167–175.

Aronson, R.B., and Precht, W.F. (2000). Evolutionary paleoecology of Caribbean coral reefs, In *Evolutionary paleoecology: The ecological context of macroevolutionary change* (W.D. Allmon and D.J. Bottjer, eds., pp. 171–234). Columbia: Columbia University Press.

Aronson, R.B., and Precht, W.F. (2001). White-band disease and the changing face of Caribbean coral reefs. *Hydrobiologia.* **460**, 25–38.

Aronson, R.B., Precht, W.F., Macintyre, I.G., and Murdoch, T.J.T. (2000). Coral bleach-out in Belize. *Nature.* **405**, 36.

Ayre, D.J., and Hughes, T.P. (2000). Genotypic diversity and gene flow in brooding and spawning corals along the Great Barrier Reef, Australia. *Evolution.* **54**, 1590–1605.

Babcock, R. (1988). Fine-scale spatial and temporal patterns in coral settlement. *Proceedings of the 6th International Coral Reef Symposium, Australia.* **2**, 635–639.

Babcock, R., and Mundy, C. (1996). Coral recruitment: Consequences of settlement choice for early growth and survivorship in two scleractinians. *J. Exp. Mar. Biol. Ecol.* **206**, 179–201.

Bak, R.P.M., and Engel, M.S. (1979). Distribution, abundance and survival of juvenile hermatypic corals (Scleractinia) and the importance of life history strategies in the parent coral community. *Mar. Biol.* **54**, 341–352.

Bak, R.P.M., and Meesters, E.H. (1998). Coral population structure: The hidden information of colony size–frequency distributions. *Mar. Ecol. Prog. Ser.* **162**, 301–306.

Bak, R.P.M., and Nieuwland, G. (1995). Long-term change in coral communities along depth gradients over leeward reefs in the Netherlands Antilles. *Bull. Mar. Sci.* **56**, 609–619.

Berkelmans, R. (2002). Time-integrated thermal bleaching thresholds of reefs and their variation on the Great Barrier Reef. *Mar. Ecol. Prog. Ser.* **229**, 73–82.

Birkeland, C. (1977). The importance of rate of biomass accumulation in early successional stages of benthic communities to the survival of coral recruits. *Proceedings of the 3rd International Coral Reef Symposium.* 16–21.

Black, K.P., Moran, P.J., and Hammond, L.S. (1991). Numerical models show coral reefs can be self-seeding. *Mar. Ecol. Prog. Ser.* **74**, 1–11.

Bosscher, H., and Meesters, E.H. (1993). Depth related changes in the growth rate of *Montastraea annularis*. *7th International Coral Reef Symposium.* **1**, 507–512.

Bruggemann, J.H., Kuyper, M.W.M., and Breeman, A.M. (1994a). Comparative analysis of foraging and habitat use by the sympatric Caribbean parrotfish *Scarus vetula* and *Sparisoma viride* (Scaridae). *Mar. Ecol. Prog. Ser.* **112**, 51–66.

Bruggemann, J.H., Van Oppen, M.J.H., and Breeman, A.M. (1994b). Foraging by the stoplight parrotfish *Sparisoma viride*. I. Food selection in different socially determined habitats. *Mar. Ecol. Prog. Ser.* **106**, 41–55.

Buss, L.W., and Jackson, J.B.C. (1979). Competitive networks: Nontransitive competitive relationships in cryptic coral reef environments. *The American Naturalist.* **113**, 223–234.

Bythell, J.C., Gladfelter, E.H., and Bythell, M. (1993). Chronic and catastrophic natural mortality of three common Caribbean reef corals. *Coral Reefs.* **12**, 143–152.

Bythell, J.C., Hillis–Starr, Z.M., and Rogers, C.S. (2000). Local variability but landscape stability in coral reef communities following repeated hurricane impacts. *Marine Ecology Progress Series.* **204**, 93–100.

Caley, M.J., Carr, M.H., Hixon, M.A., Hughes, T.P., Jones, G.P., and Menge, B.A. (1996). Recruitment and the local dynamics of open marine populations. *Annu. Rev. Ecol. Syst.* **27**, 477–500.

Carleton, J.H., and Sammarco, P.W. (1987). Effects of substratum irregularity on success of coral settlement: Quantification by comparative geomorphological techniques. *Bull. Mar. Sci.* **40**, 85–98.

Carlon, D.B., and Olson, R.R. (1993). Larval dispersal distance as an explanation for adult spatial pattern in two Caribbean reef corals. *J. Exp. Mar. Biol. Ecol.* **173**, 247–263.

Carpenter, R.C. (1985). Sea urchin mass-mortality: Effects on reef algal abundance, species composition, and metabolism and other coral reef herbivores. *Proceedings of the Fifth International Coral Reef Congress, Tahiti.* **4**, 53–59.

Carpenter, R.C. (1986). Partitioning herbivory and its effects on coral reef algal communities. *Ecol. Monogr.* **56**, 345–363.

Carpenter, R.C. (1988). Mass mortality of a Caribbean sea urchin: Immediate effects on community metabolism and other herbivores. *Proceedings of the National Academy of Sciences of the United States of America.* **85**, 511–514.

Carpenter, R.C. (1990). Mass mortality of *Diadema antillarum*. II. Effects on population densities and grazing intensity of parrotfishes and surgeonfishes. *Mar. Biol.* **104**, 79–86.

Choat, J.H., Robertson, D.R., Ackerman, J.L., and Posada, J.M. (2003). An age-based demographic analysis of the Caribbean stoplight parrotfish *Sparisoma viride*. *Mar. Ecol. Prog. Ser.* **246**, 265–277.

Chornesky, E.A. (1989). Repeated reversals during spatial competition between corals. *Ecology.* **70**, 843–855.

Chornesky, E.A., and Peters, E.C. (1987). Sexual reproduction and colony growth in the scleractinian coral *Porites astreoides*. *Biol. Bull.* **172**, 161–177.

Connell, J.H. (1978). Diversity in tropical rain forests and coral reefs. *Science.* **199**, 1302–1309.

Connell, J.H., Hughes, T.P., and Wallace, C.C. (1997). A 30-year study of coral abundance, recruitment, and disturbance at several scales in space and time. *Ecol. Monogr.* **67**, 461–488.

Connolly, S.R., and Muko, S. (2003). Space preemption, size-dependent competition, and the coexistence of clonal growth forms. *Ecology.* **84**, 2979–2988.

Cowen, R.K., Lwiza, K.M.M., Sponaugle, S., Paris, C.B., and Oldson, D.B. (2000). Connectivity of marine populations: Open or closed? *Science.* **287**, 857–859.

D'Elia, C.F., and Wiebe, W.J. (1990). Biogeochemical nutrient cycles in coral-reef ecosystems. In *Coral reefs* (Z. Dubinsky, ed., vol. 25, pp. 49–74). Elsevier Science Publishers.

De Ruyter Van Steveninck, E.D., and Bak, R.P.M. (1986). Changes in abundance of coral-reef bottom components related to mass mortality of the sea urchin *Diadema antillarum*. *Mar. Ecol. Prog. Ser.* **34**, 87–94.

De Ruyter Van Steveninck, E.D., and Breeman, A.M. (1987). Deep water populations of *Lobophora variegata* (Phaeophyceae) on the coral reef of Curaçao: Influence of grazing and dispersal on distribution patterns. *Mar. Ecol. Prog. Ser.* **38**, 241–250.

De Ruyter Van Steveninck, E.D., Van Mulekom, L.L., and Breeman, A.M. (1988). Growth inhibition of *Lobophora variegata* (Lamouroux) Womersley by scleractinian corals. *J. Exp. Mar. Biol. Ecol.* **115**, 169–178.

Diaz–Pulido, G., and McCook, L.J. (2003). Relative roles of herbivory and nutrients in the recruitment of coral reef seaweeds. *Ecology.* **84**, 2026–2033.

Done, T. (1992). Constancy and change in some Great Barrier Reef coral communities: 1980–1990. *Am. Zool.* **32**, 655–662.

Done, T.J. (1987). Simulation of the effects of *Acanthaster planci* on the population structure of massive corals in the genus *Porites*: Evidence of population resilience? *Coral Reefs.* **6**, 75–90.

Done, T.J. (1988). Simulation of recovery of predisturbance size structure in populations of *Porites* spp. damaged by the crown of thorns starfish *Acanthaster planci. Mar. Biol.* **100**, 51–61.
Done, T.J., Ogden, J.C., Wiebe, W.J., and Rosen, B.R. (1996). Biodiversity and ecosystem function of coral reefs. In *Functional roles of biodiversity: A global perspective* (H.A. Mooney, J.H. Cushman, E. Medina, O.E. Sala, and E.-D. Schulze, eds., pp. 393–429). John Wiley and Sons.
Done, T.J., and Potts, D.C. (1992). Influences of habitat and natural disturbances on contributions of massive *Porites* corals to reef communities. *Mar. Biol.* **114**, 479–493.
Dytham, C. (2000). Habitat destruction and extinctions: Predictions from metapopulation models. In *British ecological society, annual symposium 1999: Ecological consequences of habitat heterogeneity* (M.J. Hutchings, E.A. John, and A.J.A. Stewart, eds., pp. 315–327). Oxford: Blackwell Scientific Publications.
Edmunds, P.J. (2002). Long-term dynamics of coral reefs in St. John, U.S. Virgin Islands. *Coral Reefs.* **21**, 357–367.
Edmunds, P.J., and Carpenter, R.C. (2001). Recovery of *Diadema antillarum* reduces macroalgal cover and increases abundance of juvenile corals on a Caribbean reef. *Proceedings of the National Academy of Sciences of the United States of America.* **98**, 5067–5071.
Edmunds, P.J., and Witman, J.D. (1991). Effect of hurricane Hugo on the primary framework of a reef along the south shore of St. John, U.S. Virgin Islands. *Mar. Ecol. Prog. Ser.* **78**, 201–204.
Fisk, D.A., and Harriott, V.J. (1990). Spatial and temporal variation in coral recruitment on the Great Barrier Reef: Implications for dispersal hypotheses. *Mar. Biol.* **107**, 485–490.
Gardner, T.A., Cote, I.M., Gill, J.A., Grant, A., and Watkinson, A.R. (2003). Long-term region-wide declines in Caribbean corals. *Science.* **301**, 958–960.
Geister, J. (1977). The influence of wave exposure on the ecological zonation of Caribbean coral reefs. *Proc. 3rd International Coral Reef Symposium.* **1**, 23–30.
Gleason, M.G. (1996). Coral recruitment in Moorea, French Polynesia: The importance of patch type and temporal variation. *J. Exp. Mar. Biol. Ecol.* **207**, 79–101.
Goldenberg, S.B., Landsea, C.W., Mestas–Nunez, A.M., and Gray, W.M. (2001). The recent increase in atlantic hurricane activity: Causes and implications. *Science.* **293**, 474–479.
Goreau, T.J., and Macfarlane, A.H. (1990). Reduced growth rate of *Montastraea annularis* following the 1987–1988 coral-bleaching event. *Coral Reefs.* **8**, 211–215.
Grigg, R.W., Polovina, J.J., and Atkinson, M.J. (1984). Model of a coral reef ecosystem III. Resource limitation, community regulation, fisheries yield and resource management. *Coral Reefs.* **3**, 23–27.
Hall, V.R., and Hughes, T.P. (1996). Reproductive strategies of modular organisms: Comparative studies of reef-building corals. *Ecology.* **77**, 950–963.
Hamner, W.M., Jones, M.S., Carleton, J.H., Hauri, I., and Williams, D.M. (1988). Zooplankton, planktivorous fish and water currents on a windward reef face: Great Barrier Reef, Australia. *Bull. Mar. Sci.* **42**, 459–479.
Hanksi, I. (1999). *Metapopulation ecology.* Oxford: Oxford University Press.
Harrison, P.L., and Wallace, C.C. (1990). Reproduction, dispersal and recruitment of scleractinian corals. In *Ecosystems of the world 25: Coral reefs* (Z. Dubinsky, ed., pp. 133–207). Amsterdam: Elsevier.
Harvell, C.D., Mitchell, C.E., Ward, J.R., Altizer, S., Dobson, A.P., Ostfeld, R.S., and Samuel, M.D. (2002). Climate warming and disease risks for terrestrial and marine biota. *Science.* **296**, 2158–2162.
Hastings, A. (1980). Disturbance, coexistence, history, and competition for space. *Theor. Popul. Biol.* **14**, 380–395.
Hatcher, B.G., and Larkum, A.W.D. (1983). An experimental analysis of factors controlling the standing crop of the epilithic algal community on a coral reef. *J. Exp. Mar. Biol. Ecol.* **69**, 61–84.
Hay, M.E. (1984). Patterns of fish and urchin grazing on Caribbean coral reefs: Are previous results typical? *Ecology.* **65**, 446–454.

Hengeveld, R., and Hemerik, L. (2002). Biogeography and dispersal. In *British ecological society annual symposium 2001: Dispersal* (J.M.E.A. Bullock, ed., pp. 303–326). Oxford: Blackwell Scientific Publishers.
Highsmith, R.C. (1982). Reproduction by fragmentation in corals. *Mar. Ecol. Prog. Ser.* **7**, 207–226.
Highsmith, R.C., Lueptow, R.L., and Schonberg, S.C. (1983). Growth and bioerosion of three massive corals on the Belize barrier reef. *Mar. Ecol. Prog. Ser.* **13**, 261–271.
Hodgson, G. (1985). Abundance and distribution of planktonic coral larvae in Kaneohe Bay, Oahu, Hawaii. *Mar. Ecol. Prog. Ser.* **26**, 61–71.
Hoegh–Guldberg, O. (1999). Climate change, coral bleaching and the future of the world's coral reefs. *Mar. Freshw. Res.* **50**, 839–866.
Hoegh–Guldberg, O., Takabayashi, M., and Moreno, G. (1997). The impact of long-term nutrient enrichment on coral calcification and growth. *8th International Coral Reef Symposium.* **1**, 861–866.
Holm, E.R. (1990). Effects of density-dependent mortality on the relationship between recruitment and larval settlement. *Mar. Ecol. Prog. Ser.* **60**, 141–146.
Hughes, T.P. (1984). Population dynamics based on individual size rather than age: A general model with a reef coral example. *Am. Nat.* **123**, 778–795.
Hughes, T.P. (1985). Life histories and population dynamics of early successional corals. *Proceedings of the Fifth International Coral Reef Congress, Tahiti.* **4**, 101–106.
Hughes, T.P. (1989). Community structure and diversity of coral reefs: The role of history. *Ecology.* **70**, 275–279.
Hughes, T.P. (1994). Catastrophes, phase shifts, and large-scale degradation of a Caribbean coral reef. *Science.* **265**, 1547–1551.
Hughes, T.P., Baird, A.H., Dinsdale, E.A., Moltschaniwskyj, N.A., Pratchett, M.S., Tanner, J.E., and Willis, B.L. (1999). Patterns of recruitment and abundance of corals along the Great Barrier Reef. *Nature.* **397**, 59–63.
Hughes, T.P., Baird, A.H., Dinsdale, E., Moltschaniwskyj, N.A., Pratchett, M.S., Tanner, J.E., and Willis, B.L. (2000). Supply-side ecology works both ways: The link between benthic adults, fecundity, and larval recruits. *Ecology.* **81**, 2241–2249.
Hughes, T.P., and Connell, J.H. (1987). Population dynamics based on size or age? A reef-coral analysis. *Am. Nat.* **129**, 818–829.
Hughes, T.P., and Jackson, J.B.C. (1985). Population dynamics and life histories of foliaceous corals. *Ecol. Monogr.* **55**, 141–166.
Hughes, T.P., Reed, D.C., and Boyle, M.J. (1987). Herbivory on coral reefs: Community structure following mass mortalities of sea urchins. *J. Exp. Mar. Biol. Ecol.* **113**, 39–59.
Hughes, T.P., and Tanner, J.E. (2000). Recruitment failure, life histories, and long-term decline of Caribbean corals. *Ecology.* **81**, 2250–2263.
Hunte, W., and Wittenberg, M. (1992). Effects of eutrophication and sedimentation on juvenile corals. II. Settlement. *Mar. Biol.* **114**, 625–631.
Huston, M. (1985). Variation in coral growth rates with depth at Discovery Bay, Jamaica. *Coral Reefs.* **4**, 19–25.
Jackson, J.B.C., Budd, A.F., and Pandolfi, J.M. (1996). The shifting balance of natural communities? In Evolutionary paleobiology: In honor of James W. Valentine (D. Jablonski, D.H. Erwin, and J.H. Lipps, eds., pp. 89–122). Chicago: University of Chicago Press.
James, M.K., Armsworth, P.R., Mason, L.B., and Bode, L. (2002). The structure of reef fish metapopulations: Modeling larval dispersal and retention patterns. *Proc. R. Soc. Lond. Ser. B. Biol. Sci.* **269**, 2079–2086.
Johnson, C.R., and Seinen, I. (2002). Selection for restraint in competitive ability in spatial competition systems. *Proc. R. Soc. Lond. Ser. B Biol. Sci.* **269**, 655–663.
Jompa, J., and McCook, L.J. (2002). Effects of competition and herbivory on interactions between a hard coral and a brown alga. *J. Exp. Mar. Biol. Ecol.* **271**, 25–39.

Karlson, R.H., and Buss, L.W. (1984). Competition, disturbance and local diversity patterns of substratum-bound clonal organisms: A simulation. *Ecol. Model.* **23**, 243–255.

Karlson, R.H., and Jackson, J.B.C. (1981). Competitive networks and community structure: A simulation study. *Ecology* **62**, 670–678.

Knowlton, N. (1992). Thresholds and multiple stable states in coral reef community dynamics. *Am. Zool.* **32**, 674–682.

Knowlton, N. (2001). The future of coral reefs. *Proceedings of the National Academy of Sciences of the United States of America.* **98**, 5419–5425.

Knutson, T.R., Tuleya, R.E., Shen, W.X., and Ginis, I. (2001). Impact of CO_2-induced warming on hurricane intensities as simulated in a hurricane model with ocean coupling. *J. Clim.* **14**, 2458–2468.

Kojis, B.L., and Quinn, B.J. (1984). Seasonal and depth variation in fecundity of *Acropora palifera* at two reefs in Papua New Guinea. *Coral Reefs.* **3**, 165–172.

Kritzer, J.P., and Sale, P.F. (2004). Metapopulation ecology in the sea: From Levins' model to marine ecology and fisheries science. *Fish and Fisheries.* **4**, 1–10.

Largier, J.L. (2003). Considerations in estimating larval dispersal distances from oceanographic data. *Ecol. Appl.* **13**, S71–S89.

Lessios, H.A. (1988). Population dynamics of *Diadema antillarum* (Echinodermata: Echinoidea) following mass mortality in Panama. *Mar. Biol.* **99**, 515–526.

Lessios, H.A., Robertson, D.R., and Cubit, J.D. (1984). Spread of *Diadema* mass mortality through the Caribbean. *Science.* **226**, 335–337.

Levins, R. (1968). *Evolution in changing environments.* Princeton, NJ: Princeton University Press.

Levins, R. (1969). Some demographic and genetic consequences of environmental heterogeneity for biological control. *Bulletin of the Entomological Society of America.* **15**, 237–240.

Levitan, D.R. (1988). Algal–urchin biomass responses following mass mortality of *Diadema antillarum* Philippi at Saint John, U.S. Virgin Islands. *J. Exp. Mar. Biol. Ecol.* **119**, 167–178.

Lewis, J.B. (1974). The settlement behavior of planulae larvae of the hermatypic coral *Favia fragum* (esper). *J. Exp. Mar. Biol. Ecol.* **15**, 165–172.

Lewis, S.M. (1986). The role of herbivorous fishes in the organization of a Caribbean reef community. *Ecol. Monogr.* **56**, 183–200.

Liddell, W.D., and Ohlhorst, S.L. (1986). Changes in benthic community composition following the mass mortality of *Diadema* at Jamaica. *J. Exp. Mar. Biol. Ecol.* **95**, 271–278.

Lindquist, N., and Hay, M.E. (1996). Palatability and chemical defense of marine invertebrate larvae. *Ecol. Monogr.* **66**, 431–450.

Lirman, D., and Fong, P. (1997). Susceptibility of coral communities to storm intensity, duration and frequency. *8th International Coral Reef Symposium.* **1**, 561–566.

Littler, M.M., Littler, D.S., and Lapointe, B.E. (1993). Modification of tropical reef community structure due to cultural eutrophication: The southwest coast of Martinique. *7th International Coral Reef Symposium.* **1**, 335–343.

MacArthur, R.H., and Wilson, E.O. (1967). *The theory of island biogeography.* Princeton, NJ: Princeton University Press.

Maida, M., Coll, J.C., and Sammarco, P.W. (1994). Shedding new light on scleractinian coral recruitment. *J. Exp. Mar. Biol. Ecol.* **180**, 189–202.

Maida, M., Sammarco, P.W., and Coll, J.C. (1995). Effects of soft corals on scleractinian coral recruitment. I. Directional allelopathy and inhibition of settlement. *Mar. Ecol. Prog. Ser.* **121**, 191–202.

Massel, S.R., and Done, T.J. (1993). Effects of cyclone waves on massive coral assemblages on the Great Barrier Reef: Meteorology, hydrodynamics and demography. *Coral Reefs.* **12**, 153–166.

McClanahan, T.R. (1992). Resource utilization, competition, and predation: A model and example from coral reef grazers. *Ecol. Model.* **61**, 195–215.

McClanahan, T.R., Kamukuru, A.T., Muthiga, N.A., Gilagabher Yebio, M., and Obura, D. (1996). Effect of sea urchin reductions on algae, coral, and fish populations. *Conserv. Biol.* **10**, 136–154.

McGuire, M.P., and Szmant, A.M. (1997). Time course of physiological responses to NH4 enrichment by a coral–zooxanthellae symbiosis. *8th International Coral Reef Symposium.* **1**, 909–914.

Meesters, E.H., Wesseling, I., and Bak, R.P.M. (1997). Coral colony tissue damage in six species of reef-building corals: Partial mortality in relation with depth and surface area. *J. Sea Res.* **37**, 131–144.

Miller, R.J., Adams, A.J., Ogden, N.B., Ogden, J.C., and Ebersole, J.P. (2003). *Diadema antillarum* 17 years after mass mortality: Is recovery beginning on St. Croix? *Coral Reefs.* **22**, 181–187.

Moorsel, V.G.W.N.M. (1988). Early maximum growth of stong corals (Scleractinia) after settlement on artificial substrata on a Caribbean reef. *Mar. Ecol. Prog. Ser.* **50**, 127–136.

Morse, D.E., Hooker, N., Morse, A.N.C., and Jensen, R.A. (1988). Control of larval metamorphosis and recruitment in sympatric Agariciid corals. *J. Exp. Mar. Biol. Ecol.* **116**, 193–217.

Mumby, P.J. (1999a). Bleaching and hurricane disturbances to populations of coral recruits in Belize. *Marine Ecology Progress Series.* **190**, 27–35.

Mumby, P.J. (1999b). Can Caribbean coral populations be modeled at metapopulation scales? *Marine Ecology Progress Series* **180**, 275–288.

Mumby, P.J., Chisholm, J.R.M., Edwards, A.J., Clark, C.D., Roark, E.B., Andrefouet, S., and Jaubert, J. (2001). Unprecedented bleaching-induced mortality in *Porites* spp. at Rangiroa Atoll, French Polynesia. *Marine Biology.* **139**, 183–189.

Mumby, P.J., and Wabnitz, C.C.C. (2002). Spatial patterns of aggression, territory size, and harem size in five sympatric Caribbean parrotfish species. *Environmental Biology of Fishes.* **63**, 265–279.

Neumann, A.C., and Land, L.S. (1975). Lime mud deposition and calcareous algae in the bight of Abaco, Bahamas: A budget. *Journal of Sedimentary Petrology.* **45**, 763–786.

Pandolfi, J.M., Bradbury, R.H., Sala, E., Hughes, T.P., Bjorndal, K.A., Cooke, R.G., Mcardle, D., Mcclenachan, L., Newman, M.J.H., Parades, G., Warner, R.R., and Jackson, J.B.C. (2003). Global trajectories of the long-term decline of coral reef ecosystems. *Science.* **301**, 955–958.

Preece, A.L., and Johnson, C.R. (1993). Recovery of model coral communities: Complex behaviors from interaction of parameters operating at different spatial scales. In *Complex systems: From biology to computation* (D.G. Green and T. Bossomaier, eds., pp. 69–81). Amsterdam: IOS Press.

Richardson, L.L. (1998). Coral disease: What is really known? *Trends Ecol. Evol.* **13**, 438–443.

Richmond, R.H., and Hunter, C.L. (1990). Reproduction and recruitment of corals: Comparisons among the Caribbean, the tropical pacific, and the Red Sea. *Mar. Ecol. Prog. Ser.* **60**, 185–203.

Robertson, D.R. (1996). Interspecific competition controls abundance and habitat use of territorial Caribbean damselfishes. *Ecology.* **77**, 885–899.

Rogers, C.S. (1993). Hurricanes and coral reefs: The intermediate disturbance hypothesis revisited. *Coral Reefs.* **12**, 127–137.

Rogers, C.S., Fitz, H.C., Gilnack, M., Beets, J., and Hardin, J. (1984). Scleractinian coral recruitment patterns at Salt River submarine canyon, St. Croix, U.S. Virgin Islands. *Coral Reefs.* **3**, 69–76.

Rogers, C.S., Garrison, V., and Grober–Dunsmore, R. (1997). A fishy story about hurricanes and herbivory: Seven years of research on a reef in St. John. *8th International Coral Reef Symposium.* **1**, 555–560.

Ruesink, J.L. (1997). Coral injury and recovery: Matrix models link process to pattern. *J. Exp. Mar. Biol. Ecol.* **210**, 187–208.

Russ, G.R., and McCook, L.J. (1999). Potential effects of a cyclone on benthic algal production and yield to grazers on coral reefs across the central Great Barrier Reef. *J. Exp. Mar. Biol. Ecol.* **235**, 237–254.

Rylaarsdam, K.W. (1983). Life histories and abundance patterns of colonial corals on Jamaican reefs. *Mar. Ecol. Prog. Ser.* **13**, 249–260.

Sammarco, P.W. (1980). *Diadema* and its relationship to coral spat mortality: Grazing, competition, and biological disturbance. *J. Exp. Mar. Biol. Ecol.* **45**, 245–272.

Sammarco, P.W. (1991). Geographically specific recruitment and postsettlement mortality as influences on coral communities: The cross-continental shelf transplant experiment. *Limnol. Oceanogr.* **36**, 496–514.

Sammarco, P.W., and Andrews, J.C. (1988). Localized dispersal and recruitment in Great Barrier Reef corals: The Helix experiment. *Science* **239**, 1422–1424.

Sammarco, P.W., Levinton, J.S., and Ogden, J.C. (1974). Grazing and control of coral reef community structure by *Diadema antillarum* Philippi (Echinodermata: Echinoidea): A preliminary study. *J. Mar. Res.* **32**, 47–53.

Sebens, K.P. (1987). Competition for space: Effects of disturbance and indeterminate competitive success. *Theor. Popul. Biol.* **32**, 430–441.

Smith, S.R. (1992). Patterns of coral recruitment and postsettlement mortality on Bermuda's reefs: Comparisons to Caribbean and Pacific reefs. *Am. Zool.* **32**, 663–673.

Smith, S.R. (1997). Patterns of coral settlement, recruitment and juvenile mortality with depth at conch reef, Florida. *Proceedings of the 8th International Coral Reef Symposium.* **2**, 1197–1202.

Soong, K. (1993). Colony size as a species character in massive reef corals. *Coral Reefs.* **12**, 77–83.

Soong, K.Y., and Lang, J.C. (1992). Reproductive integration in reef corals. *Biol. Bull.* **183**, 418–431.

Steneck, R.S. (1988). Herbivory on coral reefs: A synthesis. *Proceedings of the 6th International Coral Reef Symposium, Australia.* **1**, 37–49.

Steven, A.D.L., and Broadbent, A.D. (1997). Growth and metabolic responses of *Acropora palifera* to long term nutrient enrichment. *8th International Coral Reef Symposium.* **1**, 867–872.

Stiger, V., and Payri, C.E. (1999). Spatial and temporal patterns of settlement of the brown macroalgae *Turbinaria ornata* and *Sargassum mangarevense* in a coral reef on Tahiti. *Mar. Ecol. Prog. Ser.* **191**, 91–100.

Stone, L. (1995). Biodiversity and habitat destruction: A comparative study of model forest and coral reef ecosystems. *Proc. R. Soc. Lond. Ser. B Biol. Sci.* **261**, 381–388.

Szmant, A.M. (1986). Reproductive ecology of Caribbean reef corals. *Coral Reefs.* **5**, 43–53.

Szmant, A.M. (1991). Sexual reproduction by the Caribbean reef corals *Montastraea annularis* and *M. cavernosa*. *Mar. Ecol. Prog. Ser.* **74**, 13–25.

Tanner, J.E. (1995). Competition between scleractinian corals and macroalgae: An experimental investigation of coral growth, survival and reproduction. *J. Exp. Mar. Biol. Ecol.* **190**, 151–168.

Tilman, D., Lehman, C.L., and Nowak, M.A. (1997). Habitat destruction, dispersal, and deterministic extinction in competitive communities. *Am. Nat.* **149**, 407–435.

Tilman, D., May, R.M., Lehman, C.L., and Nowak, M.A. (1994). Habitat destruction and extinction debt. *Nature.* **371**, 65–66.

Tomascik, T., and Sander, F. (1985). Effects of eutrophication on reef-building corals. I. Growth rate of the reef-building coral *Montastraea annularis*. *Mar. Biol.* **87**, 143–155.

Tomascik, T., and Sander, F. (1987). Effects of eutrophication on reef-building corals. III. Reproduction of the reef-building coral. *Porites porites*. *Mar. Biol.* **94**, 77–94.

Treml, E., Colgan, M., and Keevican, M. (1997). Hurricane disturbance and coral reef development: A geographic information system (GIS) analysis of 501 years of hurricane data from the Lesser Antilles. *8th International Coral Reef Symposium.* **1**, 51–546.

Van Moorsel, G.W.N.M. (1988). Early maximum growth of stony corals (Scleractinia) after settlement on artificial substrata on a Caribbean reef. *Mar. Ecol. Prog. Ser.* **50**, 127–135.

Van Rooij, J.M., De Jong, E., Vaandrager, F., and Videler, J.J. (1996a). Resource and habitat sharing by the stoplight parrotfish, *Sparisoma viride*, a Caribbean reef herbivore. *Environ. Biol. Fishes.* **47**, 81–91.

Van Rooij, J.M., Kok, J.P., and Videler, J.J. (1996b). Local variability in population structure and density of the protogynous reef herbivore *Sparisoma viride*. *Environ. Biol. Fishes.* **47**, 65–80.

Van Woesik, R. (2002). Processes regulating coral communities. *Comments on Theoretical Biology.* **7**, 201–214.
Veron, J.E.N. (1995). *Corals in space and time: The biogeography and evolution of the Scleractinia.* New York: Cornell University Press.
Victor, B.C. (1986). Larval settlement and juvenile mortality in a recruitment-limited coral reef fish population. *Ecol. Monogr.* **56**, 145–160.
Ward, S., and Harrison, P.L. (1997). The effects of elevated nutrient levels on settlement of coral larvae during the ENCORE experiment, Great Barrier Reef, Australia. *Proc. 8th International Coral Reef Symposium.* **1**, 891–896.
Warner, R.R., and Hughes, T.P. (1988). The population dynamics of reef fishes. *Proceedings of the 6th International Coral Reef Symposium, Australia.* **1**, 149–155.
Weins, J.A. (1997). Metapopulation dynamics and landscape ecology. In *Metapopulation biology: Ecology, genetics and evolution* (I. Hanski and M.E. Gilpin, eds., pp. 43–62). San Diego: Academic Press.
West, J.M., and Salm, R.V. (2003). Resistance and resilience to coral bleaching: Implications for coral reef conservation and management. *Conserv. Biol.* **17**, 956–957.
Westneat, M., and Resing, J.M. (1988). Predation on coral spawn by planktivorous fish. *Coral Reefs.* **7**, 89–92.
Wilkinson, C.R. (1987). Interocean differences in size and nutrition of coral reef sponge populations. *Science.* **236**, 1654–1657.
Williams, I.D., and Polunin, N.V.C. (2000). Large-scale associations between macroalgal cover and grazer biomass on mid-depth reefs in the Caribbean. *Coral Reefs.* **19**, 358–366.
Williams, I.D., Polunin, N.V.C., and Hendrick, V.J. (2001). Limits to grazing by herbivorous fishes and the impact of low coral cover on macroalgal abundance on a coral reef in Belize. *Marine Ecology Progress Series.* **222**, 187–196.
Willis, B.L., and Oliver, J.K. (1988). Inter-reef dispersal of coral larvae following the annual mass spawning on the great barrier reef. *Proc. 6th International Coral Reef Symposium.* **2**, 853–859.
Wolanski, E., Richmond, R.H., and McCook, L.J. (2004). A model of the effects of land-based, human activities on the health of coral reefs in the Great Barrier Reef and in Fouha Bay, Guam, Micronesia. *Journal of Marine Systems.* In press.
Woodley, J.D. (1992). The incidence of hurricanes on the north coast of Jamaica since 1870: Are the classic reef descriptions atypical? *Hydrobiologia.* **247**, 133–138.
Woodley, J.D., Chornesky, E.A., Clifford, P.A., Jackson, J.B.C., Kaufman, L.S., Knowlton, N., Lang, J.C., Pearson, M.P., Porter, J.W., Rooney, M.C., Rylaarsdam, K.W., Tunnicliffe, V.J., Wahle, C.M., Wulff, J.L., Curtis, A.S.G., Dallmeyer, M.D., Jupp, B.P., Koehl, M.A.R., Neigel, J., and Sides, E.M. (1981). Hurricane Allen's impact on Jamaican coral reefs. *Science.* **214**, 749–755.
Wooldridge, S., and Done, T.J. (2004). Learning to predict large-scale coral bleaching from past events: A bayesian approach using remotely sensed data, in-situ data, and environmental proxies. *Coral Reefs.* In press.

CHAPTER 6

Population and Spatial Structure of Two Common Temperate Reef Herbivores: Abalone and Sea Urchins

LANCE E. MORGAN and SCORESBY A. SHEPHERD

I. Introduction
II. Life History
   A. Abalone
   B. Sea Urchins
III. Larval Dispersal and Settlement
   A. Abalone
   B. Sea Urchins
   C. Summary of Larval Dispersal
IV. Population Genetics
   A. Abalone
   B. Sea Urchins
   C. Summary of Genetics
V. Spatial Variability in Adult Distributions and Demographics
   A. Abalone: Adult Habitat and Spatial Structure
   B. Sea Urchins: Adult Habitat and Spatial Structure
VI. Fishing and Management
   A. Abalone
   B. Sea Urchins
   C. Optimal Harvesting of Invertebrate Metapopulations
VII. Summary
VIII. Acknowledgment
   References

I. INTRODUCTION

The spatial organization of many benthic marine invertebrates, and its importance for the management of exploited species, have emerged as key issues in recent years. The metapopulation concept, developed to describe such spatial structures, however, needs clarification despite its common use. As Levins (1969) first described the concept, a metapopulation was depicted as a population of populations in which each individual population was subject to extinction and recolonization. Subsequent authors have variously defined metapopulation, but most theorists have maintained that in order for a metapopulation to be considered valid, these subpopulations must have a nonzero probability of extinction. Increasingly, however, empirical reviews have used a less restrictive definition of metapopulation typically invoking the concept to describe spatial variability in population dynamics of an interconnected set of population units, often termed *local populations* (Kritzer and Sale, 2004). With maximum longevities of 15 to 30 years for abalone, and possibly 100 to 200 years for red sea urchins (Ebert and Southon, 2003), local populations are likely not prone to extinction, in the absence of fishing, except in unusual cases of decimation by disease or anthropogenic disturbance (reviewed by Shepherd and Breen, 1992). Similarly, local population extinction for sea urchins is unlikely because they can exist in poor-quality habitats for years, and individual sea urchins may shrink in size as they await chance encounters with drift algae (Ebert, 1980; Edwards and Ebert, 1991; Wing et al., 2001).

For the purposes of our discussion, we define a metapopulation of benthic invertebrates as a number of relatively sedentary adult local populations linked by dispersal of a planktonic larval stage (sensu Hanski and Gilpin, 1991; Botsford et al., 1994; Gaines and Lafferty, 1995). In these situations, local birth rates may be little to largely dependent on immigrant larvae from other connected local populations (Grimm et al., 2003; Kritzer and Sale, 2004). This is not strictly in accord with the classic Levins definition, but is consistent with the use of the term in most marine applications and the literature on benthic invertebrates. It does not agree with the premise of those authors who define metapopulations based on the nonzero probability of extinction within one or more local populations (see Smedbol et al., 2002).

Sea urchins and abalone are important members of rocky subtidal environments throughout temperate oceans of the world. They share a number of characteristics, including being economically important fisheries, ecologically significant herbivores of macroalgae, having planktonic larvae, and being relatively long lived and sedentary as adults. At the same time, there are important differences in their dispersal ability that influence their spatial distribution and fishery management. Because of their high unit value, stocks of all larger species have been exploited around the world, many to commercial extinction, and the remainder to various degrees of decline. This has led to an interest in the application of metapopulation theory to these organisms in an effort to manage their fisheries better or to restore collapsed ones. Hence, our focus is necessarily on fishing as the major threat to population viability, rather than environmental or demographic stochasticity, habitat fragmentation or transience, or natural enemies as in terrestrial landscapes (see Shaffer, 1981).

Application of metapopulation theory to abalone and sea urchins is in its infancy, largely as a result of the difficulties in measuring larval dispersal and in identifying larval sources. Indeed, much of the information from which metapopulation processes are inferred is derived from studies directed at answering more general questions about larval ecology and dispersal. But first we must define the terms that we use to describe the spatial pattern at which these species occur.

At the smallest scale of patches, abalone and sea urchins each form clusters, in groups of a few to hundreds of individuals over a few to tens of square meters. Local populations are clusters of patches in a patchy seascape, which is frequently a mosaic of rock, sand, and sometimes sea grass habitats (McShane, 1995a). Continuity of a local population is inferred from continuity of habitat, and adjacent local populations are separated by unsuitable unoccupied habitat. The distinction between patch and local population is in part arbitrary, but is essentially defined by the scale of adult movement. Adults may move between patches, but not between local populations. By definition, local populations within a metapopulation are connected by larval dispersal. The three scales—patch, local population, and metapopulation—are illustrated schematically in Fig. 6-1.

We now review the literature on larval dispersal, genetics, adult populations, and spatial variability in demographic rates as evidence justifying the application of metapopulation theory to these organisms. We consider the larval behavior, dispersal patterns, and recruitment in the light of regional hydrodynamics and coastal topography, as well as genetic differences between populations and the effects of fishing, and then infer the resulting spatial structure of adult populations. Throughout this chapter we use the term *recruitment* to refer to the establishment of juveniles into adult habitats, rather than entry into the fishery. Although we focus extensively on two common Australian abalone species, greenlip (*Haliotis laevigata*) and blacklip (*H. rubra*), and sea urchins of the genus *Strongylocentrotus* in North America, we believe our results are widely applicable to other exploited haliotids and echinoids.

II. LIFE HISTORY

A. ABALONE

The existence of fisheries for abalone (genus *Haliotis*) in Australia (*H. laevigata, H. rubra,* and *H. roei*), New Zealand (*H. iris*), South Africa (*H. midae*), Japan (mainly *H. discus discus* and *H. d. hannai*), and from Alaska (*H. kamtschatkana*) to Baja California (*H. cracherodii, H. rufescens, H. sorenseni, H. corrugate,* and

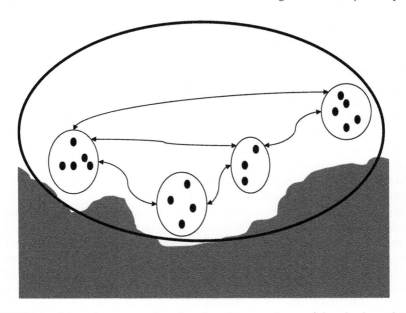

FIGURE 6-1. Schematic representation of patches, shown as clumps of dots, local populations (circles surround patches), and metapopulation (circle in bold surrounding local populations). Lines with arrows indicate connections via larval dispersal. (Morgan and Botsford, 2001. Redrawn with permission.)

H. fulgens), has prompted much research on their population biology (Shepherd et al., 1992b) and more recently their spatial pathology (Karpov et al., 2000; Shepherd and Rodda, 2001; Shepherd et al., 2001; Rogers–Bennett et al., 2002a,b). Exploited abalone species are benthic, prosobranch mollusks with sedentary and long-lived adults that live attached to rock. Maximum longevities are often 15 to 30 years and movement is limited to a few to tens of meters a year. Dispersal takes place during the pelagic larval stage lasting 5 to 7 days, rarely to 12 days in colder waters, and after settlement juveniles assume the sedentary behavior of adults.

B. SEA URCHINS

Sea urchins are commercially valuable echinoderms that are exploited in temperate waters throughout the world, primarily the North Atlantic (*S. droebachensis, Paracentrotus lividus*), North Pacific (*S. franciscanus, S. droebachensis, S. intermedius, S. nudus*. and to a lesser degree *S. purpuratus*), Chile (*Loxechinus albus*), and New Zealand (*Evechinus chloroticus*) (Andrew et al., 2002). For the most part, sea urchin fisheries show a poor history of sustained catch, and nearly all show a record of boom-and-bust landings. Sea urchins are also exploited in some tropical seas, but the focus of this review draws upon the extensive research in temperate reef systems, especially the Pacific coast of North America.

Sea urchins are relatively long lived, with maximum longevities ranging from 20 to 30 years for green sea urchins (Robinson and MacIntyre, 1997) to perhaps in excess of 100 years for red sea urchins (Ebert and Southon, 2003). Adult movement is limited to a few to several tens of meters a year. Dispersal takes place during the pelagic larval stage, lasting from a few weeks to a few months, and after settlement juveniles assume the sedentary behavior of adults.

III. LARVAL DISPERSAL AND SETTLEMENT

Sea urchins and abalone, and the vast majority of marine organisms (approximately 70%), have a dispersive planktonic larval stage. Starting with Hjort (1914), for most of the 20th century, recruitment of these larvae to a local population was considered the primary driver of population dynamics. Spatial patterns of larval dispersal and recruitment have attracted considerable attention because of their importance to management (Sinclair, 1988), and more recently to questions about the efficacy of marine reserves to manage or enhance overexploited populations (Polacheck, 1990; Dugan and Davis, 1993; Roberts, 1997). However, our ignorance of larval sources has delayed progress in determining the causes of recruitment variability. Although it is often assumed that recruitment is independent of local reproduction (Caley et al., 1996; Roughgarden et al., 1985), there is increasing evidence that physical (Sammarco and Andrews, 1988; Wolanski and Hamner, 1988; Wing et al., 1995b; Swearer et al., 1999) and behavioral (Neilson and Perry, 1990; Larson et al., 1994; Stobutzki and Bellwood, 1997) mechanisms can help maintain larvae close to their source populations.

Studies have inferred dispersal from a combination of pattern observations (e.g., size distributions or settlement) and process descriptions (e.g., circulation during planktonic period; Sale and Kritzer, 2003). Several factors influence larval dispersal, including behavior and duration of the larva, timing and duration of the spawning season, and coastal circulation. These topics as they relate to sea urchins and abalone are summarized in the following paragraphs.

A. ABALONE

1. Larval Ecology and Dispersal

Abalone larval behavior and dispersal are reviewed by McShane (1992, 1995b, 1996). Most species have a larval duration of about 5 days. During the initial larval stage of about 1 to 2 days, the trochophore tends to swim upward as a result of positive phototaxis or negative geotaxis, according to species. During the succeeding veliger stage, larvae alternately swim upward and then sink, in some species through positive geotaxis. In a detailed study by Madigan (2001) on the larval behavior of the greenlip, *H. laevigata*, and blacklip, *H. rubra*, the proportion of upward-swimming veligers declined with age so that by the fifth day veligers mostly swam downward, but with considerable variability between species, and between batches within species in swimming rate and direction. Given such variability within and between species, it is clear that species have the capacity to evolve behaviors to maximize survival, which may entail longer distance and/or short-distance dispersal. The former is an important component of fitness (Scheltema, 1986) and favors genetic variability, whereas the latter promotes local adaptations (Raimondi and Keough, 1990).

Abalone larval studies in the field show great variability in dispersal. One group of studies (Tanaka et al., 1986; Shepherd et al., 1992a; Sasaki and Shepherd, 1995) indicates larval dispersal at the scale of several kilometers, suggesting a diffusion model of dispersal. By this model, larvae are transported passively by water currents and concentrated in eddies, stagnation zones, and by coastal topography into particular habitats. Greenlip larvae may be transported several kilometers in tidal currents (Rodda et al., 1997; Shepherd et al., 1992a). In *Haliotis discus hannai*, spawning occurs during typhoons, and larvae are transported several kilometers offshore in rip currents, to be returned to the coast by wind-driven currents (Sasaki and Shepherd, 1995).

Other studies show localized larval dispersal at the scale of a few to tens of meters (Prince et al., 1987, 1988; reviews of McShane, 1992, 1995b); this is termed the *philopatric model*. In studies of blacklip abalone, larvae released in macroalgal forests on reefs were retained within the reef system, largely because of the local hydrodynamics (McShane et al., 1988; and discussed later). Several factors may then operate to minimize dispersal. Abalone larvae (both trochophores and veligers) cease swimming when they encounter turbulence or a hard surface (Crofts, 1938), a behavior possibly enhanced by the inhibiting effect of natural settlement inducers in crustose coralline algae (GABA mimics; see Yano and Ogawa, 1977, and Barlow, 1990). If crustose coralline algae do inhibit swimming, then settlement may occur soon after the trochophore encounters the substratum, as reported for a trochid larva (Heslinga, 1981; and see Prince et al., 1988; Morse, 1992).

The spawning behavior of adult abalone and the season of spawning are other factors affecting larval dispersal. For example, *H. kamtschatkana* and blacklip spawn in very calm weather (Breen and Adkins, 1980; McShane et al., 1988; Prince et al., 1987). In the case of *H. kamtschatkana*, dispersal is maximized because the abalone climbs to the top of a prominence and spawns into the current, but in the case of blacklip, dispersal is reduced in a calm embayment (Prince et al., 1987). In northern China, *H. discus hannai* migrates into shallow water for spawning presumably to increase dispersal in more turbulent conditions (Nie, 1992; Sasaki and Shepherd, 1995). In greenlip, spawning in different populations apparently occurs at different parts of the lunar cycle. Dispersal is least when larvae are released during periods of weak tidal currents, and increases when released into strong tidal currents (Shepherd and Daume, 1996). *H. discus hannai* also has the capacity for both short-and longer distance dispersal, because spawning during typhoons dispersed larvae via rip currents for some kilometers alongshore, whereas spawning during lesser storm events achieved local dispersal (Sasaki and Shepherd, 1995).

The duration of the spawning season may also play an important role in dispersal. For example, *H. cracherodii*, with a narrow summer spawning season, shows much greater genetic spatial differentiation than *H. rufescens*, which spawns throughout the year (Boolootian et al., 1962; Tegner, 1993). In the former case larvae are exposed to a much more limited range of oceanographic conditions compared with larvae of the latter species. Similarly, *H. fulgens* and *H. corrugata*, with short spawning seasons and a shorter larval life, have less dispersal ability than *H. rufescens* (Tegner and Butler, 1985; Tegner, 1993).

2. Hydrodynamics and Coastal Topography

The ability of larvae to swim vertically up or down enables them to locate themselves in any part of the water column, as shown by Sasaki and Shepherd (1995). In that study larvae, when competent to settle, became concentrated at a depth where they were returned to shore by wind-driven currents. Hence, larval behavior and timing of spawning interact with the local hydrodynamics and coastal topography to determine dispersal distance.

The strength of coastal currents, wind driven or tidal, sets an upper limit on the distance larvae can be advected. Coastal features such as bays, inlets, headlands, and small islands generate sheer zones, convergences, eddy systems with countercurrents, and stagnation zones, each of which may variously concentrate or retain larvae (Hamner and Hauri, 1977; Alldredge and Hamner, 1980; Tanaka et al., 1986; Black, 1988; Shepherd et al., 1992a; Largier, 2003). In exposed kelp habitats, nearshore currents are attenuated by swell and reef topography (Eckman, 1983; Eckman et al., 1989; Black and McShane, 1990; McShane, 1992; Jackson, 1998; Guzmán del Próo et al., 2000), so that larvae are retained near their natal reef (McShane et al., 1988; Prince, 2003). In some high wave energy conditions, offshore-flowing rip currents that weaken in deeper water are developed, giving rise to circulation cells as large as 500 m diameter that correspond with local abalone beds (Itosu and Miki, 1983). Drift tube studies mimicking larval dispersal of *H. fulgens* suggested little mixing between offshore and mainland populations off southern California, but considerable mixing alongshore over tens of kilometers (Tegner and Butler, 1985; and see Gaines et al., 2003).

With increasing distance from shore, coastal features assume less, and bottom topography more, significance (Grant and Madsen, 1979). Horseshoe vortices and wakes form around benthic reef structures of even low relief (Davis, 1986), and these reduce water movement and concentrate larvae (Shepherd et al., 1992a).

Next we consider two important aspects of larval settlement, which influence choice of settlement sites in abalone habitat.

3. Larval Behavior at Settlement

Both physical and biotic factors may influence abalone larval settlement (reviewed by Roberts, 2001). Water turbulence or strong current, and at the other extreme, sedimentary environs, may prevent larval settlement (Jonsson et al., 1991; Nash, 1992; reviews by McShane, 1993, 1995b; Shepherd and Partington, 1995). Of the many important biotic factors, Daume et al. (1999a,b) found experimentally that certain types of crustose coralline algae, some diatom films, and the presence of older conspecifics each induced settlement of greenlip larvae. The last-named effect is explicable in terms of the slime trail hypothesis of Seki and Kan-no (1981; see Slattery, 1992; reviews of McShane, 1992; Roberts, 2001), according to which conspecifics induce or favor larval settlement.

Field experiments have confirmed the facilitative role of conspecifics. Tutschulte (1976) removed adult abalone from experimental plots of $25\,m^2$ and found that recruitment failed in the plots. Conversely, Andrew et al. (1998) and T. Adams (personal communication) placed adult abalone in habitats from which they had disappeared, and found that recruitment increased many times. Note, however, that sometimes manipulations of adult spawners (but not subadults), as in the removal experiments of Prince et al. (1987, 1988), are best explained in terms of larval retention near natal sites. A conclusion from these studies is that, for some abalone species, the persistence of patches is in double jeopardy, depending both on larval availability and on larval behavior, where it is influenced by conspecifics. Other postsettlement factors, such as topographic complexity, food availability and abundance, and numbers of competitors or predators, are also important but are not considered here (reviewed by Shepherd et al., 2000; Sasaki and Shepherd, 2001).

B. Sea Urchins

1. Larval Ecology and Dispersal

Along the Pacific coast of North America, sea urchins in the genus *Strongylocentrotus* range from Baja California, to southeast Alaska, approximately 30° of latitude (Kato and Schroeter, 1985). Adult populations are neither entirely isolated patches nor are they continuous; rather, they represent a metapopulation of loosely connected local populations connected by larval dispersal (see Kritzer and

Sale, 2004). Adult populations may be entirely restricted to isolated habitat patches, (e.g., kelp beds) or more continuously distributed along larger regions of appropriate seafloor habitat. Larval dispersal patterns, influenced by coastal circulation, result in patchy settlement, which influences the abundance and population dynamics at recruitment sites (Ebert and Russell, 1988; Wing et al., 1995b; Morgan et al., 1999, 2000b; Wing et al., 2003b).

Typical of sea urchins in general, strongylocentrotids have long-lived planktonic larvae, which may link local populations across large distances. There is a seasonal pattern in the size of gonads and maturity of gametes, with development in late winter and an apparent peak in spawning during the spring (Bernard, 1977; Kato and Schroeter, 1985; Pearse and Cameron, 1991; Wing et al., 1995b). Red sea urchins spawn during the winter to early spring months, although the spawning season can be prolonged (Kato and Schroeter, 1985). Larvae of the red sea urchin may spend 7 to 19 weeks developing in the water column prior to settlement (Strathmann, 1978; Cameron and Schroeter, 1980; Strathmann, 1987; Rowley, 1989). During this time they are subject to dispersal in ocean currents that may transport them 100 to 1000 km away (Grantham et al., 2003). Green sea urchin larval growth in the water column is related to temperature and food availability (Hart and Scheibling, 1988). Because of the timing of spawning and larval development, red and purple sea urchins (*S. franciscanus* and *S. purpuratus* respectively) in the California Current system are subject to strong offshore transport following the spring transition to upwelling conditions (Ebert and Russell, 1988; Wing et al., 1995a; Morgan et al., 2000b).

Extreme recruitment variation appears to be a hallmark of sea urchin species (Lawrence, 2001). Strongylocentrotid sea urchins demonstrate episodic recruitment from California northward through British Columbia (Bernard and Miller, 1973; Pearse and Hines, 1987; Sloan et al., 1987; Ebert and Russell, 1988; Ebert et al., 1994; Wing et al., 1995a,b; Morgan et al., 2000b; Wing et al., 2003a), although more consistent annual recruitment is observed in southern California (Ebert et al., 1994). Additionally, atypical oceanographic events appear to influence recruitment (Ebert et al., 1994; Scheibling and Hennigar, 1997), including El Niño–southern oscillation events (Tegner and Dayton, 1991; Lundquist et al., 2000). Large interannual recruitment variability has also been observed in Japan for *S. nudus* (Agatsuma et al., 1998), in the Gulf of Maine for *S. droebachiensis* (Harris et al., 2001), in Chile for *Loxechinus albus*, and in Ireland for *Paracentrotus lividus* (Andrew et al., 2002, and references therein).

2. Hydrodynamics and Coastal Topography

Because sea urchin larvae are relatively small, they are often poorly sampled by shipboard collectors and cannot be reliably tagged. However, observations of spatial patterns of population size–frequency distributions and settlement have often been used to infer the influence of coastal circulation on dispersal patterns because of the relative ease in collecting these field data. Several studies in the California Current system have attempted the task of understanding the complex association between physical transport and the recruitment of red and purple sea urchins (Ebert and Russell, 1988; Ebert et al., 1994; Wing et al., 1995b; Miller and Emlet, 1997; Morgan et al., 2000b). Other meroplanktonic organisms such as crabs (Shanks, 1983; Hobbs et al., 1992; McConnaughey et al., 1992; Botsford et al., 1994, Wing et al., 1995a,b), barnacles (Connell, 1985; Gaines and Roughgarden, 1985; Gaines et al., 1985; Shanks, 1986; Farrell et al., 1991; Gaines and Bertness, 1992, 1993; Roughgarden et al., 1992), and rockfish (Ainley et al., 1993; Larson et al., 1994) have been similarly studied. Some understanding of the spatial variability in recruitment over the ranges of entire metapopulations connected by larval dispersal has been achieved for the organism being discussed (Botsford et al., 1994; Morgan et al., 2000b), but there still is very little empirical information with which to assess connectivity patterns.

For the past two decades, increased awareness and understanding of mesoscale (on the order of 100 km) variability in coastal circulation off northern California has focused attention on its implications for meroplanktonic larvae. The presence of strong seasonal upwelling led to the question of how meroplanktonic populations persist despite larvae being transported offshore (Parrish et al., 1981). Ebert and Russell (1988) sampled size distributions of purple sea urchins along the West Coast, from Oregon south to California, and concluded that recruitment is strongly influenced by coastal circulation associated with headlands. Based on the coefficients of variation of size distributions at various locations along the coast, they concluded that populations at and to the south of headlands had lower recruitment because of stronger offshore transport of larvae at these locations. Areas just to the north of these strong, offshore, upwelling jets had larger coefficients of variation in the sizes of purple sea urchins, which they concluded indicated higher recruitment. Their results, however, left unresolved questions regarding the mechanism that would lead to increasing recruitment from south of a headland to the next promontory, the method of estimating recruitment variability, and the long-term population implications of their assumptions. The explanation for lower recruitment to the south of headlands was that the features associated with headlands generally transport larvae southward in the California Current system. Although it is true that larvae would generally be transported southward in the California Current system, especially under the influence of the strong coastal current that accompanies upwelling, it does not provide a specific physical mechanism for formation of the gradient in settlement between headlands during active upwelling.

Synthesis of many more studies during the past decade indicates that the probable answer to this question of how meroplanktonic populations persist in strong upwelling regions lies in the combination of (1) the potential for larval retention in alongshore frontal zones (larval concentration areas that form offshore of upwelling jets; Richardson and Pearcy, 1977; Shenker, 1988) and "upwelling shadows" (eddies that concentrate larvae to the south of coastal promontories during upwelling events; Graham et al., 1992; Graham and Largier, 1997; Wing et al., 1998), and (2) the potential for subsequent transport of these concentrated larval patches to the coast in flow reversals, both cross-shelf and alongshore, during occasional periods of relaxation or reversal of upwelling winds (Roughgarden et al., 1992; Shanks, 1995; Wing et al., 1995a,b; Lundquist et al., 2000). Cross-shelf flows of water masses during relaxation of upwelling winds, typically lasting a few days, influence the timing of barnacle settlement in central California, producing pulses in recruitment (Farrell et al., 1991; Roughgarden et al., 1992). Larvae settled in the intertidal when the front containing larvae moved cross-shelf and collided with the shore. Upwelling relaxations also influence transport of pelagic juvenile rockfish in this region (Larson et al., 1994). Other mechanisms such as internal waves or tidal bores could play a role in the cross-shelf transport during upwelling (for a review, see Shanks, 1995).

In northern California, observations near a major promontory (Point Reyes) have shown that event-scale (3–10 days) variability in upwelling winds drives alongshore as well as cross-shelf transport of larvae. This results in similar temporal variability in the settlement of crab and sea urchin larvae, and a consistent spatial distribution of settlement along the coast (Wing et al., 1995a; Morgan et al., 2000b; Wing et al., 2003a). Observations of planktonic larval distributions and settlement on artificial collectors over space and time concurrent with measurements of hydrographic conditions have led to the conclusion that larvae are retained south of Point Reyes during upwelling winds (Wing et al., 1998), then are transported poleward and shoreward, where they settle during upwelling relaxations (Wing et al., 1995a,b).

On weekly time scales, crab (primarily *Cancer* spp.) settlement north of Point Reyes increased during periods of upwelling relaxation. Settlement was associated with a sharp increase in temperature (Wing et al., 1995a), suggesting larvae were transported poleward in the alongshore current (Morgan et al., 2000b). This coastally trapped flow may be buoyancy forced, driven in part by freshwater outflow from San Francisco Bay. More recent alongshore temperature monitoring indicates that the distance north of Point Reyes reached by this current varies, depending on the duration of the relaxation (Wing et al., 1995a). Drifter studies also support this pattern, as those released during the upwelling season generally moved south, but with occasional periods of northward movement (Davis 1985). The temporal pattern of sea urchin (*Strongylocentrotus* spp.) settlement was similar to that described for crabs, but was not as predictably tied to relaxation events, suggesting that larvae may either reside in a different water mass than crabs, or may be transported in a different manner as a result of biomechanical differences (e.g., size, swimming behavior).

Morgan et al. (2000b) tested whether spatial patterns in recruitment are caused by known aspects of coastal circulation in northern California. Their approach analyzed spatial patterns of recruitment from postsettlement observations of size distributions and density. In addition to the effects of circulation prior to settlement, postsettlement processes (e.g., space competition and predation) can also influence recruitment (Connell, 1985; Rowley, 1989; Holm, 1990). Morgan et al. (2000b) evaluated the effect of the protection of newly settled individuals beneath the spine canopy of adults (Tegner and Dayton, 1977) as an alternative explanation of the spatial pattern of recruitment. They found no correlation between adult spine canopy area or density with the abundance of small sea urchins (<30-mm test diameter) at the mesoscale (Morgan et al., 2000b). Although studies have reported higher juvenile sea urchin and juvenile abalone densities under the spine canopy of adult sea urchins (Tegner and Dayton, 1977; Rogers–Bennett and Pearse, 2001), it is not clear if this is a postsettlement effect and needs to be distinguished from larval delivery, or whether the presence of adults cues larval settlement, as noted for abalone.

Morgan et al. (2000b) interpreted their results in terms of the upwelling/relaxation mechanism observed near Point Reyes, California. This mechanism, based on alongshore flow during upwelling relaxation, explains the gradient in the observations by Ebert and Russell (1988), as well as the pattern of recruitment found in their more restricted study in northern California (see also Wing et al., 1995a,b, 1998; Lundquist et al., 2000). Studies of larval dispersal and settlement outside the California Current system similarly show the influence of coastal circulation on local population dynamics. Wing et al. (2003b) found distinct regions of reproductive sources and sinks for *Evechinus chloroticus* in New Zealand fjords.

3. Larval Behavior at Settlement

Harrold et al. (1991) discuss sea urchin recruitment in three components: larval delivery, settlement, and survival and growth of newly settled juveniles. Teasing apart these stages is difficult. Many studies have sampled early postsettlement densities of sea urchins (Ebert et al., 1994; Wing et al., 1995b; Miller and Emlet, 1997) without determining the effect of settlement on subsequent adult size–frequency distributions. Rowley (1989) found that larvae displayed no substrate settlement preference for kelp beds or barrens, and Cameron and Schroeter (1980) found that red sea urchin larvae showed no substratum selection for either adults or adult habitats. Rowley (1990) found that growth rate was higher for sea urchins settling in kelp forests than in barrens, and that this was an important factor in successful adult recruitment. It is unclear whether the association of juvenile sea urchins with adults is the result of juveniles seeking out adults or of reduced mortality of juveniles associated with adult spine canopies (Breen et al., 1985; Sloan et al., 1987).

Sea urchins may be subject to a postsettlement Allee effect (Tegner and Dayton, 1977), in which juvenile sea urchin survival is higher in areas where they can benefit from the protection of the adult spine canopy (Duggins, 1981). A study of red sea urchins at Bodega Bay in northern California showed that recruitment of juvenile sea urchins was highest in association with adults in shallow habitats, and adult sea urchins residing in shallow "rock bowl" habitats were 12 times more likely to shelter juveniles than more mobile adults in deep water (Rogers–Bennett et al., 1995). Studies have shown that spine canopies are also important for juvenile abalone survival along the coast of California (Rogers–Bennett and Pearse, 2001) and in South Africa (Day and Branch, 2000; Mayfield and Branch, 2000).

C. Summary of Larval Dispersal

Our review of larval dispersal suggests that the key to unlocking metapopulation dynamics of abalone and sea urchins rests on an understanding of their planktonic life history phases, because both recruitment patterns are strongly influenced by coastal circulation patterns. Yet there are important differences between the two groups. Abalone larval duration ranges from 5 to 12 days, whereas sea urchins range from 2 to 19 weeks. Thus, abalone disperse distances from 0.01 to 10 km, whereas sea urchins disperse distances potentially greater than 100 km. These distances fit well the log/log relation between dispersal distance and larval life as reviewed by Shanks et al. (2003). The differences in larval life between the two groups have led to different emphases and divergent findings regarding key factors, in part as a result of the different systems in which most studies have been done. For both taxa, coastal topography, swell, and current clearly impose bounds on the extent of larval dispersal, but especially for abalone biological factors (e.g., timing and duration of spawning, and larval ecology and behavior) have assumed greater importance in determining actual dispersal distances. Both sea urchin larvae and abalone larvae during their trochophore and veliger stages are dispersed passively according to local hydrodynamics. Yet the actual dispersal distance can be maximized or minimized, or kept variable, by intrinsic behavioral mechanisms both of the spawning adults and of the larva in any given water body.

Despite the variability, two generalizations emerge from abalone studies. First, larval dispersal distances tend to be greater when spawners are farther from the coast; conversely, dispersal distances tend to be lower when spawners are close to islands, elevated reefs, or indented coastlines with bays and headlands. From studies of sea urchins, it appears that their longer planktonic lives promote longer distance dispersal, and that recruitment to local populations is likely dependent on immigration of larvae from other local populations subject to the environmental influences and coastal circulation in a given year. Abalone demonstrate two distinct models of dispersal—diffusive and philopatric—in comparison with sea urchins, which at present appear to be almost entirely diffusive except in rare situations when larvae are constrained by circulation patterns to isolated fjords or sounds.

The optimal dispersal distance in terms of fitness will involve tradeoffs, and vary according to the local habitat and conditions (Hamilton and May, 1977; Roff, 2002). According to Iwasa and Roughgarden (1985), selection favors local settlement in source habitats, whereas Sasaki and Shepherd (1995) emphasized the selective advantages of mechanisms that favored high variability of dispersal distance.

Intriguingly, recent evidence, primarily from fish (e.g., Swearer et al., 1999; Cowan et al., 2003; Taylor and Hellberg, 2003), suggests that local retention of larvae is more important than previously recognized, and that long-distance transport from distant "source" populations is likely insufficient to sustain marine populations over demographic timescales. The consequences of this will have a large bearing on metapopulation models. Not surprisingly, given the lack of methods for tracking larvae, there is currently no means by which to understand the linkages between local populations. Should advances (such as microelemental analysis or genetics) allow larval tracking, we would still be hampered by the lack of constancy between years in larval survival and transport. Together, these studies leave many of the questions concerning larval connectivity open, and point to a need for continued research aimed at identifying larval sources and destinations. This question continues to be one of the outstanding challenges in marine ecology (Palumbi and Warner, 2003).

IV. POPULATION GENETICS

Given the difficulties in actually determining the realized dispersal distances and pathways of larval dispersal, scientists have sought to infer patterns of dispersal from genetics. A variety of genetic methods allow for estimation of differences among populations, and thereby an indirect estimate of population connectivity. Many studies have focused on sea urchin and abalone population genetics, and have led to new insights regarding dispersal.

A. Abalone

The genetic population structure has been examined for a number of abalone, variously with allozymes, mtDNA, and nuclear DNA, with no one method as yet demonstrably superior (Bossart and Powell, 1998; Conod et al., 2002; reviewed by Withler, 2000). Here we shall simply summarize authors' interpretative conclusions in terms of the estimated neighborhood size (N_m) or region of complete genetic mixing. Although species with small N_m are generally inferred to have small metapopulation ranges, and species with very large N_m have correspondingly large metapopulation ranges (Palumbi, 2003), there are numerous exceptions.

The intertidal species *H. cracherodii* and *H. roei* have N_m values of approximately 6 to 15 km (Hamm and Burton, 2000; Hancock, 2000), and for both species this corresponds to the patchy coastal intertidal reef systems that commonly stretch from a few to approximately 10 km along the coast, separated by stretches of sandy beaches or unsuitable habitat (R.S. Burton and B. Hancock respectively, personal communication). Two other shallow-water species, *H. diversicolor* and *H. discus hannai*, showed fixed genetic differences between neighboring populations approximately 35 km apart, suggesting a complete lack of genetic exchange at this scale (Hara and Kikuchi, 1992; Hara and Fujio, 1992; Jiang et al., 1995). Yet in blacklip, *H. rubra*, a species with philopatric larvae, Elliott et al. (2002), confirming Brown and Murray's (1992) results, found little differentiation and a very high N_m value of 500 km. These authors suggested a broad-scale isolation-by-distance model, in which recruitment was mainly local, but in some years influxes of larvae from distant reefs could strongly influence allelic frequencies and mask fine scale differentiation.

In contrast, some other species show very large values of N_m. Zuñiga et al. (2000) found little genetic differentiation in *H. fulgens* in populations separated by as much as 100 km, and in the South African *H. midae*, two major stocks were homogeneous over hundreds of kilometers (Evans et al., 2004). In British Columbia, *H. kamtschatkana* was genetically homogeneous over 800 km of coast, an unsurprising finding given a larval life of 8 days and the strong tidal currents flowing through the maze of islands (Withler et al., 2003). In California, Burton and Tegner (2000) suggested little genetic differentiation in *H. rufescens,* but later studies (R.S. Burton, personal communication) have indicated some local differentiation of stocks. Two deeper water species, the greenlip *H. laevigata* and the Mexican *H. corrugata* had anomalously small N_m values near to zero, as a result of sampling populations at intervals of 30 to 50 km (Brown and Murray, 1992; del Rio–Portilla and González–Aviles 2000). Sampled populations were concluded to be genetically distinct, reflecting, on a geographical scale, the genetic distinctness of even neighboring populations usually separated by tens of kilometers (Shepherd and Brown, 1993). However, even close populations isolated by unfavorable currents were quite distinct (del Rio–Portilla and González–Aviles, 2000; Guzmán del Próo et al., 2000).

B. Sea Urchins

Attempts to understand population genetics from different genetic markers have yielded some success (Palumbi, 1995; Edmands et al., 1996; Mladenov et al.,

1997; Moberg and Burton, 2000; Flowers et al., 2002), although the results at times have shown conflicting patterns. Palumbi (1995) reported on *S. purpuratus* from the west coast of North America, where very little differentiation of mtDNA was observed over 2500 km from southern California to Washington. A sea urchin with a relatively similar life history, *Echinometra* sp., from the tropics showed large differentiation over distances on the order of 100 km. These contrasting patterns may reflect variation in currents and circulation along a linear coastline (west coast of North America), compared with an island archipelago (subtropical and tropical Pacific). Linear coastlines enable a steppingstone pattern of connectivity to emerge, whereas the complex circulation around island archipelagos may well mask any easy understanding of population connectivity. Presumably high larval dispersal is also reflected by a lack of genetic differentiation in *E. chloroticus* along the New Zealand coast. One exception is the genetically distinct population in Doubtful Sound, where restricted larval transport occurs (Mladenov et al., 1997).

Edmands et al. (1996) used both allozyme and mtDNA sequencing for purple sea urchins and found pronounced population structure, especially using allozymes. Moberg and Burton (2000), using allozyme electrophoresis, found patterns of genetic heterogeneity, but with no pattern of geographical affinity for red sea urchins in California. Northern and southern populations were not genetically differentiated at any of the six loci studied. Neighboring populations were often more distinct than geographically distant populations. They also tested within-population genetic variability and patterns of recruitment by analyzing recruits, subadults, and adults from sites in northern California. These comparisons also revealed significant spatial and temporal differentiation among size-stratified samples. This pattern was observed in both of 2 years sampled, and suggests that the concept of a geographically well mixed larval pool is unlikely in northern California. These patterns do not support the steppingstone hypothesis of connectivity, and point, rather, toward "sweepstakes" recruitment, or high variance in reproductive success (Hedgecock, 1994). For free-spawning invertebrates, the unpredictable nearshore environment can lead to extreme variance in reproductive success. In any given spawning event, if only a small subset of the adult population contributes to the larval cohort, then reproduction may be viewed as a sweepstakes, with the likelihood for successful spawning being subject to chance events.

A more recent study of the purple sea urchin tested this "sweepstakes" hypothesis for sea urchins throughout California (Flowers et al., 2002). In this study researchers sampled mtDNA haplotype numbers and diversity of recently settled recruits (1–14 days old). They found little evidence of reduced genetic variation in the recruits relative to the diversity estimated from a previously reported sample of 145 *S. purpuratus* adults. Different cohorts of recruits were in some cases mildly differentiated from each other. These researchers concluded that sweepstakes reproduction can retain many successful fertilization events and that there are likely multiple "winners."

C. SUMMARY OF GENETICS

Studies of the population genetics of sea urchins and abalone suggest that they have very different metapopulation structures. Abalone studies provide a picture of an inherently variable metapopulation structure between species. Intertidal or shallow-water species, such as *H. cracherodii*, *H. diversicolor*, *H. roei*, and *H. discus hannai*, tend to show distinct genetic differences at a scale of a few kilometers of coastline, often corresponding to the scale at which reefs are found, whereas deeper water species such as greenlip, *H. laevigata*, *H. corrugata*, and *H. rufescens*, with populations that may extend over tens of square kilometers of rocky substratum, show genetic differentiation at a larger scale. Species such as the blacklip, *H. rubra*, and *H. fulgens* often show little genetic differentiation but may still belong to the former group as a result of mainly localized dispersal. This dichotomy between intertidal/shallow subtidal species and those in deeper water parallels our earlier broad division of abalone into those with philopatric larvae and those with diffusive larvae. The former will tend to have smaller, and the latter will have larger, metapopulation areas. These trends, of course, may be modified by habitat, seascape features, and evolved behavior.

Sea urchins, with much longer lived larvae and longer reproductive seasons compared with abalone, appear to have higher levels of homogeneity across local populations, although coastal topography may also interact such that species along long stretches of linear coastline show less population segregation than species found in islands, fjordlands, or areas with irregular coastlines. Presumably, while genetic homogeneity may occur throughout large geographical ranges, it is not clear whether it is maintained by limited dispersal between adjacent local populations (i.e., limited distance or steppingstone dispersal; Fig. 6-2) that occurs relatively frequently, or rather through infrequent long-distance dispersal. The data reviewed here are consistent with Bohonak's (1999) review of larval dispersal of 27 groups across many phyla, which found that increased dispersal was associated with decreased population differentiation.

V. SPATIAL VARIABILITY IN ADULT DISTRIBUTIONS AND DEMOGRAPHICS

We turn now to the influence that local habitat quality has on metapopulation dynamics, briefly describing the processes that lead to variability in adult spatial distributions and demographics. To understand metapopulation dynamics, it is

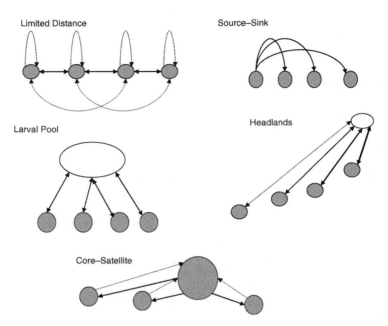

FIGURE 6-2. Proposed models for abalone and sea urchin metapopulation connectivity (based on Carr and Reed, 1993; and Morgan and Botsford, 2001). Models assume local populations do not exchange adults, but are linked by larval dispersal. Arrows indicate direction and hypothesized level of contributed larva. The headlands model describes a hypothesized scenario to explain empirical settlement patterns near promontories in the California Current system, where coastal circulation delivers more larvae to local populations near these retention features than to those more distant. Open circles represent planktonic larval patches (source of larva unknown), solid circles indicate adult populations (possible sources and destinations of larva). Core–satellite, limited distance, source–sink, and larval pool models describe various types of connectivity inferred for abalone populations.

important to identify reproductive sources and sinks (Morgan and Botsford, 2001). This is especially important to proposals for spatial management of abalone and sea urchins, and the discussion of marine reserves (Botsford et al., 1993; Quinn et al., 1993; Pfister and Bradbury, 1996; Morgan et al., 1999).

The spatial distribution of abalone and sea urchin populations is determined by the patchiness of rocky habitat and the differential arrival of larvae. Physical and biotic qualities of the habitat then determine the growth, reproduction, fecundity, and survival of a species, and ultimately its metapopulation dynamics. We have previously defined three scales—patch, local population, and metapopulation—that we use to describe the spatial pattern of abalone and sea urchins, and we here summarize the processes occurring at each scale. The fertilization of eggs occurs at the finest scale of the patch, and declines exponentially with distance between individuals (Levitan et al., 1992; Babcock and Keesing, 1999). Patches are maintained by aggregative, or food- or shelter-seeking behavior (Cameron and Schroeter, 1980; Breen et al., 1985; Shepherd, 1986b) at sites fixed by the presence of crevices, sand boundaries, or other critical habitat features (McShane, 1995a; Prince et al., 1998). Patches may become vacant, for example, after a fishing episode, and reoccupied as animals move from surrounding habitat patches into them. Disease outbreaks may also lead to localized die-offs in sea urchins and abalone (Shepherd and Breen, 1992). It is clearly inappropriate to term such patches local populations, which we have defined as clusters of patches in a patchy seascape, separated from other local populations by unsuitable, unoccupied habitat (Fig. 6-1). Adults can move distances of meters to tens of meters between patches, but not between local populations.

A. ABALONE: ADULT HABITAT AND SPATIAL STRUCTURE

Adult abalone habitats are on rocky reefs within the photic zone, with nearshore species (e.g., blacklip) usually on reefs of high relief and deeper water species (e.g., greenlip) on offshore reefs of low relief (Shepherd, 1973). A very extensive body of literature has accumulated on the growth, survival, and fecundity of abalone, and has been reviewed by Day and Fleming (1992) and Shepherd and Breen (1992). Variation in growth between individuals, patches, local populations, and seasonally is common, and has been variously attributed to qualities of the habitat, such as food supply (often amount of algal drift), food type, water movement, crevice abundance, as well as density of the species (Dixon and Day, 2004).

Populations of stunted individuals, with slower growth, lower fecundity, and likely higher mortality compared with nearby "normal" populations, are a well-known phenomenon among exploited species (e.g., Emmett and Jamieson, 1988; Nash, 1992; Wells and Mulvay, 1995; Worthington and Andrew, 1998; Prince, 2003). Such stunted populations may occur in marginal or sink habitats and may contribute little to egg production (Rodda et al., 2002). A striking example of a mapped mosaic of patches of normal and stunted greenlip, which vary according to food availability, is given by Prince (2003). The natural mortality of abalone is equally as variable as growth, but as yet few correlations have been found, other than density and latitude (Shepherd and Breen, 1992).

Many greenlip, *H. laevigata*, populations occur from 10 to 25m in depth as clusters of local populations linked by larval dispersal and typically separated by distances of 20 to 50km (Shepherd and Brown, 1993). In contrast, blacklip, *H. rubra*, populations are largely constrained to nearshore high-relief reefs, forming interconnected steppingstonelike local populations (Prince and Hilborn, 1998; Prince et al., 1998; Prince, 2003). Comparison of the spatial distribution of many greenlip populations at a reef scale before and after stocks had declined or collapsed (Shepherd and Baker, 1998; Shepherd and Rodda, 2001; Shepherd et al., 2001) gave information on range contraction, from which it was possible to infer the original metapopulation structure, based on likely larval transport distance and the distance from neighboring populations. Here we give examples of the spatial structure of two well-studied abalone populations: greenlip in Waterloo Bay, South Australia, and blacklip on George III Reef, Tasmania (Fig. 6-3).

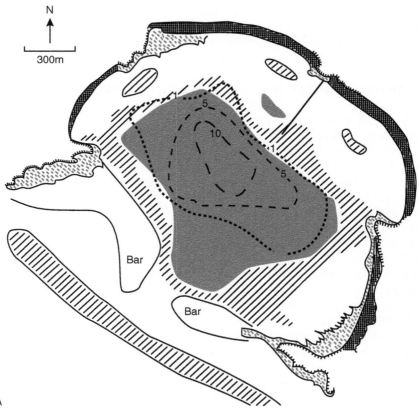

A

FIGURE 6-3. Spatial structure of a greenlip and a blacklip metapopulation in southern Australia. (A) Waterloo Bay, showing core local population and five surrounding local populations occupied by greenlip abalone. Isolines represent recruit densities (2+ age class) in numbers per $1000\,m^2$ measured in 1979, when population was in decline. The areas with light diagonal lines show local populations now extinct, and spatial contraction of the core local population, after the collapse of the population between 1979 and 1982. Dark shading shows occupied habitat in 1998. (B) George III Reef, Tasmania, occupied by blacklip abalone, showing central *Durvillaea* and outer *Macrocystis* zones (narrow, intervening *Phyllospora* zone not shown), with one core and one satellite local population (from Prince, 1989).

Chapter 6 Abalone and Sea Urchins 225

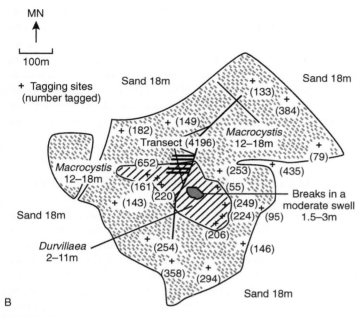

B

FIGURE 6-3. *Continued*

The Waterloo Bay population originally occupied approximately 2 km², including the central local population and six satellite local populations at the margins and outside the Bay (Shepherd and Baker, 1998). Abalone were clustered mainly in crevices in patches of a size that increased with density (Shepherd and Brown, 1993; Shepherd and Partington, 1995). Recruitment was highest in the central local population and declined toward the margins of the Bay (Fig. 6-3A). With increasing size, subadults migrated toward the Bay's entrance, where patch sizes numbered in the hundreds (Shepherd, 1986a). Under intense fishing, marginal local populations collapsed, and the central local population contracted in range by 40% toward the center and entrance to the Bay, which, from water movement studies (Shepherd and Womersley, 1981), were likely larval source habitats.

The blacklip population on the isolated George III Reef complex (Fig. 6-3B), covering approximately 0.35 km², occurred in three distinct habitat types: a shallow bull kelp (*Durvillaea*) zone, surrounded by a narrow *Phyllospora* zone, and an outer deep *Macrocystis* zone, with extensive sand patches (Fig. 6-3B; Prince, 1989; Prince et al., 1998). The population is likely self-recruiting, like many other Tasmanian blacklip populations (Prince, 1989). Adult blacklip lived in patches in all three zones, with no movement between the western and core local populations, and average net movement within the latter of 40 to 60 m over 2 years, mainly toward the shallow *Durvillaea* zone, where recruitment was highest, but declined with depth. Under intense fishing, the population contracted toward the shallow *Durvillaea* zone, and when fishing pressure relaxed, the population reestablished itself in the outer zone. The stock is best viewed as an unstructured metapopulation as defined by Hanski (1991), with a single satellite local population, although J.D. Prince (personal communication) inferred that abalone within each zone were mainly self-recruiting, resulting from the strong nexus between adult and recruit density.

B. Sea Urchins: Adult Habitat and Spatial Structure

Adult sea urchins are distributed on rocky reefs where attached and drift macroalgae occur. The degree to which these rocky habitats are broken into discrete patches limits adult movement. At the patch scale, individuals aggregate to spawn. Individual sea urchin reproductive fitness depends on the quality of the food and habitat, and (as witnessed by the fishery), there can be great variability between patches in morphology (McShane and Anderson, 1977; Rogers–Bennett et al., 1995;) and reproductive fitness (i.e., gonad quality; Holland et al., 1967; Vadas, 1977; Larson et al., 1980).

In many places throughout the world "barrens" (areas devoid of macroalgae) occur in response to sea urchin overgrazing (Breen and Mann, 1976; Harrold and Pearse, 1987; Estes and Duggins, 1995). These areas are characterized by high-density populations of sea urchins that are able to persist with very low nutritional input by feeding on microalgal films and corallines. Sea urchin barrens are notable in a metapopulation context because they result in the local persistence of sink habitats, where growth and reproductive rates decline, and where, without recruitment from other populations, the population declines. Sea urchins in barrens are often nutritionally stressed and unable to develop mature gonads, but are able to maintain themselves over long periods without growth (Wing et al., 2001). There exists a rich body of literature on the topic of sea urchin barrens that we do not discuss here (see Estes and Duggins, 1995; Scheibling, 1996).

Individual variation in growth, metabolism, mortality, and reproduction at the patch and local population scales may result from differences in life history traits (Russell 1987), genetics (Edmands et al., 1996; Moberg and Burton, 2000), depth (Rogers–Bennett et al., 1995), climate variation (e.g., ENSO; Dayton and Tegner, 1990; Tegner and Dayton, 1991; Tegner et al., 1997), density (Levitan, 1989; Lundquist, 2000), body size (Levitan, 1991), and wave action (Denny and Shibata, 1989; Edwards and Ebert, 1991). Growth of sea urchins is typically seasonal (Walker, 1981) and dependent on the availability of fleshy macroalgae (Ebert, 1968; Vadas, 1977).

Patch habitat quality influences the distribution, size, and density of sea urchins. Studies based on size–frequency distributions show large variation in density and abundance at both patch and local population scales (e.g., Lundquist, 2000; Kalvass et al., 1990, 1991; Kalvass and Taniguchi, 1993). Presumably these differences reflect interannual variation in settlement, growth, and mortality (including fishing mortality; Morgan et al., 2000a). A study of red sea urchins at Bodega Bay in northern California showed that there are morphologically distinct sea urchins in shallow (5 m) and deep (23 m) habitats. Shallower sea urchins had significantly heavier gonads and occurred at higher densities, where food was more abundant (Rogers–Bennett et al., 1995). In one study of green sea urchins along the coast of Maine, two sympatric growth morphs were identified (Vadas et al., 2002). It is hypothesized that these morphs resulted from differential survival and postsettlement migration, with the smaller, slower growing morphs likely moving from deeper waters where they first settled.

Several studies have looked at variations in size–frequency distributions between sites in an attempt to assess the variability in demographic rates among local populations (Russell, 1987; Ebert et al., 1999; Morgan et al., 2000a). In the most geographically comprehensive study of growth and survival, Ebert et al. (1999) studied *S. franciscanus* at 18 sites from California to Alaska. They found no north–south differences among the local populations sampled, and suggested that no latitudinal differences in food availability existed. In contrast with growth rates, annual survival rates were correlated with latitude (higher in the north), and disease and temperature-related stress increased mortality to the south (Ebert et al., 1999). Russell (1987) also described clinal variation in purple sea urchin recruitment and longevity that mirrored the patterns in red sea urchins. That is, higher recruitment rates in the southern portion of the range and greater longevity to the north. An examination of growth, mortality, and recruitment data suggests that sea urchins have great plasticity in life history traits in response to environmental variables.

In a more geographically restricted study, Morgan et al. (2000a) estimated growth and mortality rates for local populations of red sea urchins from tagged/recaptured individuals at three sites in northern California. The estimated growth and mortality parameters differed among sites, leading to substantially different estimates of yield per recruit from these sites. No predictable spatial pattern was observed in the sites over a range of 100 km, and individual growth rates within a site were as variable as those between sites.

In summary, growth rate estimates for red sea urchins from multiple studies along the west coast of North America do not support a predictable pattern of variation (Ebert and Russell, 1992; Ebert et al., 1999; Morgan et al., 2000a; Rogers–Bennett et al., 2003). Growth studies in other parts of the world also indicate between-patch habitat quality leading to varying growth and reproductive output. Wing et al. (2003b) observed large differences in the abundance, growth, gonad development, and larval settlement of the sea urchin E. *chloroticus* at nine sites in the Doubtful Thompson Sound complex, New Zealand. The observed variation in growth was attributed to nutritional history. The results of growth studies suggest that habitat quality leads to large patch scale variation in growth rates that will influence egg-per-recruit analyses, and the value of particular habitats as larval sources (see Rogers–Bennett et al., 1985; Morgan et al., 2000a). Unfortunately, no clear predictive spatial patterns are currently apparent.

VI. FISHING AND MANAGEMENT

Not surprisingly, metapopulation theory has tended to resonate with ecologists and fisheries managers because it offers a conceptual framework that accounts for spatial variability in catch rates and productivity. The ability to delineate source and sink populations (Pulliam, 1988; Carr and Reed, 1993; Allison et al., 1998; Morgan and Botsford, 2001; Lipcius et al., 2005) is necessary for the design of spatial management schemes and is useful in detecting recruitment overfishing.

A. Abalone

Intense fishing leads to differential decline of abalone patches and ultimately local populations, because fishers target the largest patches and then progressively smaller ones (Shepherd and Partington, 1995; Prince, 2003). Continual reaggregation of abalone and fisher behavior, which tends to target the high-density patches (Prince and Hilborn, 1998; Prince, 2003), together ensure that local populations with strong recruitment to the adult population are fished first, followed by those with lower recruitment. Local populations in habitats with poor or sporadic recruitment, although fished less often, recover far more slowly and may eventually go extinct.

In an analysis of 23 inshore greenlip stocks, Shepherd et al. (2001) found that the resilience of individual stocks to intense fishing was determined by the extent of larval retention (estimated from coastal topographic "closure indices") within the metapopulations' spatial extent. Thus, stocks bounded to a greater extent by land or emergent reef had consistently higher recruitment and declined more slowly under intense fishing than those in open water. The ranges of stocks on open coasts contracted from deeper offshore waters toward the coast, where they persisted near headlands, the lee side of islands, in inlets, or near distinct bottom features like reefs of high relief (Fig. 6-3). Others contracted upstream toward presumed source populations. In every studied case, contraction was from marginal, deep-water, or distant subpopulations toward the core areas, which mainly supported fishing, (e.g., Fig. 6-3). A strong inference from these studies is that core areas are larval sources and they export larvae to the surrounding areas. Without internal metapopulation structure as proposed, we would not expect a common pattern of decline, such as preferential demise of downstream populations or patches, and local populations would simply fragment and disappear haphazardly.

The causes of differential recruitment failure in a population are less clear. One mechanism—the Allee effect (i.e., the effect of low adult density on fertilization success—has been suggested for recruitment failure, and a critical mean threshold density of 0.15 to 0.3 adults/m^2 has been proposed from empirical data (Shepherd and Partington, 1995; Shepherd and Rodda, 2001; Shepherd et al., 2001; cf. Prince et al., 1987, 1988). Another mechanism may be the fragmentation effect, well-known from many terrestrial studies to increase the extinction risk of populations (Brown, 1984, 1995; Gaston and Lawton, 1990; Hanski, 1991; Rodriguez, 2002). In many long-term studies, the number of extinct habitat patches and/or local populations increased over time under intense fishing (Shepherd and Brown, 1993; Shepherd and Partington, 1995; Shepherd and Rodda, 2001; Shepherd et al., 2001). In these cases, larvae arriving at extinct patches would have lacked settlement cues, and then suffered higher mortality by failing to settle, or settling in inferior sites. Other nonmutually exclusive explanations for patch extinction include poor larval retention or poor juvenile survival in marginal habitats. In contrast, Prince (2003) has presented a highly localized recruitment hypothesis to explain progressive patch extinction. He has argued that microstocks functionally exist at the scale of the patch, and that stock–recruit relations essentially occur at this scale. This hypothesis puts the scale of the metapopulation much smaller than we have suggested.

B. Sea Urchins

Sustainable management of sea urchin fisheries requires maintaining sufficient reproductive capacity, which necessitates not just knowledge of the impacts of fishing on reproduction at each location, but also the impact of that reproduction on recruitment to other locations. Understanding the spatial pattern of larval dispersal is an important aspect of a complete understanding of the role of recruitment dynamics in determining sustainable catch (Botsford et al., 1999; Morgan and Botsford, 2001).

At low population levels, sea urchins demonstrate Allee effects in recruitment from both a predispersal (Levitan et al., 1992) and a postdispersal mechanism (Tegner and Dayton, 1977). Predispersal Allee effects occur in broadcast spawning species because fertilization success is dependent on the density and proximity of other spawners. Successful fertilization is hypothesized to decline rapidly below threshold adult densities (Pennington, 1985; Levitan et al., 1992; Lundquist, 2000), although recent model projections suggest that threshold effects may be too weak to be detected in stock–recruitment relationships (Lundquist and Botsford, 2004).

Because larger sea urchins provide larger sheltering areas, the effect of fisheries is to remove the most appropriate juvenile shelters for small urchins as well as abalone (Tegner and Dayton, 1977; Tegner, 1989; Mayfield and Branch, 2000). Predation experiments show that red sea urchins greater than 90 mm test diameter are essentially free from predation by spiny lobsters, the main predator in southern California (Tegner and Levin, 1983). Postdispersal Allee effects result from too few adults present to provide a spine canopy refuge from predation for newly settled sea urchins (Tegner and Dayton, 1977). Similarly, Rogers–Bennett and Pearse (2001) discuss the importance of sea urchin spine canopies on the survival of juvenile abalone.

Several studies have discussed sea urchin management based on spatial closures (Botsford et al., 1993; Quinn et al., 1993; Pfister and Bradbury, 1996; Botsford et al., 1999; Morgan et al., 1999). Accounting for larval dispersal in no-fishing reserve design has been a difficult task (but see Allison et al., 1998; Morgan and Botsford, 2001; and Lipcius et al., 2001). Few studies address spatially explicit reserve designs because of the difficulties in empirically diagnosing larval dispersal pathways, the number of significant uncertainties, and the inability to generalize from one species to another. Lipcius et al. (2001) showed that knowledge of dispersal patterns and connectivity could optimize spiny lobster reserve placement in the Bahamas. Morgan and Botsford (2001) looked at the role different forms of larval dispersal might have on the management of a sea urchin fishery (Fig. 6-2). They evaluated marine reserves in the context of hypothetical but realistic dispersal models for sea urchins and found that identification of larval sources was an important aspect of management. Without explicit knowledge of larval sources and sinks, placing 25% of the population in a marine reserve was the best precautionary strategy. The pattern of larval dispersal could account for a large variation in the performance of a reserve system, especially when larval source populations were left open to fishing and larval sinks were placed in no-fishing reserves.

In summary, then, overfishing has multiple effects. Reduction of adults may inhibit larval settlement and subsequent survival. The export of larvae to distant local populations or to those in an inferior habitat is reduced under fishing, and this is followed by the progressive extinction of habitat patches and the fragmentation of local populations. In the late stage of collapse, only a few core local populations with high productivity persist. A metapopulation framework that accounts for variation in local population productivity, including larval sources and sinks, can help to maintain catch rates and population persistence.

C. Optimal Harvesting of Invertebrate Metapopulations

Once it is established that a species demonstrates spatial variability in population dynamics (can be considered a metapopulation) then management should also take on a spatial dimension. Given the decline or collapse of abalone and sea urchin fisheries around the world, robust spatial harvesting strategies have received some attention (Quinn et al., 1993; Botsford et al., 1993; Dugan and Davis, 1993; Botsford et al., 1999). MaCall (1990) and Gerber et al. (2003) have examined the optimal placement of reserves in model metapopulations. In the context of abalone fisheries, Tuck and Possingham (1994, 2000), modeled two local populations in a metapopulation under various relative source–sink scenarios involving larval transfer between them. Under conservative (i.e., precautionary) management, harvesting of sink local populations and protecting source local populations will maximize the yield, whereas reverse strategies will entail a high risk of local collapse. If the metapopulation structure is not recognized, source local populations would be overharvested and sinks underharvested.

Other studies have addressed conceptual issues for exploited metapopulations, many in the context of optimal siting of reserves, but here we cite only those relevant to sedentary species with locally dispersing larvae. Crowder et al. (2000) and Lipcius et al. (2005) reviewed recent literature and concluded that reserves in source habitats were superior to those in sinks. The protection of sink habitats has been explored by a few authors. Howe et al. (1991) pointed out that they have distinct benefits in a metapopulation context. They augment total population size, provide a more diverse gene pool, and may extend the survival of declining metapopulations, especially in the case of "leaky" sinks, in which back dispersal may occur and may contribute to overall stability of the population (Dias, 1996).

Where the metapopulation structure is not known, some (Tegner, 1993; Wallace, 1998; Morgan and Botsford, 2001; Shanks et al., 2003) have recommended the systematic placement of reserves in both productive and less productive habitats to protect 20% of the spawning stock in each significant metapopulation. Others (Crowder et al., 2000; Rogers–Bennett et al., 2002b) have suggested use of the pragmatic distinction between productive and marginal habitats, which is more easily recognizable than sources and sinks, as surrogates in deciding reserve placement. In this context, productive habitats are those with strong recruitment, in which persistent fishing is sustained. However, marginal habitats may still be useful as sources for adult transplants to replenish depleted neighboring populations (e.g., Rodda et al., 2002).

Empirical studies of reserve benefits are rare. Gell and Roberts (2003) document several cases in which species protected in marine reserves showed enhanced fishery landings, and MaCall (1990) gave examples of upstream reserves, which maintained exploited downstream clam beds. Baker et al. (1996) cited examples of their successful use to enhance or sustain abalone stocks in Australia (Wells and Keesing, 1990), Mexico (Shepherd et al., 1998), and New Zealand (Ballantine, 1989). However, few of these studies examined the question of optimal siting of a reserve, although the small reserve at La Natividad, Mexico (Shepherd et al., 1998), was centrally placed in likely source habitat and was believed by cooperative biologists to have contributed to the survival of the local fishery.

There are several studies of reserve failures. Shepherd and Brown (1993) described the collapse of a reserve population resulting from fishing of a core local population outside the reserve boundary, but this was a problem of a reserve too small to protect a marginally viable metapopulation. Similarly, reserves in marginal habitats have (in two examples) failed to reestablish collapsed mollusk populations (Heslinga et al., 1984; McCay, 1988, cited in Lipcius et al., 2005).

Because of coastal hydrodynamics, local habitat quality can, at the same time, be very high and have no relationship to local recruitment as a result of the decoupling of local gamete production from larval supply and recruitment. Management strategies for sea urchins have been modeled to reflect this, and in the extreme case, segments of these marine metapopulations can be demographic sinks where local mortality outpaces larval production. In these areas, recruits survive but do not produce successful larvae, and the patch persistence depends entirely on immigration from productive larval sources elsewhere in the metapopulation. If source areas in these metapopulations produce a large number of successful larvae, it is possible that a high proportion of the regional population network can be maintained within habitats with a local reproductive deficit. In such systems, assumptions about the productivity and resilience of populations to fishing pressure, drawn from estimates of local demographic parameters, can be particularly misleading (Morgan and Botsford, 2001). Model studies suggest metapopulations with spatial structure and reproductive Allee effects may suddenly collapse under widespread habitat degradation or fishing pressure that results in the loss or decline of source populations (Quinn et al., 1993; Morgan and Botsford, 2001).

Field evidence is mounting that establishing marine reserves results in localized increases of both abundance and body size of previously overexploited populations (e.g., Gell and Roberts, 2003; Halpern, 2003). It is now recognized that metapopulation dynamics (especially source–sink dynamics) have a profound influence over the efficacy of marine reserves to conserve populations (Crowder et al., 2000; Lipcius et al., 2001; Botsford et al., 2003), but scientists have not yet successfully categorized the different pathways by which population or metapopulation structure may drive reserve success. The vagaries of the biotic and physical environment are so great that the potential for a single metapopulation model to describe adequately multiple species with different life history traits, including species such as abalone and sea urchins, is likely small, and our discussion highlights important considerations for marine reserve proponents.

VII. SUMMARY

Can we come to a comprehensive statement of metapopulation dynamics for both sea urchins and abalone in temperate reefs of the world? We have already noted (see Section I) that the classic Levins' extinction-centered model of local populations is inappropriate for many marine invertebrates, at least over timescales of years to even decades. Although we cannot empirically document metapopulation structure under the classic extinction–recolonization Levins model, sea urchins and abalone do fit more contemporary definitions of metapopulations (Kritzer and Sale, 1994). We suggest that, for sea urchins and abalone, a definition that emphasizes larval dispersal between adult local populations with varying habitat quality is appropriate. It is possible that if ongoing studies extend to timescales that are appropriate to the longevity of abalone and sea urchin population dynamics (decades to centuries), a more classic extinction–recolonization metapopulation model will be revealed for these species.

Our review of a great body of empirical research agrees with the work of Botsford et al. (1994), which demonstrated through modeling the critical effects of variability in the spatial and temporal components of larval dispersal on population fluctuations of meroplanktonic organisms such as abalone and sea urchins. The importance of transport from these findings was greater than simply its influence on recruitment. Rather, it was shown to be critical to the underlying population dynamics. Hence, models that emphasize spatiotemporal variability in population dynamics and connections through dispersal are more realistic for abalone and sea urchins (Harrison, 1991). Despite current difficulties in tagging larvae, its application to larvae may provide one avenue through which to measure larval dispersal and may lead to an understanding of the internal metapopulation dynamics of these species.

The spatial configuration, including pathology, of abalone and sea urchin stocks, referred to earlier, suggests that they may variously conform to one of several metapopulation types, according to habitat availability, hydrodynamics, and the extent of larval dispersion. Our illustration of common metapopulation types (Fig. 6-2)—core–satellite, source–sink, limited distance, and larval pool models—may apply to many species. Greenlip abalone often appear to fit the first two models (e.g., Fig. 6-3A for Waterloo Bay). Such populations typically occur in extensive, spatially dispersed habitat patches, comprising one or more central local populations, which contribute larvae to subsets of a common pool. The larvae disperse widely enough to settle potentially anywhere within a central or nearby local populations. Where alongshore currents have a net direction of flow during spawning, then upstream local populations will be relative sources and downstream local populations will be relative sinks. Many other species, such as *H. corrugata, H. fulgens* (Tegner, 1992), *H. kamtschatkana, H. midae,* and *H. rufescens,* probably conform to one or the other of these two models, although knowledge of their spatial distribution patterns is anecdotal or lacking.

The larval pool and limited distance models (Fig. 6-2) seem more typical of blacklip, *H. rubra, H. discus hannai, H. roei, H. cracherodii, H. diversicolor,* and other shallow-water species with philopatric larvae. Among these species, local populations, or in the extreme case of philopatry, patches (Prince, 2003), are largely self-recruiting, although leakage of larvae to neighboring local populations, depending on local currents, is likely.

Based on field studies of settlement in northern California, Morgan and Botsford (2001) hypothesized an additional metapopulation dispersal model: headlands (Fig. 6-2). This model focuses on the spatial variability observed at the local population scale around upwelling cells in the California Current system. The degree to which this or other models fit other sea urchins will need to be evaluated based on experience with coastal circulation and hydrography in the region of study. Sea urchins in this region may well exhibit limited dispersal model structure, whereas other sea urchin populations have been described in terms of source–sink models (Wing et al., 2003b).

Although abalone and sea urchins appear more diverse in their metapopulation structure than previously thought (e.g., Allison et al., 1998), a word of caution is pertinent (Grimm et al., 2003). The internal dynamics of populations are still unproved and rest on inductive inference, without the benefit of experimental manipulation (except fishing). The "patchy population" scenario (Harrison, 1991) with interconnected patches within ecologically independent, unstructured local populations may be more common than we suppose (Prince, 2003). The field is ripe for empirical testing of larval connections between local populations.

VIII. ACKNOWLEDGMENT

The authors thank many colleagues, especially Drs. Ron Burton, Lou Botsford, Rob Day, Cameron Dixon, Nick Elliott, Jeremy Prince, Jason Tanner, Stephen Mayfield, Steve Palumbi, Jake Kritzer, and Peter Sale, for valued suggestions and discussions, and for reviewing the manuscript. Tony Olsen assisted with preparation of Figure 6-3.

REFERENCES

Agatsuma, Y., Nakao, S., Motoya, S., Tajima, K., and Miyamoto, T. (1998). Relationships between year-to-year fluctuations in recruitment of juvenile sea urchins *Strongylocentrotus nudus* and seawater temperature in southwestern Hokkaido. *Fisheries Science.* **64**, 12.

Ainley, D.G., Sydeman, W.J., Parrish, R.H., and Lenarz, W.H. (1993). Oceanic factors influencing distribution of young rockfish (*Sebastes*) in central California: A predator's perspective. *CalCOFI Report.* **34**, 133–139.

Alldredge, A.L., and Hamner, W.M. (1980). Recurring aggregation of zooplankton by a tidal current. *Estuarine Coastal Mar. Sci.* **10**, 31–37.

Allison, G.W., Lubchenco, J., and Carr, M.H. (1998). Marine reserves are necessary but not sufficient for marine conservation. *Ecol. Appl.* **8 (suppl.)**, S79–S92.

Andrew, N.L., Agatsuma, Y., Ballesteros, E., Bazhin, A.G., Creaser, E.P., Barnes, D.K.A., Botsford, L.W., Bradbury, A., Campbell, A., Dixon, J.D., Einarsson, S., Gerring, P.K., Herbert, K., Hunter, M., Hur, S.B., Johnson, C.R., Juinio-Menez, M.A., Kalvass, P., Miller, R.J., Moreno, C.A., Palleiro, J.S., Rivas, D., Robinson, S.M.L., Schroeter, S.C., Steneck, R.S., Vadas, R.L., Woodby, D.A., and Xiaoqi, Z. (2002). Status and management of the world sea urchin fisheries. *Oceanogr. Mar. Biol. Ann. Rev.* **40**, 343–425.

Andrew, N.L., Worthington, D.G., Brett, P.A., and Bentley N. (1998). Interaction between the abalone fishery and sea urchins in New South Wales. Final report to Fisheries Research and Development Corporation. Sydney, Australia: Fisheries Research Institute.

Babcock, R., and Keesing, J.K. (1999). Fertilization biology of the abalone *Haliotis laevigata*: Laboratory and field studies. *Can. J. Fish. Aquat. Sci.* **56**, 1668–1678.

Baker, J.L., Shepherd, S.A., and Edyvane, K. (1996). The use of marine fishery reserves to manage benthic fisheries, with emphasis on the South Australian abalone fishery. In *Developing Australia's representative system of marine protected areas* (R. Thackway, ed., pp. 103–113). Canberra, Australia: Department of Environment.

Ballantine, W. (1989). Marine reserves: Lessons from New Zealand. *Progress in Underwater Science.* **13**, 1–14.

Barlow, L.A. (1990). Electrophysiological and behavioural responses of larvae of the red abalone (*Haliotis rufescens*) to settlement-inducing substances. *Bull. Mar. Sci.* **46**, 537–554.

Bernard, F.R. (1977). Fishery and reproductive cycle of the red sea urchin, *Strongylocentrotus franciscanus*, in British Columbia. *J. Fish. Res. Bd Can.* **34**, 604–610.

Bernard, F.R., and Miller, D.C. (1973). Preliminary investigation on the red sea urchin resources of British Columbia *Strongylocentrotus franciscanus* (Agassiz). *Fish. Res. Bd Canada Technical Report.* **400**, 1–35.

Black, K.P. (1988). The relationship of reef hydrodynamics to variations in numbers of planktonic larvae on and around coral reefs. *Proc. 6th Int. Coral Reef Symp.* **2**, 125–130.

Black, K.P., and McShane, P.E. (1990). Influence of surface gravity waves on wind-driven circulation in intermediate depths on an exposed coast. *Aust. J. Mar. Freshw. Res.* **41**, 353–363.

Bohonak, A.J. (1999). Dispersal, gene flow, and population structure. *Quart. Rev. Biol.* **74**, 21–63.

Boolootian, R.A., Farmanfarmaian, A., and Giese, A.C. (1962). On the reproductive cycle and breeding habits of two western species of *Haliotis*. *Biol. Bull.* **122**, 183–193.

Bossart, J.L., and Powell, D.P. (1998). Genetic estimates of population structure and gene flow: Limitations, lessons and new directions. *Trends in Evolution and Ecology.* **13**, 202–206.

Botsford, L.W., Micheli, F., and Hastings, A. (2003). Principles for the design of marine reserves. *Ecol. Appl.* **13 (suppl.)**, S25–S31.

Botsford, L.W., Moloney, C.L., Hastings, A., Largier, J.L., Powell, T.M., Higgins, K., and Quinn, J.F. (1994). The influence of spatially and temporally varying oceanographic conditions on meroplanktonic metapopulations. *Deep-Sea Research II.* **41**, 107–145.

Botsford, L.W., Morgan, L.E., Lockwood, D., and Wilen, J. (1999). Marine reserves and management of the northern California red sea urchin fishery. **CalCOFI Report 40**, 87–93.

Botsford, L.W., Quinn, J.F., Wing, S.R., and Brittnacher, J.G. (1993). Rotating spatial harvest of a benthic invertebrate, the red sea urchin *Strongylocentrotus franciscanus*. In *Proceedings of the International Symposium on Management Strategies of Exploited Fish Populations* (G. Kruse, D.M. Eggers,

R.J. Marasco, C. Pautzke, and T.J. Quinn II, eds., pp. 408–429). Fairbanks: University of Alaska.

Breen, P.A., and Adkins, B.E. (1980). Spawning in a British Columbia population of northern abalone *Haliotis kamtschatkana*. *The Veliger.* **23**, 177–179.

Breen, P.A., Carolsfeld, W., and Yamanaka, K.L. (1985). Social behaviour or juvenile red sea urchins *Strongylocentrotus franciscanus* (Agassiz). *J. Exp. Mar. Biol. Ecol.* **92**, 45–61.

Breen, P.A., and Mann, K.H. (1976). Destructive grazing by sea urchins in eastern Canada. *J. Fish. Res. Bd Canada.* **33**, 1278–1283.

Brown, J.H. (1984). On the relationship between abundance and distribution of species. *Am. Nat.* **124**, 253–279.

Brown, J.H. (1995). *Macroecology.* Chicago: University of Chicago Press.

Brown, L.D., and Murray, N.D. (1992). Genetic relations within the genus *Haliotis*. In *Abalone of the world: Biology, fisheries and culture* (S.A. Shepherd, M.J. Tegner, and S.A. Guzmán del Próo. eds., pp. 19–23). Oxford: Blackwell.

Burton, S.R., and Tegner, M.J. (2000). Enhancement of red abalone *Haliotis rufescens* stocks at San Miguel Island: Reassessing a success story. *Mar. Ecol. Prog. Ser.* **202**, 303–308.

Caley, M.J., Carr, M.H., Hixon, M.A., Hughes, T.P., Jones, G.P., and Menge, B.A. (1996). Recruitment and the local dynamics of marine populations. *Ann. Rev. Ecol. Syst.* **7**, 477–500.

Cameron, R.A., and Schroeter, S.C. (1980). Sea urchin recruitment: Effect of substrate selections on juvenile distribution. *Mar. Ecol. Prog. Ser.* **2**, 243–247.

Carr, M.H., and Reed, D.C. (1993). Conceptual issues relevant to marine harvest refuges: Examples from temperate reef fishes. *Can. J. Fish. Aquat. Sci.* **50**, 2019–2028.

Connell, J. (1985). The consequences of variation in initial settlement vs. postsettlement mortality in rocky intertidal communities. *J. Exp. Mar. Biol. Ecol.* **93**, 11–45.

Conod, N., Bartlett J.P., Evans, B.S., and Elliott, N.G. (2002). Comparison of mitochondrial and nuclear DNA analyses of population structure in a blacklip abalone *Haliotis rubra* Leach. *Mar. Freshwater Res.* **53**, 711–718.

Cowen, R.K., Lwiza, K.M.M., Sponaugle, S., Paris C.B., and Olson D.B. (2003). Connectivity in marine populations: Open or closed? *Science.* **287**, 857–859.

Crofts, D.R. (1938). Development of *Haliotis tuberculata*. *Phil. Trans. R. Soc. Ser. B.* **228**, 219–267.

Crowder, L.B., Lyman, S.J., Figueira W.F., and Priddy J. (2000). Source-sink population dynamics and the problem of siting marine reserves. *Bull. Mar. Sci.* **66**, 799–820.

Daume, S., Brand-Gardner, S., and Woelkerling, W.J. (1999a). Preferential settlement of abalone larvae: Diatom films vs. nongeniculate coralline algae. *Aquaculture.* **174**, 243–254.

Daume, S., Brand-Gardner, S., and Woelkerling, W.J. (1999b). Settlement of abalone larvae (*Haliotis laevigata* Donovan) in response to nongeniculate coralline red algae (Corallinales, Rhodophyta). *J. Exp. Mar. Biol. Ecol.* **234**, 125–143.

Davis, J.A. (1986). Boundary layers, flow microenvironments and stream benthos. In *Limnology in Australia* (P. de Dekker and W.D. Williams, eds., pp. 303–312). Dordrecht: Junk.

Davis, R.E. (1985). Drifter observations of coastal surface currents during CODE: The statistical and dynamical views. *J. Geophys. Res.* **90**, 4756–4772.

Day, E., and Branch, G.M. (2000). Evidence for a positive relationship between juvenile abalone *Haliotis midae* and the sea urchin *Parechinus angulosus* in the South-western Cape, South Africa. *S. Afr. J. Mar. Sci.* **22**, 145–156.

Day, R.W., and Fleming, A. (1992). The determinants and measurement of abalone growth. In *Abalone of the world: Biology, fisheries and culture* (S.A. Shepherd, M.J. Tegner, and S.A. Guzmán del Próo, eds., pp. 141–168). Oxford: Blackwell.

Dayton, P.K., and Tegner, M.J. (1990). Ecological consequences of the 1982–82 El Niño to marine life. In *Bottoms below troubled waters: Benthic impacts of the 1982–84 El Niño in the temperate zone* (P.W. Glynn, ed., pp. 433–472). Amsterdam: Elsevier.

Del Río–Portilla, M., and González-Aviles, J.G. (2000). Population genetics of the yellow abalone, *Haliotis corrugata*, in Cedros and San Benito Islands: A preliminary survey. *J. Shellfish Res.* **20**, 765–770.

Denny, M.W., and Shibata, M.F. (1989). Consequences of surf-zone turbulence for settlement and external fertilization. *Am. Nat.* **134**, 859–889.

Dias, P.C. (1996). Sources and sinks in population biology. *Trends in Evolution and Ecology.* **11**, 326–330.

Dixon, C.D., and Day, R.W. (2004). Growth responses in emergent greenlip abalone to density reductions and translocations. *J. Shellfish Res.* **23**, 1223–1228.

Dugan, J.E., and Davis, G.E. (1993). Applications of marine refugia to coastal fisheries management. *Can. J. Fish. Aquat. Sci.* **50**, 2029–2042.

Duggins, D.O. (1981). Interspecific facilitation in a guild of benthic marine herbivores. *Oecologia.* **48**, 157–163.

Ebert, T.A. (1968). Growth rates of the sea urchin *Strongylocentrotus purpuratus* related to food availability and spine abrasion. *Ecology.* **49**, 1075–1091.

Ebert, T.A. (1980). Flexible growth of sea urchin jaws: An example of plastic resource allocation. *Bull. Mar. Sci.* **30**, 467–474.

Ebert, T.A., Dixon, J.D., Schroeter, S.C., Kalvass, P.E., Richmond, N.T., Bradbury, W.A., and Woodby, D.A. (1999). Growth and mortality of red sea urchins *Strongylocentrotus franciscanus* across a latitudinal gradient. *Mar. Ecol. Prog. Ser.* **190**, 189–209.

Ebert, T.A., and. Russell, M.P. (1988). Latitudinal variation in size structure of the west coast purple sea urchin: a correlation with headlands. *Limnol. Oceanogr.* **33**, 286–294.

Ebert, T.A., and Russell, M.P. (1992). Growth and mortality estimates for red sea urchin *Strongylocentrotus franciscanus* from San Nicolas Island, California. *Mar. Ecol. Prog. Ser.* **81**, 31–41.

Ebert, T.A., and Southon, J.R. (2003). Red sea urchins (*Strongylocentrotus franciscanus*) can live over 100 years: Confirmation with A-bomb 14carbon. *Fish. Bull.* **101**, 915–922.

Ebert, T.A., Schroeter, S.C., and Dixon, J.D. (1994). Settlement patterns of red and purple sea urchins (*Strongylocentrotus franciscanus* and *S. purpuratus*) in California, USA. *Mar. Ecol. Prog. Ser.* **111**, 41–52.

Eckman, J.E. (1983). Hydrodynamic processes affecting benthic recruitment. *Limnol. Oceanogr.* **28**, 241–257.

Eckman, J.E., Duggins, D.O., and Sewell, A.T. (1989). Ecology of understorey kelp environments. Effects of kelps on flow and particle transport near the bottom. *J. Exp. Mar. Biol. Ecol.* **129**, 173–187.

Edmands, S., Moberg, P.E., and Burton, R.S. (1996). Allozyme and mitochondrial DNA evidence of population subdivision in the purple sea urchin *Strongylocentrotus purpuratus*. *Mar. Biol.* **126**, 443–450.

Edwards, P.B., and Ebert, T.A. (1991). Plastic responses to limited food availability and spine damage in the sea urchin *Strongylocentrotus purpuratus* (Stimpson). *J. Exp. Mar. Biol. Ecol.* **145**, 205–220.

Elliott, N.G., Bartlett, J., Evans, B., Officer, R., and Haddon, M. (2002). *Application of molecular genetics to the Australian abalone fisheries: Forensic protocols for species identification and blacklip stock structure.* FRDC report 1999/164. Hobart: CSIRO Marine Research and Fisheries Research and Development Corporation.

Emmett, B., and. Jamieson, G.S. (1988). An experimental transplant of northern abalone, *Haliotis kamtschatkana*, in Barkley Sound, British Columbia. *Fish. Bull.* **87**, 95–104.

Estes, J.A., and Duggins, D.O. (1995). Sea otters and kelp forests in Alaska: Generality and variation in a community ecological paradigm. *Ecol. Monogr.* **65**, 75–100.

Evans, B.S., Sweijd, N.A., Bowie, R.C.K., Cook, P.A., and Elliott, N.G. (2004). Population genetic structure of the perlemoen *Haliotis midae* in South Africa: Evidence of range expansion and founder effects. *Mar. Ecol. Prog. Ser.* **270**, 163–172.

Farrell, T.M., Bracher, D., and Roughgarden, J. (1991). Cross-shelf transport causes recruitment to intertidal populations in central California. *Limnol. Oceanogr.* **36**, 279–288.

Flowers, J.M., Schroeter, S.C., and Burton, R.S. (2002). The recruitment sweepstakes has many winners: Genetic evidence from the sea urchin *Strongylocentrotus purpuratus*. *Evolution.* **56**, 1445–1453.

Gaines, S.D., and Bertness, M.D. (1992). Dispersal of juveniles and variable recruitment in sessile marine species. *Nature.* **360**, 579–580.

Gaines, S.D., and Bertness, M.D. (1993). The dynamics of juvenile dispersal: Why field ecologists must integrate. *Ecology.* **74**, 2430–2435.

Gaines, S., Brown, S., and Roughgarden, J. (1985). Spatial variation in larval concentrations as a cause of spatial variation in settlement for the barnacle, *Balanus glandula*. *Oecologia.* **67**, 267–272.

Gaines, S.D., Gaylord, B., and Largier, J.L. (2003). Avoiding current oversights in marine reserve design. *Ecol. Appl.* **13 (suppl.)**, S32–S46.

Gaines, S.D., and Lafferty, K.D. (1995). Modeling the dynamics of marine species: The importance of incorporating larval dispersal. In *Ecology of marine invertebrate larvae* (L. McEdward, ed., pp. 389–412). New York: CRC Press.

Gaines, S.D., and Roughgarden, J. (1985). Larval settlement rate: A leading determinant of structure in an ecological community of the marine intertidal zone. *Proc. Natl Acad. Sci. USA.* **82**, 3707–3711.

Gaston, K.J., and Lawton, J.H. (1990). Effects of scale and habitat on the relationship between regional distribution and local abundance. *Oikos.* **58**, 329–335.

Gell, F.R., and Roberts, C.M. (2003). Benefits beyond boundaries: The fishery effects of marine reserves. *Trends in Ecology and Evolution.* **18**, 448–455.

Gerber, L.R., Botsford, L.W., Hastings, A., Possingham, H.P., Gaines, S.D., Palumbi S.R., and Andelman, S. (2003). Population models for marine reserve designs: A retrospective and prospective synthesis. *Ecol. Appl.* **13**, S47–S64.

Graham, W.M., Field, J.G., and Potts, D.C. (1992). Persistent "upwelling shadows" and their influence on zooplankton distributions. *Mar. Biol.* **114**, 561–570.

Graham, W.M., and Largier, J.L. (1997). Upwelling shadows as nearshore retention sites: The example of northern Monterey Bay. *Cont. Shelf Res.* **17**, 509–532.

Grant, W.D., and Madsen, O.S. (1979). Combined wave and current interaction with a rough bottom. *J. Geophysical Res.* **84**, 1797–1808.

Grantham, B.A., Eckert, G.L., and Shanks, A.L. (2003). Dispersal potential of marine invertebrates in diverse habitats. *Ecol. Appl.* **13**, S108–S116.

Grimm, V., Reise, K., and Strasser, M. (2003). Marine metapopulations: A useful concept? *Helgol. Mar. Res.* **56**, 222–228.

Guzmán del Próo, S.A., Salinas, F., Zaytsez, O., Belmar-Pérez, J., and Carillo-Laguna, J. (2000). Potential dispersion of reproductive products and larval stages of abalone (*Haliotis* spp.) as a function of the hydrodynamics of Bahia Tortuga, Mexico. *J. Shellfish Res.* **19**, 869–881.

Halpern, B. (2003). The impact of marine reserves: Do reserves work and does reserve size matter? *Ecol. Appl.* **13**, S117–S137.

Hamilton, W.D., and May, R.M. (1977). Dispersal in stable habitats. *Nature (London).* **269**, 578–581.

Hamm, D.E., and Burton, R.S. (2000). Population genetics of black abalone, *Haliotis cracherodii*, along the central California coast. *J. Exp. Mar. Biol. Ecol.* **254**, 235–247.

Hamner, W.M., and Hauri, I.R. (1977). Fine-scale surface currents in the Whitsunday Islands, Queensland, Australia: Effect of tide and topography. *Aust. J. Mar. Freshwat. Res.* **28**, 333–359.

Hancock, B. (2000). Genetic subdivision of Roe's abalone, *Haliotis roei*, Grey (Mollusca: Gastropoda), in south-western Australia. *Mar. Freshwater Res.* **52**, 679–687.

Hanski, I. (1991). Single-species metapopulation dynamics: Concepts, models and observations. *Biol. J. Linn. Soc.* **42**, 17–38.

Hanski, I., and Gilpin, M. (1991). Metapopulation dynamics: Brief history and conceptual domain. *Biol. J. Linn. Soc.* **42**, 3–16.

Hara, M., and Fujio, Y. (1992). Geographic distribution of isozyme genes in natural abalone. *Bull. Tohoku Natl Fish. Res. Inst.* **54**, 115–124.

Hara, M., and Kikuchi, S. (1992). Genetic variability and population structure in the abalone *Haliotis discus hannai*. *Bull. Tohoku Natl Fish. Res. Inst.* **54**, 107–114.

Harris, L.G., Williams, C.T. Chester, C.M., Tyrrell, M., Sisson, C., and Chavanich, S. (2001). Declining sea urchin recruitment in the Gulf of Maine: Is overfishing to blame? In *Echinoderms 2000: Proceedings of the 10th International Echinoderm Conference* (M.F. Barker, ed., pp. 439–444). Rotterdam: Balkema.

Harrison, S. (1991). Local extinction in a metapopulation context: An empirical evaluation. *Biol. J. Linn. Soc.* **42**, 73–88.

Harrold C., Lisin, S., Light, K., and Tudor, S. (1991). Isolating settlement from recruitment of sea urchins. *J. Exp. Mar. Biol. Ecol.* **147**, 81–94.

Harrold, C., and Pearse, J.S. (1987). The ecological role of echinoderms in kelp forests. In *Echinoderm studies* (M. Jangoux and J.M. Lawrence, eds., vol. 2., pp. 137–233). Rotterdam: Balkema.

Hart, M.W., and Scheibling, R.E. (1988). Heat waves, baby booms, and the destruction of kelp beds by sea urchins. *Mar. Biol.* **99**, 167–176.

Hedgecock, D. (1994). Temporal and spatial genetic structure of marine animal populations in the California current. *CalCOFI Rep.* **35**, 73–81.

Heslinga, G.A. (1981). Larval development, settlement and metamorphosis of the tropical gastropod *Trochus niloticus*. *Malacologia.* **20**, 249–257.

Heslinga, G.A., Orak, O., and Ngiramengior, M. (1984). Coral reef sanctuaries for trochus snails. *Mar. Fish. Rev.* **46**, 73–80.

Hjort, J. (1914). Fluctuations in the great fisheries of northern Europe viewed in the light of biological research. *Rapp. P. V. Reun. Cons. Perm. Int. Explor. Mer.* **20**, 1–228.

Hobbs, R.C., Botsford, L.W., and Thomas, A. (1992). Influence of hydrographic conditions and wind forcing on the distribution and abundance of Dungeness crab, *Cancer magister*, larvae. *Can. J. Fish. Aquat. Sci.* **49**, 1379–1388.

Holland, L.Z., Giese, A.C., and Phillips, J.H. (1967). Studies on the perivisceral coelomic fluid protein concentration during seasonal and nutritional changes in the purple sea urchin. *Comp. Biochem. Physiol.* **21**, 361–371.

Holm, E.R. (1990). Effects of density-dependent mortality on the relationship between recruitment and larval settlement. *Mar. Ecol. Prog. Ser.* **60**, 141–146.

Howe, R.W., Davis, G.J., and Mosca, V. (1991). The demographic significance of "sink" populations. *Biol. Cons.* **57**, 239–256.

Itosu, C., and Miki, M. (1983). Field observation of wave-induced circulation in ormer fishing ground. *Bull. Japan. Soc. Sci. Fish.* **49**, 339–346.

Iwasa, Y., and Roughgarden, J. (1985). Evolution in a metapopulation with space-limited subpopulations. *IMA J. Math. Appl. Med. Biol.* **2**, 93–107.

Jackson, G.A. (1998). Currents in the high drag environment of a coastal kelp stand off California. *Cont. Shelf Res.* **17**, 1913–1928.

Jiang, L., Wu, W.L., and Huang, P.C. (1995). The mitochondrial DNA of Taiwan abalone *Haliotis diversicolor* Reeve, 1846 (Gastropoda, Archaeogastropoda: Haliotidae). *Molec. Mar. Biol. Biotech.* **4**, 353–364.

Jonsson, P.R., Andre, C., and Lindegrath, M. (1991). Swimming behaviour of marine bivalve larvae in a flume boundary-layer flow: Evidence for near-bottom confinement. *Mar. Ecol. Prog. Ser.* **79**, 67–76.

Kalvass, P., and Taniguchi, I. (1993). Relative abundance and size composition of red sea urchin, *Strongylocentrotus franciscanus*, populations along the Mendocino County coast, 1991. *Marine Resources Administrative Report* No. 93-1. State of California, Department of Fish and Game.

Kalvass, P., Taniguchi, I., and Buttolph, P. (1990). Relative abundance and size composition of red sea urchin, *Strongylocentrotus franciscanus*, populations along the Mendocino and Sonoma County coasts, August 1988. *Marine Resources Administrative Report* No. 90-1. State of California, Department of Fish and Game.

Kalvass, P., Taniguchi, I., Buttolph, P., and DeMartini, D. (1991). Relative abundance and size composition of red sea urchin, *Strongylocentrotus franciscanus*, populations along the Mendocino and Sonoma County coasts, 1989. *Marine Resources Administrative Report* No. 91-3. State of California, Department of Fish and Game.

Karpov, K., Haaker, P.L., Taniguchi, I., and Rogers–Bennett, L. (2000). Serial depletion and the collapse of the California abalone fishery. *Can. Spec. Publ. Fish. Aquat. Sci.* **130**, 11–21.

Kato, S., and Schroeter, S.C. (1985). Biology of the red sea urchin, *Strongylocentrotus franciscanus*, and its fishery in California. *Mar. Fish. Rev.* **47**, 1–20.

Kritzer J.P., and Sale, P.F. (2004). Metapopulation ecology in the sea: From Levins' model to marine ecology and fisheries science. *Fish Fisheries.* **5**, 131–140.

Largier, J.L. (2003). Considerations in estimating larval dispersal distances from oceanographic data. *Ecol. Appl.* **13**, S71–S89.

Larson, B.R., Vadas, R.L., and Kesar, K. (1980). Feeding and nutritional ecology of the sea urchin *Strongylocentrotus droebachiensis* in Maine, USA. *Mar. Biol.* **59**, 49–62.

Larson, R.J., Lenarz, W.H., and Ralston, S. (1994). The distribution of pelagic juvenile rockfish of the genus *Sebastes* in the upwelling region off central California. *CalCOFI Rep.* **35**, 175–221.

Lawrence, J.H. (2001). *Edible sea urchins: Biology and ecology.* Amsterdam: Elsevier.

Levins, R. (1969). Some demographic and genetic consequences of environmental heterogeneity for biological control. *Bull. Entomol. Soc. Amer.* **15**, 237–240.

Levitan, D.R. (1989). Density-dependent size regulation in *Diadema antillarum*: Effects on fecundity and survivorship. *Ecology.* **70**, 1414–1424.

Levitan, D.R. (1991). Influence of body size and population density on fertilization success and reproductive output in a free spawning invertebrate. *Biol. Bull. (Woods Hole).* **18**, 261–268.

Levitan, D.R., Sewell, M.A., and Chia, F.S. (1992). How distribution and abundance influence fertilization success in the sea urchin *Strongylocentrotus franciscanus*. *Ecology.* **73**, 248–254.

Lipcius, R.N., Crowder, L.B., and Morgan, L.E. (2005) Conservation of populations and metapopulations by marine reserves: Caveats, conceptual framework, and an opportunity for optimality. In *Marine conservation biology: The science of maintaining the sea's biodiversity* (E.A. Norse and L.B. Crowder, eds. pp 328–345). Covelo: Island Press.

Lipcius, R.N., Stockhausen, W.T., and Eggleston, D.B. (2001). Marine reserves for Caribbean spiny lobster: Empirical evaluation and theoretical metapopulation recruitment dynamics. *Mar. Freshwater Res.* **52**, 1589–1598.

Lundquist, C.J. (2000). *Effects of density dependence and environment on recruitment of coastal invertebrates.* PhD dissertation. Davis: University of California.

Lundquist, C.J., and Botsford, L.W. (2004). Model projections of the fishery implications of the Allee effect in broadcast spawners. *Ecol. Appl.* **14**, 929–941.

Lundquist, C.J., Botsford, L.W., Morgan, L.E., Diehl, J.A., Lee, T., Lockwood, D.R., and Pearson, E.L. (2000). Effects of El Niño and La Niña on local invertebrate settlement in northern California. *CalCOFI Rep.* **41**, 167–176.

MaCall, A.D. (1990). *Dynamic geography of marine fish populations.* Seattle: Washington Sea Grant Program.

Madigan, S.M. (2001). *Larval and juvenile biology of the abalone Haliotis laevigata and Haliotis rubra.* PhD thesis. Adelaide: Flinders University.

Mayfield, S., and Branch, G.M. (2000). Interrelations among rock lobsters, sea urchins, and juvenile abalone: Implications for community management. *Can. J. Fish. Aquat. Sci.* **57**, 2175–2185.

McCay, B.J. (1988). Muddling through the clam beds: Cooperative management of New Jersey's hard clam spawner sanctuaries. *J. Shellfish Res.* **79**, 327–340.

McConnaughey, R.A., Armstrong, D.A., Hickey, B.M., and Gunderson, D.R. (1992). Juvenile Dungeness crab (*Cancer magister*) recruitment variability and oceanic transport during the pelagic larval phase. *Can. J. Fish. Aquat. Sci.* **49**, 2028–2034.

McShane, P.E. (1992). Early life history of abalone: A review. In *Abalone of the world: Biology, fisheries and culture* (S.A. Shepherd, M.E. Tegner, and S.A. Guzmán del Próo, eds., pp 120–138). Oxford: Blackwell.

McShane, P.E. (1993). Evidence for localized recruitment failure in the New Zealand abalone *Haliotis iris* (Mollusca: Gastropoda). In Proceedings of the Second International Temperate Reef Symposium (C.N. Battershill, D.R. Schiel, G.P. Jones, R.G. Creese, and A.B. McDiarmid, eds., pp. 145–150). Wellington: NIWA.

McShane, P.E. (1995a). Estimating the abundance of abalone: The importance of patch size. *Mar. Freshwater Res.* **46**, 657–662.

McShane, P.E. (1995b). Recruitment variation in abalone: Its importance to fisheries management. *Mar. Freshwater Res.* **46**, 555–570.

McShane, P.E. (1996). Recruitment processes in abalone (*Haliotis* spp). In *Survival strategies in early life history stages of marine resources* (Y. Watanabe, Y. Yamashita, and Y. Oozeki, eds., pp. 315–324). Rotterdam: Balkema.

McShane, P.E., and Anderson, O.F. (1997). Resource allocation and growth rates in the sea urchin *Evechinus chloroticus* (Echinoidea: Echinometridae). *Mar. Biol.* **128**, 657–663.

McShane, P.E., Black, K.P., and Smith, M.G. (1988). Recruitment processes in *Haliotis rubra* (Mollusca: Gastropoda) and regional hydrodynamics in southeastern Australia imply localized dispersal of larvae. *J. Exp. Mar. Biol. Ecol.* **124**, 175–203.

Miller, B.A., and Emlet, R.B. (1997). Influence of nearshore hydrodynamics on larval abundance and settlement of sea urchins *Strongylocentrotus franciscanus* and *S. purpuratus* in the Oregon upwelling zone. *Mar. Ecol. Prog. Ser.* **148**, 83–94.

Mladenov, P.V., Allibone, R.M., and Wallis, G.P. (1997). Genetic differentiation in the New Zealand sea urchin *Evechinus chloroticus* (Echinodermata: Echinoidea). *N.Z. J. Mar. Freshw. Res.* **31**, 261–269.

Moberg, P.E., and Burton, R.S. (2000). Genetic heterogeneity among adult and recruit red sea urchins, *Strongylocentrotus franciscanus*. *Mar. Biol.* **136**, 773–784.

Morgan, L.E., and Botsford, L.W. (2001). Managing with reserves: Modeling uncertainty in larval dispersal for a sea urchin fishery. In *Spatial processes and management of marine populations* (G.H. Kruse, N. Bez, A. Booth, M.W. Dorn, S. Hills, R.N. Lipcius, D. Pelletier, C. Roy, S.J. Smith, and D. Witherell, eds., pp. 667–684). Fairbanks: University of Alaska Sea Grant.

Morgan, L.E., Botsford, L.W., Lundquist, C.J., and Quinn, J.F. (1999). The potential of no-take marine reserves to sustain the red sea urchin (*Strongylocentrotus franciscanus*) fishery in northern California. *Bull. Tohoku Natl Fish. Res. Inst.* **62**, 83–94.

Morgan, L.E., Botsford, L.W., Wing, S.R., and Smith, B.D. (2000a). Spatial variability in growth and mortality of the red sea urchin, *Strongylocentrotus franciscanus* (Agassiz), in northern California. *Can. J. Fish. Aquat. Sci.* **57**, 980–992.

Morgan, L.E., Wing, S.R., Botsford, L.W., Lundquist, C.J., and Diehl, J.M. (2000b). Spatial variability in red sea urchin (*Strongylocentrotus franciscanus*) recruitment in northern California. *Fish. Oceanogr.* **9**, 83–98.

Morse, D.E. (1992). Molecular mechanisms controlling metamorphosis and recruitment in abalone larvae. In *Abalone of the world: Biology, fisheries and culture* (S.A. Shepherd, M.J. Tegner, and S.A. Guzmán del Próo, eds., pp. 107–119). Oxford: Blackwell.

Nash, W.J. (1992). An evaluation of egg-per-recruit analysis as a means of assessing size limits for blacklip abalone (*Haliotis rubra*) in Tasmania. In *Abalone of the world: Biology, fisheries and culture* (S.A. Shepherd, M.J. Tegner, and S.A. Guzmán del Próo, eds., pp. 318–340). Oxford: Blackwell.

Neilson, J.D., and Perry, R.I. (1990). Diel vertical migrations of marine fishes: An obligate or facultative process? *Adv. Mar. Biol.* **26**, 115–168.

Nie, Z.-Q. (1992). A review of abalone culture in China. In Abalone of the world: Biology, fisheries and culture (S.A. Shepherd, M.J. Tegner, and S.A. Guzmán del Próo, eds., pp. 592–602). Oxford: Blackwell.
Palumbi, S.R. (1995). Using genetics as an indirect estimator of larval dispersal. In Ecology of marine invertebrate larvae (L. McEdward, ed., pp. 369–387). New York: CRC Press.
Palumbi, S.R. (2003). Population genetics, demographic connectivity, and the design of marine reserves. Ecol. Appl. 13, S146–S158.
Palumbi, S.R., and Warner, R.R. (2003). Why gobies are like hobbits. Science. 299, 51–52.
Parrish, R.H., Nelson, C.S., and Bakun, A. (1981). Transport mechanisms and reproductive success of fishes in the California current. Biol. Oceanogr. 1, 175–203.
Pearse, J.S., and Cameron, R.A. (1991). Echinodermata: Echinoidea. In Reproduction of marine invertebrates. Vol. VI. Echinoderms and lophophorates (A.C. Giese, J.S. Pearse, and V.B. Pearse, eds., pp. 513–662). Pacific Grove, CA: Boxwood Press.
Pearse, J.S., and Hines, A.H. (1979). Expansion of a central California kelp forest following the mass mortality of sea urchins. Mar. Biol. 51, 83–91.
Pennington, J.T. (1985). The ecology and fertilisation of echinoid eggs: The consequences of sperm dilution, adult aggregation, and synchronous spawning. Biol. Bull. 169, 417–430.
Pfister, C.A., and Bradbury, A. (1996). Harvesting red sea urchins: Recent effects and future predictions. Ecol. Appl. 6, 298–310.
Polacheck, T. (1990). Year around closed areas as a management tool. Nat. Resource Mod. 4, 327–354.
Prince, J.D. (1989). The fisheries biology of the Tasmanian stocks of Haliotis rubra. PhD dissertation. Tasmania: Department of Zoology, University of Tasmania.
Prince, J.D. (2003). The barefoot ecologist goes fishing. Fish Fisheries. 4, 359–371.
Prince, J.D., and Hilborn, R. (1998). Concentration profiles and invertebrate fisheries management. In Proceedings of the North Pacific Symposium on Invertebrate Stock Assessment and Management (G.S. Jamieson and A. Campbell, eds.) Can. Spec. Publ. Fish. Aquat. Sci. 125, 187–196.
Prince, J.D., Sellers, T.L., Ford, W.B., and Talbot, S.R. (1987). Experimental evidence for limited dispersal of haliotid larvae (genus Haliotis) (Mollusca: Gastropoda). J. Exp. Mar. Biol. Ecol. 106, 243–263.
Prince, J.D., Sellers, T.L., Ford, W.B., and Talbot, S.R. (1988). Confirmation of a relationship between the localized abundance of breeding stock and recruitment for Haliotis rubra Leach (Mollusca: Gastropoda). J. Exp. Mar. Biol. Ecol. 122, 91–104.
Prince, J.D., Walters, C., Ruiz–Avila, R., and Sluczanowski, P. (1998). Territorial user's rights and the Australian abalone (Haliotis sp.) fishery. In Proceedings of the North Pacific Symposium on Invertebrate Stock Assessment and Management (G.S. Jamieson and A. Campbell eds). Can. Spec. Publ. Fish. Aquat. Sci. 125, 367–375.
Pulliam, H.R. (1988). Sources, sinks, and population regulation. Am. Nat. 132, 652–661.
Quinn, J.F., Wing, S.R., and Botsford, L.W. (1993). Harvest refugia in marine invertebrate fisheries: Models and applications to the red sea urchin, Strongylocentrotus franciscanus. Am. Zool. 33, 537–550.
Raimondi, P.T., and Keough, M.J. (1990). Behavioural variability in marine larvae. Aust. J. Ecol. 15, 427–437.
Richardson, S.L., and Pearcy, W.G. (1977). Coastal and oceanic fish larvae in an area of upwelling off Yaquina Bay, Oregon. Fish. Bull. 75, 125–145.
Roberts, C.M. (1997). Connectivity and management of Caribbean coral reefs. Science. 278, 1454–1457.
Roberts, R. (2001). A review of settlement cues for larval abalone (Haliotis spp.) J. Shellfish Res. 20, 571–586.
Robinson, S.M., and MacIntyre, A.M. (1997). Ageing and growth of the green sea urchin, Bull. Aquaculture Assoc. Canada 97, 56–60.

Rodda, K.R., Keesing, J.K., and Foureur, B.L. (1997). Variability of larval settlement of abalone on larval collectors. *Molluscan Res.* **18**, 253–264.

Rodda, K.R., Mayfield, S., and Shepherd, S.A. (2002). Translocation of greenlip abalone in South Australia. *South Australian Research and Development Institute Report* (33 pp). Adelaide: South Australia Research and Development Institute.

Rodriguez, J.P. (2002). Range contraction in declining North American bird populations. *Ecol. Appl.* **12**, 238–248.

Roff, D. (2002). *Life history evolution.* Sunderland, MA: Sinauer Associates.

Rogers-Bennett, L., Bennett, W.A., Fastenau, H.C., and Dewees, C.M. (1995). Spatial variation in red sea urchin reproduction and morphology: Implications for harvest refugia. *Ecol. Appl.* **5**, 1171–1180.

Rogers-Bennett, L., Haaker, P.L., Huff, T.O., and Dayton, P.K. (2002a). Estimating baseline abundances of abalone in California for restoration. *CalCOFI Rep.* **43**, 97–111.

Rogers-Bennett, L., Haaker, P.L., Karpov, K.A., and Kushner, D.J. (2002b). Using spatially explicit data to evaluate marine protected areas for abalone in southern California. *Conserv. Biol.* **16**, 1308–1317.

Rogers-Bennett, L., and Pearse, J.S. (2001). Indirect benefits of marine protected areas for juvenile abalone. *Conserv. Biol.* **15**, 642–647.

Rogers-Bennett, L., Rogers, D.W., Bennett, W.A., and Ebert, T.A. (2003). Modeling red sea urchin (*Strongylocentrotus franciscanus*) growth using six growth functions. *Fish. Bull.* **101**, 614–626.

Roughgarden, J., Iwasa, Y., and Baxter, C. (1985). Dynamics of a metapopulation with space-limited subpopulations. *Ecology.* **66**, 54–67.

Roughgarden, J., Pennington, J.T., Stoner, D., Alexander, S., and Miller, K. (1992). Collisions of upwelling fronts with the intertidal zone: The cause of recruitment pulses in barnacle populations of central California. *Acta Oecologia.* **12**, 35–51.

Rowley, R.J. (1989). Settlement and recruitment of sea urchins (*Strongylocentrotus* spp.) in a sea urchin barren ground and a kelp bed: Are populations regulated by settlement or postsettlement processes? *Mar. Biol.* **100**, 485–494.

Rowley, R.J. (1990). Newly settled sea urchins in a kelp bed and urchin barren ground: A comparison of growth and mortality. *Mar. Ecol. Prog. Ser.* **62**, 229–240.

Russell, M.P. (1987). Life history traits and resource allocation in the purple sea urchin *Stongylocentrotus purpuratus* (Stimpson). *J. Exp. Mar. Biol. Ecol.* **108**, 199–216.

Sale, P.F., and Kritzer, J.P. (2003). Determining the extent and spatial scale of population connectivity: Decapods and coral reef fishes compared. *Fish. Res.* **65**, 153–172.

Sammarco, P.W., and Andrews, J.C. (1988). Localized dispersal and recruitment in Great Barrier Reef corals: The Helix experiment. *Science.* **239**, 1422–1424.

Sasaki, R., and Shepherd, S.A. (1995). Larval dispersal and recruitment of *Haliotis discus hannai* and *Tegula* spp. on Miyagi coasts, Japan. *Mar. Freshwater Res.* **46**, 519–529.

Sasaki, R., and Shepherd, S.A. (2001). Ecology and postsettlement survival of the Ezo abalone, *Haliotis discus hannai*, on Miyagi coasts, Japan. *J. Shellfish Res.* **20**, 619–626.

Scheibling, R.E. (1996). The role of predation in regulating sea urchin populations in eastern Canada. *Oceanologica Acta.* **19**, 421–430.

Scheibling, R.E., and Hennigar, A.W. (1997). Recurrent outbreaks of disease in sea urchins *Strongylocentrotus droebrachiensis* in Nova Scotia: Evidence for a link with large-scale meteorological and oceanographic events. *Mar. Ecol. Prog. Ser.* **152**, 155–165.

Scheltema, R.S. (1986). On dispersal and planktonic larvae of benthic invertebrates: An eclectic overview and summary of problems. *Bull. Mar. Sci.* **39**, 290–322.

Seki, T., and Kan-no, H. (1981). Induced settlement of the Japanese abalone, *Haliotis discus hannai*, veliger by the mucous trails of the juvenile and adult abalones. *Bull. Tohoku Reg. Fish. Lab.* **43**, 29–36.

Shaffer, M.L. (1981). Minimum population sizes for species conservation. *Bioscience.* **31**, 131–134.
Shanks, A.L. (1983). Surface slicks associated with tidally forced internal waves may transport pelagic larvae of benthic invertebrates and fishes shoreward. *Mar. Ecol. Prog. Ser.* **13**, 311–315.
Shanks, A.L. (1986). Tidal periodicity in the daily settlement of intertidal barnacle larvae and an hypothesized mechanism for the cross-shelf transport of cyprids. *Biol. Bull.* **170**, 429–440.
Shanks, A.L. (1995). Mechanisms of cross-shelf dispersal of larval invertebrates and fish. In *Ecology of marine invertebrate larvae* (L.R. McEdward, ed., pp. 323–367). Boca Raton, FL: CRC Press.
Shanks, A.L., Grantham, B.A., and Carr M.H. (2003). Propagule dispersal distance and the size and spacing of marine reserves. *Ecol. Appl.* **13**, S159–S169.
Shenker, J.M. (1988). Oceanographic associations of neustonic larval and juvenile fishes and Dungeness crab megalopae off Oregon. *Fish. Bull.* **86**, 299–319.
Shepherd, S.A. (1973). Studies on southern Australian abalone (genus *Haliotis*). I. Ecology of five sympatric species. *Aust. J. Mar. Freshw. Res.* **24**, 217–257.
Shepherd, S.A. (1986a). Movement of the South Australian abalone, *Haliotis laevigata*, in relation to crevice abundance. *Aust. J. Ecol.* **11**, 295–302.
Shepherd, S.A. (1986b). Studies on southern Australian abalone (genus *Haliotis*). VII. Aggregative behaviour of *H. laevigata* in relation to spawning. *Mar. Biol.* **90**, 231–236.
Shepherd, S.A., and Baker, J.L. (1998). Biological reference points in an abalone (*Haliotis laevigata*) fishery. In *Proceedings of the North Pacific Symposium on Invertebrate Stock Assessment and Management* (G.S. Jamieson and A. Campbell, eds.). *Can. Spec. Publ. Fish. Aquat. Sci.* **125**, 235–245.
Shepherd, S.A., and Breen, P.A. (1992). Mortality in abalone: Its estimation, variability and causes. In *Abalone of the world: Biology, fisheries and culture* (S.A. Shepherd, M.J. Tegner, and S.A. Guzmán del Próo, eds., pp. 276–304). Oxford: Blackwell.
Shepherd, S.A., and Brown, L.D. (1993). What is an abalone stock? Implications for the role of refugia in conservation. *Can. J. Fish. Aquat. Sci.* **50**, 2001–2009.
Shepherd, S.A., and Daume, S. (1996). Ecology and survival of juvenile abalone in a crustose coralline habitat in South Australia. In *Survival strategies in early life history stages of marine resources* (Y. Watanabe, Y. Yamashita, and Y. Oozeki, eds., pp. 297–313). Rotterdam: Balkema.
Shepherd, S.A., Lowe, D., and Partington, D. (1992a). Studies on southern Australian abalone (genus *Haliotis*). XIII. Larval dispersal and recruitment. *J. Exp. Mar. Biol. Ecol.* **164**, 247–260.
Shepherd, S.A., and Partington, D. (1995). Studies on southern Australian abalone (genus *Haliotis*). XVI. Recruitment, habitat and stock relations. *Mar. Freshwater Res.* **46**, 475–492.
Shepherd, S.A., Preece, P.A., and White, R.W.G. (2000). Tired nature's sweet restorer? Ecology of stock enhancement in Australia. In *Workshop on rebuilding abalone stocks in British Columbia* (A. Campbell, ed.). *Can. Spec. Publ. Fish. Aquat. Sci.* **130**, 84–97.
Shepherd, S.A., and Rodda, K.R. (2001). Sustainability demands vigilance: Evidence for serial decline of the greenlip abalone fishery and a review of management. *J. Shellfish Res.* **20**, 829–841.
Shepherd, S.A., Rodda, K.R., and Vargas, K.M. (2001). A chronicle of collapse in two abalone stocks with proposals for precautionary management. *J. Shellfish Res.* **20**, 843–856.
Shepherd, S.A., Tegner, M.J., and Guzmán del Próo, S.A., eds. (1992b). *Abalone of the world: Biology, fisheries and culture.* Oxford: Blackwell.
Shepherd, S.A., Turrubiates, J.R., and Hall, K. (1998). Decline of the abalone fishery at La Natividad, México: Overfishing or climate change? *J. Shellfish Res.* **17**, 839–846.
Shepherd, S.A., and Womersley, H.B.S. (1981). The algal and seagrass ecology of Waterloo Bay, South Australia. *Aquat. Bot.* **11**, 305–371.
Sinclair, M. (1988). *Marine populations: An essay on population regulation and speciation.* Washington Sea Grant. Seattle: University of Washington Press.
Slattery, M. (1992). Larval settlement and juvenile survival in the red abalone (*Haliotis rufescens*): An examination of inductive cues and substratum selection. *Aquaculture.* **102**, 143–153.

Sloan, N.A., Lauridsen, C.P., and Harbo, R.M. (1987). Recruitment characteristics of the commercially harvested red sea urchin, *Strongylocentrotus franciscanus*, in southern British Columbia. *Fish. Res.* **5**, 55–69.
Smedbol, R.K., McPherson, A., Hansen, M.M., and Kenchington, E. (2002). Myths and moderation in marine "metapopulations?" *Fish Fisheries.* **3**, 20–35.
Stobutzki, I.C., and Bellwood, D.R. (1997). Sustained swimming abilities of the late pelagic stages of coral reef fishes. *Mar. Ecol. Prog. Ser.* **149**, 35–41.
Strathmann, R.R. (1978). The length of pelagic period in echinoderms with feeding larvae from the northeastern Pacific. *J. Exp. Mar. Biol. Ecol.* **34**, 23–27.
Strathmann, R.R. (1987). *Reproduction and development of marine invertebrates of the northern Pacific coast.* Seattle: University of Washington Press.
Swearer, S.E., Caselle, J.E., Lea, D.W., and Warner, R.R. (1999). Larval retention and recruitment in an island population of a coral-reef fish. *Nature.* **402**, 799–802.
Tanaka, K., Tanaka, T., Ishida, O., and Oba, T. (1986). On the distribution of swimming and deposited larvae of nursery ground of abalone at the southern coast off Chiba Prefecture. *Bull. Jpn Soc. Sci. Fish.* **52**, 1525–1532.
Taylor, M.S., and Hellberg, M.E. (2003). Genetic evidence for local retention of pelagic larvae in a Caribbean reef fish. *Science.* **299**, 107–109.
Tegner, M.J. (1989). The feasibility of enhancing red sea urchin, *Strongylocentrotus franciscanus*, stocks in California: An analysis of the options. *Mar. Fish. Rev.* **51**, 1–22.
Tegner, M.J. (1992). Brood-stock transplants as an approach to abalone stock enhancement. In *Abalone of the world: biology, fisheries and culture* (S.A. Shepherd, M.J. Tegner, and S.A. Guzmán del Próo, eds., pp. 461–473). Oxford: Blackwell.
Tegner, M.J. (1993). Southern California abalones: Can stocks be rebuilt using harvest refugia? *Can. J. Fish. Aquat. Sci.* **50**, 2010–2918.
Tegner, M.J., and Butler, R.A. (1985). Drift-tube study of the dispersal potential of green abalone (*Haliotis fulgens*) larvae in the southern California Bight: Implications for recovery of depleted populations. *Mar. Ecol. Prog. Ser.* **26**, 73–84.
Tegner, M.J., and Dayton, P.K. (1977). Sea urchin recruitment patterns and implications for commercial fishing. *Science.* **196**, 324–326.
Tegner, M.J., and Dayton, P.K. (1991). Sea urchins, El Niños, and the long term stability of southern California kelp forest communities. *Mar. Ecol. Prog. Ser.* **77**, 49–63.
Tegner, M.J., Dayton, P.K. Edwards, P.B., and Riser, K.L. (1997). Large-scale, low-frequency oceanographic effects on kelp forest succession: A tale of two cohorts. *Mar. Ecol. Prog. Ser.* **146**, 117–134.
Tegner, M.J., and Levin, L.A. (1983). Spiny lobsters and sea urchins: Analysis of a predator–prey interaction. *J. Exp. Mar. Biol. Ecol.* **73**, 125–150.
Tuck, G.N., and Possingham, H.P. (1994). Optimal harvesting strategies for a metapopulation. *Bull. Mathematical Biol.* **56**, 107–127.
Tuck, G.N., and Possingham, H.P. (2000). Marine protected areas for spatially structured exploited stocks. *Mar. Ecol. Prog. Ser.* **192**, 89–101.
Tutschulte, T.C. (1976). The comparative ecology of three sympatric species. PhD dissertation. San Diego: University of California.
Vadas, R.L. (1977). Preferential feeding: An optimization strategy in sea urchins. *Ecol. Monogr.* **47**, 337–371.
Vadas, R.L., Sr., Smith, B.D., Beal, B., and Dowling, T. (2002). Sympatric growth morphs and size bimodality in the green sea urchin (*Strongylocentrotus droebachiensis*). *Ecol. Monogr.* **72**, 113–132.
Walker, M.M. (1981). Influence of season on growth of the sea urchin *Evechinus chloroticus*. *N.Z. J. Mar. Freshw. Res.* **15**, 201–205.
Wallace, S.S. (1998). Evaluating the effects of three forms of marine reserve on northern abalone populations in British Columbia, Canada. *Conserv. Biol.* **13**, 882–887.

Wells, F.E., and Keesing, J.K. (1990). Population characteristics of the abalone, *Haliotis roei*, on intertidal platforms in the Perth metropolitan area. *J. Malac. Soc. Aust.* **11**, 65–71.

Wells, F.E., and Mulvay, P. (1995). Good and bad fishing areas for *Haliotis laevigata*: A comparison of population parameters. In *Progress in abalone fisheries research* (S.A. Shepherd, R.W. Day, and A.J. Butler, eds.). *Mar. Freshw. Res.* **46**, 689–695.

Wing, S.R., Botsford, L.W., Largier, J.L., and Morgan, L.E. (1995a). Spatial structure of relaxation events and crab settlement in the northern California upwelling system. *Mar. Ecol. Prog. Ser.* **28**, 199–211.

Wing, S.R., Botsford, L.W., Largier, J.L., and Ralston, S. (1998). Meroplankton distribution and circulation associated with a coastal retention zone in the northern California upwelling zone. *Limnol. Oceanogr.* **43**, 1710–1721.

Wing, S.R., Botsford, L.W., Morgan, L.E., Diehl, J.M., and Lundquist, C.J. (2003a). Inter-annual variability in larval supply to populations of three invertebrate taxa in the northern California Current. *Estuar. Coast. Shelf Sci.* **57**, 859–872.

Wing, S.R., Gibb, M.T., and Lamare, M.D. (2003b). Reproductive sources and sinks within a sea urchin, *Evechinus chloroticus*, population of a New Zealand fjord. *Mar. Ecol. Prog. Ser.* **248**, 109–123.

Wing, S.R., Lamare, M.D., and Vasques, J. (2001). Population structure of sea urchins (*Evechinus chloroticus*) along gradients in benthic productivity in the New Zealand fjords. In *Echinoderms 2000* (M.F. Barker, ed., pp. 56–75). Rotterdam: Balkema.

Wing, S.R., Largier, J.L., Botsford, L.W., and Quinn, J.F. (1995b). Settlement and transport of benthic invertebrates in an intermittent upwelling zone. *Limnol. Oceangr.* **40**, 316–329.

Withler, R.E. (2000). Genetic tools for identification and conservation of exploited abalone (*Haliotis* spp.) species. In *Workshop on rebuilding abalone stocks in British Columbia* (A. Campbell, ed.). *Can. Spec. Publ. Fish. Aquat. Sci.* **130**, 101–110.

Withler, R.E., Campbell, A., Li, S., Brouwer, D., Supernault, K.J., and Miller, K.M. (2003). Implications for high levels of genetic diversity and weak population structure for the rebuilding of northern abalone in British Columbia, Canada. *J. Shellfish Res.* **22**, 839–847.

Wolanski, E., and Hamner, W. (1988). Topographically-controlled fronts in the ocean, and their influence on the distribution of organisms. *Science.* **241**, 177–181.

Worthington, D., and Andrew, N.L. (1998). Small-scale variation in demography and its implications for an alternative size limit in the fishery for blacklip abalone (*Haliotis rubra*) in New South Wales, Australia. *Can. Spec. Publ. Fish. Aquat. Sci.* **125**, 341–348.

Yano, I., and Ogawa, Y. (1977). Effects of light, temperature and hydrostatic pressure on vertical distribution of abalone larvae, *Haliotis gigantea*. *Bull. Tokai Reg. Fish. Res. Lab.* **91**, 19–27.

Zuñiga, G., Guzmán del Próo, S.A., Cisnero, R., and Rodriguez, G. (2000). Population genetic analysis of the abalone *Haliotis fulgens* (Mollusca: Gastropoda) in Baja California, Mexico. *J. Shellfish Res.* **19**, 853–859.

CHAPTER 7

Rocky Intertidal Invertebrates: The Potential for Metapopulations within and among Shores

MARK P. JOHNSON

I. Introduction
II. Patch Models
III. Within-Shore Metapopulations
IV. Metapopulations at Different Scales
V. Measured Scales of Variability
VI. Summary
   References

I. INTRODUCTION

The rocky intertidal has traditionally acted as an interface between ecologists from different disciplines. This role is reflected in spatial ecology, for which a variety of metapopulation models have been explicitly or implicitly examined on rocky shores. Metapopulation models (related to Levins' [1969] patch-occupancy approach) are cited in early theoretical work on rocky shore patch mosaics (Levin and Paine, 1974). However, much of the patch dynamic literature that developed in association with the landmark paper by Paine and Levin (1981) did not take an explicitly metapopulation-oriented approach. The Levins metapopulation model has not been widely accepted as a description for marine systems (Grimm et al., 2003). In a database search carried out for this review, only a small fraction (0.2%) of intertidal studies also had a reference to metapopulations in the title, key words, or abstract. The separation of research fields is perhaps surprising, because metapopulation biology and intertidal ecology share an appreciation of dispersal as a key theme. Indeed, larval ecology is central to population and community studies in marine ecology (Underwood and Fairweather, 1989;

Connolly and Roughgarden, 1999b). Despite wide interest in the variability of larval supply (Shkedy and Roughgarden, 1997; Shanks et al., 2000), a metapopulation perspective is not necessarily used by researchers.

This review will follow the lead of other authors (e.g., Hanski, 1997) by starting with the Levins model. Although there are conceptual similarities, links between the patch-based Levins model and patch dynamics have generally not been emphasized in the literature. Because patch dynamics are often described within individual shores, it is possible that metapopulations may exist on individual shores. The evidence for such "within-shore" metapopulations can be evaluated with reference to a limited number of criteria. Moving to larger spatial scales is likely to contradict the formal criteria for a patch-based metapopulation. Few, if any, authors have applied a Levins patch model to networks of separate shores (by considering the presence or absence of populations on individual shores as a balance of extinction–colonization processes). Researchers have preferred more complex models to describe the influence of spatial processes on population dynamics (e.g., Bascompte et al., 2002). Complex spatially structured models can still be considered within the metapopulation framework if a broad definition of "metapopulation" is accepted. In a description of the evolving usage of the terms in spatial ecology, Hanski and Simberloff (1997) suggested that any set of local populations linked by dispersal could be considered to be a metapopulation. Complex spatial models with a range of processes occurring at different scales can therefore be considered to represent metapopulations, as long as dispersal is a key feature.

Complexity in spatial models arises partly from the ways in which population dynamics can be described and partly from the links that can be specified between local populations. Analyses of models frequently suggest that linking by dispersal has fundamental consequences for the sizes and dynamics of populations. The complexity of models may, however, restrict the generality of conclusions. Further applications of the metapopulation approach are likely to need a better understanding of the relevant scales of demographic processes, based on field studies at a range of different spatial scales. Without such studies, there is a danger that the construction of theoretical models involves an arbitrary choice of scales.

Many populations of invertebrates may be best described by metapopulation processes at two levels: within and among shores. Few, if any, species have been comprehensively evaluated within this framework. Alternatively, a metapopulation approach may not be sufficient or necessary in describing large-scale spatial dynamics. Environmental forcing by climate or nearshore oceanography may be more relevant to population variability. Given the potential complexities of spatial population dynamics, the most tractable applications of metapopulations to rocky shores may involve within-shore metapopulations and comparisons between species with different degrees of dispersal.

II. PATCH MODELS

The Levins (1969) metapopulation model is a patch dynamic model that describes colonization and extinction processes in a landscape of patches. There is no requirement that the patches are fixed in time and space (like oceanic islands). Patches can be defined as gaps in a homogenous reference background, with no restrictions on location, size, or persistence (Levin and Paine, 1974). Using an inclusive definition for patches has unified different areas of ecology. Patch dynamics has provided a general description for a variety of terrestrial and marine systems (Watt, 1947; Pickett and White, 1985; Levin et al., 1993). The asynchronous turnover of patches can create a heterogeneous landscape in which the possibilities of species coexistence are far larger than would be possible in a homogenous environment (Horn and McArthur, 1972; Levin and Paine, 1974). In particular, species competing to occupy patches can coexist where there are tradeoffs between competitive ability and the colonization of unoccupied patches (Levins and Culver, 1971; Nee and May, 1992).

The patch dynamic perspective has been widely adopted to describe rocky shore communities (Underwood, 2000). It is unusual to find a description of a rocky shore system that does not invoke a spatiotemporal mosaic of patches. This viewpoint emphasizes that variability in time and space may be important for ecosystem structure and function: Patches are more than ecological "noise" (Steele et al., 1993). Patches may be generated by holes in a monoculture (e.g., "bare rock" in mussel beds; Paine and Levin, 1981) or patches may be clumps of individuals within a matrix formed by a different species or substrate (e.g., patches of algae interspersed by barnacles; Hartnoll and Hawkins, 1985). Individual boulders can also constitute patches within a habitat (Sousa, 1979). However, despite some conceptual links between Levins' models and the patch dynamic literature, patch dynamic studies tend not to focus on metapopulation issues. A more normal context for discussing patch dynamic processes is when emphasizing the creation of spatiotemporal heterogeneity and the role of disturbance in ecological systems (Sousa, 1979; Paine and Levin, 1981).

Given the early conceptual links between patch dynamics and Levin's metapopulation model, it is perhaps surprising that the two approaches have not been more formally linked. Patch dynamics literature tends to emphasize habitat diversity within a landscape whereas Levins' original model contains a population of identical patches. This conflict of emphasis may have discouraged researchers from making formal links between the two approaches. However, it can be relatively straightforward to extend Levins–type patch occupancy models to include disturbance, successional processes, substrate heterogeneity, and spatial interactions (e.g., Caswell and Etter, 1993; Gyllenberg and Hanski, 1997; Johnson, 2000c; Amarasekare and Possingham, 2001; Ellner and Fussmann, 2003; Hastings, 2003). These recent developments of theory may provide some of the clearest examples of metapopulation dynamics. Although the likelihood of Levins–type metapopulations has been questioned on empirical and theoretical grounds (Harrison, 1991; Ellner and Fussmann, 2003), the most likely metapopulation examples may be where species occupy transitory patches, which are subsequently lost through succession (Harrison and Taylor, 1997). Patches of bare space in mussel mosaics are lost through succession to mussel dominance (Paine and Levin, 1981). Could the dynamics of species in many rocky shore (patch dynamic) systems therefore be described using metapopulation theory? Both Hanski (1997) and Grimm et al. (2003) suggest criteria for answering such a question. These criteria can be summarized as follows:

1. Patches contain local populations rather than ephemeral aggregations formed by the movement of individuals within a continuous population.
2. The dynamics of local populations should not be synchronous.
3. The occupants of a patch face a considerable risk of local extinction.
4. Patches are linked by dispersal with the possibility of new local populations becoming established in suitable, currently unoccupied patches of habitat.

These metapopulation criteria can be reconciled with patch dynamics approaches. An example is the persistence of the sea palm *Postelsia palmaeformis* on mussel-dominated (*Mytilus californianus*) shores on the Pacific coast of North America (Paine, 1988). Patches of sessile adult algae are found in wave-induced gaps in the mussel bed. Patches are asynchronous and persist for different periods of time before extinction (which occurs through lateral spread of the mussel bed). Drifting individuals can establish *Postelsia* patches in newly created gaps in the mussel bed. Paine (1988) emphasizes that the persistence of *Postelsia* depends on the dynamics of its resource: bare space. Such sentiments are echoed by an emphasis on the availability of suitable habitat in the metapopulation literature (e.g., Hanski et al., 1996).

The links between patch dynamics and patch-based metapopulations are becoming more widely recognized (Amarasekare and Possingham, 2001). Development of these links seems likely to provide new empirical evidence for metapopulation dynamics (Harrison and Taylor, 1997). For rocky shore ecologists, patch dynamics has provided a framework for understanding how the interactions between species-specific traits and availability of suitable habitat contribute to mosaic patterns on the shore (e.g., Sousa, 1984; Menge et al., 1993). Metapopulation theory can complement this understanding by providing a framework for studying the demographics of occupied and unoccupied patches. In the *Postelsia* example, a structured model that incorporates patch age or size may be required (Paine and Levin, 1981; Paine, 1988). For both structured and unstructured models, a metapopulation model allows evaluations of the shorewide persistence of organisms. Persistence of species in a patch dynamic system can be greater or less than expected from an unstructured model, depending on the details of patch demography (Johnson, 2000c; Hastings, 2003). Modeling techniques such as sensitivity tests can be used to compare the importance of different features of a system and to interpret system-level statistics such as the fraction of unoccupied patches. For example, a shorewide metapopulation may be more affected by the extent of the shore or by climate-driven changes in fecundity and recruitment events. Models provide mechanisms for the synthesis and comparison of elements in a system. This integrated perspective has the potential to provide tractable predictions for community responses to changes in the disturbance regime on shores (e.g., changes in wave climate).

III. WITHIN-SHORE METAPOPULATIONS

A restriction on Levins–type metapopulations, which does not occur with the looser patch dynamic concept, is that there are essentially only two scales of interest. Patches are assumed to have dynamics that can be reduced to occupied or not, whereas the "regional" scale describes the net change in occupancy in a network of suitable patches. It is decisions on the regional scale that are problematic (Fig. 7-1). The regional scale could refer to a collection of patches within a shore. However, species with extended periods of planktonic development may be demographically reliant on larvae produced at different shores. Although this phenomenon can be incorporated into patch-occupancy models (as a propagule "rain"; Gotelli, 1991), the influence of an external source of larvae may make shore-scale metapopulation models redundant. For example, recruitment to a shore may vary sufficiently from year to year that local populations have synchronous dynamics (in violation of condition 2 listed earlier). Such an external forcing of dynamics would not necessarily prevent a patch dynamic approach. For example, the structure of patches may still be relevant to foraging patterns and trophic interactions. However, with events occurring across a range of scales potentially compromising a metapopulation approach, careful evaluation of the Hanski (1997) and Grimm et al. (2003) criteria becomes relevant if the concept is to be useful.

Dispersal is the key process that determines the spatial scales and structure of populations. A metapopulation is not possible without dispersal between local populations. Too much dispersal between local populations means that the details of local dynamics are not important. If shores are to represent the "regional" level in a metapopulation, the dispersal of individuals should be sufficient to link patches within a shore, but not sufficient to synchronize the dynamics on different shores. Within-shore metapopulation structures may therefore be more common in species with restricted or absent planktonic dispersal stages. Somewhat counterintuitively, rock pool plankton may represent good candidates for

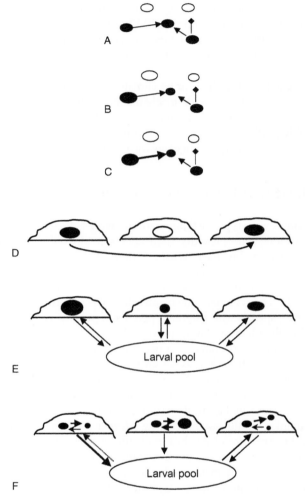

FIGURE 7-1. Examples of metapopulation structures at different scales. Dispersal can be homogenous across the metapopulation (as in A, D, or E) or local populations may have more complex links (e.g., distances between patches or hydrography may affect dispersal in C or F). (A) Patch occupancy within shore. Patches identical (classic or Levins–type metapopulation. (B) Patch occupancy. Patch quality may differ as a result of age, chance of extinction, and so forth. (C) Patches contain populations with dynamics linked by the number of dispersing individuals. (D) Shore occupancy model. Shores considered identical. (E) Shores contain populations. Populations are linked by contribution to a common larval pool. (F) Shores contain multiple populations. Shores contribute to a larval pool, but the larval pool is not redistributed homogenously (e.g., local currents increase or decrease the expected larval supply).

within-shore metapopulations. Despite a planktonic lifestyle, many rock pool species are poor dispersers, because they have behavioral and structural characteristics that aid retention in pools. Some rock pool species are rarely or never found in open water samples (Dethier, 1980; Jonsson, 1994). Ciliates and dinoflagellates are known to form benthic resting stages that restrict washout during high tides (Jonsson, 1994). Rock pool copepods such as *Tigriopus brevicornis* have hooks and serrations that may improve resistance to dislodgment by wave action (McAllen, 2001). Studies on a related species, *T. californicus*, have shown that dispersal between shores is unlikely: "Private" alleles and large differences in allozyme frequencies were maintained between adjacent rocky outcrops over decadal timescales (Burton, 1997).

Studies of *T. californicus* have assumed that a metapopulation structure exists, based on the observation of extinction events (Dybdahl, 1994). Other authors have questioned whether the drying out of upper shore pools necessarily leads to extinction, because *Tigriopus* spp. may take refuge within algae (McAllen, 1999) and there is some evidence for a desiccation-resistant state (Fraser, 1936; Ranade, 1957; Powlik and Lewis, 1996). Population surveys of different pools are, however, required to evaluate the majority of the metapopulation criteria suggested by Hanski (1997) and Grimm et al. (2003). The results of such a survey for *T. brevicornis* in pools on the Isle of Man are presented by Johnson (2001). There was clear evidence for persistent metapopulation structure over the course of a generation (Table 7-1). Populations in separate pools maintained distinct structures (based on the comparative abundance of different life history stages), and the dynamics of pool populations were not synchronous.

Although *T. brevicornis* (and probably other *Tigriopus* spp.) seem to persist in within-shore metapopulations formed from local populations in individual pools, there is considerable heterogeneity in pool quality. High shore pools can provide refuges from predation (Dethier, 1980) and from washout during storms and exceptionally high tides. Lower shore pools are less susceptible to drying out. Even with the documented resistance to desiccation, local population extinction is likely in high shore pools if they are isolated in dry conditions for periods exceeding 1 week (Powlik and Lewis, 1996). An important feature of the quality (in terms of suitability for *Tigriopus*) of individual pools is its variability in time. Washout is more likely in winter, whereas drying out of pools is more likely in summer. Hence, replenishing or recolonizing upper shore pools is more important in a season when lower shore pools are experiencing relatively benign conditions. Conversely, upper shore pools are more likely to restock storm-washed lower shore pools at a time when the influence of desiccation is reduced. Seasonal variations of *Tigriopus* metapopulations may track the changes in pool quality (as observed by Johnson, 2001). Hence, the internal structure of the copepod metapopulation does not consist of fixed sources or sinks of individuals: A sink in one season may be a source at another time of the year. These ideas

TABLE 7-1. Evaluation of the Structure of *Tigriopus brevicornis* Populations within Tide Pools

Criterion	Criterion Met?	Evidence
1. Patches contain local populations rather than ephemeral aggregations formed by the movement of individuals within a continuous population.	Yes	Differences in population structure between separate pools maintained over time. Pools are not just random samples from a shorewide population structure.
2. The dynamics of local populations should not be synchronous.	Yes	Population synchrony decays with distance between pairs of pools. The majority of correlations between pools were weak.
3. The occupants of a patch face a considerable risk of local extinction.	Yes	There are changes in occupation status of specific pools on different dates.
4. Patches are linked by dispersal.	Yes	There is evidence for migration of adults and copepodites during spring tides.

Criteria are given in the text and are based on Hanski (1997) and Grimm et al. (2003).

can make a contribution to the wider debate on metapopulations. A metapopulation of pools at the same height would be predicted to become extinct before a metapopulation using the same number of potentially suitable pools at a range of heights. A parallel prediction has recently been demonstrated in butterflies (McLaughlin et al., 2002). Population extinction occurred in homogenous habitat, but a heterogeneous habitat buffered the influence of weather and facilitated population persistence. Metapopulations with variable patch quality may be better conserved by including a variety of sites, not just those that appear optimal at any one time (Johnson, 2001).

A different group of invertebrates may also represent candidate within-shore metapopulations. Many species are direct developers: They have no planktonic dispersal stage. Grantham et al. (2003) estimate that 15% of the invertebrate fauna on rocky shores in the northwest Pacific have direct development. Among specific taxonomic groups, the prevalence of direct development may be higher. More than one third of Molluscan taxa collected by Johnson et al. (2001) had direct development. For these species, dispersal between shores is possible through rafting or drifting adults (Highsmith, 1985). The small numbers of individuals moving by such methods of dispersal seem unlikely to affect population dynamics above the shore scale. Genetic studies of direct developers often emphasize differences between individuals over short distances within shores (Johannesson, 2003). This implies that individual movement is relatively unimportant and that separate local populations may exist. Johnson et al. (2001) found that mollusks with direct development were more patchy within shores than species with planktonic dispersal. These observations suggest a potential within-shore metapopulation structure for direct developers. Local populations could consist of aggregations based on colonization events, weakly linked by the movements of adults. The proposed population structure may not function as a metapopulation if there is strong exogenous forcing of population dynamics. For example, patches could vary together if a predator forages at a larger spatial scale than the patch size, or if weather causes synchronous variation in mortality. However, there is good evidence that at least one species with direct development persists as a within-shore metapopulation. Breeding aggregations of the whelk Nucella (Thais) lamellosa represent local populations because individuals generally return to breed in the same group each year (Spight, 1974). There are asynchronous dynamics in different aggregations and there is a nonzero risk of local population extinction. Individuals generally do not move more than a few meters, but there is some dispersal between aggregations. This dispersal allows the reestablishment of populations following local extinctions (Spight, 1974). Although Spight (1974) did not use the term *metapopulation*, N. (Thais) lamellosa meets all the criteria to be considered as one.

As mentioned earlier, patch dynamic processes may provide a context for metapopulation dynamics if a species is dependent on a certain type of patch. For example, patches of fucoid algae may represent suitable habitat to species such as N. lapillus and Littorina fabalis, which are otherwise rare on smooth rock platforms, unlikely to move far as adults, and have direct development (Hawkins et al., 1992; Thompson et al. 1996). N. lapillus and L. fabalis could persist in networks of Fucus patches, with patch turnover causing local extinctions. This type of within-shore metapopulation of direct developers has been simulated by Johnson et al. (1998). Simulations mimicked processes in the Fucus–barnacle mosaic on moderately exposed northeast Atlantic shores (Hartnoll and Hawkins, 1985). Patches of algae become established in areas with low local densities of limpets. Limpets move into algal patches as they grow, changing the spatial pattern of grazing on the shore and creating new low grazing areas where algae can establish. Probably as a result of the increased local density of limpets, patches last for a finite period (approximately 5 years; Southward, 1956), because adult plants are undercut by abrasion and juvenile algae are removed by grazing. The interaction between algae and grazers produces a dynamic mosaic of patches of different ages. It is this temporally and spatially variable resource that the species dependent on algal patches need to track.

Changes in the spatiotemporal characteristics of a patch mosaic may affect the within-shore persistence of dependent species (as stressed in the patch dynamic literature; e.g., Sousa, 1984). The survival of patch dependent species was simulated in the model used for investigating the *Fucus* mosaic to examine the potential links between population persistence and patch dynamics (Johnson et al., 1998). In these simulations, limpets could drive changes in patch characteristics, both in terms of patch length scale (measured by spatial autocorrelation) and patch turnover times (measured by temporal autocorrelation). Patches were smaller and less persistent when limpets changed the position of their home scar more frequently. Limpets that tended to aggregate into clumps increased algal patch sizes, unless the limpets were also likely to move into patches of *Fucus*. Not surprisingly, a tendency for limpets to move into clumps of *Fucus* increased algal patch turnover rates (Johnson et al., 1998). In an extension of the Levins patch model approach, sites on the simulation grid occupied by algae represented the habitat available to direct developers at any one time. A site containing algae could be occupied by a local population of direct developers or could be empty. Extinction of local populations occurred when an occupied algal site reverted to bare rock. Colonization of sites was a local process, reflecting the limited mobility of adults. Algal sites that did not contain a population of direct developers had a probability of becoming occupied if adjacent sites were also occupied. Persistence of direct developers at large scales was therefore a metapopulation process, dependent on a balance between patch colonization and extinction.

Differences in the spatiotemporal dynamics of the underlying *Fucus* mosaic affected the simulated metapopulations of direct developers (Fig. 7-2). When limpets clumped but avoided algae, the *Fucus* mosaic was stable with well-defined patches. The majority of sites occupied by algae also contained algal-dependent direct developers. In contrast, when limpets clumped but were indifferent to the presence of algae, the mosaic of *Fucus* patches was less well defined. Under these conditions there was a greater proportion of suitable sites (containing algae) that did not contain direct developers. These results demonstrate the potential interactions between patch dynamic processes and within-shore metapopulations. Limpet behavior varies with environmental conditions and between species, so variation in patch dynamics may affect the persistence of metapopulations of direct developers on different shores. Other processes can affect patch dynamics, such as disturbance frequency, surface topography and artificial structures. These factors also have the potential to affect the persistence of metapopulations on shores.

A reasonable proportion of invertebrate taxa are not highly mobile as adults and receive relatively small numbers of recruits from separate shores. Even for species with planktonic larvae, local recruitment may be more important than recruits from other locations. The nudibranch *Adalaria proxima* has a short lived lecithotrophic larvae, but local populations (approximately 100 m apart) are genetically distinct and appear to retain genetic structure over time (Lambert et al., 2003). Intertidal populations of *A. proxima* occur on small scales (reproduction occurs by fertilizing an attached egg mass so breeding adults must be within crawling distance) and

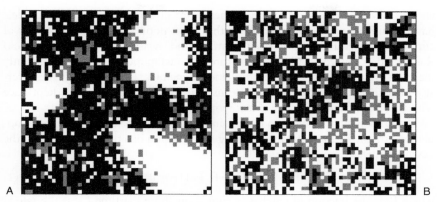

FIGURE 7-2. (A, B) Alternative spatial patterns in simulated *Fucus* mosaics. Gray cells indicate that a site is occupied by *Fucus*. Where a *Fucus*-occupied site also contains a local population of direct developers, the cell is colored black. The simulations represent 25 m² of a moderately exposed shore in the northeast Atlantic (for details see Johnson et al., 1998). Parameters are the same in both panels except (A) limpets aggregate but avoid *Fucus* and (B) limpets aggregate but are indifferent to *Fucus*. The predicted abundance of *Fucus* is approximately the same in both simulations, but patch occupancy of direct developers is higher when *Fucus* patches are more defined.

are subject to extinction and recolonization processes (Lambert et al., 2003). Hence there is evidence that even intertidal invertebrates with planktonic larvae can be structured as within-shore metapopulations.

As yet, the concept of within-shore metapopulations is novel and only a limited number of potential examples exist. Further testing of this framework will involve greater characterization of the spatial and temporal scales of processes on rocky shores, including the extent and frequency of colonization and extinction/disturbance events (e.g., Kelaher and Rouse, 2003). Metapopulation theory offers one route to extending the ecological truism that "things vary in space and time" to a more analytical framework that integrates spatiotemporal variability into population dynamics. In particular, this approach emphasizes how small-scale stochastic patch events can be studied using deterministic models at larger spatial scales.

IV. METAPOPULATIONS AT DIFFERENT SCALES

Levins–type patch-occupancy models have not been applied at scales above the shore scale. Such an approach would require individual shores to be occupied or not, with the metapopulation consisting of a collection of shores linked by dispersal. This minimal level of structure may have some heuristic use, but most authors have preferred to include further levels of detail. For example, models have included population structure within local populations and the effects of currents on larval dispersal (Roughgarden and Iwasa, 1986; Alexander and Roughgarden, 1996). Adding structure to metapopulation models greatly increases the range of potential population dynamics. Unfortunately, the diversity of potential model structures means that observations have not kept pace with theoretical investigations of metapopulation dynamics.

The fundamental changes to population dynamics associated with a metapopulation structure can be illustrated using modification of a discrete version of the Levins model:

$$P_{i,t+1} = cP_{i,t}(H - P_{i,t}) + (1-m)P_{i,t} \qquad (1)$$

where $P_{i,t}$ is the fraction of patches occupied on shore i at time t, c is the rate at which unoccupied patches are colonized, H is the fraction of the total habitat on each shore consisting of available patches, and m is the rate at which occupied patches become extinct. A simple way to link separate shores is to assume that the supply of potential colonists comes from a larval pool, such that each shore receives the same fraction of the larval pool (colonization occurs in proportion to the average of P across shores). Numerical solutions for a series of shores can be given by iterating Equation 1, given a vector of initial occupancy levels.

The simple model demonstrates the changes in dynamics that occur when local populations are linked by dispersal (Fig. 7-3). Linking shores allows metapopulations to persist on shores where they would go extinct in the absence of a larval pool ("sinks"). The average population size is, however, larger across independent metapopulations than in the linked array. The overall dynamics are also different. For example, recovery from a low within-shore occupancy is slower when shores are linked. Roughgarden and Iwasa (1986) analyzed a similar metapopulation model in more depth, showing how spatial variation in average larval output could lead to alternative stable states for the metapopulation.

The relative simplicity of Equation 1 is deceptive. There are many ways in which processes at different scales may be influencing dynamics (Fig. 7-1). A fundamental problem with choosing scales for a metapopulation model is that there is usually little evidence that the scales are appropriate. Levels of organization may reflect mathematical constraints, and they may be arbitrary when compared with the scales of processes in natural systems. In the model of Roughgarden and Iwasa (1986), the smallest scale is that of a "local" population. It might be intuitive to think of the local population as a shore. This implies that demographic processes are defined at the shore scale. However, density-dependent mortality is thought to occur at smaller scales because it is associated with loss of crowded (hummocked) barnacles or local predation (Roughgarden et al., 1985; Roughgarden and Iwasa, 1986; Possingham et al., 1994). There may be no clear justification for the scaling up of local density-dependent processes. Small-scale

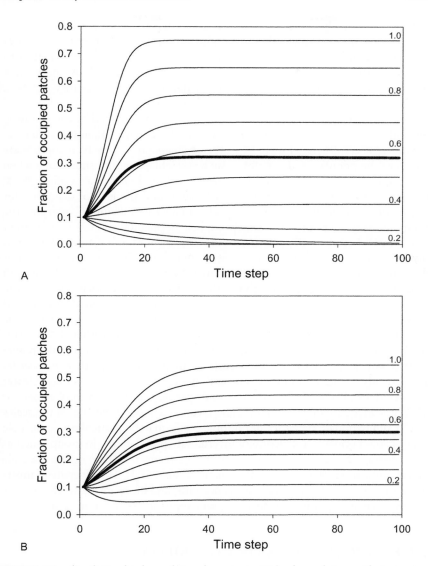

FIGURE 7-3. (A, B) Simulated population dynamics in 10 local populations with Levins–type dynamics. The available fraction of habitat in each local population (H) varies between 0.1 and 1. The average level of occupancy across 10 populations is shown as a thicker line for 10 independent local populations (A) and 10 populations that share a common larval pool (B). The dynamics show trajectories from an initial occupancy of 0.1 in each local population. Patch dynamic parameters are the same for all local populations ($c = 0.4$, $m = 0.1$). Average occupancy across all populations in (A) is 0.32 with an average occupancy in (B) of 0.3.

oscillations are likely to be averaged out at the larger (shore) scales (Johnson, 2000a). Even at small scales, the oscillations predicted for high settlement rates in Roughgarden et al. (1985) and related models may require unrealistic constraints on the density-dependent processes (Johnson 2000a). A shore scale model based on patch scale observations of density dependence may not be a useful basis for a between-shores metapopulation model.

Some of the problems of model structure may be exaggerated by the relatively low adult mobility of intertidal invertebrates. A shore may contain individuals in very different conditions of wave exposure or emersion. It may be inappropriate to average the population dynamics of such assemblages at the shore scale when demographic rates are likely to vary greatly within shores (e.g., Caffey, 1985; Bertness et al., 1991). This is not a trivial matter and the lack of definition for demographic parameters at relevant scales makes it difficult to interpret the rapidly growing literature on spatially explicit population models (e.g., Gerber et al., 2003). For example, parameters used by Gaylord and Gaines (2000) can only be "based loosely" on barnacles, even though this is "one of the few cases where sufficient data exist to allow parameter estimation. (p. 773)" Even with barnacles, the parameter estimation is difficult. The current models scale up from small-scale studies to simulation grid scales (approximately 550 m in Gaylord and Gaines, 2000). However, small-scale studies could include processes such as the aggregation of whelks on patches of high recruitment (Fairweather, 1988). This process probably cannot be scaled up to simulation grid scales because the movements of whelks are not at the same scale. Demographic rates other than settlement are also held constant in current models. In part this allows clearer interpretations of single processes such as dispersal, but this also emphasizes our lack of knowledge about the variance and mean of demographic rates at the scales used in simulations. The issues of multiple scales may be particularly applicable to the spatial population structure of invertebrate populations. In groups such as fish, breeding aggregations or homing to natal streams may help define the scale of local populations (Hill et al., 2002; Rhodes et al., 2003). The diversity of invertebrate dispersal modes, reproductive allocations, and life spans, however, implies that no single metapopulation structure will apply across different taxa.

V. MEASURED SCALES OF VARIABILITY

Given the potential pitfalls of arbitrary scale choices in metapopulation models, there is a need for better definition of the relevant scales in the field. Levin (1992) stresses that there is no "correct" scale, but spatial statistics can be used to define the scales where significant variation exists (e.g., Underwood and Chapman, 1996). Identification of the scales of variability in population density is the first step in identifying demographically important processes. Rocky shore invertebrate species have been shown to have a high degree of variation in abundance at small scales (less than a meter) with additional variation at mesoscales (100m) and macroscales (>1 km; Underwood and Chapman, 1996; Hyder et al., 1998; Benedetti-Cecchi, 2001). This suggests that a metapopulation model should contain these scales if it is to describe all the relevant spatial processes.

Studies of spatial pattern invoke hypotheses for the causes of variation at different scales. Hydrodynamics are likely to be involved at all scales, from small-scale turbulence to alongshore transport at mesoscales (Pineda, 1994). At scales more than 1 km, larval concentrations will be affected by phenomena such as fronts, upwelling, currents, and internal waves (Shanks et al., 2000; Gaylord and Gaines, 2000). Because many intertidal invertebrates are sessile or have limited mobility as adults, individual behavior is likely to contribute most to small-scale pattern (although larval behavior may influence interactions with hydrographic processes at other scales). Topography is likely to influence spatial pattern and larval settlement behavior (e.g., Crisp and Barnes, 1954). Predation may be involved at a number of scales, depending on the mobility of the predators. Climate variation is likely to affect populations at large scales. Some variation at the kilometer scale may reflect the orientation of shores (e.g., the effects of wind on larval supply; Hawkins and Hartnoll, 1982; Bertness et al., 1996).

A useful technique for defining relevant scales is to examine the scales of population dynamic synchrony (Koenig, 1999). Burrows et al. (2002) applied this approach to time series collected from shores in Shetland. The majority of species had synchronous population dynamics on shores separated by distances of up to 30 km. Synchrony did not tend to decline with distance. The direct developer *N. lapillus* had asynchronous dynamics on different shores, but there was no general contrast between species with planktonic dispersal and direct developers. Synchrony between shores implies a link through a common pool of larvae or through environmentally mediated disturbance effects occurring across all shores. Burrows et al. (2002) stressed that that a wider range of coastline configurations should be examined with this approach, because most of their sites were in the same semi-enclosed area.

Processes at the edge of species ranges may offer further opportunities for defining the spatial scales of populations. Characteristics of range expansion in invasive species indicate the alongshore supply from established populations (McQuaid and Phillips, 2000; Armonies, 2001). Similar processes may be relevant as species ranges change as a result of climate change. A metapopulation approach may aid in understanding processes at species boundaries (Lennon et al., 1997). Bascompte et al (2002) were able to make robust predictions for persistence thresholds in finite patch networks with spatial variability in demographic parameters. Changes in site occupancy at the edge of species' ranges could be predicted using these techniques. For example, some species appear to be expanding along gradients of diminishing habitat availability (Herbert et al.,

2003). An integration of habitat availability and climate-driven changes in demography would allow a prediction of the extent of potential range expansion. As an alternative, species may have boundaries set purely by climatic thresholds or hydrographic features.

Genetic markers can be used to identify the geographic structure of populations and the sources of recruits (Avise, 1994; Drouin et al., 2002). Wider issues, such as the influence of metapopulation structure on genetics and evolution (e.g., Pannell and Charlesworth, 2000), are beyond the scope of this chapter, because these processes are not restricted to intertidal invertebrates. Molecular methods can be used to discriminate between alternative population structures (Jolly et al., 2003). Isolation-by-distance slopes indicate a wide variation in the mean dispersal distances of marine invertebrates (Kinlan and Gaines, 2003). There are, however, limitations to the use of genetic makers as tools for investigating demographic processes. For example, a relatively small number of migrants can remove the genetic differences between populations. This may make genetic similarity between local populations hard to interpret (Mora and Sale, 2002). Exchange of individuals between sites could be extensive or demographically unimportant.

Observations of pattern are unlikely to discriminate between competing hypotheses about process. The scaling of some processes can be investigated experimentally (e.g., Fernandes et al., 1999). Further information can be obtained by making direct measurements of demographic rates. Jenkins et al. (2000) emphasize that barnacle recruitment is variable at all scales from meters to kilometers, but mortality does not vary significantly between shores separated by approximately 5 km (Jenkins et al., 2001). Such results imply that the assumption of a common mortality across shores (used in the model for Fig. 7-3) may be reasonable in some cases.

The issues associated with defining relevant spatial scales for populations demonstrate how metapopulation ecology (in the broadest sense) interfaces with most areas of intertidal population ecology. It is tempting to feel that intertidal invertebrates must be organized as metapopulations. Islands represent patches of suitable habitat and coastlines are also divisible into suitable and unsuitable (e.g., sedimentary) shores. However, not every patchy system is a metapopulation (Harrison and Taylor, 1997). By moving beyond Levins–type patch-occupancy models, there is a danger that metapopulation ecology degrades into a "theory of everything" with little value (see comments by Hanski and Simberloff, 1997). The model for Figure 7-3 suggests a technical definition: A metapopulation is a parsimonious description for a set of populations linked by dispersal in which the most influential demographic processes are best characterized at different spatial scales. This definition is not meant as a replacement for the criteria of Hanski (1997) and Grimm et al. (2003); however, it does introduce the importance of explicitly defining scales. A metapopulation therefore exists when populations are supplied from a larval pool (a large-scale process), but mortality or carrying capacity is set at the shore level. This is a pragmatic definition because there is sufficient evidence to suggest that the interaction of processes with different scales can alter population dynamics when compared with populations in which demographic processes vary at the same scale (Roughgarden and Iwasa, 1986; Fig. 7-3). The definition also distinguishes a metapopulation from a single patchy population. The size of a patchy population defines the larval pool but, although demographic processes may vary among sites, movement of individuals smooths this out to produce uniformity at the same scale as the larval pool. The focus on process given in the previous metapopulation definition reflects the concerns of Thomas and Kunin (1999): Because real populations will not fit smoothly into categories such as "mainland–island," it is better to categorize by process.

The examples outlined in this section show that studies defining scales of interest are relatively recent. Further empirical work is required to ensure that metapopulation models reflect the appropriate interactions between scales in the systems under investigation.

VI. SUMMARY

Intertidal invertebrates typically have patchy distributions, linked by the dispersal of larvae or by individual rafting and floating. By the broader definition given earlier, many invertebrates are likely to have metapopulation structures. Dispersing individuals are affected by hydrodynamics across a range of scales, whereas processes such as growth and mortality vary at scales below the dispersal scale. Rocky shore ecologists have frequently emphasized how trophic interactions structure communities. Interacting species may, however, have different dispersal scales, and this has consequences for community ecology and evolution (Kinlan and Gaines, 2003). Furthermore, species-specific larval biology can cause species to respond to hydrodynamic forcing in different ways (Wing et al., 2003). Spatially explicit population models have emphasized how processes at different spatiotemporal scales can produce patterns such as traveling waves and cyclic or chaotic dynamics. Topographic features, potentially including coastlines, can generate traveling waves of population density by a number of mechanisms (Johnson, 2000b; Sherratt et al., 2003). Hence, metapopulation dynamics may be an underrecognized source of spatiotemporal variability in coastal ecosystems.

Although metapopulation models help define the potential dynamics of a system, spatially extensive fine-scale data are required to determine the relevant scales of processes and to discriminate between competing hypotheses. In the case of rocky shore invertebrates, this information does not yet seem to be available for any species. Time series of adult density, mortality, and recruitment are needed from within and among shores to examine unambiguously the scales of synchrony. Synchronous dynamics among sites could be driven by internal dynamics linked by dispersal and/or correlated environmental forcing (Botsford et al., 1994). Although the resolution of metapopulation dynamics is a formidable undertaking, there are clear justifications for this research from the potential applications to climate change prediction, marine reserve, and fishery research (Botsford et al., 1994; Gerber et al., 2003).

The data requirements for an integration of oceanography, habitat heterogeneity, and population ecology in metapopulation studies may restrict the number and generality of case studies developed. Although a metapopulation structure may be appropriate, many populations may be approximated by simpler approaches. For example, at larger scales there may be stock–recruit relationships that permit some prediction of population dynamic processes (Hughes et al., 2000). Metapopulation structure suggests some linkage between adults and recruits at some scale in the system. The alternative to this is that environmental stochasticity creates demographically open populations at all scales. Variation in recruitment may be predicted from a clearer understanding of larval biology and its interactions with forcing by climate and nearshore oceanography (Gaines et al., 2003; Gilg and Hilbish, 2003). As yet, the demographic "openness" of invertebrate populations has generally been demonstrated at small scales (within shores; e.g., Hyder et al., 2001). There are few, if any, studies that establish the extent of adult–recruit relationships at large scales in a framework that also evaluates the role of large-scale hydrographic processes. In part this gap may be addressed by the increase in large-scale studies of dispersal and/or recruitment (e.g., Connolly and Roughgarden, 1999a). However, one trend is for settlement and postsettlement processes to be studied separately. This creates a debate that appears, at times, to be rather artificial (the "pre- vs. postsettlement" debate). To really understand any potential metapopulation structure, reproduction, settlement, and mortality need to be studied together across different scales.

The relative degree of population closure for interacting species may affect the scales over which processes such as upwelling or nutrient enrichment act on community structure (e.g., Menge et al., 2003). For example, predator populations may increase in upwelling areas resulting from propagule retention and increases in food supply (Menge et al., 2003). However, if predators are demographically linked over larger scales than their prey, the "top-down" effects may be dependent on a certain scale of upwelling. If upwelling areas are too small, predator propagules could be lost in relatively greater proportions than their prey, preventing predator populations from responding to the local increase in productivity (but potentially affecting communities adjoining the upwelling area). Metapopulation biology could help predict over what scales trophic links will be most strongly emphasized, including the different potential responses of systems to episodic events at different scales. Witman et al. (2003) describe the response of predators to a large-scale mussel recruitment event. Metapopulation ecology could examine whether there was a critical patch (recruitment event) scale below which the observed community response would not occur. These issues are yet to be explored fully in theory. There could, however, be fundamental consequences for our understanding and definition of large-scale environmental forcing of ecosystems.

Beyond the challenges in defining the spatial structure of populations across different scales, there are two areas where metapopulation ecology may be particularly relevant to rocky intertidal invertebrates. First, there is relatively good evidence that some species are structured as within-shore populations. Species with direct development and rock pool specialists can be dependent on relatively well-defined patches and fulfill the criteria to be considered metapopulations. There is scope for more integration between patch dynamic and metapopulation studies. Tractable experiments to test metapopulation hypotheses at the within-shore scale should be possible. Artificial habitat patches can be created (e.g., Quinn et al., 1989) and shores with different patch dynamics can be compared. Such experiments have direct application to the conservation of intertidal species (particularly where there are analyses of metapopulation persistence and sensitivity). Threats particularly relevant to metapopulations include alterations to the extent and continuity of shores with the construction of artificial structures such as seawalls (Chapman and Bulleri, 2003), and where harvesting in the intertidal can change the structure of patches used by species (Monteiro et al., 2002).

Second, intertidal invertebrate species vary greatly in life history characteristics. This provides scope for a range of comparative studies. The population dynamics of species with planktonic development and direct developers would be predicted to differ in ways that reflect the underlying differences in spatial structure of populations. Population stability may vary between dispersal modes and at different scales. Eckert (2003) found that time series of planktonic dispersers were more stable than those for direct developers. A greater degree of planktonic dispersal between sites may have smoothed out local population fluctuations (Eckert, 2003). Comparisons between developmental modes can also be used to examine recovery from disturbances. Reed et al. (2000) suggested that species with higher planktonic dispersal potential were first to recover following a disturbance associated with El Niño. The conclusions about recovery were made using time series collected at different scales, which may have influenced the results. However, the studies of Reed et al. (2000) and Eckert (2003) suggest that differences in dispersal affect spatiotemporal population dynamics. Such studies emphasize that an understanding of metapopulation structure may be relevant to predictions of population stability and recovery.

The metapopulation approach can be integrated into studies of the population ecology of intertidal invertebrates at a number of different points. This integration is an ongoing process. It seems likely that a critical evaluation of metapopulation structure is capable of shedding new light on familiar studies in addition to informing future research on intertidal population ecology.

REFERENCES

Alexander, S.E., and Roughgarden, J. (1996). Larval transport and population dynamics of intertidal barnacles: A coupled benthic/oceanic model. *Ecol. Monogr.* **66**, 259–275.

Amarasekare, P., and Possingham, H. (2001). Patch dynamics and metapopulation theory: The case of successional species. *J. Theor. Biol.* **209**, 333–344

Armonies, W. (2001). What an introduced species can tell us about the spatial extension of benthic populations. *Mar. Ecol. Prog. Ser.* **209**, 289–294.

Avise, J.C. (1994). *Molecular markers, natural history and evolution.* London: Chapman and Hall.

Bascompte, J., Possingham, H., and Roughgarden, J. (2002). Patchy populations in stochastic environments: Critical number of patches for persistence. *Am. Nat.* **159**, 128–137.

Benedetti-Cecchi, L. (2001). Variability in abundance of algae and invertebrates at different spatial scales on rocky sea shores. *Mar. Ecol. Prog. Ser.* **215**, 79–92.

Bertness, M.D., Gaines, S.D., Bermudez, D., and Sanford, E. (1991). Extreme spatial variation in the growth and reproductive output of the acorn barnacle *Semibalanus balanoides*. *Mar. Ecol. Prog. Ser.* **75**, 91–100.

Bertness, M.D., Gaines, S.D., and Wahle, R.A. (1996). Wind driven settlement patterns in the acorn barnacle *Semibalanus balanoides*. *Mar. Ecol. Prog. Ser.* **137**, 103–110.

Botsford, L.W., Moloney, C.L., Hastings, A., Largier, J.L., Powell, T.M., Higgins, K., and Quinn, J.F. (1994). The influence of spatially and temporally varying oceanographic conditions on meroplanktonic metapopulations. *Deep Sea Res. II.* **41**, 107–145.

Burrows, M.T., Moore, J.J., and James, B. (2002). Spatial synchrony of population changes in rocky shore communities in Shetland. *Mar. Ecol. Prog. Ser.* **240**, 39–48.

Burton, R.S. (1997). Genetic evidence for long-term persistence of marine invertebrate populations in an ephemeral environment. *Evolution.* **51**, 993–998.

Caffey, H.M. (1985). Spatial and temporal variation in settlement and recruitment of intertidal barnacles. *Ecol. Monogr.* **55**, 313–332.

Caswell, H., and Etter, R.J. (1993). Ecological interactions in patchy environments: From patch occupancy models to cellular automata. In *Patch dynamics* (S.A. Levin, T.M. Powell, and J.H. Steele, eds., pp. 93–109). Berlin: Springer-Verlag.

Chapman, M.G., and Bulleri, F. (2003). Intertidal seawalls: New features of landscape in intertidal environments. *Landscape Urban Plann.* **62**, 159–172.

Connolly, S.R., and Roughgarden, J. (1999a). Increased recruitment of northeast Pacific barnacles during the 1997 El Nino. *Limnol. Oceanogr.* **44**, 466–469.

Connolly, S.R., and Roughgarden, J. (1999b). Theory of marine communities: Competition, predation, and recruitment-dependent interaction strength. *Ecol. Monogr.* **69**, 277–296.

Crisp, D.J., and Barnes, H. (1954) The orientation and distribution of barnacles at settlement with particular reference to surface contour. *J. Anim. Ecol.* **23**, 142–162.

Dethier, M.N. (1980). Tidepools as refuges: Predation and the limits of the harpacticoid copepod *Tigriopus californicus* (Baker). *J. Exp. Mar. Biol. Ecol.* **42**, 99–111.

Drouin, C.-A., Bourget, E., Tremblay, R. (2002). Larval transport processes of barnacle larvae in the vicinity of the interface between two genetically different populations of *Semibalanus balanoides*. *Mar. Ecol. Prog. Ser.* **229**, 165–172.

Dybdahl, M.F. (1994). Extinction, recolonization, and the genetic structure of tidepool copepod populations. *Evol. Ecol.* **8**, 113–124.

Eckert, G.L. (2003). Effects of the planktonic period on marine population fluctuations. *Ecology.* **84**, 372–383.

Ellner, S.P., and Fussmann, G. (2003). Effects of successional dynamics on metapopulation persistence. *Ecology.* **84**, 882–889.

Fairweather, P.G. (1988). Consequences of supply-side ecology: Manipulating the recruitment of intertidal barnacles affects the intensity of predation upon them. *Biol. Bull.* **175**, 349–354.

Fernandes, T.F., Huxham, M., and Piper, S.R. (1999). Predator caging experiments: A test of the importance of scale. *J. Exp. Mar. Biol. Ecol.* **241**, 137–154.

Fraser, J.H. (1936). The occurrence, ecology and life history of *Tigriopus fulvus* (Fischer). *J. Mar. Biol. Ass. UK.* **20**, 523–536.

Gaines, S.D., Gaylord, B., and Largier, J.L. (2003). Avoiding current oversights in marine reserve design. *Ecol. Appl.* **13**, S32–S46.

Gaylord, B., and Gaines, S.D. (2000). Temperature or transport? Range limits in marine species mediated solely by flow. *Am. Nat.* **155**, 769–789.

Gerber, L.R., Botsford, L.W., Hastings, A., Possingham, H.P., Gaines, S.D., Palumbi, S.R., and Andelman, S. (2003). Population models for marine reserve design: A retrospective and prospective synthesis. *Ecol. Appl.* **13**, S47–S64.

Gilg, M.R., and Hilbish, T.J. (2003). The geography of marine larval dispersal: Coupling genetics with fine-scale physical oceanography. *Ecology.* **84**, 2989–2998.

Gotelli, N.J. (1991). Metapopulation models: The rescue effect, the propagule rain, and the core–satellite hypothesis. *Am. Nat.* **138**, 768–776.

Grantham, B.A., Eckert, G.L., and Shanks, A.L. (2003). Dispersal potential of marine invertebrates in diverse habitats. *Ecol. Appl.* **13**, S108–S116.

Grimm, V., Reise, K., and Strasser, M. (2003). Marine metapopulations: A useful concept? *Helgol. Mar. Res.* **56**, 222–228.

Gyllenberg, M., and Hanski, I. (1997). Habitat deterioration, habitat destruction, and metapopulation persistence in a heterogenous landscape. *Theor. Popul. Biol.* **52**, 198–215.

Hanski, I. (1997). Metapopulation dynamics: From concepts and observations to predictive models. In *Metapopulation biology: Ecology, genetics and evolution* (I. Hanski and M.E. Gilpin, eds., pp. 69–91). London: Academic Press.

Hanski, I., Moilanen, A., and Gyllenberg, M. (1996). Minimum viable metapopulation size. *Am. Nat.* **147**, 527–541.

Hanski, I., and Simberloff, D. (1997). The metapopulation approach, its history, conceptual domain and application to conservation. In *Metapopulation biology: Ecology, genetics and evolution* (I. Hanski and M.E. Gilpin, eds., pp. 5–26). London: Academic Press.

Harrison, S. (1991). Local extinction in a metapopulation context: An empirical evaluation. *Biol. J. Linn. Soc.* **42**, 73–88.

Harrison, S., and Taylor, A.D. (1997). Empirical evidence for metapopulation dynamics. In *Metapopulation biology: Ecology, genetics and evolution* (I. Hanski and M.E. Gilpin, eds., pp. 27–42). London: Academic Press.

Hartnoll, R.G., and Hawkins, S.J. (1985). Patchiness and fluctuations on moderately exposed rocky shores. *Ophelia.* **24**, 53–63.

Hastings, A. (2003). Metapopulation persistence with age-dependent disturbance or succession. *Science.* **301**, 1525–1526.

Hawkins, S.J., and Hartnoll, R.G. (1982). Settlement patterns of *Semibalanus balanoides* (L.) in the Isle of Man (1977–1981). *J. Exp. Mar. Biol. Ecol.* **62**, 271–283.

Hawkins, S.J., Hartnoll, R.G., Kain-Jones, J.M., Norton, T.A. (1992). Plant–animal interactions on hard substrata in the north-east Atlantic. In *Plant–animal interactions in the marine benthos* (D.M. John, S.J. Hawkins, J.H. Price, eds., pp. 1–32). Systematics Association special volume no. 46. Oxford: Clarendon Press.

Herbert, R.J.H., Hawkins, S.J., Sheader, M., and Southward, A.J. (2003). Range extension and reproduction of the barnacle *Balanus perforatus* in the eastern English Channel. *J. Mar. Biol. Assoc. UK.* **83**, 73–82.

Highsmith, R.C. (1985). Floating and algal rafting as potential dispersal mechanisms in brooding invertebrates. *Mar. Ecol. Prog. Ser.* **25**, 169–179.

Hill, M.F., Hastings, A., and Botsford. L.W. (2002). The effects of small dispersal rates on extinction times in structured metapopulation models. *Am. Nat.* **160**, 389–402.

Horn, H.S., and MacArthur, R.H. (1972). Competition among fugitive species in a harlequin environment. *Ecology.* **53**, 749–752.

Hughes, T.P., Baird, A.H., Dinsdale, E.A., Moltschaniwskyj, N.A., Pratchett, M.S., Tanner, J.E., and Willis, B.L. (2000). Supply-side ecology works both ways: The link between benthic adults, fecundity, and larval recruits. *Ecology.* **81**, 2241–2249.

Hyder, K., Åberg, P., Johnson, M.P., and Hawkins, S.J. (2001). Models of open populations with space-limited recruitment: Extension of theory and application to the barnacle *Chthamalus montagui*. *J. Anim. Ecol.* **70**, 853–863.

Hyder, K., Johnson, M.P., Hawkins, S.J., and Gurney, W.S.C. (1998). Barnacle demography: Evidence for an existing model and spatial scales of variation. *Mar. Ecol. Prog. Ser.* **174**, 89–99.

Jenkins, S.R., Aberg, P., Cervin, G., Coleman, R.A., Delany, J., Della Santina, P., Hawkins, S.J., LaCroix, E., Myers, A.A., Lindegarth, M., Power, A.M., Roberts, M.F., and Hartnoll, R.G. (2000). Spatial and temporal variation in settlement and recruitment of the intertidal barnacle *Semibalanus balanoides* (L.) (Crustacea: Cirripedia) over a European scale. *J. Exp. Mar. Biol. Ecol.* **243**, 209–225.

Jenkins, S.R., Aberg, P., Cervin, G., Coleman, R.A., Delany, J., Hawkins, S.J., Hyder, K., Myers, A.A., Paula, J., Power, A.M., Range, P., and Hartnoll, R.G. (2001). Population dynamics of the intertidal barnacle *Semibalanus balanoides* at three European locations: Spatial scales of variability. *Mar. Ecol. Prog. Ser.* **217**, 207–217.

Johannesson, K. (2003). Evolution in *Littorina*: Ecology matters. *J. Sea. Res.* **49**, 107–117.

Johnson, M. (2000a). A re-evaluation of density dependent population cycles in open systems. *Am. Nat.* **155**, 36–45.

Johnson, M.P. (2000b). Scale of density dependence as an alternative to local dispersal in spatial ecology. *J. Anim. Ecol.* **69**, 536–540.

Johnson, M.P. (2000c). The influence of patch demographics on metapopulations, with particular reference to successional landscapes. *Oikos.* **88**, 67–74.

Johnson, M.P. (2001). Metapopulation dynamics of *Tigriopus brevicornis* (Harpacticoida) in intertidal rock pools. *Mar. Ecol. Prog. Ser.* **211**, 215–224.

Johnson, M.P., Allcock, A.L., Pye, S.E., Chambers, S.J. and Fitton, D.M. (2001). The effects of dispersal mode on the spatial distribution patterns of intertidal molluscs. *J. Anim. Ecol.* **70**, 641–649.

Johnson, M.P., Burrows, M.T., and Hawkins, S.J. (1998). Individual based simulations of the direct and indirect effects of limpets on a rocky shore *Fucus* mosaic. *Mar. Ecol. Prog. Ser.* **169**, 179–188.

Jonsson, P.R. (1994). Tidal rhythm of cyst formation in the rock pool ciliate *Strombidium oculatum* Gruber (Ciliophora, Oligotrichida): A description of the functional biology and an analysis of the tidal synchronization of encystment. *J. Exp. Mar. Biol. Ecol.* **175**, 77–103.

Jolly, M.T., Viard, F., Weinmayr, G., Gentil, F., Thiebaut, E., and Jollivet, D. (2003). Does the genetic structure of *Pectinaria koreni* (Polychaeta: Pectinariidae) conform to a source–sink metapopulation model at the scale of the Baie de Seine? *Helgol. Mar. Res.* **56**, 238–246.

Kelaher, B.P., and Rouse, G.W. (2003). The role of colonization in determining spatial patterns of *Proscoloplos bondi* sp. nov (Orbiniidae: Annelida) in coralline algal turf. *Mar. Biol.* **143**, 909–917.

Kinlan, B.P., and Gaines, S.D. (2003). Propagule dispersal in marine and terrestrial environments: A community perspective. *Ecology.* **84**, 2007–2020.

Koenig, W.D. (1999). Spatial autocorrelation of ecological phenomena. *Trends Ecol. Evol.* **14**, 22–26.

Lambert, W.J., Todd, C.D., and Thorpe, J.P. (2003). Genetic population structure of two intertidal nudibranch molluscs with contrasting larval types: Temporal variation and transplant experiments. *Mar. Biol.* **142**, 461–471.

Lennon, J.J., Turner, J.R.G., and Connell, D. (1997). A metapopulation model of species boundaries. *Oikos.* **78**, 486–502.

Levin, S.A. (1992). The problem of pattern and scale in ecology. *Ecology.* **73**, 1943–1967.
Levin, S.A., and Paine, R.T. (1974). Disturbance, patch formation and community structure. *Proc. Nat. Acad. Sci. U S A.* **71**, 2744–2747.
Levin, S.A., Powell, T.M., and Steele, J.H. (1993). *Patch dynamics.* Berlin: Springer-Verlag.
Levins, R. (1969). Some demographic and genetic consequences of environmental heterogeneity for biological control. *Bull. Entomol. Soc. Am.* **15**, 237–240.
Levins, R., and Culver, D. (1971). Regional coexistence of species and competition between rare species. *Proc. Nat. Acad. Sci. U S A.* **68**, 1246–1248.
McAllen, R. (1999). *Enteromorpha intestinalis:* A refuge for the supralittoral rockpool harpacticoid copepod *Tigriopus brevicornis. J. Mar. Biol. Assoc. UK.* **79**, 1125–1126.
McAllen, R. (2001). Hanging on in there: Position maintenance by the high-shore rockpool harpacticoid copepod *Tigrioptis brevicornis. J. Nat. His.* **35**, 1821–1829.
McLaughlin, J.F., Hellmann, J.J., Boggs, C.L., and Ehrlich, P.R. (2002). The route to extinction: Population dynamics of a threatened butterfly. *Oecologia.* **132**, 538–548.
McQuaid, C.D., and Phillips, T.E. (2000). Limited wind-driven dispersal of intertidal mussel larvae: In situ evidence from the plankton and the spread of the invasive species *Mytilus galloprovincialis* in South Africa. *Mar. Ecol. Prog. Ser.* **201**, 211–220.
Menge, B.A., Farrell, T.M., Olson, A.M., Vantamelen, P., and Turner, T. (1993). Algal recruitment and the maintenance of a plant mosaic in the low intertidal region on the Oregon coast. *J. Exp. Mar. Biol. Ecol.* **170**, 91–116.
Menge, B.A., Lubchenco, J., Bracken, M.E.S., Chan, F., Foley, M.M., Freidenburg, T.L., Gaines, S.D., Hudson, G., Krenz, C., Leslie, H., Menge, D.N.L., Russell, R., and Webster, M.S. (2003). Coastal oceanography sets the pace of rocky intertidal community dynamics. *Proc. Natl. Acad. Sci. USA.* **100**, 12229–12234.
Monteiro, S.M., Chapman, M.G., and Underwood, A.J. (2002). Patches of the ascidian *Pyura stolonifera* (Heller, 1878): Structure of habitat and associated intertidal assemblages. *J. Exp. Mar. Biol. Ecol.* **270**, 171–189.
Mora, C., and Sale, P.F. (2002). Are populations of coral reef fish open or closed? *Trends Ecol. Evol.* **17**, 422–428.
Nee, S., and May, R.M. (1992). Dynamics of metapopulations: Habitat destruction and competitive coexistence. *J. Anim. Ecol.* **61**, 37–40.
Paine, R.T. (1988). Habitat suitability and local population persistence of the sea palm *Postelsia palmaeformis. Ecology.* **69**, 1787–1794.
Paine, R.T., and Levin, S.A. (1981). Intertidal landscapes: Disturbance and the dynamics of pattern. *Ecol. Monogr.* **51**, 145–178.
Pannell, J.R., and Charlesworth, B. (2000). Effects of metapopulation processes on measures of genetic diversity. *Phil. Trans. R. Soc. Lond. B.* **355**, 1851–1864.
Pickett, S.T.A., and White, P.S. (1985). *The ecology of natural disturbance and patch dynamics.* London: Academic Press.
Pineda, J. (1994). Spatial and temporal patterns in barnacle settlement rate along a southern California rocky shore. *Mar. Ecol. Prog. Ser.* **107**, 125–138.
Possingham, H., Tuljapurkar, S.D., Roughgarden, J., and Wilks, M. (1994). Population cycling in space limited organisms subject to density dependent predation. *Am. Nat.* **143**, 563–582.
Powlik, J.J., and Lewis, A.G. (1996). Desiccation resistance in *Tigriopus californicus* (Copepoda: Harpacticoida). *Estuar. Coast. Shelf. Sci.* **43**, 521–532.
Quinn, J.F., Wolin, C.L., and Judge, M.L. (1989). An experimental-analysis of patch size, habitat subdivision, and extinction in a marine intertidal snail. *Conserv. Biol.* **3**, 242–251.
Ranade, M.R. (1957). Observations on the resistance of *Tigriopus fulvus* (Fischer) to changes in temperature and salinity. *J. Mar. Biol. Assoc. UK.* **36**, 115–119.
Reed, D.C., Raimondi, P.T., Carr, M.H., and Goldwasser, L. (2000). The role of dispersal and disturbance in determining spatial heterogeneity in sedentary organisms. *Ecology.* **81**, 2011–2026.

Rhodes, K.L., Lewis, R.I., Chapman, R.W., and Sadovy, Y. (2003). Genetic structure of camouflage grouper, *Epinephelus polyphekadion* (Pisces: Serranidae), in the western central Pacific. *Mar. Biol.* **142**, 771–776.

Roughgarden, J., and Iwasa, Y. (1986). Dynamics of a metapopulation with space-limited subpopulations. *Theor. Popul. Biol.* **29**, 235–261.

Roughgarden, J., Iwasa, Y., and Baxter, C. (1985). Demographic theory for an open marine population with space-limited recruitment. *Ecology.* **66**, 54–67.

Shanks, A.L., Largier, J., Brink, L., Brubaker, J., and Hooff, R. (2000). Demonstration of the onshore transport of larval invertebrates by the shoreward movement of an upwelling front. *Limnol. Oceanogr.* **45**, 230–236.

Sherratt, J.A., Lambin, X., and Sherratt, T.N. (2003). The effects of the size and shape of landscape features on the formation of traveling waves in cyclic populations. *Am. Nat.* **162**, 503–513.

Shkedy, Y., and Roughgarden, J. (1997). Barnacle recruitment and population dynamics predicted from coastal upwelling. *Oikos.* **80**, 487–498.

Sousa, W.P. (1979). Disturbance in marine intertidal boulder fields: The nonequilibrium maintenance of species diversity. *Ecology.* **60**, 1225–1239.

Sousa, W.P. (1984). Intertidal mosaics: Patch size, propagule availability, and spatially-variable patterns of succession. *Ecology.* **65**, 1918–1935.

Southward, A.J. (1956). The population balance between limpets and seaweeds on wave beaten rocky shores. *Ann. Rep. Mar. Biol. Stat. Port Erin.* **68**, 20–29.

Spight, T.M. (1974). Sizes of populations of a marine snail. *Ecology.* **55**, 712–729.

Steele, J.H., Carpenter, S.R., Cohen, J.E., Dayton, P.K., and Ricklefs, R.E. (1993). Comparing terrestrial and marine ecological systems. In *Patch dynamics* (S.A. Levin, T.M. Powell, and J.H. Steele, eds., pp. 1–12). Berlin: Springer-Verlag.

Thomas, C.D., and Kunin, W.E. (1999). The spatial structure of populations. *J. Anim. Ecol.* **68**, 647–657.

Thompson, R.C., Wilson, B.J., Tobin, M.L., Hill, A.S., and Hawkins, S.J. (1996). Biologically generated habitat provision and diversity of rocky shore organisms at a hierarchy of spatial scales. *J. Exp. Mar. Biol. Ecol.* **202**, 73–84.

Underwood, A.J. (2000). Experimental ecology of rocky intertidal habitats: What are we learning? *J. Exp. Mar. Biol. Ecol.* **250**, 51–76.

Underwood, A.J., and Chapman, M.G. (1996). Scales of spatial patterns of distribution of intertidal invertebrates. *Oecologia.* **107**, 212–224.

Underwood, A.J., and Fairweather, P.G. (1989). Supply-side ecology and benthic marine assemblages. *Trends Ecol. Evol.* **4**, 16–20.

Watt, A.S. (1947). Pattern and process in the plant community. *J. Ecol.* **35**, 1–22.

Wing, S.R., Botsford, L.W., Morgan, L.E., Diehl, J.M., Lundquist, C.J. (2003). Inter-annual variability in larval supply to populations of three invertebrate taxa in the northern California Current. *Estuar. Coast. Shelf. Sci.* **57**, 859–872.

Witman, J.D., Genovese, S.J., Bruno, J.F., McLaughlin, J.W., Pavlin, B.I. (2003). Massive prey recruitment and the control of rocky subtidal communities on large spatial scales. *Ecol. Monogr.* **73**, 441–462.

CHAPTER 8

Metapopulation Dynamics of Coastal Decapods

MICHAEL J. FOGARTY and LOUIS W. BOTSFORD

I. Introduction
II. Decapod Life Histories
III. Identifying Decapod Metapopulations
IV. Case Studies
   A. American Lobster (*Homarus americanus*)
   B. Blue Crab (*Callinectes sapidus*)
   C. Dungeness Crab (*Cancer magister*)
   D. Pink Shrimp (*Pandalus borealis*)
V. Discussion
VI. Summary
VII. Acknowledgment
   References

I. INTRODUCTION

Marine decapod crustaceans, including lobsters, crabs, and shrimp, support important fisheries from tropical to polar seas. The high unit value of decapods and their longstanding importance in coastal fisheries have provided powerful incentives for scientific study of their ecology and population dynamics. As for many marine species, high levels of population variability in space and time are hallmarks of decapod populations. Key aspects of the life history characteristics of the Decapoda make them important candidates for the study of metapopulation dynamics. Most lobsters, crabs, and shrimp exhibit well-defined spatial structures linked to habitat preferences and requirements. Dispersal during the meroplanktonic larval stages is potentially extensive for many (extending thousands of kilometers in some cases), whereas juvenile and adult movements are typically (but not exclusively) more restricted. In general, exploitation rates of commercially fished decapods are very high. Hence, the influence of harvesting on their demography, abundance, and distribution patterns must be carefully considered in any evaluation of decapod metapopulations. The practical implications of metapopulation dynamics for heavily exploited decapod crustaceans are considerable. Global landings of decapod crustaceans have increased steadily during the last five decades in response to increased demand (for more information, go to www.fao.org). Increased exploitation rates may place many decapods at risk. Where metapopulation structures exist, the linkage between local population structure and spatial patterns of fishing intensity will critically define the stability and resilience of decapods to harvesting.

Classic metapopulation theory is framed by the dynamic interplay of extinction and recolonization events (Levins, 1970; reviewed in Hanski, 1999), but here we will not confine ourselves to that type of dynamic. Rather, we are interested in how other characteristics such as stability and variability are affected by the metapopulation structure. In this chapter, we adopt the more general definition of a metapopulation as a number of populations distributed over space, linked through dispersal processes (e.g., Botsford et al., 1994; Hanski and Simberloff, 1997; Kritzer and Sale, 2004). Hence, the defining characteristics that we require of metapopulations are simply (1) that juveniles and adults be distributed over space to the degree that some individuals would not encounter others (i.e., the benthic populations are not well mixed) and (2) there is interchange among populations. We focus in particular on exchange of larvae.

For the Decapoda, there is little evidence of extinction–recolonization processes in local marine populations, although the implications of widespread loss of essential habitats for many decapods (sea grass meadows, mangrove forests, coral reefs, etc.) through natural and anthropogenic disturbances are clear. Several decapod stocks have been brought to the brink of commercial if not biological extinction as a consequence of overexploitation and habitat loss and degradation. However, currently, no marine decapods appear in the World Conservation Union (IUCN) Red List of Threatened Species (for more information, go to www.redlist.org), although threatened freshwater decapods (dominated by crayfish species) are identified. When information is available, we address issues related to extinction probabilities for decapods under more restrictive metapopulation definitions (see Smedbol et al., 2002, for discussion within a fisheries context).

We structure our synthesis around case studies of intensively studied marine decapod species, interpreting key aspects of their population behavior in terms of metapopulation dynamics. The evidence we use to adduce decapod metapopulation structure is as mentioned earlier: dispersal in at least one life history stage and some level of spatial segregation. We characterize decapod metapopulations using inferences based on genetic structure and other information on interchange among local populations. We also examine characteristics such as variation in the spatial distribution over time and the relative degree of synchrony in population fluctuations. To understand these fluctuations, we initially focus on life history and single population dynamics, the pattern of the interconnecting larval dispersal, then the variability in distribution of populations over space. The topic—

single population dynamics—involves the levels of stability, variability, and persistence that would occur in a population that was isolated and well mixed, rather than part of a metapopulation. To describe the consequent metapopulation dynamics, we indicate how single population dynamics, spatial distribution, and larval dispersal interact to produce greater or less variability in abundance, stability, synchrony over space, and persistence.

When contrasting single population and metapopulation dynamics, we make use of single population results regarding stability and persistence, which are well-known, and metapopulation results, which are not fully developed (Botsford and Hastings, this volume). To avoid repetition, we briefly review qualitative aspects of the former here. Population stability in marine decapods has been explored primarily in the context of cycles occurring in population data (Ricker, 1954; Botsford and Wickham, 1978; Botsford, 1997; Higgins et al., 1997a). Briefly, cycles resulting from density dependence in recruitment are more likely to occur when there is a combination of (1) a steeper negative slope in recruitment survival (or in a Ricker–type stock–recruitment relationship) and (2) a narrow adult age distribution. The period of the cycles is typically near twice the mean age in the population, and higher periods and chaotic behavior are unlikely to occur as long as there are several adult age classes (Botsford, 1992). Adding explicit accounts of environmental variability to models of these populations can lead to a variance in abundance that is dominated by frequencies near the frequency of cycles (e.g., McCann et al., 2003).

For persistence, we follow the description in the fishery literature of the persistence of age-structured populations with density-dependent recruitment (Sissenwine and Shepherd, 1987). In this deterministic description, populations will persist as long as the lifetime egg production exceeds the inverse of the slope of the egg–recruit relationship at its origin. Thus, steeper stock–recruit relationships and unfished (i.e., broader) age structures are more persistent. This *replacement* result is similar to requiring R_0 be greater than 1.0 in linear age-structured models in general ecology, the difference being that the threshold extinction value in populations with a planktonic stage is highly uncertain. Benchmark thresholds in the range of a lifetime egg production of 35 to 50% of natural lifetime egg production have been used (Mace and Sissenwine, 1993; Myers et al., 1999). Adding explicit consideration of environmental variability to models does not change the nature of the population behavior, but does make the population growth rate more negative (Tuljapurkar, 1982).

A central issue of interest at the metapopulation level is, then, how these characteristics—stability, variability, and persistence—are changed by the presence of several individual populations and larval exchange between them. In addition, other characteristics, such as synchrony over space, become important. Seeking an understanding of synchrony requires an examination of environmental variability and its contribution to spatio-temporal fluctuations. Unfortunately, we do not have a complete understanding of stability, synchrony, and persistence in metapopulations (although we give one example of metapopulation structure changing stability here). As described in Botsford and Hastings (2005), we are only beginning to understand how metapopulation structure affects persistence. Botsford and Hastings (2005) explain progress in terms of the way in which single population concept of persistence as individuals replacing themselves by reproduction through their life (i.e., over age), for metapopulations can be viewed as individuals replacing themselves over space.

To place our conclusions in a general context, we first present an overview of life histories of the Decapoda with implications for metapopulation dynamics.

II. DECAPOD LIFE HISTORIES

Decapod crustaceans differ from many other marine taxa in critical aspects of their reproductive biology and early life history dynamics. Mating systems involving courtship and insemination, and the prominence of external egg brooding by the female are important features of decapod life histories. Decapods share with other marine taxa potentially extensive larval dispersal patterns and specific habitat preferences. At later life stages, generally circumscribed juvenile and adult movement patterns are common in many crab, lobster, and shrimp species, although long-range migration of adults of some taxa are well-known. These features hold direct relevance to decapod metapopulation dynamics.

The classic life history tradeoff of egg size and energy content and fecundity is well represented in the Decapoda. Clawed lobsters typically produce thousands (*Nephrops*) to tens of thousands (*Homarus*) of large eggs (approximately 1.5 mm in diameter), whereas some brachyuran crabs produce clutches of as many as three million much smaller eggs (approximately 0.3–0.4 mm in diameter; Cobb et al., 1997). Egg development in decapods generally occurs during an incubation phase in which the ova are brooded by the female and are not subject to passive dispersal in the plankton. Certain peneaid shrimp do, however, release their eggs directly without an intervening brood phase. For many marine decapods, the naupliar phase of the life history is passed entirely during this incubation period. The eggs may be held by the female for extensive periods (e.g., 6–10 months for the Norway lobster *Nephrops norvegicus*, and 9–11 months for the American lobster *Homarus americanus*) before hatching into the first larval stage. Decapods differ from some other crustacean taxa in undergoing a fixed number of larval instars, although exceptions do occur. For example, the blue crab *Calinectes sapidus* and the stone crab *Menippe mercenaria* both have been observed to display slight variations in the number of zoeal stages, with related changes in the overall duration of the larval phase (Sastry, 1983). Decapod larvae are planktotrophic—an important feature that allows relatively extended larval stage durations and dispersal potential.

Chapter 8 Metapopulation Dynamics of Coastal Decapods

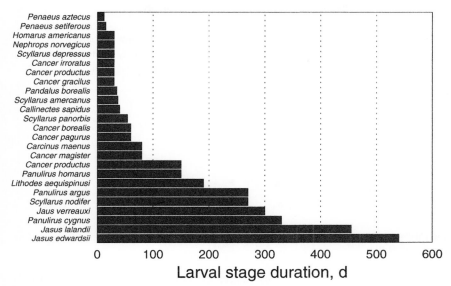

FIGURE 8-1. Larval stage duration of selected decapods.

The duration of the larval phase differs markedly among marine decapod groups (Fig. 8-1). In addition, the length of the larval stage is dependent on temperature and other factors such as food supply (Sastry, 1983). Relative stage durations of the egg and larval phases play a critical role in overall dispersion potential during the early life stages. Contrasts among selected clawed lobsters (*Homarus* and *Nephrops*), spiny lobsters (*Panulirus* and *Jasus*), and brachyuran crabs (*Cancer*) with respect to maximum fecundity, egg size, brood period, larval stage duration, and postlarva size indicate distinct differences with important implications for dispersal potential and larval survival (Cobb et al., 1997). Lobster taxa offer a particularly vivid example. The overall duration of the egg and larval stages of the clawed lobsters of the genus *Homarus* and of the spiny lobsters of the genus *Panulirus* is on the order of 1 year or more. However, the relative duration of the egg and larval stages differs markedly (Fig. 8-2). For *Homarus americanus*, the 10 to 12-month incubation period is followed by an approximate 1-month larval phase. In contrast, the incubation period for many panulirid lobsters is 2 to 3 months followed by a 6 to 18-month larval stage. The dispersal potential of *Homarus americanus* during the larval stage is on the order 100 to 200 km (Fogarty, 1983), whereas that for many panulirids is an order of magnitude higher (Phillips and Sastry, 1980). For example, *P. argus* zoea (phyllosomes) disperse on a pan-Caribbean scale. Phyllosomes of the western rock lobster *P. cygnus* have been located 1500 km from the hatching areas off the west coast of Australia, and the dispersal range of *P. gracilis* in the Pacific has been estimated at as much as 4000 km (Cobb and Wang, 1985). Both lobster groups differ sub-

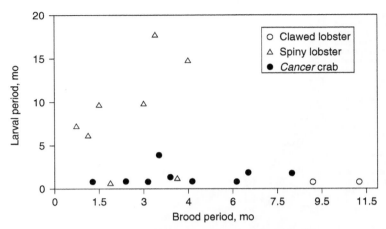

FIGURE 8-2. Duration of the egg brood and larval phase for three decapod taxa. Compiled by Cobb et al. (1997).

stantially from crabs of the genus *Cancer* in terms of life history traits (Cobb et al., 1997). Fecundity is much higher in *Cancer*, and egg and postlarval size are much smaller compared with the lobster taxa. In addition, the larval period is substantially shorter in most *Cancer* species than for most spiny lobster species, whereas the brood period is much shorter than for clawed lobsters (Fig. 8-2). In general, we expect the dispersal potential to be relatively low and planktonic mortality to be high for *Cancer* relative to lobster species (see Cobb et al., 1997).

Behavioral responses play a critical role in dispersal in many marine invertebrates (Young, 1995). Consideration of larval behavior strongly shapes our perspectives on transport and the relative probabilities of interchange among local populations, often altering views based on consideration of larval duration and mesoscale hydrography alone (e.g., Hedgecock, 1986; McConaugha, 1992). For example, early phyllosome stages of *P. cygnus* are photopositive and are transported cross-shelf in surface waters from the hatching locations on the nearshore continental shelf of Western Australia in the divergence created by the Leeuwen Current. Later phyllosome stages are photonegative and are transported in countercurrents at depth back to the coast, where settlement of the postlarval puerulus stage occurs. Active swimming by the pueruli is an important component of movement toward suitable settlement sites in sea grass meadows on the coast (Phillips et al., 1994). Similar observations have been made for the American lobster (Katz et al., 1994)

Metamorphosis from the final zoeal stage to the postlarval form of most decapods is accompanied by both morphological and behavioral changes related to habitat selection and settlement. Many decapods are shelter dependent during the early juvenile stages after settlement and dispersal is relatively limited. For a number of decapods, movements and directed migrations increase markedly during the adult phase of the life history and may be a mechanism for later interchange among populations. For panulirid lobsters, mass migration of adults in well-defined queues more than hundreds of kilometers or more have been documented.

III. IDENTIFYING DECAPOD METAPOPULATIONS

Identification of metapopulation structures in the marine environment holds important challenges and in many ways has lagged developments in terrestrial and freshwater systems in which sampling is often more tractable (see the review by Sale and Kritzer, 2003). Our examination of decapod metapopulations focuses initially on identification of spatial structure and larval exchange, but also includes inferences derived from an integrated evaluation of spatiotemporal patterns in abundance (or, in some cases, catches as a proxy for abundance), oceanography and larval biology, juvenile and adult movements, and information on genetic structure. We examine evidence for spatial structure of juveniles and adults, related to habitat preferences and other factors, and consider the relative coherence in population fluctuations in spatially separated groups of conspecifics. Metapopulations are generally more likely to persist if their subpopulations are driven by independent environmental variability (see Hanski, 1999) and this consideration also has important implications for conservation and management (Earn et al., 2000). For exploited decapod crustaceans, the situation is complicated by the potential effects of harvesting in which, for example, trends in population levels over relatively large spatial domains can be imposed by large-scale removals by the fishery. In other instances, large-scale environmental forcing may impose common trends in animal population dynamics (Post and Forchhammer, 2002).

Because spatial scales of adult movement in the examples described herein are much less than likely scales of larval dispersal, knowing the pattern of larval dispersal is a key element to understanding metapopulation dynamics completely. Although we will describe general tendencies of larval movement, unfortunately we do not know the dispersal patterns at the scales necessary to describe metapopulation connectivity. However, research on this important aspect continues on several fronts, including larval tracking, coupled physical/biological models, and genetic analysis.

Difficulties in sampling and tracking the dispersive larval stages of decapods are substantial, but not insurmountable. The potential development of predictive indices of recruitment has motivated many efforts to monitor decapod larval or postlarval abundance and distribution (e.g., Scarratt, 1973; Phillips et al., 1994; Wing et al., 1995a, b). For decapods, recently developed approaches for artificially marking and tracking marine fish larvae and the use of natural tracers (cf., Thorrold et al., 2002) are not generally possible throughout the entire larval period. Successful tracking on shorter timescales for crab larvae has proved possible using elemental fingerprinting (DiBacco and Levin, 2000; DiBacco and Chadwick, 2001), and various marking methods have been tried (Anastasia et al., 1998). Molting processes in decapods and other crustaceans preclude the use of the approach taken for fish, because the hard parts that might retain distinctive geochemical signatures of source locations or that may allow marking using artificial tracers are lost at each molt. Despite these difficulties, a substantial body of information has been accrued on decapod dispersal based on larval distributions in relation to spawning locations and ontogenetic changes in the spatial distribution of successive larval stages.

The development of coupled biological–physical models used to track potential dispersal trajectories (e.g., Johnson and Hess, 1990; Katz et al., 1994; Incze and Naime, 2000; Cowen et al., 2000; Pederson et al., 2003; Harding et al., in press) has further served to delineate possible pathways of population interchange. These models provide a means for examining the consequences for dispersal of interactions between known circulation and larval behavior.

Information on genetic structure can provide important checks on inferences concerning decapod metapopulations derived from consideration of factors such as larval stage duration, distribution, and oceanographic transport mechanisms. Again, the importance of decapods to commercial fisheries has resulted in extensive efforts to define genetic structure, often in an attempt to define management units. Evidence of strong genetic separation among groups would be taken as an indication of distinct populations rather than a metapopulation structure for these groups (e.g., Palumbi, 1995; Hellberg et al., 2002; Palumbi, 2003). Conversely, evidence for genetic homogeneity implies some level of interchange (possibly at very low levels). In some instances, however, genetic homogeneity can reflect historical chance events rather than frequent mixing (see Hedgecock, 1986, for a review). Information on overall levels of genetic variation can also provide insights into metapopulation structures (Hanski, 1999). Under some metapopulation constructs, reduced genetic diversity is expected (Hellberg et al., 2002). Dispersal distances have been estimated from genetic data, using the rate of change in F_{ST} (or an equivalent measure) with distance to estimate dispersal distances based on modeling results (Kinlan and Gaines, 2003). Following that approach, it was found that marine benthic invertebrates had a wider range of dispersal distances than either macroalgae or fish.

It is now widely appreciated that information on factors such as larval stage duration can only provide insights into dispersal *potential* and that *realized* dispersal can be very different. Nonetheless, Shanks et al. (2003) have demonstrated useful relationships between duration and dispersal distance. As more sensitive measures of genetic structure have been devised (e.g., assays of nuclear DNA), further evidence of finer scale population differentiation of marine organisms has emerged. Such a result points to the possibility of oceanographic retention mechanisms or perhaps the interplay of larval behavior and hydrography, and may redirect our understanding of the scales at which metapopulations operate.

In some instances, the economic importance of depleted decapods has led to remediation efforts that also might provide insights into aspects of metapopulation dynamics such as patch occupancy and colonization rates. For example, transplant programs designed to reestablish submerged aquatic vegetation in estuarine systems with respect to blue crabs (e.g., Orth et al., 1996) or the establishment of artificial reefs for lobsters (e.g., Cobb et al., 1997; Castro, 2003) can provide insights into colonization processes that parallel issues in metapopulation theory related to extinction and recolonization. In other cases, the inadvertent introduction of decapod species and patterns of population expansion may provide information on dispersal patterns. For example, the green crab *Carcinus maenas*, introduced from Europe to the east coast of North America in the 18th century has now been introduced to the west coast, and tracking the dispersion of this species could be important in understanding colonization events (Grosholz, 2002; Jamieson et al., 2002).

IV. CASE STUDIES

A. AMERICAN LOBSTER (*HOMARUS AMERICANUS*)

The American lobster, *Homarus americanus*, is distributed in the western North Atlantic from Labrador to Cape Hatteras in estuarine waters to the edge of the continental shelf. The highest lobster densities are typically found in structurally complex habitats with abundant shelter (Cooper and Uzmann, 1980; Lawton and Lavalli, 1995). The spatial mosaic of lobster populations is strongly defined by substrate type and lobster habitat preferences. Lobsters have supported important commercial fisheries in the northeastern United States and the maritime provinces of Canada since the early 19th century, with a large-scale coherence in overall landings dominated by a dramatic increase during the last two decades (Fig. 8-3; Fogarty, 1995; Fogarty and Gendron, 2004). The recent increase in abundance and landings appears to be linked to changes in lobster distribution patterns, with increased utilization of soft-substrate environments (reviewed in Fogarty and Gendron, 2004). The apparent change in habitat utilization may have resulted in an increase in overall carrying capacity for lobsters under current environmental conditions. Reduction in fish predators through overharvesting has been suggested as one potential mechanism underlying an increase in lobster recruitment. Reduction in predatory fish populations has been linked to higher abundance levels of other crustacean populations in the northwest Atlantic (e.g.,

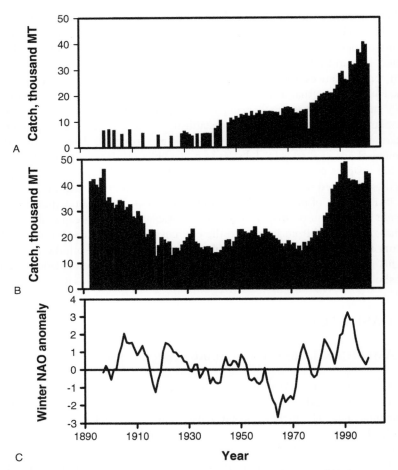

FIGURE 8-3. Landings of American lobster in (a) Canada (b) United States and (c) the winter North Atlantic Oscillation index (smoothed) [Fogarty and Gendron (2005)].

Witman and Sebens, 1992; Worm and Myers, 2003; see Section II.B.4). A potential role of declining cod predation on lobster populations has also been posited (Jackson et al., 2001), although Hanson and Lanteigne (2000) concluded that changes in predation rates by cod on lobster in the Gulf of St. Lawrence could not account for changes in landings in this region. Other predators, however, may still play an important role in this dynamic.

Changes in large-scale climatic conditions in the North Atlantic have also been identified and may have played a role in the increase. For example, the winter

North Atlantic Oscillation (NAO) index has exhibited a generally increasing trend from the early 1970s to the mid 1990s (Fig. 8-3). Changes in the NAO potentially affect wind fields, temperature patterns, and precipitation over the North Atlantic. In turn, these factors can potentially affect larval transport, individual growth, and ecosystem productivity. It is likely that the increase in lobster abundance reflects the interplay of a number of fishery-related, abiotic, and biotic factors (Fogarty, 1995). One or more of these factors may be dominant during different time periods.

Life History and Single Population Dynamics

The American lobster is among the largest benthic invertebrate species throughout its range. It has been suggested, not without controversy, that lobsters are keystone predators in these systems. In the absence of harvesting, lobsters are long lived, with estimated life spans in excess of 50 years (Cooper and Uzmann, 1980) and they can attain weights as much as 20 kg. The long life span implies a low natural mortality rate, with estimates of 0.05 to 0.15 per year typically used in population models. Fertilization occurs during egg extrusion, and the eggs undergo a 10 to 12-month incubation period. Hatching occurs during late spring and early summer, with geographical differences controlled by temperature regimes. A brief prelarval stage at hatching is succeeded by three zoeal stages followed by the postlarval settlement stage. The larval phase of the life history lasts approximately 3 to 4 weeks, depending on temperature. The different larval stages exhibit different phototactic responses, which may influence dispersal patterns. The first zoeal stage and the early phase of the postlarval stage are strongly photopositive and the larvae are transported in surface currents. The second and third zoeal stages are moderately photonegative (Fogarty, 1983). Differential distribution within the water column at different stages may provide for retention mechanism through transport in countercurrent systems. Larvae in the postlarval stage preferentially settle in cobble and other habitats where shelter is readily available, and they remain cryptic for about the first 2 years of life.

Populations of American lobster appear to be unexpectedly persistent in the face of high removals by fishing. Both single population and metapopulation explanations have been advanced to explain the apparently high resilience of American lobster stocks to exploitation. Single population explanations point to the nature of the stock–recruitment relationship (Caddy, 1986; Fogarty and Idoine, 1986; Ennis and Fogarty, 1997). Examination of the relationship between lobster population egg production and subsequent recruitment in Arnolds Cove, Newfoundland, indicated a strongly asymptotic functional relationship with a steep slope at the origin (Ennis and Fogarty, 1997). Fogarty and Idoine (1986) found a similar relationship between stage IV larval production and a proxy for subsequent recruitment in Northumberland Strait. Relationships of this type are consistent with high resilience to exploitation and may be a critical factor in the long-term persistence of lobster populations under intensive exploitation (see the description of single population dynamics in Section I of this chapter and in Botsford and Hastings, this volume). The benchmark level of fraction of natural lifetime egg production required for persistence in the lobster fishery is 10%, rather than the more typical 35%, in part because of the steep slope of the egg–recruit relationship at the origin. Metapopulation explanations have involved the existence of spatial refugia and larval source–sink dynamics (Anthony and Caddy, 1980; Fogarty, 1998). Larval subsidies from less intensively exploited components of an overall metapopulation may play a critical role in resilience to exploitation (Fogarty, 1998).

Larval Dispersal

The potential importance of larval transport in American lobster and larval subsidies to heavily exploited inshore lobster groups has elicited considerable interest (Fogarty, 1995; Fogarty, 1998). It has been suggested that heavily exploited inshore lobster stocks off Nova Scotia (Harding et al., 1983; Harding and Trites, 1988) and New England (Katz et al., 1994) receive larval input from offshore areas based on consideration of distribution and abundance of larvae and residual circulation patterns. American lobster larvae originating on the edge of the continental shelf off southern New England could reach coastal waters using a combination of directed swimming and drift in prevailing wind-driven surface currents (Katz et al., 1994). Examination of the distribution patterns of successive larval stages in synoptic samples along an inshore–offshore transect indicated that later larval stages were more common closer to shore, suggesting that shoreward transport of larvae may be important. Furthermore, lobsters from the middle and outer continental shelf in this region are known to undertake extensive seasonal migrations, with a shoreward component prior to release of the larvae (see Fogarty, 1995, for a review). This seasonal directional movement in the southern New England region would substantially reduce the dispersal distance for linkage between inshore and offshore lobster groups.

Incze and Naime (2000) examined potential larval dispersal patterns in the Gulf of Maine in a numerical hydrodynamic transport model coupled with a simple lobster cohort model. The timing and location of hatching strongly influenced both the duration of the larval phase and the transport patterns in these simulations. The general cyclonic circulation pattern in the gulf dominated transport processes, with an apparent net removal of lobster larvae from eastern Maine and a strong larval subsidy to mideastern Maine from "upstream" locations in eastern Maine. Prevailing onshore winds during the larval development time also result in strong linkages between inshore and offshore locations in the Gulf of Maine. Inshore and offshore groups are also linked through adult migration in the southern New England region and off Nova Scotia (see review by Fogarty, 1995).

Harding et al. (in press) used a three-dimensional numerical hydrodynamic model to evaluate potential source areas for stage IV lobster larvae found off Browns Bank and off Georges Bank in the Gulf of Maine. Field studies in the vicinity of Georges Bank indicated that first and second stage larvae were most common on the bank whereas the third and fourth stages were primarily found off-bank. Furthermore, the condition index of later stage larvae found on the bank was lower than those off-bank. Results of the particle-tracking exercise indicated that the most probable source areas were known hatching locations off Cape Cod and in the vicinity of Penobscot Bay on the Maine coast. Harding et al. (in press) noted, however, that the off-bank location of later stage larvae near Georges Bank may in fact reflect directed swimming of larvae originating on the bank or the effects of transient mesoscale hydrographic events, rather than a more distant source location.

Metapopulation Dynamics

The findings to date regarding larval transport are generally consistent with the view provided by research on the genetic structure of lobster. Relatively low levels of genetic variation have typically been observed in American lobster populations. Tracey et al. (1975) reported an average proportion of heterozygous loci per individual of 3.8%. Low genetic diversity is consistent with metapopulation dynamics (Smedbol et al., 2002), but other mechanisms can be readily invoked. Comparisons of the genetic structure of inshore and offshore lobster populations have generally shown no significant differences (Barlow and Ridgway, 1971; Odense and Annand, 1978; Kornfield and Moran, 1990; Harding et al., 1997). However, Tracey et al. (1975) did report differences between inshore lobsters off Massachusetts and samples collected in an offshore canyon area on Georges Bank. Differences were also reported between inshore New England and the Gulf of St. Lawrence (Tracey et al., 1975). Based on a reanalysis of the data of Tracey et al. (1975), finer scale differences among inshore and among offshore locations have also been suggested (Burton, 1983; Shaklee, 1983). Differences between inshore and offshore lobsters based on analyses of morphometric characteristics and incidence of parasites have been used to adduce that inshore and offshore groups are separate (see Fogarty, 1995).

Monitoring of postlarval settlement in structurally complex habitats (small cobbles, shell, or gravel) has been carried out in a number of locations in the Gulf of Maine and in southern New England (Fig. 8-4; R. Wahle, Bigelow Laboratory for Ocean Sciences, personal communication). Settlement declined through the late 1990s at the longest running sampling locations in New Brunswick, Maine, and Rhode Island, followed by a consistent increase

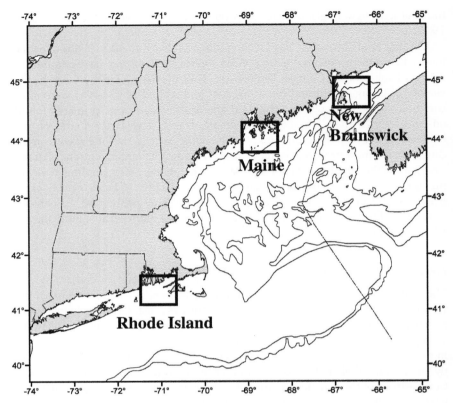

FIGURE 8-4. Location of lobster settlement sample sites used in this analysis.

(Fig. 8-5). In these habitats, a moderate to high level of coherence among the three locations is evident. The correlation between the New Brunswick, Maine, and Rhode Island sites is approximately 0.6 to 0.7. The overall level of coherence in settlement presumably is indicative of the role of large-scale environmental forcing and possible interchange among locations. The Rhode Island site differs notably in not having exhibited the dramatic increase evident in Maine and New Brunswick during 2001 to 2003. The northern sites are part of the same oceanographic domain in the Gulf of Maine.

Consideration of linkages between inshore and offshore lobster populations through migration and larval dispersal has provided important insights into elements of the response of inshore lobster populations to exploitation. Fogarty (1998) examined the potential importance of dispersal processes in a simple discrete space–discrete time model to explore implications for the stability and resilience of lobster populations. The model represented two areas—inshore and

Chapter 8 Metapopulation Dynamics of Coastal Decapods 285

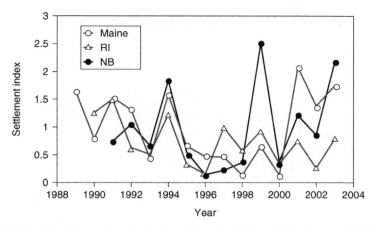

FIGURE 8-5. Trends in post-larval settlement of American lobster in New Brunswick (NB), mid-Coast Maine (Maine), and Rhode Island (RI) [R. Wahle, personal communication].

offshore—with differential patterns of fishing mortality within each. It was shown that even relatively small rates of exchange potentially confer substantial resilience to exploitation if a source area experiences lower exploitation rates.

Potentially high levels of fishing mortality can be sustained in the inshore population if exploitation rates in the offshore region remain low, even at low levels of larval subsidy from offshore (Fig. 8-6). Similarly, the subsidy provided by the migration of adults from inshore to offshore permits higher levels of offshore fishing mortality when inshore exploitation rates are low to moderate than for the case of independent populations. The maximum yield occurs at low to moderate levels of fishing mortality on both subpopulations (Fig. 8-6).

It should be noted that the higher the level of subsidy (fraction of larval production exported) provided from offshore to inshore subpopulations, the more vulnerable the offshore group to exploitation. Transport (loss) of larvae from the offshore to inshore regions results in a reduction in the slope of the recruitment curve at the origin. In the absence of any return migration of adults from inshore to offshore in particular, the offshore stock is at increasingly greater risk of overexploitation with increasing levels of larval transport to the inshore group. This suggests that the uncertainty associated with levels of larval subsidy provided from the offshore stock might require greater caution in exploiting this source population.

B. BLUE CRAB (*CALLINECTES SAPIDUS*)

The blue crab (*Callinectes sapidus*) is broadly distributed in tropical, subtropical, and temperate waters in the western Atlantic from Nova Scotia to Uruguay. The

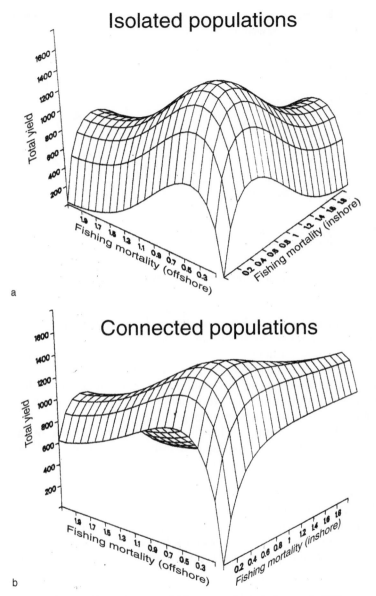

FIGURE 8-6. Lobster yield as a function of fishing mortality in inshore and offshore grounds for the case of no interchange between areas (upper) and linkage through larval dispersal and adult migration (lower) [Fogarty (1998)].

blue crab supports valuable commercial fisheries, with the highest landings derived from major estuaries and embayments on the Atlantic coast of the United States and from the Gulf of Mexico. The estuarine-dependent life history plays a dominant role in the population structure of blue crabs throughout its range. Examination of population structure and the role of oceanographic features in larval transport has focused on the Mid- and South-Atlantic Bights and the Gulf of Mexico. Accordingly, our overview of possible metapopulation dynamics emphasizes these regions.

Life History and Single Population Dynamics

The blue crab is an important predator in estuarine systems, regulating prey populations of benthic infaunal organisms in many areas. Mating occurs in the low-salinity waters of the upper reaches of estuaries after the pubertal molt by the female. Females then migrate to the higher salinity waters of the estuary mouth, where hatching occurs after an approximate 2-week incubation period by the female. Two to three broods are extruded by an individual female during the reproductive season. The larvae are transported from the estuary mouth to continental shelf waters, where a sequence of seven zoeal stages develops. The duration of the larval stages is 29 to 31 days at 25°C. Reinvasion of the estuaries occurs with entry by the megalopae to nursery grounds in the lower estuary dominated by sea grass meadows, where shelter is abundant.

The longevity of blue crabs in the absence of exploitation has been estimated at 2 to 8 years based on life history considerations and evidence from mark and recapture experiments (reviewed by Fogarty and Lipcius, in press). Taking a life span of 4 to 6 years in an unexploited population as the most plausible would imply an instantaneous natural mortality rate of approximately 0.5 to 0.75. The pubertal molt for females appears to be a terminal stage, although a small fraction of females undergo one or more further molts. Males continue to grow throughout the life span and can attain weights as much as 1 kg.

It has long been suggested that blue crab landings exhibit cyclical patterns on multidecadal timescales (e.g., Burkenroad, 1946; Fig. 8-7). Spectral analysis of Chesapeake Bay blue crab landings for the period 1922 to 1976 indicated peaks in the spectral density at periods of 8.6 to 10.7 and 18.0 years (Hurt et al., 1979). Comparisons with the power spectrum of air temperature and precipitation records from Philadelphia and with tidal forcing suggested that the 8.6 and 18.0-year periodicities in blue crab landings correspond approximately to components of the tidal cycle. The 10.7-year cycle in landings was roughly related to a dominant period in the temperature cycle. Cyclical patterns in blue crab landings in Florida (Prochaska and Taylor, 1982) and recruitment in Louisiana (Guillory, 1997) on subdecadal timescales have also been identified (Fogarty and Lipcius, in press). As noted in Section I, cycles may also occur at periods determined by

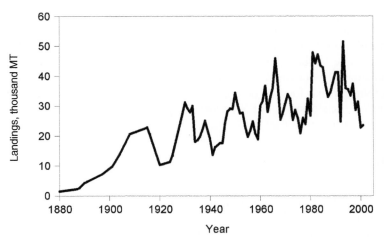

FIGURE 8-7. Landings of blue crab in Chesapeake Bay (uncorrected for changes in reporting systems over time) [Fogarty and Lipcius, in press].

endogenous mechanisms. This possibility of endogenous cycles in combination with overcompensatory population mechanisms in the blue crab has not been extensively explored, but cannot be discounted. Cannibalism, known to be important in blue crab populations (Darnell, 1959; Tagatz, 1968; Laughlin, 1982; Moksnes et al., 1997), can generate overcompensatory dynamics leading to cycles as described earlier.

Considerable attention has been devoted to determining the parent–progeny relationship for the blue crab in Chesapeake Bay (reviewed in Fogarty and Lipcius, in press). These studies have repeatedly demonstrated evidence for density-dependent controls during the prerecruit stages of the blue crab. The Ricker (1954) stock–recruitment model, which is consistent with compensatory control through cannibalism, has widely applied to blue crab populations. Blue crab recruitment is highly variable, but numerous attempts to link recruitment to environmental forcing mechanisms have provided equivocal results. Attention has centered on factors affecting transport of the meroplanktonic stages and on physical factors affecting basic physiology. Statistically significant environmental effects have been reported for extended Ricker recruitment models with explicit consideration of stream flow, wind speed and direction, incident solar radiation, temperature, and salinity (Applegate, 1983; Tang, 1985). However, an updated analysis by Rugulo et al. (1998) found no significant environmental effects, and Lipcius and Van Engel (1990) indicated that the improvement in model fit in Tang's analysis relative to a simple Ricker model was not significant. Qualitative comparisons and correlative studies, often using landings as a proxy for recruitment, have provided some indication of the importance of environmental factors such as wind speed and direction affecting transport mechanisms (Johnson et al., 1984; Johnson and Hester, 1989). However, in an updated analysis, Watkins (1995) found no significant relationship between landings and wind stress or between landings and river discharge during the period of larval occurrence. In estuaries in Florida, significant correlations between river discharge, and both landings (Wilber, 1994) and abundance (Livingston, 1991) have been reported.

Larval Dispersal

Transport mechanisms for blue crab larvae are dominated by baroclinic flows driven by estuarine processes and cross-shelf transport driven by wind fields. In the Mid-Atlantic Bight blue crab zoea are transported onto the shelf after hatching at the mouth of the estuary at the nocturnal high tide (Fig. 8-8). The larvae are neustonic throughout the period of larval development and do not appear to undergo diurnal vertical migrations or change vertical distribution with ontogeny (McConaugha et al., 1983; Epifanio et al., 1984; McConaugha, 1988; Epifanio, 1988a; Epifanio et al., 1989; Epifanio, in press). Retention of larvae in the Mid-Atlantic Bight, despite the southward-flowing nearshore currents, appears to be related to episodic reversals of flow patterns related to wind events and the occurrence of extended periods of northward-flowing water in the midshelf region off Chesapeake Bay during late summer when blue crab zoeae are most common in that area (Boicourt, 1982).

Transport of megalopae back to the estuaries has been linked to southerly wind events over the continental shelf. Goodrich et al. (1989) and Little and Epifanio (1991) demonstrated a relationship between episodic settlement of blue crab megalopae in Chesapeake and Delaware Bays with southward wind. Downwelling circulation driven by southward wind events would result in cross-shelf transport of *C. sapidus* megalopae into estuaries. Southerly winds over the Mid-Atlantic Bight are typically related to passage of low-pressure systems through the region, particularly in autumn, when the larvae are present in the water column (Epifanio, in press).

In the northern Gulf of Mexico, settlement of megalopae has been related to periods of onshore wind events (Morgan et al., 1996). Relationships between megalopal settlement and lunar phase have been suggested (van Montfrans et al., 1990; Boylan and Wenner, 1992; Olmi, 1995), but these may be principally related to the spring/neap cycle (Epifanio, in press). Within the estuary, the megalopae appear to utilize selective tidal stream transport to move up-estuary. By moving to the surface during the flood phase and to the bottom during the ebb phase, the megalopae are transported within the estuary by riding only flood currents. Settlement and metamorphosis of the megalopal stage appears to be mediated by chemical cues related to salinity, and other signals related to estuarine conditions, including floral composition.

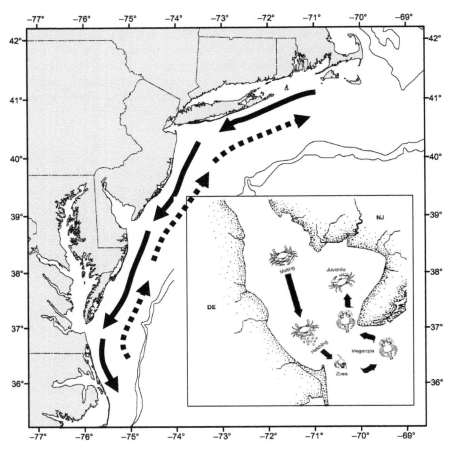

FIGURE 8-8. Mean surface currents in the Mid-Atlantic Bight in summer driven by baroclinic (solid arrows) and wind-driven forcing (dashed arrows). The inset shows the distribution of blue crab life history stages and the transport of larvae from the Delaware Bay estuary to the continental shelf and return of the megalopal stage to the estuary (after Epifanio, in press).

Application of numerical hydrodynamic models has shown some promise as a synthetic tool in understanding patterns of larval dispersal and in predicting settlement events (Johnson, 1985; Johnson and Hess, 1990). Variations in the winds over the continental shelf were found to have the greatest effect on larval transport. For Delaware Bay, model results indicated the importance of upwelling-favorable (northward) winds for larval retention near the parent estuary and that downwelling-favorable (southward) wind events later in the larval season control cross-shelf transport and subsequent recruitment to juvenile habitat within the estuary (Garvine et al., 1997).

Metapopulation Dynamics

Results of genetic studies appear to be generally consistent with the observations of dispersal mechanisms of blue crabs, particularly for the larval stages. Examination of the broad-scale genetic structure of the blue crab within US waters from New York through Texas indicated substantial patchiness in genetic structure, apparently related to the formation of larval aggregations with distinct genetic structure and the subsequent settlement patterns of these groups. Furthermore, clinal variation at one locus was discerned on the Atlantic coast and attributed to selective forces (although there was no evidence of clinal structure in the Gulf of Mexico). Examination of mtDNA haplotypes in samples collected at 14 locations distributed from New York to Mexico showed relatively high levels of diversity (Murphy et al., unpublished). North–south clinal patterns of mtDNA haplotype diversity were found in the Atlantic, and east–west gradients were identified in the Gulf of Mexico. Overall results from these broad-scale studies are consistent with a mechanism in which mixing is sufficient to prevent full population differentiation, but rates of exchange among geographically distant stocks were relatively low.

On finer spatial scales, no significant differences in genetic structure in comparisons of samples collected in Chesapeake Bay and nearby Chincoteaque Bay were found. Conflicting evidence concerning levels of genetic structure in the Gulf of Mexico has, however, been reported. Kordos and Burton (1993) reported significant differences along the Texas coast, whereas at a somewhat larger spatial scale Berthelemy–Okazaki and Okazaki (1997) found no significant differences over space in the Gulf.

The interplay of baroclinic and wind-driven flows on the continental shelf strongly influences the distribution patterns of the zoeal stages and their proximity to different estuarine systems prior to reentry and settlement of the megalopae. The possibility of mixing of larvae hatched in different estuaries is high, given the dominant and variable role of wind forcing in transport and retention of the zoeal stages during summer and early autumn. Although it has been assumed that there is a relatively strong degree of philopatry for blue crabs spawned in the larger estuarine systems, interchange among smaller adjacent estuaries and river systems appears to be particularly likely. This general pattern is consistent with the observation of sufficient exchange to infer panmixia, while allowing for the potential for some degree of separation through genetic drift or if selective forces affect the survival of larvae and later stages in different regions.

Examination of abundance trends derived from research vessel surveys shows a general decline during the 1990s in the Gulf of Mexico from Alabama through Texas (Fig. 8-9; Fogarty and Lipcius, in press). The relative roles of exploitation and environmental forcing in the decline have not been delineated, although excessive fishing pressure is clearly an important component of the population

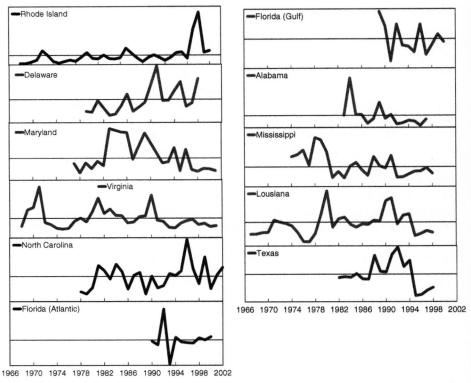

FIGURE 8-9. Trends in relative abundance of blue crab normalized by subtracting the mean and dividing by the standard deviation from Atlantic coast and Gulf coast states.

trends. The situation on the Atlantic coast of the United States is more complex and appears to reflect changing environmental conditions and the effects of harvesting. Populations have increased at the northern end of the Mid-Atlantic Bight in Rhode Island waters and in Delaware Bay, most likely in response to warming temperatures and a lower probability of winter mortality. In Chesapeake Bay, a general decline during the last decade has been evident with increasing fishing mortality rates (Fig. 8-9).

When wind conditions are not favorable for local retention of zoeal stages near the natal estuaries, the most probable exchange of larvae is from north to south in the Mid-Atlantic Bight (Fig. 8-8) and from east to west in the Gulf of Mexico. Under these conditions, "upstream" populations can provide important subsidies to adjacent estuaries, but are themselves particularly vulnerable under combinations of high exploitation rates and sustained transport patterns affecting the successful reinvasion of the natal estuaries.

Chapter 8 Metapopulation Dynamics of Coastal Decapods 293

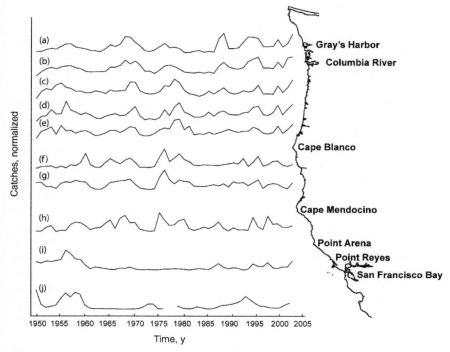

FIGURE 8-10. Dungeness crab catches normalized by subtracting the mean and dividing by the standard deviation from (a) Grays Harbour, Willapa Bay, and Columbia River, Washington; (b) Columbia River, Oregon; (c) Tillamook and Garibaldi; (d) Newport and Depoe Bay; (e) Coos Bay and Winchester Bay; (f) Brookings, Gold Beach, and Port Orford; (g) Eureka and Crescent City; (h) Fort Bragg; (i) Bodega Bay and San Francisco; and (j) Monterey and Morro Bay. (Pacific Coast Marine Fisheries Commission.)

C. DUNGENESS CRAB (*CANCER MAGISTER*)

Dungeness crab (*Cancer magister*) are found on sandy bottoms as far south as Baja, California, Mexico, and as far north as the Priboloff Islands (Jensen and Armstrong, 1987), but we focus on the metapopulation found in the California Current, from Monterey, California, in the south to the Straits of San Juan de Fuca. The variation in abundance and recruitment of Dungeness crab are reasonably well-known (Fig. 8-10) because of several characteristics of the fishery: (1) It is a male-only fishery with a minimum size limit and (2) it is very intense with some annual harvest rates in excess of 0.9 (Methot and Botsford, 1982). This produces a recruitment dominated catch, which is a good index of recruit abundance.

Life History and Single Population Dynamics

The life history of Dungeness crab is also well-known. There are several studies of growth rate (e.g., Botsford, 1984; Wainwright and Armstrong, 1993). In northern California, male crabs enter the fishery primarily at ages 3, 4, and 5 years. There is a latitudinal cline in growth rate, with slower growth to the north, but this may be the result of differences in temperature (Wainwright and Armstrong, 1993). Natural mortality is poorly known, but a value of 0.3 per year is often assumed (Botsford and Wickham, 1978).

The larval phase is reasonably well studied, but dispersal patterns are still poorly known. Eggs hatch near the beginning of the calendar year, and five zoeal stages last several months followed by a megalopal stage with about a month's duration. From plankton samples of late larval stages over several years, mortality was estimated to be 0.0061 per day in the larval stage (Hobbs et al., 1992). Diel vertical migratory behavior begins during the last zoeal stage and is strongest in the megalopal stage (Hobbs and Botsford, 1992; Jamieson et al., 1989). The diet of the larval stages is not known, but analysis of nitrogen isotope content indicates they are omnivorous (Kline, 2002). After the larval stage, some juvenile Dungeness crab settle nearshore whereas others settle or migrate into bays and estuaries (Armstrong et al., 2003; Roegner et al., 2003). The availability of large bays and estuaries varies along the coast (e.g., San Francisco Bay to the south and Grays Harbor, Willipa Bay, and the Columbia River Estuary to the north; Fig. 8-10).

Recruitment varies in a cyclical fashion at every point along the coast in the California Current, except the southern end of the fished distribution, where the population collapsed to low levels during the late 1950s (Fig. 8-10). Explanation of this cyclical behavior has attracted considerable research attention. Numerous studies (reviewed in Botsford et al., 1989; and Botsford and Hobbs, 1995) have led to the consensus that it is likely the result of a combination of endogenous population dynamic mechanisms and exogenous environmental forcing, but not predator–prey mechanisms with salmon or humans as the predator (Botsford et al., 1983).

The endogenous mechanism is compensatory density-dependent recruitment with cannibalism (Botsford and Hobbs, 1995), an egg predator worm (Hobbs and Botsford, 1989), and density-dependent fecundity (McKelvey et al., 1980) as possible biological mechanisms. Assessment of the relative likelihood of these three biological mechanisms has been based on conditions for cyclical behavior developed from age-structured models as described in Section I. This assessment has involved (1) comparing the period of the cycles with that expected from models of each mechanism and (2) determining whether each mechanism caused a steep enough slope of the recruitment survival function to cause cycles (see reviews in Botsford et al., 1989; and Botsford and Hobbs, 1995). Essentially, any of the three mechanisms can cause cycles of the observed period. The slope of the recruit-

Chapter 8 Metapopulation Dynamics of Coastal Decapods

ment survival function has been determined only for the egg predator worm, and it cannot by itself cause the cycles, but may contribute to them. The effect of cannibalism on recruitment survival has not been quantified, and density-dependent fecundity has not been directly observed. An interesting feature of the stability analysis was that single sex harvest appears to make populations less stable than they otherwise would be (Botsford, 1997).

Potential sources of environmental forcing have been assessed through computation of the covariability of the catch record with various environmental time series, including various upwelling indices (Peterson, 1973; Botsford and Wickham, 1975; Johnson et al., 1986), alongshore transport (McConnaughey et al., 1994), and general warm/cool conditions in the California Current (Botsford and Lawrence, 2001). For the most part these environmental mechanisms influence the larval phase. Not surprisingly, the various driving variables identified are highly correlated in the California Current, where temperature and sea level are high, and upwelling index and southward transport are low during El Niños, and the reverse is true during La Niñas.

Throughout these investigations of endogenous mechanisms and environmental forcing, there has also been some attention to their combined effect through (1) comparison of the autocorrelation functions of the environmental and the population time series (Botsford and Wickham, 1975), (2) simulation of environmentally forced density-dependent populations (Botsford, 1986), and (3) analysis of the timescales of variability in environmentally forced nonlinear populations (McCann et al., 2003). The essential result common to these investigations is that a population that is deterministically stable, but close to producing cycles, can produce cycles when influenced by non-cyclic environmental forcing.

To summarize these studies of cyclical behavior of single isolated populations, it is quite likely that a mechanism such as cannibalism is strong enough in these populations to be at least close to causing cycles in a deterministic model, and that the noncyclical influence of warm/cool conditions in the California Current on the larval phase is strong enough to drive the cyclical covariability (Botsford and Lawrence, 2002; McCann et al., 2003). The variation from warm to cool conditions in the California Current includes variation from weak to strong upwelling winds and weak to strong southward transport, in addition to the change in temperature (see Botsford and Lawrence, 2002, for details of covariability).

Persistence is not a concern over most of the range of Dungeness crab because only males can be legally landed in the fishery. With single-sex management, as long as there are adequate males to fertilize a high percentage of females, lifetime egg production will remain high despite the intensive fishery. Assessment of northern California females indicates their fertilization rate has not been reduced by the fishery (Hankin et al., 1997).

The collapse of the central California population of Dungeness crab in the late 1950s (see (i) and (j) in Fig. 8-10) has drawn considerable attention regarding the cause of its lack of persistence. Purported causes of the collapse of the central California population include both single population and metapopulation explanations. The single population explanations are (1) an environmental shift such as an increase in temperature, which laboratory studies indicated led to greater egg mortality (Wild, 1980); (2) a decrease in estuarine survival; (3) an egg predator worm (Wickham, 1986); and (4) the presence of multiple equilibria through endogenous population dynamics (Botsford, 1981). The multiple-equilibria mechanism involves a density-dependent increase in the growth rate of juveniles caused by the reduction in density by fishery removals. The greater fraction of recruits reaching larger sizes with greater cannibalistic tendencies upsets the balance between density-dependent mortality and reproduction, locking the population into a lower equilibrium (cf., Collie et al., 2004).

The metapopulation explanation is related to the environmental changes in the late 1950s. In the California Current, the increase in temperature suggested by Wild et al. (1980) to be a possible cause of greater egg mortality would also be associated with stronger alongshore northward (or weaker southward) flow. As noted by Gaylord and Gaines (2001), strong advection can lead to local extinction upstream of the distribution of meroplanktonic species.

Larval Dispersal

There is a growing appreciation for the fact that larval dispersal patterns are a key element in the dynamics of marine metapopulations. However, the dispersal patterns of Dungeness crab larvae are poorly known. Dungeness crab larvae are released into the northward-flowing nearshore current—the Davidson Current. At some time during their larval phase (typically March or April), the *spring transition* occurs in the California Current (Strub et al., 1987) and strong northerly, upwelling winds begin, leading to a southward offshore flow (Parrish et al., 1984). Development rate is temperature dependent, which could lead to greater survival through the larval period in warmer years (if mortality rate does not also increase with temperature), and differing patterns of development of larvae between the north and south (Moloney et al., 1994). The former is the result of changing growth rate during a stage with high mortality, and the latter results from temperature declining during the spring in the south, where upwelling is strongest, and increasing in the north from heating. We have some knowledge of larval distribution during the later stages from 5 years of Russian ichthyoplankton cruises (Hobbs et al., 1992). The cross-shelf distribution is consistent with that predicted by wind-driven flows in an advection diffusion model with vertically migrating, temperature-dependent larvae (Hobbs et al., 1992; Botsford et al., unpublished). Larvae disperse offshore during the early larval periods, but are returned to shore by the onshore component of winds acting on vertically migrating larvae.

The alongshore transport of Dungeness crab larvae presented a perplexing question: Why are these larvae not swept to the south after the spring transition, beyond the range of Dungeness crab, by the strong, southward upwelling jet (cf., Parrish et al., 1984)? A possible answer is that they are retained in retention areas associated with promontories in at least the southern part of their range (i.e., Point Reyes, Cape Mendocino, Cape Blanco [Fig. 8-10], and possibly other smaller ones). The potential for this mechanism has been studied at Point Reyes, where invertebrate larvae, including *Cancer* species, appear to be retained in the lee of the point (Wing et al., 1998) during active upwelling winds, they are returned to the north of the Point, where they settle during relaxation of the upwelling winds (Wing et al., 1995a, b). It is possible that a retention mechanism exists at other promontories, accompanied by northward transport during relaxation (cf., Morgan et al., 2000; Miller and Emmet, 1997). This would not only serve to explain persistence, but could also set the spatial scale of larval dispersal at the distance between promontories (i.e., several hundred kilometers; Botsford et al., 1998).

Metapopulation Behavior

As noted earlier, genetic variability and relatedness over space provide important information on the potential for metapopulation structure in decapods. Techniques have been developed for examining several microsatellite loci of Dungeness crab, but to our knowledge there are no published studies of genetic variability over space (Jensen and Bentzen, 2004; Toonen et al., 2004). However, several genetic studies of other meroplanktonic species have focused on the possibility of a rentention mechanism as proposed by Wing et al. (1995a, b; 1998) for Point Reyes. The combination of retention during actively upwelling and subsequent northward flow during relaxation may occur at other coastal promontories to the north, although these areas have not been studied as intensively. This could provide for a degree of persistence in each embayment (i.e., Point Reyes–Point Arena, Point Arena–Cape Mendocino, Cape Mendocino–Cape Blanco, and possibly others at finer spatial scales). This mechanism has been tested by comparing genetic relatedness within embayments to relatedness between embayments.

One study sampled populations of the intertidal porcelain shore crab, *Petrolisthes cinctipes*, at 12 locations from Neah Bay, Washington, south to Morro Bay, California (Toonen and Grosberg, submitted). Female *P. cinctipes* produce brood larvae then release them from February through mid April, and they spend at least several weeks in the plankton before competency. As a result, some of them will be in the plankton during the upwelling period, which begins at the spring transition, typically between March 15 and April 15 (Strub et al., 1987). Toonen and Grosberg (submitted) did not find greater relatedness within embayments compared with between embayments. Thus, if these promontories do provide retention zones for *P. cinctipes*, they are not "leak proof." A low percentage of leakage beyond each promontory would lead to the genetic homogeneity obtained, yet the persistence mechanism could be ecologically operable. A significant conclusion regarding spatial structure was that Monterey Bay and Morrow Bay differed from all other sites (i.e., from Dillon Beach north). The study also found that the relationship between relatedness and distance that did not peak at the origin, peaked in the range of hundreds of kilometers. Another genetic study of the acorn barnacle, *Balanus glandula*, in the California Current obtained a similar result—a strong cline between Monterey and Cape Mendocino (Sotka et al., 2004).

The spatial structure of the Dungeness crab metapopulation, the single population dynamics, and larval dispersal patterns may combine to produce characteristics of behavior that would not be seen if they were a number of single isolated populations. Here we examine several possible examples.

Relative variability appears to vary along the coast, with the coefficient of variation being greater to the south (Higgins et al., 1997b). The proposed explanations of this tend to depend on heterogeneities of habitat, rather than a metapopulation effect per se. One possibility is that the upwelling winds are strongest to the south, so that their effects on variability would be greater there. A second possibility is that large estuaries, which would be substantial nursery grounds for juvenile crabs, are more common in the north (Armstrong et al., 2003). Abundance in these areas tends to vary less than abundance along the outer coast, reaching a constant abundance possibly resulting from within-cohort cannibalism (Fernandez, 1999).

Information on the nature of connectivity among populations, and hence persistence, can be obtained from examination of covariability. Although the cyclical nature of population segments reflected in the Dungeness crab catch records appear visually to be synchronous (Fig. 8-9), analysis of their covariability over space leads to a different view. Substantial correlations exist only over several hundred kilometers (Botsford et al., 1998).

Specific metapopulation dynamic effects have been investigated through modeling by examining the changes in expected population behavior that result from increasing the connectivity between adjacent populations from zero to higher values using an idealized dispersal pattern. Increasing the width of a Gaussian-shaped larval distribution pattern tended to synchronize cyclical populations along a coastline if the compensatory density dependence occurred prior to larval dispersal (e.g., density-dependent fecundity), but when the density dependence occurred after the dispersal phase (e.g., cannibalism), it caused adjacent bands of populations to be out of phase (Fig. 8-11; Botsford et al., 1998). The width of the banding increased with the spatial scale of dispersal (i.e., the width of the Gaussian dispersal). The fact that the correlation scales in the catch data and the

Chapter 8 Metapopulation Dynamics of Coastal Decapods 299

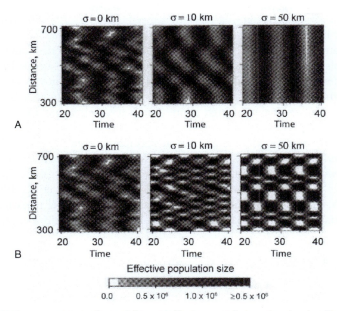

FIGURE 8-11. Spatiotemporal variability in effective population size (as it affects density-dependent recruitment survival) for different values of width of the Gaussian-shaped larval dispersal function, s, when density-dependent recruitment is predispersal (A) and postdispersal (B).

potential spatial scales of larval dispersal inferred from the spacing of promontories along the coast are similar indicates this mechanism may be operative at some level.

To summarize metapopulation behavior, the cyclical, near-synchronous variability is likely a combination of deterministic, endogenous mechanisms and stochastic, exogenous mechanisms. We have probably determined all the possible driving mechanisms, but we do not know exactly where this metapopulation system lies in parameter space. The system must include density-dependent recruitment (e.g., cannibalism) to be as periodic as it appears. Individual populations, however, may not be deterministically stable; they may be driven by variability in survival on 3 to 10-year timescales (i.e., the dominant timescale of warm/cool variability in the California Current, which is closely related to El Niño). The observed spatial covariability (decorrelation scales on the order of several hundred kilometers) could be cause by spatial covariability in the environmental forcing variable, and the lack of covariability over long spatial scales could be the result of the presence of the spatial patterning mechanism described earlier associated with postdispersal density-dependent recruitment and a dispersal scale set by the retention/relaxation mechanism described in Wing et al. (1995a, b; 1998).

Explanation of the collapse of the central California Dungeness crab population may also involve a metapopulation component. The shift in the late 1950s to higher temperature would suggest a tendency for greater northward flow in the California Current (or less southward flow). Because this area is near the southern end of the range of high abundance, such a shift could contribute to the decline (e.g., as in Gaylord and Gaines, 2001).

Application of metapopulation concepts and full accounting of spatial aspects can also aid in the assessment of influences of environmental variables. For example, when the proposal that interannual variability in alongshore flow could drive cycles in Dungeness crab populations (McConnaughey et al., 1994) was tested with a metapopulation model, it was found to lead either to extinction or to cycles that were out of phase at each end of the range (Botsford et al., 1998).

D. PINK SHRIMP (*PANDALUS BOREALIS*)

The pink shrimp *Pandalus borealis* is an arcto-boreal distributed Caridean decapod supporting valuable commercial fisheries in high-latitude systems. Although originally described as having a circumpolar distribution with populations in both the North Atlantic and North Pacific, it has been proposed that the Pacific form comprises a separate sibling species, *P. eous* (formerly considered in some sources to be a subspecies, *P. borealis eous*), on the basis of morphological differences in the adult and sizes differences in the larval stages (Squires, 1992). Genetic evidence is equivocal on this point (Bergstrom, 2000). Because of the unresolved taxonomic status, we will structure our case study on the Atlantic populations, although, when appropriate, we will draw general lessons from the Pacific concerning population structure and dispersal patterns.

Pink shrimp are distributed on soft substrates (principally soft muds and silty sands), with highest concentrations on sediments characterized by high organic carbon content (Shumway et al., 1985). *P. borealis* inhabits a broad depth range from 9 to 1450m on the continental shelves and slopes of the North Atlantic (Shumway et al., 1985). Commercially exploited concentrations are typically found at 50 to 300m in depth. Pink shrimp are eurythermal, occurring in temperature ranges of approximately 1.5 to 12.0°C and are generally considered to be stenohaline, preferring relatively high-salinity waters.

P. borealis forms dense aggregations, often characterized as schools. Typically, well-defined spatial segregation patterns by sex and life stage are observed and exploited in commercial fisheries. The shrimp are harvested using both otter trawls and traps in different parts of the range. Pink shrimp are also important components of the diet of a number of commercially important fish populations including Atlantic cod (*Gadus morhua*), redfish (*Sebastes* spp.), and whiting (*Merluccius* spp.).

Life History and Single Population Dynamics

Pink shrimp are protandric hermaphrodites and particular attention has been given to the potential role of environmental and fishery-related factors on the timing of sex reversal (Charnov, 1982). The life history is divided into male, transitional, and female forms (with subdivisions including mature and immature males); four transitional substages; and three principal female groupings. Typically, most 1 to 2-year-old pink shrimp breed as males. A relatively small but variable fraction develop directly as primary females, and a early-maturing class that reproduces as females at about 1.5 years is also recognized. These stage durations are temperature dependent, however. The time required to reach the mature female stage varies inversely with temperature, ranging from approximately 6 years at the northern extent of the range and 4 years at the southern end (Shumway et al., 1985).

Mature females incubate from 500 to 3500 eggs, depending on body size, averaging approximately 2000 eggs. The duration of the incubation period, during which the embryonic stages pass, ranges from 4.5 to 10 months, depending on water temperature. Egg loss during the incubation phase has been attributed to cannibalism by other pink shrimp and attrition resulting from mechanical disturbance. Six larval stages are generally recognized (although higher numbers have been reported). In culture systems, the duration of the pelagic larval stage ranges from approximately 70 days at 4°C to about 32 days at 10°C. Thus, the potential, at least, for broad dispersal of the larval stages is clear.

The effects of fishery removals on the demographic structure of the populations and the implications for sustainable harvesting practices offer particularly interesting tests of sex allocation theory. Most of the attention on pink shrimp population biology has centered on this aspect of their dynamics. It has been proposed that *P. borealis* exhibits a form of phenotypic or environmental sex determination and that the proportion of early-maturing females may provide a sensitive indicator of demographic change. In particular, it is hypothesized that the frequency of early-maturing females in the population is inversely related to the abundance of older females. The effects of removal by the fishery is particularly relevant in this regard. Jensen reported a concomitant decline in the mean size of pink shrimp in the Skagerrak as a result of exploitation, and an increase in the proportion of early-maturing females in the catches. Charnov (1981, 1982) evaluated these changes in terms of life history theory in which adaptation to changing demographic structures occurs as a compensatory mechanism. Latitudinal clines in life span and in the demographic structure have also been reported, with populations characterized by shorter life spans exhibiting higher frequencies of early maturing females (reviewed in Charnov, 1982). Charnov (1982) suggested that the proportion of early-maturing females decreases with life span, which is in turn controlled by growth and mortality (both of which vary with latitude).

Bergstrom (1992, 1997) suggested, however, that natural selection at the northern extent of the range acts against primary and early maturing females, and that these genotypes are maintained in higher latitude systems through larval drift and adult migration from other systems (Bergstrom, 2000). If this underlying mechanism is operative, it has important ramifications for consideration of shrimp metapopulation structures.

Larval Dispersal

The larvae of *P. borealis* undergo diel vertical migrations, moving toward surface waters at night (Ouellet and Lefaivre, 1994). During daylight hours, highest concentrations of larvae coincided with subsurface chlorophyll—a maxima associated with the pycnocline in the Gulf of St. Lawrence. Vertical swimming at speeds of 1 per centimeter have been reported (Weinberg, 1982). The role of larval behavior in dispersal has not been firmly established. However, the observed diel vertical migrations suggest that the larvae may be subject to current–countercurrent systems affecting dispersal patterns. Shumway et al. (1985) suggested that year class strength of *P. borealis* is set during the pelagic larval phase, implying the critical importance of dispersal and survival processes in the early life stages.

Horstedt and Smidt (1956) reported that influx of larvae into the fjords of West Greenland is critical to persistence of pink shrimp populations in these systems. In the Gulf of Maine, it has been inferred that larvae released in nearshore environments are retained in these environments through the early juvenile stages (Shumway et al., 1985). Pederson et al. (2003) demonstrated relatively wide dispersal for pink shrimp in West Greenland waters. However, no linkage with specific water mass characteristics was observed. Drifter studies were used to infer a net northward dispersal of larvae of 3.1 km per day with a potential overall dispersal range of 200 to 400 km.

Pedersen et al. (2003) used a numerical hydrodynamic modeling approach to predict the dispersal pattern of pink shrimp larvae from known spawning locations in the Barents Sea. A Lagrangian particle-tracking model was used to follow simulated larvae over a 60-day developmental period. Simulated dispersal patterns for 3 years were checked against observed settlement locations for 1996 to 1998. The model indicated that a dispersal range of as much as 330 km from the source location was possible in the Barents Sea. In the 3 years simulated, the mean dispersal distance varied from 74 to 122 km in response to interannual variation in physical forcing. A major determinant of interannual variability in hydrographic conditions was related to Atlantic water inflow into the Barents Sea. Higher dispersal rates in 1998 were related to higher inflow rates, possibly related to the NAO. In contrast, low predicted dispersal in 1996 was related to reduced hydrodynamic forcing. In 1996, reduced predicted transport to the Kola coast was apparently reflected in poor recruitment to this region in this year.

In each of the 3 years examined, the probability distribution of distance displaced was bimodal, apparently reflecting patterns related to whether the larvae were entrained in local gyres. Anticyclonic gyres are observed in association with submarine banks in the Barents Sea, and larvae can be retained in these mesoscale features. Accordingly, in some areas of the Barents Sea, local larval retention is possible and may play an important role in defining metapopulation structures. In the Gulf of Alaska and Bering Sea, Ivanov (1969) reported a similar patterns for *P. eous* in which gyre systems resulted in the retention of larvae and the replenishment of shrimp beds. The relatively broad dispersal potential indicated in the modeling for the Barents Sea was deemed consistent with the overall lack of genetic differentiation in this region. Pedersen et al. (2003) concluded that pink shrimp in the Barents Sea are not completely dependent on larval subsidies from outside the region as suggested by Lysy (1981; as cited in Bergstrom, 2000), but that substantial regional interchange of larvae is possible. Again, mesoscale oceanographic retention features may provide a fundamental mechanism for formation of pink shrimp metapopulations with superimposed overall flow fields providing the linkage among population components.

Metapopulation Structure

Bergstrom (2000) speculated that localized stocks of pandalid shrimp are recruited from large regional larval pools. The role of hydrographic conditions was hypothesized to be greatest during the first two larval stages, when active swimming is relatively weak, but with diminishing impact on later larval stages. This model has clear implications for understanding metapopulation structure of pink shrimp, because it would imply well-defined population structures with the potential for mixing from larvae drawn from the regional larval pool.

Adult pink shrimp also undergo extensive migrations, and strong linkages between inshore and offshore population components have been inferred. The well-defined ontogenetic spatial distribution patterns are critical in this regard. Adult female shrimp in many areas move from deeper offshore locations to release their larvae in nearshore waters. In areas such as the Gulf of Maine, the juveniles remain inshore for the first 2 years of life. During their second winter, the mature males move offshore and remain in deeper waters until their fourth winter of life. The ovigerous females, after releasing their larvae, return to offshore habitats.

Large-scale movements of *P. borealis*, particularly in the northern extent of the range, are related to water mass movements. Off West Greenland, Smidt (1981) reported that offshore stocks are linked to inshore stocks when intrusions of deep, warm, water masses result in onshore transport from offshore to inshore locations. Teigsmark (1983) indicated that shrimp groups related to different water mass characteristics could be identified in the Barents Sea and tracked to different locations.

Analyses of population genetic structure of pink shrimp in the northeast Atlantic indicate no population differentiation in oceanic environments (Rasmussen et al., 1993; Martinez et al., 1997; Drengstig et al., 2000). Martinez et al. (1997) found no significant genetic differences among pink shrimp throughout the Barents Sea. Nor were differences between Spitsbergen and Barents samples found. However, Martinez et al. (1997) did find evidence of genetic divergence between Spitsbergen–Barents and northern Norwegian fjords. Complementary results based on allozyme studies were reported by Drengstig and Fevolden (1997) and Drengstig et al. (2000). Lysy (1981; as cited in Bergstrom, 2000) had earlier suggested that female shrimp along the northern Norwegian coast supply recruits to the Barents Sea as a whole. Sevigny et al. (2000) found no evidence of genetic differentiation from pink shrimp samples in the northwest Atlantic from Labrador to the Gulf of St. Lawrence. Collectively, these genetic results are consistent with field observations indicating relatively wide dispersal potential and interchange among local populations.

An examination of trends in pink shrimp biomass at nine locations in the North Atlantic (Fig. 8-12) reveals relatively high coherence among the northernmost populations (Worm and Myers, 2003). In the western North Atlantic, large-scale increases in biomass are evident during the last two decades from Labrador to the eastern Scotian Shelf (Fig. 8-13). A population increase off Iceland is also evident. Population trajectories for the Gulf of Maine, Barents Sea, and in the Skagerrak show less pronounced and more variable increases since the mid 1980s. In a meta-analysis of these data, Worms and Myers (2003) indicated strong inverse relationships between shrimp and cod biomass in the regions exhibiting rapid shrimp population increases. Cod is known to prey on pink shrimp, and a top-down control on shrimp production is postulated. In contrast, tests for temperature effects for these stocks indicated little effect in those populations apparently dominated by top-down predation controls. However, in the more southerly populations, the linkage between shrimp biomass and cod abundance was less clear, and the apparent temperature effect was stronger. Worm and Myers (2003) concluded that temperature does have the potential to mediate the predation effect in these southern groups.

Strong temperature controls on shrimp production in the Gulf of Maine have previously been explored in a number of studies (see Shumway et al., 1985, and Bergstrom, 2000, for an overview). In this southernmost population, landings and abundance of shrimp are strongly inversely related, as might be expected. It has been suggested that the mechanism underlying the temperature effect in the Gulf of Maine is related to susceptibility to a peridinean parasite (Shumway et al., 1985).

The apparently strong determinants of shrimp production through either top-down controls or bottom-up forcing complicate the interpretation of the available time series of biomass with respect to synchrony and its implications for

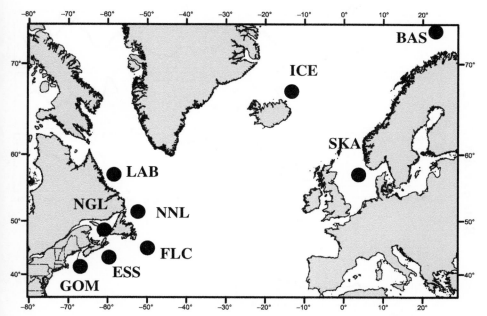

FIGURE 8-12. Map of pink shrimp populations used in trend analysis (after Worm and Myers, 2003) including Barents Sea (BAS), Iceland (ICE), the Skagerrak (SKA), Labrador (LAB), northern Newfoundland (NNL), northern Gulf of St. Lawrence (NGL), Flemish Cap (FLC), eastern Scotian Shelf (ESS), and Gulf of Maine (GOM).

metapopulation dynamics. The northernmost populations exhibit a high level of synchrony, presumably resulting from the strong top-down controls identified by Worm and Myers (2003). The apparent lack of synchrony, however, between the Scotian Shelf and Gulf of Maine, coupled with considerations of the general southward transport between these regions, opens the possibility that a rescue effect could be an important factor in maintaining the southernmost population in the northwest Atlantic.

V. DISCUSSION

The case studies summarized clearly illustrate the importance of demographic interactions over space. Decapod populations cannot be considered to be single, well-mixed populations (on ecological timescales), nor can they be considered to be a number of independent populations. The characteristics of single population behavior of most interest are persistence and stability. Being part of a

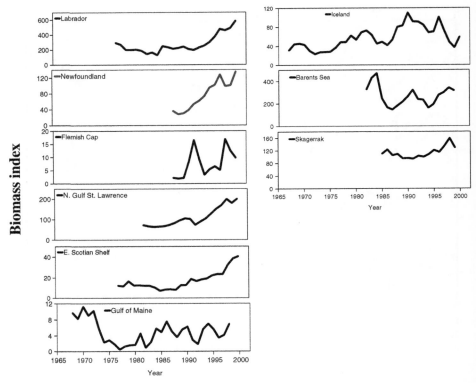

FIGURE 8-13. Trends in relative abundance of pink shrimp in the North Atlantic (Worm and Myers, 2003).

metapopulation likely changes both, but in the cases presented it is typically not possible to separate the single population effect from the metapopulation contributions completely. For example, in the case of the American lobster, local stock–recruitment relationships seem to have steep slopes at the origin, indicating persistent populations, but one cannot dismiss the potential effect of larval subsidy on the unexpectedly high resilience of lobsters to high harvest.

With regard to population stability, the metapopulation structure can actually introduce a new kind of behavior: When density-dependent recruitment occurs after settlement, "Turing–type" instabilities can arise (Turing, 1952). These have been shown to arise in other ecological systems with nonlinear dynamics and age structure (Hastings, 1992). For our purposes here, a Turing–type instability can be viewed simply as spatial patterns in abundance that occur when there are density-dependent effects that occur with a time and space lag. In the Dungeness crab example, note that this effect did not occur when density-dependent recruit-

Chapter 8 Metapopulation Dynamics of Coastal Decapods 307

ment was predispersal (i.e., directly and immediately controlled), but rather only when it was postdispersal (i.e., controlled by density at another location, at a time lag, the time to maturation). This phenomenon arose in the modeling and analysis of cycles in the Dungeness crab populations, and although it provides a convenient connection between scales of dispersal and variability, it is not yet clear how strong it is in the actual populations.

Another important metapopulation feature relevant to decapods is the potential for losses of persistence for populations at the edges of the range of a metapopulation resulting from changes in advection. There are several examples from simulations of metapopulations that underscore the sensitivity of persistence to alongshore flows (Botsford et al., 1998; Gaylord and Gaines, 2001). It is not clear how much such a mechanism affects specific cases, such as the decline of the central California population of Dungeness crab, but the possibility clearly needs to be considered. Advection along a metapopulation structure can also interact with population shifts resulting from other environmental effects as proposed, for example, in recent research on snow crab (*Chinocetes opilio*) in the Bering Sea. Analysis of annual surveys conducted by the US National Marine Fisheries Service since 1975 has indicated that mature females migrate from the middle domain of the intermediate shelf to the shelf edge by tracking near-bottom temperature (Ernst et al., in press). Snow crab larvae presumably settle and grow to maturity in the middle domain, a possible retention zone lacking strong currents (Orensanz et al., 2005). A recent (1975–1979) warming in the Bering Sea led to a northward contraction of the mature female's range. Yet during subsequent years of cool temperatures, recruitment did not expand back to the midshelf areas to the south. This apparent "environmental ratchet effect" may be the result of increased cod predation by an expanded cod stock, but a second possibility is of greater interest in terms of metapopulation behavior: That snow crab larvae cannot reinvade the lost areas because they are downstream in terms of the dominant patterns of circulation (Orensanz et al., submitted).

The variability in reproductive parameters among decapods identified by Cobb et al. (1997) led to the question of whether we would expect consequent differences in metapopulation behavior. How would differences in (1) the duration of the larval stage, (2) the number of eggs, and (3) the sizes of eggs and larvae affect metapopulation dynamics? Making the most obvious, parsimonious assumptions regarding their effects, (1) greater duration of the larval stage in a diffusive environment would be expected to lead to a broader spatial pattern of larval dispersal; (2) a greater number of eggs would increase the magnitude of the larval settlement, but not the spatial pattern; and (3) greater size of larvae in an environment with presumed size-dependent mortality would also lead to greater magnitude of settlement. The population dynamic effects of greater settlement would be greater persistence and potentially greater spatial synchrony, unless spatial patterning was caused by strong density dependence as in Figure 8-11.

We can view potential differences in the range of dispersal in terms of dispersal distances ranging from insignificant to intermediate to so large as to form a common larval pool that would supply a constant supply of potentially settling larvae. The first, extreme case would consist of independent populations, hence an increase in larval dispersal distance leads to greater metapopulation behavior. Based on the results in Botsford et al. (2001; see also Botsford and Hastings, 2005), populations with shorter dispersal distances can persist better on widely separated patches of habitat, whereas populations with longer dispersal distances, and the same fecundity, are more likely to require a certain fraction of the coastline be viable habitat to persist. With regard to stability and synchrony, as demonstrated by Figure 8-11, broader dispersal patterns are more likely to lead to spatial patterning if the conditions for it exist (i.e., postdispersal density dependence).

The increasing global interest in the effects of exploitation on decapod species leads naturally to the question of how metapopulation structure affects persistence of commercially fished decapods. The possibility of serial depletion, sequentially fishing down subpopulations within a metapopulation, rather than fishing all subpopulations at once, provides perhaps the largest difference in vulnerability between single populations and metapopulations. In a metapopulation being serially depleted, catch-per-unit-effort declines much more slowly than in a single well-mixed population (Prince and Hilborn, 1998), declining only when the last population is being fished to low levels. In such cases, the application of single species models results in parameter estimates that are essentially biased high, until it is too late. Orensanz et al. (1998) present a cogent analysis of serial depletion within and among metapopulations of North Pacific decapod species: red king crab (*Paralithodes camtschaticus*), shrimp (primarily *Pandalus borealis*), tanner crab (*Chionoecetes bairdi*), and brown king crab (*Lithodes aquispina*). They note that decapods can be vulnerable to diminishing persistence through exploitation even with male-only fishing (e.g., the decline in fertilization rate in the red king crab), and note the underappreciated effects that natural refuges have played in keeping species persistent under exploitation (see the quote from Carl Walters in Orensanz et al., 1998). These suggest that spatial management through marine protected areas may be appropriate for management of decapod fisheries.

The significant disparity between adult movement and larval movement makes decapod metapopulations ideal candidates for spatial management through marine reserves. Populations in reserves have the potential to provide larval subsidies to populations outside reserves, without the losses of adults seen in more mobile species, in which individuals are caught in the fishery when they swim out of reserves. Although not configured to represent an MPA explicitly, the results from the two-patch model for the American lobster (Fogarty, 1998) are directly relevant to the case in which fishing mortality in one area is set to zero, as in a

Chapter 8 Metapopulation Dynamics of Coastal Decapods

no-take marine reserve, and suggest that the potential benefits of protection of some segment of the population could be significant. The key consideration would rest in identifying important source areas for protection, possibly defined by habitat or other characteristics, while fishing would continue in areas receiving larval subsidies from the closed areas. Specification of the size of the areas required for protection would require further study. In this case, as in most other reserve designs, uncertainty in dispersal patterns is a critical limitation (Morgan and Botsford, 2001). Although MPAs have been specifically established in few instances for decapods, some examples can be cited. In Chesapeake Bay, a historical spawning sanctuary for protection of adult female crabs in the lower bay has been in place seasonally (June 1–September 15) for several decades. Recently, the historical spawning sanctuary was expanded by about 200%, including a deep-water dispersal corridor in water depths more than 11 m (Seitz et al., 2001; Lipcius et al., 2001, 2003), effective June 2000. The combined sanctuary and dispersal corridor comprises 172,235 ha in the lower bay and main stem of Chesapeake Bay. The sanctuary and corridor system has been shown to be highly selective for adult female crabs (Seitz et al., 2001; Lipcius et al., 2001, 2003). The historical spawning sanctuary was estimated to provide protection to 11 to 22% of the adult female population (mean, 16%; Seitz et al., 2001b), because of the high exploitation rates in the bay and the vulnerability of females prior to the closure. Inclusion of the expanded sanctuary and deep-water dispersal corridor is estimated to increase the protection afforded to mature females to approximately 50 to 70% (Lipcius et al., 2003).

Year-round closures for American lobster management have been established in Bonavista Bay, Newfoundland (Rowe, 2001, 2002). Two small reserves were established in 1997; increases in male population density and mean size were observed at both sites (Rowe, 2002). An increase in female size and the proportion of ovigerous females was observed at one of the sites. Restricted movements of lobsters resulted in considerable protection afforded to individuals within the closed areas (Rowe, 2001). These results suggest that the broader application of year-round fishery closures could be beneficial as a tool in lobster management, particularly when direct controls on fishing mortality are difficult to implement.

Some interesting contrasts are also afforded by comparing the life history characteristics of decapods and teleost fish in the context of metapopulation dynamics (see Sale and Kritzer, 2003). A few key distinguishing characteristics deserve to be highlighted. First, the prevalence of brooding behavior during the egg stage for decapods has few analogs among teleosts. The relative time spent in the brood phase versus the dispersive larval phase differs markedly among decapods (Fig. 8-1), and the relevant contrast with teleosts is restricted to the meroplanktonic stages. The range of larval durations among the decapods substantially exceeds that of teleosts. The longest decapod larval stage durations occur principally in the spiny lobsters. Other decapod taxa exhibit shorter larval stage durations and more restricted potential dispersal.

Among the teleosts, coral reef fish typically have relatively short larval stage durations (Sale and Kritzer, 2003; Kritzer and Sale, this volume), and restricted larval stage durations occur among some temperate rocky reef fishes whereas others are viviparous with no larval dispersive phase (see Gunderson and Vetter, this volume).

In general, benthic invertebrates tend to differ from teleosts in terms of larval settlement and habitat selection, and the ability to delay metamorphosis if suitable substrate conditions are not encountered (Bradbury and Snelgrove, 2001). The end points of the larval dispersal process, therefore, can be substantially different in decapods and fish. Although generalizations always hold exceptions, these attributes may result in greater patchiness in the settlement stages of decapods. The extent to which this may also result in a higher probability of metapopulation formation as defined in this chapter cannot currently be assessed. Coral reef fishes would appear to hold greater affinities with decapods in this regard than other fishes.

In general, fish larvae are capable of higher swim speeds than most benthic invertebrates (see reviews by Bradbury and Snelgrove, 2001; and Sale and Kritzer, 2003). The role of active dispersal in larval fish distribution is undoubtedly important in defining the spatial structure of fish populations, also with important implications for metapopulation structure. The well-defined homing behavior of some fishes does not appear to have an analog in decapods (or at least has not been reported). Accordingly, the aspects of metapopulation dynamics documented by Jones (this volume) for anadromous and some estuarine-dependent fishes is not found among the decapods. In particular, comparable evidence for extinction events in local populations shown for anadromous fish does not currently exist for decapods (see Jones, this volume).

VI. SUMMARY

Observed spatial patterns of benthic life stages, larval dispersal mechanisms, genetic structure, and the level of synchrony in fluctuations in abundance are consistent with underlying metapopulation structures for the decapods considered here. Although there is sufficient evidence to conclude that metapopulation structure and dynamics are important in decapod populations, and that marine reserves may be particularly appropriate for species with their life history, we are far from a complete understanding of decapod metapopulation dynamics. The dynamic behavior of metapopulations depends on tendencies in local single population behavior and the connectivity among these populations. The former has been the focal point of much research, but the latter is not well-known. Although we describe larval movement here in general terms, we need better descriptions of larval dispersal patterns to understand metapopulation dynamics well enough to be able to use that understanding in the management of decapod resources. Persistence is the key behavior for conservation of decapods, and we need to know the many possible paths of connectivity over space that provide the replacement necessary for decapod metapopulations to persist.

VII. ACKNOWLEDGMENT

The authors are grateful for the constructive reviews provided by J. Stanley Cobb, Jake Kritzer, and Peter Sale. Rick Wahle made unpublished lobster settlement data available to the authors. Betty Holmes kindly provided GIS applications.

REFERENCES

Anastasia, J.R., Morgan, S.G., and Fisher, N.S. (1998). Tagging crustacean larvae: Assimilation and retention of trace elements. *Limn. Oceanogr.* **43**, 362–368.

Anthony, V.C., and Caddy, J.F., eds. (1980). Proceedings of the Canada–U.S. workshop on status of assessment science for N.W. Atlantic lobster (*Homarus americanus*) stocks. *Can. Tech. Rept. Fish. Aquat. Sci.* **932**, 186.

Applegate, A.J. (1983). *An environmental model predicting the relative recruitment success of the blue crab,* Callinectes sapidus, *in Chesapeake Bay, Virginia*. MS thesis. Williamsburg, VA: College of William and Mary.

Armstrong, D.A., Roper, C., and Gunderson, D. (2003). Estuarine production of juvenile Dungeness crab (*Cancer magister*) and contribution to the Oregon–Washington coastal fishery. *Estuaries*. **26**, 1174–1186.

Barlow, J., and Ridgway, G.J. (1971). Polymorphisms of esterase isozymes in the American lobster (*Homarus americanus*). *J. Fish. Res. Board Can.* **28**, 15–21.

Bergstrom, B.I. (1992). *Demography and sex change in pandalid shrimps*. PhD dissertation. Goteborg: Goteborg University, Faculty of Natural Sciences.

Bergstrom, B.I. (1997). Do protandric shrimp have an environmental sex determination? *Mar. Biol.* **128**, 397–407.

Bergstrom, B.I. (2000). *The biology of* Pandalus: *Advances in marine biology* (vol. 38, pp. 57–245). London: Academic Press.

Berthelemy-Okazaki, N.J., and Okazaki, R.K. (1997). Population genetics of the blue crab *Callinectes sapidus* from the northwestern Gulf of Mexico. *Gulf Mex. Sci.* **15**, 35–39.

Boicourt, W.C. (1982). Estuarine larval retention mechanisms on two scales. In *Estuarine comparisons* (V.S. Kennedy, ed., pp. 445–457). New York: Academic Press.

Botsford, L.W. (1981). The effects of increased individual growth rates on depressed population size. *Am. Nat.* **117**, 38–63.

Botsford, L.W. (1984). Effect of individual growth rates on expected behavior of the northern California Dungeness crab (*Cancer magister*) fishery. *Can. J. Fish. Aquat. Sci.* **41**, 99–107.

Botsford, L.W. (1986). Population dynamics of the Dungeness crab (*Cancer magister*). *Can. J. Fish. Aquat. Sci.* **92**, 140–153.

Botsford, L.W. (1992). Further analysis of Clark's delayed recruitment model. *Bulletin of Math. Biol.* **54**, 275–293.

Botsford, L.W. (1997). Dynamics of populations with density-dependent recruitment and age structure. In *Structured population models in marine, terrestrial, and freshwater systems* (S. Tuljapurkar and H. Caswell, eds., pp. 371–408) New York: Chapman and Hall.

Botsford, L.W., Armstrong, D.A., and Shenker, J.M. (1989). Oceanographic influences on the dynamics of commercially fished populations. In *Coastal oceanography of Washington and Oregon* (M.R. Landry and B.M. Hickey, eds., pp. 511–565). Amsterdam.

Botsford, L.W., and Hobbs, R.C. (1995). Recent advances in the understanding of cyclical behavior of Dungeness crab (*Cancer magister*) populations. *ICES Mar. Sci. Symp.* **199**, 157–166.

Botsford, L.W., and Lawrence, C.A. (2002). Patterns of co-variability among California current chinook salmon, coho salmon, Dungeness crab, and physical oceanographic conditions. *Prog. Oceanogr.* **53**, 283–305.

Botsford, L.W., Methot, R.D., Jr., and Johnston, W.E. (1983). Effort dynamics of the northern California Dungeness crab (*Cancer magister*) fishery. *Can. J. Fish. Aquat. Sci.* **40**, 337–346.

Botsford, L.W., Moloney, C.L., Hastings, A., Largier, J.L., Powell, T.M., Higgins, K., and Quinn, J.F. (1994). The influence of spatially and temporally varying oceanographic conditions on meroplanktonic metapopulations. *Deep-Sea Res. II.* **41**, 107–145.

Botsford, L.W., Moloney, C.L., Largier, J.L., and Hastings, A. (1998). Metapopulation dynamics of meroplanktonic invertebrates: The Dungeness crab (*Cancer magister*) as an example. In *Proceedings of the North Pacific Symposium on Invertebrate Stock Assessment and Management* (G.S. Jamieson and A. Campbell). *Can. Spec. Publ. Fish. Aquat. Sci.* **125**, 295–306.

Botsford, L.W., and Wickham, D.E. (1975). Correlation of upwelling index and Dungeness crab catch. *Fish. Bull. US.* **73**, 901–907.

Botsford, L.W., and Wickham, D.E. (1978). Behavior of age-specific, density-dependent models and the northern California Dungeness crab (*Cancer magister*) fishery. *J. Fish. Res. Board Can.* **35**, 833–843.

Boylan, J.M., and Wenner, E.L. (1992). Settlement of brachyuran megalopae in a South Carolina, USA, estuary. *Mar. Ecol. Progr. Ser.* **97**, 237–246.

Bradbury, I.R., and Snelgrove, P.V.R. (2001). Contrasting larval transport in demersal fish and benthic invertebrates: The roles of behaviour and advective processes in determining spatial pattern. *Can. J. Fish. Aquat. Sci.* **58**, 811–823.

Burkenroad, M.D. (1946). Fluctuations in abundance of marine animals. *Science.* **103**, 684–686.

Burton, R.S. (1983). Protein polymorphism and genetic differentiation of marine invertebrate populations. *Mar. Biol. Lett.* **4**, 193–206.

Caddy, J.F. (1986). Modelling stock–recruitment processes in Crustacea: Some practical and theoretical perspectives. *Can. J. Fish. Aquat. Sci.* **43**, 2330–2344.

Castro, K.M. (2003). Assessing the impact of habitat and stock enhancement for the American lobster (*Homarus americanus*), in Narragansett Bay, Rhode Island. PhD dissertation. Kingston, RI: University of Rhode Island.

Charnov, E.L. (1981). Sex reversal in *Pandalus borealis*: Effect of a shrimp fishery. *Mar. Biol. Lett.* **2**, 53–57.

Charnov, E.L. (1982). *The theory of sex allocation: Monographs in population biology.* Princeton, NJ: Princeton University Press.

Cobb, J.S., and Wang, D. (1985). Fisheries biology of lobsters and crayfish. In *The biology of crustacea* (D.E. Bliss, ed., vol. 10, pp. 167–247). Orlando, FL: Academic Press.

Cobb, J.S, Booth, J.D., and Clancy, M. (1997). Recruitment strategies in lobsters and crabs: A comparison. *Mar. Freshwater Res.* **48**, 797–806.

Collie, J.S., Richardson, K., and Steele, J.H. (2004). Regime shifts: Can ecological theory illuminate the mechanisms? *Prog. Oceanogr.* **60**, 281–302.

Cooper, R.A., and Uzmann, J.R. (1980). Ecology of juvenile and adult *Homarus*. In *The biology and management of lobsters* (J.S. Cobb and B.F. Phillips, eds., vol. 1, pp. 215–276). New York: Academic Press.

Cowen, R.K., Lwiza, K.M.M., Sponagaule, S., Paris, C.B., and Olson, D.B. (2000). Connectivity of marine populations: Open or closed? *Science*. **287**, 857–859.

Darnell, R.M. (1959). Studies of the life history of the blue crab (*Callinectes sapidus* Rathbun) in Louisiana waters. *Trans. Amer. Fish. Soc.* **88**, 294–304.

DiBacco, C., Chadwick, D.B. (2001). Assessing the dispersal and exchange of brachyuran larvae between regions of San Diego Bay, California and nearshore coastal habitats using elemental fingerprinting. *J. Mar. Res.* **59**, 53–78.

DiBacco, C., Levin, L.A. (2000). Development and application of elemental fingerprinting to track the dispersal of marine invertebrate larvae. *Limnol. Oceanogr.* **45**, 871–880.

Drengstig, A., and Fevolden, S.E. (1997). Genetic structuring of *Pandalus borealis*, in the North Atlantic. 1. Allozyme studies. Int. Council Explor. Sea C.M. 1997/AA:3.

Drengstig, A., Fevolden, S.E., Garland, P.E., and Aschan, M.M. (2000). Genetic structure of the deep sea shrimp (*Pandalus borealis*) in the north-east Atlantic based on allozyme variation. *Aquat. Living Res.* **13**, 121–128.

Earn, D.J., Lecin, S.A., and Rohani, P. (2000). Coherence and conservations. *Science*. **290**, 1360–1364.

Ennis, G.P., and Fogarty, M.J. (1997). Recruitment overfishing reference point for the American lobster, *Homarus americanus*. *Mar. Freshwat. Res.* **48**, 1029–1034.

Epifanio, C.E. (1988a). Dispersal strategies of two species of swimming crab on the continental shelf adjacent to Delaware Bay. *Mar. Ecol. Progr. Series*. **49**, 243–248.

Epifanio, C.E. (In press). Biology of larvae. In *The biology of the blue crab*, Callinectes sapidus (V.S. Kennedy, ed.) Maryland Sea Grant Press.

Epifanio, C.E., Masse, A.K., Garvine, R.W. 1989. Transport of blue crab larvae by surface currents off Delaware Bay, USA. Marine ecology progress series. Oldendorf. Vol. 54, no. 1–2, pp. 35–41.

Epifanio, C.E., Valenti, C.C., and Pembroke, A.E. (1984). Dispersal and recruitment of blue crab larvae in Delaware Bay, U.S.A. *Can. J. Fish Aquat. Sci.* **62**, 250–268.

Ernst, B., Orensanz, J.M., and Armstrong, D.A. (In press). Spatial dynamics of female snow crab (*Chionoecetes opilio*) in the eastern Bering Sea. *Can. J. Fish. Aquat. Sci.*

Fernandez, M. (1999). Cannibalism in Dungeness crab *Cancer magister*. Effects of predator–prey size ratio, density, and habitat type. *Mar. Ecol. Progr. Ser.* **182**, 221–230.

Fogarty, M.J. (1983). Distribution and relative abundance of American lobster (*Homarus americanus*) larvae: A review. In *Distribution and relative abundance of American lobster* (Homarus americanus) *larvae: New England investigations during 1974–1979* (M.J. Fogarty, ed., pp. 3–8). NOAA Technical report SSR-F-775. Seattle: National Marine Fisheries Service.

Fogarty, M.J. (1995). Populations, fisheries, and management. In *The biology of the American lobster* (J. Factor, ed., pp. 111–137). San Diego: Academic Press.

Fogarty, M.J. (1998). Implications of larval dispersal and directed migration in American lobster stocks: Spatial structure and resilience. *Can. Spec. Publ. Fish. Aquat. Sci.* **125**, 273–283.

Fogarty, M.J., and Gendron, L. (2004). Biological reference points for American lobster populations: Limits to exploitation and the precautionary approach. *Can. J. Fish. Aquat. Sci.* **61**, 1392–1403.

Fogarty, M.J., and Idoine, J.S. (1986). Recruitment dynamics in an American lobster (*Homarus americanus*) population. *Can. J. Fish. Aquat. Sci.* **48**, 2368–2376.

Fogarty, M.J., and Lipcius, R. (In press). Population dynamics and fisheries. In *The biology of the blue crab*, Callinectes sapidus (V.S. Kennedy, ed.) Maryland Sea Grant Press.

Garvine, R.W., Epifanio, C.E., Epifanio, C.C., and Wong, K.C. (1997). Transport and recruitment of blue crab larvae: A model with advection and mortality. *Est. Coast. Shelf Sci.* **45**, 99–111.

Gaylord, B., and Gaines, S.D. (2001). Temperature or transport? Range limits in marine species mediated solely by flow. *Am. Nat.* **155**, 769–789.

Goodrich, D.M., van Montfrans, J., and Orth, R.J. (1989). Blue crab megalopal influx to Chesapeake Bay: Evidence for a wind-driven mechanism. *Est. Coast. Shelf Sci.* **29**, 247–260.

Grosholz, E. (2002). Ecological and evolutionary consequences of coastal invasions. *Trends Ecol. Evol.* **17**, 22–27.

Guillory, V. (1997). Long-term trends in abundance and recruitment of blue crab, according to 30 years of fishery independent data. *Proc. Louisiana Acad. Sci.* **61**, 36–42.

Hankin, D.G., Butler, T.H., Wild, P.W., and Xue, Q. (1997). Does intense fishing on males impair mating success of female Dungeness crabs? *Can. J. Fish Aquat. Sci.* **54**, 655–669.

Hanski, I. (1999). *Metapopulation ecology.* Oxford: Oxford University Press.

Hanski, I., and Simberloff, D. (1997). The metapopulation approach, its history, conceptual domain and application to conservation. In *Metapopulation biology* (I. Hanski and M. Gilpin, eds., pp. 5–26). San Diego: Academic Press.

Hanson, J.M., and Lanteigne, M. (2000). Evaluation of Atlantic cod predation on American lobster in the southern Gulf of St. Lawrence, with comments on other potential fish predators. *Trans. Amer. Fish. Soc.* **129**, 13–29.

Harding, G.C., and Trites, R.W. (1988). Dispersal of *Homarus americanus* larvae in the Gulf of Maine from Browns Bank. *Can. J. Fish. Aquat. Sci.* **45**, 416–425.

Harding, G.C., Drinkwater, K.F., Hannah, C.G., Pringle, J.D., Prena, J., Loder, J.W., Pearre, S., Jr., and Vass, W.P. (In press). Larval lobster (*Homarus americanus*) distribution and drift in the vicinity of the Gulf of Maine offshore banks and their probable origins. *Mar. Ecol. Prog. Ser.*

Harding, G.C., Drinkwater, K.F., and Vass, W.P. (1983). Factors influencing the size of American lobster (*Homarus americanus*) stocks along the Atlantic coast of Nova Scotia, Gulf of St. Lawernce, and Gulf of Maine: A new synthesis. *Fish. Oceanogr.* **14**, 112–137.

Harding, G.C., Kenchington, E.L., Bird, C.J., Pezzack, D.S., and Landry, D.C. (1997). Genetic relationships among subpopulations of the American lobster (*Homarus americanus*) as revealed by random amplified polymorphic DNA. *Can. J. Fish. Aquat. Sci.* **54**, 1762–1771.

Hastings, A. (1992) Age dependent dispersal is not a simple process: Density dependence, stability, and chaos. *Theoretical Population Biology.* **41**, 388–400.

Hedgecock, D. (1986). Is gene flow from pelagic larval dispersal important in the adaptation and evolution of marine invertebrates? *Bull. Mar. Sci.* **39**, 550–564.

Hellberg, M.E., Burton, R.S., Neigel, J.E., Palumbi, S.R. (2002). Genetic assessment of connectivity among marine populations. *Bull. Mar. Sci.* **70**, 273–290.

Higgins, K., Hastings, A., and Botsford, L.W. (1997a). Density dependence and age structure: Nonlinear dynamics and population behavior. *Amer. Nat.* **149**, 247–269.

Higgins, K., Hastings, A., Sarvela, J.N., and Botsford, L.W. (1997b). Stochastic dynamics and deterministic skeletons: Population behavior of Dungeness crab. *Science.* **276**, 1431–1435.

Hobbs, R.C., and Botsford, L.W. (1989). Dynamics of an age-structured prey with density- and predation-dependent recruitment: The Dungeness crab and a nemertean egg predator worm. *Theor. Pop. Biol.* **36**, 1–22.

Hobbs, R.C., and Botsford, L.W. (1992). Diel vertical migration and timing of metamorphosis of larval Dungeness crab *Cancer magister. Mar. Biol.* **112**, 417–428.

Hobbs, R.C., Botsford, L.W. and Thomas, A. (1992). Influence of hydrographic conditions and wind forcing on the distribution and abundance of Dungeness crab, *Cancer magister,* larvae. *Can. J. Fish. Aquat. Sci.* **49**, 1379–1388.

Horstedt, S.A., and Smidt, E. (1956). The deep sea prawn (*Pandalus borealis* Kr.) in Greenland waters. *Medd. Danmarks Fiskeri Havund.* **1**, 118.

Hurt, P.R., Libby, L.M., Pandolfi, L.J., and Levine, L.H. (1979). Periodicities in blue crab population of Chesapeake Bay. *Climatic Change.* **2**, 75–78.

Incze, L.S., and Naime, C.E. (2000). Modelling the transport of lobster (*Homarus americanus*) larvae and postlarvae in the Gulf of Maine. *Fish. Oceanogr.* **9**, 99–113.

Ivanov, B.G. (1969). Biology of the northern shrimp (*Pandalus borealis* Kr.) in the Gulf of Alaska and the Bering Sea. *FAO Fish. Rept.* **57**, 700–810.

Jackson, J.B.C., et al. (2001). Historical overfishing and the recent collapse of coastal ecosystems. *Science.* **293**, 629–638.

Jamieson, G.S., Foreman, M.G.G., Cherniawsky, J.Y., and Levings, C.D. (2002). European green crab (*Carcinus maenas*) dispersal: The Pacific experience. In *Crabs in cold water regions: Biology, management, and economics* (A.J. Paul, E.G. Dawe, R. Elner, G.S. Jamieson, G.H. Kruse, R.S. Otto, B. Sainte-Marie, T.C. Shirley. pp. 561–576). Alaska Sea Grant College Program. AK-SG-02-01. Fairbanks, AK: Alaska Sea Grant.

Jamieson, G.S., Phillips, A.C., and Hugget, W.S. (1989). Effects of ocean variability on the abundance of Dungeness crab larvae. *Can. Spec. Publ. Fish. Aquat. Sci.* **108**, 305–325.

Jensen, G.C., and Armstrong, D.A. (1987). Range extensions of some northeastern pacific Decapoda. *Crustaceana.* **52**, 215–217.

Jensen, P.C., and Bentzen, P. (2004). Isolation and inheritance of microsatellite loci in the Dungeness crab (Brachyura: Cancridae: *Cancer magister*). *Genome.* **47**, 325–331.

Johnson, D.R. (1985). Wind forced dispersion of blue crab larvae from Chesapeake Bay. *Cont. Shelf. Res.* **4**, 733–745.

Johnson, D.F., Botsford, L.W., Methot, R.D., Jr., and Wainwright, T.C. (1986). Wind stress and cycles in Dungeness crab (*Cancer magister*) catch off California, Oregon, and Washington. *Can. J. Fish. Aquat. Sci.* **43**, 838–845.

Johnson, D.F., and Hess, K.W. (1990). Numerical simulations of blue crab larval dispersal and recruitment. *Bull. Mar. Sci.* **46**, 195–213.

Johnson, D.R., and Hester, B.S. (1989). Larval transport and its association with recruitment of blue crabs to Chesapeake Bay. *Est. Coast. Shelf Sci.* **28**, 459–472.

Johnson, D.R., Hester, B.S., and McConaugha, J.R. (1984). Studies of a wind mechanism influencing the recruitment of blue crabs in the Middle Atlantic Bight. *Cont. Shelf Res.* **104**, 265–273.

Katz, C.H., Cobb, J.S., and Spaulding, M. (1994). Larval behavior, hydrodynamic transport, and potential offshore recruitment in the American lobster, *Homarus americanus. Mar. Ecol. Prog. Ser.*

Kinlan, B.P., and Gaines, S.D. (2003). Propagule dispersal in marine and terrestrial environments: A community perspective. (A.J. Paul, E.G. Dawe, R. Elner, G.S. Jamieson, G.H. Kruse, R.S. Otto, B. Sainte-Marie, T.C. Shirley). *Ecology.* **84**, 2007–2020. Fairbanks, AK: Alaska Sea Grant.

Kline, T.C., Jr. (2002). Relative trophic position of *Cancer magister* megalopae. In *Crabs in cold water regions: Biology, management, and economics* (pp. 645–649). Alaska Sea Grant College Program. AK-SG-02-01.

Kordos, L.M., and Burton, R.S. (1993). Genetic differentiation of Texas Gulf coast populations of the blue crab *Callinectes sapidus. Mar. Biol.* **117**, 227–233.

Kornfield, I. and Moran, P. (1990). Genetics of population differentiation in lobsters. In *Life history of the American lobster: Proceedings of a workshop November 29–30, 1989, Orono, Maine* (I. Kornfield, ed., pp. 23–24). Orono, ME: Lobster Institute.

Kritzer, J.P., and Sale, P.F. (2004). Metapopulation ecology in the sea: From Levin's model to marine ecology and fisheries science. *Fish. Fisher.* **5**, 131–140.

Laughlin, R.A. (1982). Feeding habits of the blue crab, *Callinectes sapidus* Rathbun, in the Apalachicola estuary, Florida. *Bull. Mar. Sci.* **32**, 807–822.

Lawton, P., and Lavalli, K. (1995). Postlarval, juvenile, adolescent, and adult ecology. In *Biology of the lobster*, Homarus americanus (J. Factor, ed., pp. 47–88). San Diego: Academic Press.

Levins, R. (1970). Extinction. *Lect. Notes Math.* **2**, 75–107.

Lipcius, R., and Van Engel, W.A. (1990). Blue crab population dynamics in Chesapeake Bay: Variation in abundance (York River, 1972–1989) and stock–recruit functions. *Bull. Mar. Sci.* **46**, 180–194.

Lipcius, R.N., Seitz, R.D., Goldsborough, W.J., Montane, M.M., and Stockhausen, W.T. (2001). A deep-water dispersal corridor for adult female blue crabs in Chesapeake Bay. In *Spatial processes and management of marine populations* (G.H. Kruse, N. Bez, A. Booth, M.W. Dorn, S. Hills, R.N. Lipcius, D. Pelletier, C. Roy, S.J. Smith, and D. Witherell, eds., pp. 643–666). University of Alaska Sea Grant publication AK-SG-01-02. Fairbanks, AK: Alaska Sea Grant.

Lipcius, R.N, Stockhausen, W.T., Seitz, R.D., and Geer, P.J. (2003). Spatial dynamics and value of a marine protected area and dispersal corridor for the blue crab spawning stock in Chesapeake Bay. *Bull. Mar. Sci.* **72**, 453–470.

Little, K.T., and Epifanio, C.E. (1991). Mechanism for the re-invasion of the estuary by two species of brachyuran megalopae. *Mar. Ecol. Prog. Ser.* **68**, 235–242.

Livingston, R.J. (1991). Historical relationships between research and resource management in the Apalachicola River estuary. *Ecol. Appl.* **1**, 361–382.

Mace, P.M., and Sissenwine, M.P. (1993). How much spawning per recruit is enough? In *Risk evaluation and biological reference points for fisheries management* (S.J. Smith, J.J. Hunt, and D. Rivard, eds.). *Can. Spec. Pub. Fish. Aquat. Sci.* **120**, 101–118.

Martinez, I., Skjerdal, T., Dreyer, B., and Aljanabi, S.M. (1997). Genetic structuring of *Pandalus borealis* in the North Atlantic. 2. RAPD analysis. *Int. Council Explor. Sea C.M.* 1997/T:24.

McCann, K.S., Botsford, L.W., and Hastings, A. (2003). Differential response of marine populations to climate forcing. *Can. J. Fish. Aquat. Sci.* **60**, 971–985.

McConaugha, J.R. (1988). Export and reinvasion of larvae as regulators of estuarine decapod populations. *Am. Fish. Soc. Symp.* **3**, 90–103.

McConaugha, J.R. (1992). Decapod larvae: Dispersal, mortality, and ecology. A working hypothesis. *Amer. Zool.* **32**, 512–523.

McConaugha, J.R., Johnson, D.F., Provenzano, A.J., and Maris, R.C. (1983). Seasonal distribution of larvae of *Callinectes sapidus* (Crustacea: Decapoda) in the waters adjacent to Chesapeake Bay. *J. Crust. Biol.* **3**, 582–591.

McConnaughey, R.A., Armstrong, D.A., Hickey, B.M., and Gunderson, D.R. (1994). Juvenile Dungeness crab (*Cancer magister*) recruitment variability and oceanic transport during the pelagic larval phase. *Can. J. Fish. Aquat. Sci.* **49**, 2028–2044.

McKelvey, R.D., Hankin, D., Yanosko, K., and Snygg, C. (1980). Stable cycles in multistage recruitment models: An application to the northern California Dungeness crab (*Cancer magister*) fishery. *Can. J. Fish. Aquat. Sci.* **37**, 2323–2345.

McMillan-Jackson, A.L., and Bert, T.M. (2004). Mitochondrial DNA variation and population genetic structure of the blue crab Callinectes sapidus in the eastern United States. *Mar. Biol.* **145**, 769–777.

Methot, R.D., Jr., and Botsford, L.W. (1982). Estimated preseason abundance in the California Dungeness crab (*Cancer magister*) fisheries. *Can. J. Fish. Aquat. Sci.* **39**, 1077–1083.

Miller, B.A., and Emlet, R.B. (1997). Influence of nearshore hydrodynamics on larval abundance and settlement of sea urchins *Strongylocentrotus franciscanus* and *S. purpuratus* in the Oregon upwelling zone. *Mar. Ecol. Prog. Ser.* **148**, 83–94.

Moksnes, P.O., Licius, R.N., Pihl, L., and van Montfrans, J. (1997). Cannibal–prey dynamics in young juveniles and post-larvae of the blue crab. *J. Exp. Mar. Ecol.* **215**, 157–187.

Moloney, C.L., Botsford, L.W., and Largier, J.L. (1994). Development, survival and timing of metamorphosis of planktonic larvae in a variable environment: The Dungeness crab as an example. *Mar. Ecol. Prog. Ser.* **113**, 61–79.

Morgan, L.E., and Botsford, L.W. (2001). Managing with reserves: modeling uncertainty in larval dispersal for a sea urchin fishery. In (G.H. kruse, N. Bez, A. Booth, M.W. Dorn, S. Hills, R.N. Lipcius, D. Pelletier, C. Roy, S.J. Smith, and D. Witherill, Eds.). Proceedings of the Symposium on Spatial Processes and Management of Marine Populations, University of Alaska Sea Grant, Fairbanks AK pp. 667–684.

Morgan, L.E., Wing, S.R., Botsford, L.W., Lundquist, C.J., and Diehl, J.M. (2000). Spatial variability in red sea urchin (*Strongylocentrotus franciscanus*) recruitment in northern California. *Fish. Oceanogr.* **9**, 83–98.

Morgan, S.G., Zimmer–Faust, R.K., Heck, K.L., Jr., and Coen, L.D. (1996). Population regulation of blue crabs in the northern Gulf of Mexico: Postlarval supply. *Mar. Ecol. Prog. Ser.* **133**, 73–88.
Myers, R.A., Bowen, K.G., and Barrowman, N.J. (1999). Maximum reproductive rate of fish at low population sizes. *Can. J. Fish. Aquat. Sci.* **56**, 2404–2419.
Odense, P.H., and Annand, C. (1978). Isoenzyme systems of an inshore and offshore lobster population. *Int. Coun. Explor. Sea Doc.* C.M. 1978/K:15.
Olmi, E.J., III. (1995). Ingress of blue crab megalopae in the York River, Virginia, 1987–1989. *Bull. Mar. Sci.* **57**, 753–780.
Orensanz, J.M., Armstrong, J., Armstrong, D., and Hilborn, R. (1998). Crustacean resources are vulnerable to serial depletion: The multifaceted decline of crab and shrimp fisheries in the Greater Gulf of Alaska. *Rev. Fish Biol. Fisher.* **8**, 117–176.
Orensanz, J.M., Ernst, B., Armstrong, D.A., Stabeno, P., and Livingston, P. (2005). Contraction of the geographic range of distribution of snow crab (*Chionoecetes opilio*) in the eastern Bering Sea: An environmental ratchet? *CALCOFI Reports.* **45**, 65–79.
Orth, R.J., van Montfrans, J., Lipcius, R.N., and Metcalf, K.S. (1996). Utilization of seagrass habitat by the blue crab, *Callinectes sapidus*, in Chesapeake Bay: A review. In *Sea grass biology: Proceedings of an international workshop* (J. Kuo, R.C. Phillips, D.I. Walker, and H. Kirkmanm, eds., pp. 213–224).
Ouellet, P., and Lefaivre, D. (1994). Vertical distribution of northern shrimp *Pandalus borealis* larvae in the Gulf of St. Lawrence: Implications for trophic interactions and transport. *Can. J. Fish. Aquat. Sci.* **51**, 123–132.
Palumbi, S.R. (1995). *Using genetics as an indirect estimator of larval dispersal. Ecology of marine invertebrate larvae.* Boca Raton, FL: CRC Press.
Palumbi, S.R. (2003). Population genetics, demographic connectivity, and the design of marine reserves. *Ecol. Appl.* **13 (suppl.)**, S146–S158.
Parrish, R.H., Nelson, C.S., and Bakun, A. (1984). Transport mechanisms and reproductive success of fishes in the California Current. *Biol. Oceanogr.* **1**, 175–203.
Pedersen, O.P., Aschan, M., Rasussen, T., Tande, K.S., and Slagstad, D. (2003). Larval dispersal and mother populations of *Pandalus borealis* investigated by a Lagrangian particle-tracking model. *Fish. Res.* **65**, 173–190.
Peterson, W.T. (1973). Upwelling indices and annual catches of Dungeness crab, *Cancer magister*, along the west coast of the United States. *Fish. Bull. U.S.* **71**, 902–910.
Phillips, B.F., Cruz, R., Brown, R.S., and Caputi, N. (1994). Predicting catch of spiny lobster fisheries. In *Spiny lobster management* (B.F. Phillips, J.S. Cobb, and J. Kittaka, eds., pp. 285–301). *Fish.* Oxford: News Books.
Phillips, B.F., and Sastry, A.N. (1980). Larval ecology. In *The biology and management of lobsters* (J.S. Cobb and B.F. Phillips, eds., pp. 11–57, vol. 2). New York: Academic Press.
Post, E., and Forchhammer, M.C. (2002). Synchronization of animal population dynamics by large-scale climate. *Nature.* **420**, 168–171.
Prince, J., and Hilborn, R. (1998). Concentration profiles and invertebrate fisheries management. In *Proceedings of the North Pacific Symposium on Invertebrate Stock Assessment and Management* (G.S. Jamieson and A. Campbell, eds.). *Can. Spec. Publ. Fish. Aquat. Sci.* **125**, 187–196.
Prochaska, F.J., and Taylor, T.G. (1982). Cyclical and seasonal effort-yield functions for the Florida west coast blue crab fishery. In (H.M. Perry and W.A. Van Engel, eds., pp. 187–194). *Proc. Blue Crab Coll.*, October 18–19, 1979. Gulf Coast Marine Fisheries Commission. *Gulf states mar. fish. comm. pub.* 7. Ocean Springs, MS.
Rasmussen, T., Thollesson, M., and Nilssen, E.M. (1993). Preliminary investigation on the population genetic differentiation of the deep water prawn, *Pandalus borealis*, Kroyer from Northern Norway and Barents Sea. *Int. Council. Explor. Sea C.M.* 1993/K:11.
Ricker, W.E. (1954). Stock and recruitment. *J. Fish. Res. Board Can.* **11**, 559–623.

Roegner, G.C., Armstrong, D.A., Hickey, B.M., and Shanks, A.L. (2003). Ocean distribution of Dungeness crab megalopae and recruitment patterns to estuaries in southern Washington state. *Estuaries.* **26**, 1058–1070.

Rowe, S. (2001). Movement and harvesting mortality of American lobsters (*Homarus americanus*) tagged inside and outside no-take reserves in Bonavista Bay, Newfoundland. *Can. J. Fish. Aquat. Sci.* **58**, 1336–1346.

Rowe, S. (2002). Population parameters of American lobster inside and outside no-take reserves in Bonavista Bay, Newfoundland. *Fish. Res.* **56**, 167–175.

Rugulo, L.J., Knotts, K.S., Lange, A.M., and Crecco, V.A. (1998). Stock assessment of the Chesapeake Bay blue crab (*Callinectes sapidus* Rathbun). *J. Shellfish. Res.* **17**, 493–518.

Sale, P.F., and Krtizer, J.P. (2003). Determining the extent and spatial scale of population connectivity: Decapods and coral reef fishes compared. *Fish. Res.* **65**, 153–172.

Sastry, A.N. (1983). Pelagic larval ecology and development. In *The biology of crustacea* (D.E. Bliss, ed., vol. 7, pp. 214–282). *Behav. Ecol.* New York: Academic Press.

Scarratt, D.J. (1973). Abundance, survival, and vertical and diurnal distribution of lobster larvae in Northumberland Strait, 1962–63, and their relationship with commercial stocks. *J. Fish. Res. Board Can.* **30**, 1819–1824.

Seitz, R.D., Lipcius, R.N., Stockhausen, W.T., and Montane, M.M. (2001). Efficacy of blue crab spawning sanctuaries in Chesapeake Bay. In *Spatial processes and management of marine populations* (G.H. Kruse, N. Bez, A. Booth, M.W. Dorn, S. Hills, R.N. Lipcius, D. Pelletier, C. Roy, S.J. Smith, and D. Witherell, eds., pp. 607–626). University of Alaska Sea Grant, AK-SG-01-02, Fairbanks, AK: Alska Sea Grant.

Shaklee, J.B. (1983). The utilization of enzymes as genetic markers in fisheries management and conservation. Isozymes. *Curr. Top. Biol. Med. Res.* **11**, 214–247.

Shanks, A.L., Grantham, B.A., and Carr, M.H. (2003). Propagule dispersal distance and the size and spacing of marine reserves. *Ecol. Appl.* **13**, S159–S169.

Shumway, S.E., Perkins, H.C., Schick, D.F., and Stickney, A.P. (1985). Synopsis of biological data of the pink shrimp, *Pandalus borealis* Kroyer 1838. FAO fisheries synopsis no. 144. NOAA Technical Report. *Nat. Mar. Fish. Serv.* Seattle: National Marine Fisheries Service.

Sissenwine, M.P., and Shepherd, J.G (1987). An alternative perspective on recruitment overfishing and biological reference points. *Can. J. Fish. Aquat. Sci.* **44**, 913–918.

Sivigny, J.-M., Savard, L., and Parsons, D.G. (2000). Genetic characterization of the northern shrimp, *Pandalus borealis*, in the northwest Atlantic using electrophoresis of enzymes. *J. Northw. Atl. Fish. Sci.* **27**, 161–175.

Smedbol, R.K., McPherson, A., Hansen, M.M., and Kenchington, E. (2002). Myths and moderation in marine metapopulations. *Fish Fisher.* **3**, 20–35.

Smidt, E. (1981). Environmental conditions and shrimp stocks at Greenland. In *Proceedings of the International Pandalid Shrimp Symposium, Kodiak, Alaska, University of Alaska February 13–15, 1979* (T. Frady, ed., pp. 391–392). Sea Grant Reports 3. Kodiak, AK: University of Alaska.

Sotka, E.E., Wares, J.P., Barth, J.A., Grosberg, R.K., and Palumbi, S.R. (2004). Strong genetic clines and geographical variation in gene flow in the rocky intertidal barnacle *Balanus glandula*. *Molecular Ecology.* **13**, 2143–2156.

Squires, H.J. (1992). Recognition of *Pandalus eous* Makarov, 1935 as a Pacific species not a variety of the Atlantic *Pandalus borealis* Kroyer 1838 (Decapoda Caridea). *Crust.* **63**, 257–262.

Strub, P.T., Allen, J.S., Huyer, A., and Smith, R.L. (1987). Large-scale structure of the spring transition in the coastal ocean off western North America. *J. Geophys. Res.* **92**, 1527–1544.

Tagatz, M.E. (1968). Biology of the blue crab, *Callinectes sapidus* Rathbun, in the St. Johns River, Florida. *Fish. Bull.* **67**, 17–33.

Tang, Q. (1985). Modification of the Ricker stock recruitment model to account for environmentally induced variation in recruitment with particular reference to the blue crab fishery in Chesapeake Bay. *Fish. Res.* **3**, 13–21.

Teigsmark, G. (1983). Populations of the deep-sea shrimp, *Pandalus borealis* (Kroyer) in the Barents Sea. *Fiskiderectoratet Skrifter, Serie Havundersokelser* **17**, 377–430.

Thorrold, S.R., Jones, G.P., Hellberg, M.E., Burton, R.S., Swearer, S.E., Neigel, J.E., Morgan, S.G., Warner, R.R. (2002). Quantifying larval retention and connectivity in marine populations with artificial and natural markers. *Bull. Mar. Sci.* **70 (suppl.)**, 291–308.

Toonen, R.J., Locke, M., and Grosberg, R.K. (2004). Isolation and characterization of polymorphic microsatellite loci from the Dungeness crab *Cancer magister*. *Molecular Ecology Notes.* **4**, 30–32.

Tracey, M.L., Nelson, K., Hedgecock, D., Schlesser, R.A., and Pressick, M.L. (1975). Biochemical genetics of lobsters: Genetic variation and structure of American lobster (*Homarus americanus*) populations. *J. Fish. Res. Board Can.* **32**, 2091–2101.

Tuljapurkar, S.D. (1982). Population dynamics in variable environments. II. Correlated environments, sensitivity analysis and dynamics. *Theor. Pop. Biol.* **21**, 114–140.

Turing, A. (1952). The chemical basis of morphogenesis. *Philos. Trans. R. Soc. London B. Biological Sci.* **237**, 37–72.

van Montfrans, J., Peery, C.A., and Orth, R.J. (1990). Daily, monthly and annual settlement patterns by *Callinectes sapidus* and *Neopanope sayi* on artificial collectors deployed in the York River, Virginia. *Bull. Mar. Sci.* **46**, 214–228.

Wainwright, T.C., and Armstrong, D.A. (1993). Growth patterns in the Dungeness crab (*Cancer magister* Dana): Synthesis of data and comparison of models. *J. Crust. Biol.* **13**, 36–50.

Watkins, E. (1995). *The relationship between estuarine hydrodynamics and larval recruitment in fish and crustacean populations: A review and case study*. MS thesis. Baltimore, MD: Johns Hopkins University.

Weinberg, R. (1982). Studies on the influence of temperature, salinity, light, and feeding rate on laboratory reared larvae of deep sea shrimp, *Pandalus borealis* Kroyer 1838. *Meeresforschung*. **29**, 136–153.

Wickham, D.E. (1986). Epizootic infestations of nemertean brood parasites on commercially important crustaceans. *Can. J. Fish. Aquat. Sci.* **43**, 2295–2302.

Wilber, D.H. (1994). The influence of Apalachicola River flows on blue crab, *Callinectes sapidus*, in north Florida. *Fish. Bull.* **92**, 180–188.

Wild, P.W. (1980). Effects of seawater temperature on spawning, egg development, hatching success, and population fluctuations of the Dungeness crab, *Cancer magister*. *CalCOFI Rep.* **XXI**, 115–120.

Wing, S.R., Botsford, L.W., Largier, J.L., and Morgan, L.E. (1995b). Spatial structure of relaxation events and crab settlement in the northern California upwelling region. *Mar. Ecol. Prog. Ser.* **128**, 199–211.

Wing, S.R., Botsford, L.W., Ralston, S.V., and Largier, J.L. (1998). Meroplanktonic distribution and circulation in a coastal retention zone of the northern California upwelling system. *Limnol. Oceanogr.* **43**, 1710–1721.

Wing, S.R., Largier, J.L., Botsford, L.W., and Quinn, J.F. (1995a). Settlement and transport of benthic invertebrates in an intermittent upwelling region. *Limnol. Oceanogr.* **40**, 316–329.

Witman, J.D., and Sebens, K.P. (1992). Regional variation in fish predation intensity: A historical perspective on the Gulf of Maine. *Oecology.* **90**, 305–315.

Worm, B., and Myers, R.A. (2003). Meta-analysis of cod-shrimp interactions reveals top-down control in oceanic food webs. *Ecology.* **84**, 162–173.

Young, C.M. (1995). Behavior and locomotion during the dispersal phase of larval life. In *Ecology of marine invertebrate larvae* (L. McEdward, ed., pp. 249–278). Boca Raton, FL: CRC Press.

CHAPTER 9

A Metapopulation Approach to Interpreting Diversity at Deep-Sea Hydrothermal Vents

MICHAEL G. NEUBERT, LAUREN S. MULLINEAUX, and M. FORREST HILL

I. Introduction
II. Vent Systems as Metapopulations
  A. Dynamics and Distribution of Vent Habitat
  B. Dispersal and Colonization
III. Species Interactions
IV. Biogeography and Diversity
V. Metapopulation Models for Vent Faunal Diversity
  A. A Null Model
  B. Facilitation
VI. Summary
References

I. INTRODUCTION

Hydrothermal vent communities inhabit the deep sea in locations where volcanic activity causes hot, chemical-rich fluids to exit the seafloor. These surprisingly productive systems were first discovered in 1977 during a research expedition to the Galapagos Rift in the eastern Pacific ocean. Investigators found giant clams, more than 25 cm in length, and red-plumed tubeworms with 3-m-long tubes. The abundance, size, and novelty of organisms astonished biologists, and challenged the prevailing view of the deep sea as a sparsely populated desert inhabited mostly by small individuals. Subsequent investigation has revealed that the vent habitat is patchy and transient, making the ecosystem well suited to a metapopulation framework. Our goal in this chapter is to provide an introduction to the natural history of vent ecosystems in the deep sea (>1 km water depth) and to give an example of how metapopulation models can be used to address one class of questions about species diversity patterns at vents. We frequently use vents along the East Pacific Rise as examples because they are the systems most familiar to us.

The discovery of productive communities at vents perplexed biologists because the flux of photosynthetically-derived food particles from the sea surface was considered insufficient to support them. However, early studies revealed the prevalence of microbial chemoautotrophy and suggested that it might supplant photosynthesis as the basis of the food web. Microbes use reduced chemicals such as hydrogen sulfide and methane in the vent fluids as their energy source for fixing organic carbon (reviewed in Karl, 1995). Many of the microbes are free-living, either as plankton or on the seafloor, but some exist in close associations with metazoans. A prominent example is the symbiotic association between vestimentiferan tubeworms and bacteria. The tubeworms have no mouth or gut, and use instead an internal organ called a trophosome filled with chemoautotrophic bacteria. The bacteria are so numerous that they can comprise 15% of the body weight of the tubeworm (Fisher et al., 1988). Other species (e.g., mollusks) have bacteria associated with their gills, and some crustaceans and polychaetes support epizootic bacteria on their outer surfaces.

Vent fluids provide the reduced chemicals needed for chemoautotrophy, but they can exit the seafloor at extreme temperatures (>400°C) and with high concentrations of trace metals (Cu, Co, Zn, Pb) that are potentially toxic to many organisms (Childress and Fisher, 1992). In cases where undiluted fluids exit the seafloor, they immediately mix with oxygenated seawater, precipitating dark ferric sulfides, giving the vents the name *black smokers*. Some of the precipitates form directly on the seafloor and develop into large chimneys. In other cases, where hydrothermal fluids have mixed with ambient (2 to 4°C) seawater in the subsurface, they leak out of cracks and crevices at temperatures of a few degrees or tens of degrees above ambient. Diffuse venting usually is clear, but has been called *shimmering water* because the elevated temperatures, and consequent density anomalies, refract light used by observers. Areas of diffuse venting often are associated with high-temperature smokers and a decrease in temperature with distance from the main conduits. Because vent fluids are heated, they are lighter than seawater and may rise to levels of hundreds of meters above the seafloor before they entrain enough seawater to become neutrally buoyant. These plumes have been important in locating new vents, because they are geographically more extensive, and therefore easier to find, than the vent orifices.

The majority of invertebrates inhabiting deep-sea vents can live only in vent habitats. More than 440 species have been observed at vents, most of which are endemic species (Tunnicliffe et al., 1998). New species still are being described at a rapid rate (Fig. 9-1). Taxonomic groups that are common in the deep sea, such as polychaetes, mollusks, and crustaceans, are also common at vents, but in very different forms (see review in Van Dover, 2000). Vestimentiferan tubeworms are among the best known of Pacific vent species because of their large

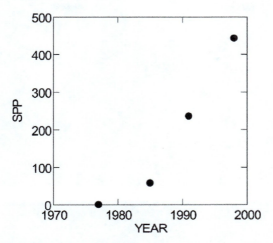

FIGURE 9-1. Number of vent species described since the discovery of vent ecosystems. Total vent species described at the time of publication of major reviews by Newman (1985), Tunnicliffe (1991) and Tunnicliffe et al. (1998).

size, bright-red plumes, and unusual morphology (Fig. 9-2A). Originally thought to be a new phylum, these tubeworms are now considered to be highly specialized polychaetes in the family Siboglinidae (Rouse, 2001). They live in the diffuse flows at temperatures as high as 30°C. Other polychaetes in the family Alvinellidae (e.g., Pompeii worms) inhabit smoker chimneys in the Pacific and are exposed regularly to temperatures of 40°C (Desbruyères et al., 1998) and perhaps higher. Numerous smaller polychaete species live in the interstices of tubes, shells, and rock, functioning as scavengers, predators, and suspension feeders. Small tube-building polychaetes in the family Serpulidae (feather-duster worms) form extensive fields outside the boundaries of detectable venting (Fig. 9-2B), capturing bacteria and organic particles as they drift away from the vents. The giant clams of the eastern Pacific have red hemoglobin-filled tissues to capture hydrogen sulfide from vent fluids. Their gills are greatly modified to support endosymbiotic bacteria rather than suspension feed, and they insert their foot directly into the vent fluids to absorb sufficient hydrogen sulfide for their endosymbionts. Mussels occur in the diffuse fluids at many vents (Fig. 9-2C), sometimes forming beds 50 cm thick and extending many tens of meters. Juvenile mussels feed on suspended matter, but older individuals appear to get most of their nutrition from bacteria in their gills. Grazing limpets are very common, covering not only the basaltic surfaces, but also the tubes and shells of other

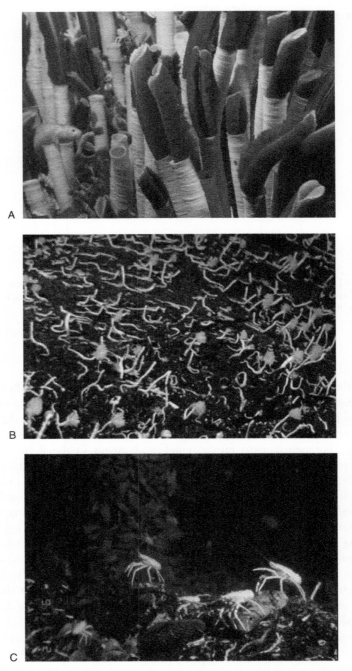

FIGURE 9-2. Hydrothermal vent fauna on East Pacific Rise: (A) thicket of vestimentiferan tubeworms providing habitat for bythograeid crabs and fish; (B) field of suspension-feeding serpulid polychaetes; (C) vent mussels (with yellowish periostracum covering shell) and galatheid crabs. (see color plate)

sessile invertebrates. They appear to subsist on benthic bacteria, just as related shallow-water species graze on benthic algae. A large predatory snail species occurs in the eastern Pacific, but its role in structuring the communities is unknown. Galatheid and bythograeid crabs function as scavengers and, when given the opportunity, predators. Other mobile predators include fish and octopods, some of which are endemic to vents, and others that are foraging opportunistically. Vent faunas in biogeographic provinces other than the eastern Pacific differ substantially in species composition; processes potentially responsible for this variation are discussed in Section IV.

Vents have now been located at volcanically active sites in all ocean basins (Fig. 9-3). Marine volcanic activity is concentrated along mountain ranges that form as a consequence of seafloor spreading at the boundaries of tectonic plates. Along these midocean ridges, magma rises up to the surface of the Earth's crust, heating fluids within the crustal rocks and episodically erupting onto the seafloor. The ridges are linear features, and constrain the distribution of most venting to the narrow margins of crustal plates. Exceptions are found in volcanically active back-arc and fore-arc basins (e.g., Lau basin in the western Pacific), where subduction of the Earth's crust generates volcanism, and on seamounts such as Loihi, which appears destined to become the southernmost island of Hawaii. Vents and their associated communities occur as discrete patches, separated from each other by areas of seafloor that are not exposed to hydrothermal fluids. An individual vent may comprise an isolated black smoker, an area of diffuse flow, or a combination of focused and diffuse venting. Vents may be separated by less than 1 km, forming clusters along an active segment of a ridge (e.g., near 9°N along the East Pacific Rise; Fig. 9-4), but these clusters are often separated from each other by hundreds to thousands of kilometers. Although our understanding of the spatial scales of vent distributions is limited by the resolution and extent of geological mapping (i.e., the absence of vents in many regions may be the result of lack of surveys), it is clear that vents are patchy, and this patchiness occurs over a wide range of spatial scales.

Like terrestrial volcanic and hydrothermal systems, deep-sea vents are transient. The life spans of vents and their spacing along a ridge both depend on the magmatic, tectonic, and hydrologic processes that result in release of hydrothermal fluids (reviewed in Fornari and Embley, 1995). The fluids enter the seafloor as ambient seawater, seeping downward into the crustal rocks. They are heated by magma (either directly by contact or indirectly through thermal conduction) and interact with the rocks, gaining some chemicals and losing others. Eventually, the hot, buoyant fluids seep back up through the crust. Conduits up to the seafloor may be formed abruptly through volcanic eruptions or by tectonic cracking and diking events. These conduits may close or become rerouted as a result of tectonic shifts, geochemical clogging (e.g., mineral precipitation as temperatures cool), lava flows, or basaltic-pillow collapses and rockfalls (Haymon et al., 1993).

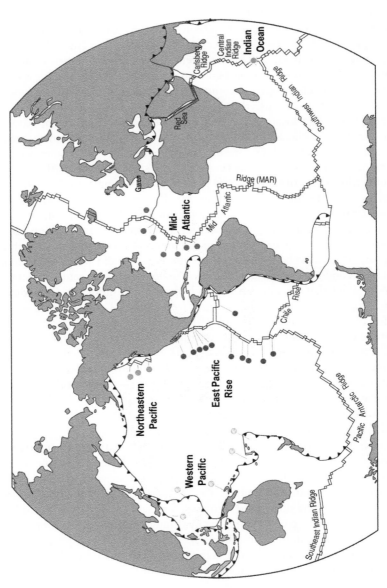

FIGURE 9-3. Global distribution of known hydrothermal vent communities on mid-ocean ridges and back-arc spreading centers. Biogeographic provinces are distinguished by color: Western Pacific (lightest gray); Northeastern Pacific (gray); East Pacific Rise (black); Mid-Atlantic Ridge (dark gray, including 3 northernmost vents near Azores); and Indian Ocean (light gray). Map redrawn from Van Dover et al. (2002).

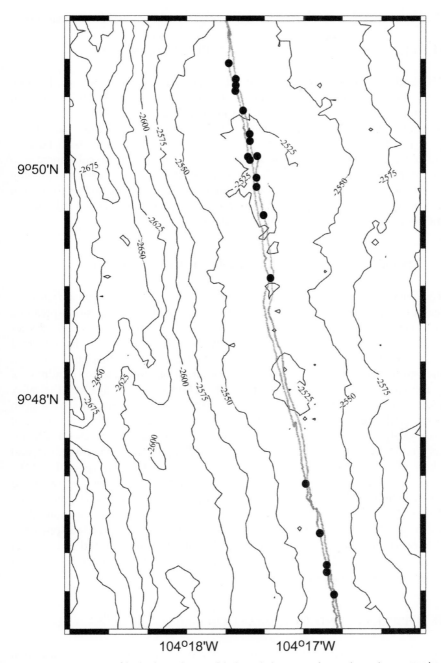

FIGURE 9-4. Spacing of hydrothermal vents (black circles) in vent clusters located near 9°50'N and 9°47'N on the East Pacific Rise. From K. von Damm, unpublished data.

II. VENT SYSTEMS AS METAPOPULATIONS

The study of vent ecology is well suited to a metapopulation approach because of the patchy distributions and transient nature of the habitat, and the endemism of component species. Relatively little is known about the life histories of species, dynamics of populations, or structure of communities, due to their recent discovery and remote locations. Although this lack of information could be considered a drawback, it does mean that theory can contribute substantially to explaining patterns, identifying important parameters and processes, and guiding future empirical research. In this section we summarize information on the spatial and temporal scales of physical and ecological processes that are relevant to metapopulation modeling of vent ecosystems.

A. Dynamics and Distribution of Vent Habitat

The dynamics of venting on the deep seafloor vary among ridge systems with different spreading rates. Vent habitats on the Mid-Atlantic Ridge (a slow-spreading ridge) are sparse, and known vents are separated by as much as a thousand kilometers (Fig. 9-3). Dating studies at one vent, TAG, show that venting has occurred over 10,000 years at this location (Lalou et al., 1993). During that time, however, the venting appears to have ceased and restarted on timescales of hundreds of years. The ridge crest is relatively broad on the Mid-Atlantic Ridge, and the valley walls can be 50 m high or more. The observed vent communities have been discovered over a wide range of depths (850–3500 m), and in a variety of geological, topographic, and oceanographic settings (Van Dover, 1995). Chemical composition of the vent fluids varies, depending on the geological setting, and likely influences the composition of colonizing species (Desbruyères et al., 2001). In contrast, vents on the East Pacific Rise (a fast-spreading ridge) are separated by a few hundred kilometers at most, and often occur in clusters (Haymon et al., 1991). The vents have life spans as short as years to decades (Haymon et al., 1993; MacDonald et al., 1980). Geologists do not fully understand the processes that control the spacing and transience of vents, but the rate and character of magma delivery to the crust appear to be associated with the spreading rate of a ridge and the dynamics of its venting.

B. Dispersal and Colonization

Given the patchiness and transience of the vent environment, it is clear that migration is essential for vent species to maintain their populations and geographic ranges. Because adults of most vent species are attached or have minimal migratory capabilities, it also seems likely that vent species disperse through the water via larval stages. However, the dispersal mechanisms and their population consequences have not been easy to predict. For instance, Lutz et al. (1980) predicted that most vent species should have larvae that feed in the water to disperse long distances, but an examination of egg sizes and larval shell types did not support the idea. Subsequent studies showed that reproduction and larval type of vent species were highly constrained by phylogenetic affinity (reviewed in Tyler and Young, 1999). In other words, species retained the life history attributes of their shallow-water ancestors, even if they weren't optimal for the vent environment.

To disperse successfully, a larva must have a sufficiently long life span to survive the transport interval between neighboring vent habitats, and must be able to locate a suitable habitat for settlement. Because larvae of most vent species are small (Berg and Van Dover, 1987) and appear to be relatively weak swimmers, dispersal rates and trajectories are strongly influenced by oceanic circulation. Circulation in the vicinity of mid-ocean ridges is controlled by topography, hydrography, Coriolis effects, turbulent mixing, and regional forcing. Near the ridge, channeling or blocking by topography may intensify flow parallel to the ridge. Coriolis effects (from the Earth's rotation) also contribute to flow along topography or to gyres and eddies in the water column (Cummins and Foreman, 1998). Tidal and longer period oscillations are often prominent at ridge sites (Lavelle and Cannon, 2001), and these can drive substantial vertical flows up and down the ridge flanks, as well as contribute to localized mean flows parallel to the ridge. In general, currents that parallel the ridge axis are expected to facilitate larval exchange among vent communities.

A direct estimate of larval life span has been obtained for only one species, the vent tubeworm *Riftia pachyptila*. Physiological studies of cultured larvae revealed that the larvae can live on the order of 30 to 40 days without feeding (Marsh et al., 2001). If the larvae are transported as passive, neutrally buoyant particles, they have the potential to disperse distances up to 100 km in typical currents near 9 to 10°N along the East Pacific Rise (Mullineaux et al., 2002). However, the ridge at this latitude experiences sustained periods of cross-axis flow, which might sweep larvae off the ridge and away from suitable habitat. Currents near East Pacific Rise vents at 13°N are aligned along the ridge axis for much longer intervals than at 9°N (Chevaldonné et al., 1997), and *Riftia* larvae appear capable of dispersing up to 200 km along the ridge (Mullineaux et al., 2002). In both regions of the East Pacific Rise, vent spacing is on the order of tens of kilometers, suggesting that larvae should be able to disperse effectively between neighboring vents. A similar approach was used to estimate dispersal potential of alvinellid polychaetes on the East Pacific Rise (Chevaldonné et al., 1997). Larval life spans of the alvinellids, as inferred from egg size and life histories of related shallow species, were quite short and allowed dispersal distances on the order of tens of kilometers. These species appear to have sufficiently long larval life spans to disperse between neighboring vents on the East Pacific Rise, but not to survive transport across ocean basins (e.g., between the East Pacific Rise and western Pacific vents) or between disjunct ridge systems (e.g., East Pacific Rise to northeastern Pacific ridge system). Because larval life span has been measured for *Riftia* only, the question of how long-distance transport influences vent ecology remains open.

Other potential topographic and oceanographic barriers to dispersal include the transform faults that separate ridges into segments (e.g., MacDonald, 1982; Fig. 9-3) and provide deep passages between abyssal basins. Although detailed oceanographic observations from discontinuities in ridges are sparse, examples from the Atlantic suggest that the flows across the ridges are often persistent and strong (Ledwell et al., 2000; Mercier and Speer, 1998). Larvae caught in these flows would likely be lost from that ridge system permanently, because no mechanism for return is known.

However, retention of larvae may occur on a local scale near a source vent. The buoyant hydrothermal plume sets up a circulation cell as a consequence of the Coriolis effect (e.g., Helfrich and Speer, 1995), which could retain particles and larvae near their source populations. The circulation cell is expected to extend roughly 1 km from the vent, with focused upward velocities at the center, diffuse downward velocities at the perimeter, and inward velocities along the bottom (Fig. 9-5). Any larvae that sank, swam downward, or had a positive tropic response to vent fluids would tend to accumulate near the vent. Such a retention mechanism would inhibit dispersal, but potentially enhance settlement back into the local population (Mullineaux and France, 1995).

Population genetic studies are useful for inferring dispersal and have constrained the spatial scales over which colonists appear to be exchanged. Most species of tubeworms and mollusks show evidence of extensive gene flow on the 100 km scale typical of a ridge segment, Craddock et al., 1997; Vrijenhoek, 1997; Won et al., 2003), but punctuated by population bottlenecks that might be the result of frequent extinctions. Surprisingly, there is little evidence for "isolation by distance" for any species except on large (thousands of kilometers) scales or across topographic barriers. Along the East Pacific Rise, neighboring populations of the polychaete *Alvinella pompejana* vary genetically, but show no increase in variation with separation distance (Jollivet et al., 1995). Populations of the amphipod *Ventiella sulfuris* are genetically similar at vents within a ridge segment, but substantially different between the East Pacific Rise and the Galapagos spreading center (France et al., 1992). These results indicate that larvae disperse extensively within segments of a mid-ocean ridge, but not across transform faults and other large topographic discontinuities in the ridge system (Tunnicliffe et al., 1998). The lack of documentation for the expected "isolation by distance" pattern has led researchers to speculate that vent larvae exist as a well-mixed, ever-present pool of potential colonists (Van Dover, 2000; Vrijenhoek, 1997). Although this

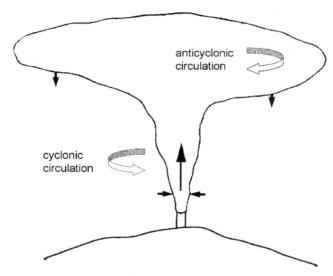

FIGURE 9-5. Diagram of flow in a buoyant hydrothermal vent plume, showing anticyclonic circulation at the neutrally buoyant level and cyclonic circulation below. Redrawn from Helfrich and Speer (1995).

may be a valid generalization on evolutionary timescales, sampling in the water column has demonstrated that abundance and species composition of vent larvae, like their shallow relatives, are highly variable on ecological timescales (Mullineaux et al., 2001).

The dynamic nature of vent habitats will likely influence dispersal success and population genetic structure. Theoretical considerations indicate that introducing variation of vent positions into a metapopulation model (to simulate shifting of hydrothermal activity along a ridge axis) can prevent gene flow and genetic drift from reaching equilibrium (Jollivet et al., 1999). When vent spacing was allowed to vary temporally in simulations, individual populations rarely experienced the prolonged isolation that led to extinction. Similarly, variance in vent positions facilitated gene flow and allowed population genetic structure to become homogeneous in a metapopulation that would have experienced genetic isolation if the vent positions had been static.

Despite potential dispersal barriers, many vent species colonize new habitat rapidly (Micheli et al., 2002; Shank et al., 1998; Tunnicliffe et al., 1997), indicating that they are prompt and efficient dispersers. The expected consequences of unimpeded dispersal and subsequent genetic exchange would be to homogenize species composition and genetic structure across vents on a ridge. However, these studies all were conducted in fields of active venting, where a new habitat was typically within 10 km of established vent communities. In contrast, at Loihi seamount, an isolated vent south of Hawaii, only a few individual metazoan colonists have been detected during the past 10 years, and no vent-endemic populations appear to have become established (Grigg, 1997). It is not clear whether the cause of sparse recruitment is unsuitable habitat or restricted supply of colonists.

III. SPECIES INTERACTIONS

Until recently, vent communities have been assumed to be structured by the extreme physical and chemical features of their environment (Luther et al., 2001; Sarrazin et al., 1999), but in situ, manipulative studies have revealed an important role of species interactions. Facilitation, competition, and predation influence the patterns of zonation, succession, and species composition at vents (Levesque et al., 2003; Micheli et al., 2002; Mullineaux et al., 2003). The physicochemical environment clearly sets the extreme limits of species' occurrences at vents, but interactions with other species influence their distributions within these limits.

Succession is particularly apparent and important in systems in which episodic disturbances induce large changes in availability of primary substratum or food. On ridges, the catastrophic perturbations that create new vents or obliterate existing ones occur on timescales comparable to the species' generation times, making disturbance and primary succession dominant characteristics of the ecosystem. The physical and chemical nature of the habitat continues to be dynamic after eruptions (Butterfield et al., 1997; von Damm, 2000), and changes in the flux and composition of hydrothermal fluids affect chemoautotrophic-based production. The responses of colonizing species at deep-sea vents to these perturbations, either directly through physiological tolerances and nutritional requirements or indirectly through biological interactions, are largely unknown.

Previous studies of invertebrate colonization at hydrothermal vents have documented a distinct and consistent successional sequence over the life span of vents on the East Pacific Rise (Fustec et al., 1987; Hessler et al., 1988). The initial visibly dominant sessile metazoan in sites with moderate-temperature fluids (i.e., <30°C, not high-temperature black smokers) appears to be the small vestimentiferan tubeworm *Tevnia jerichonana* (Shank et al., 1998). This species then is replaced by the larger tubeworm *Riftia pachyptila*, usually over a period of less than one year. Later in the sequence, the mussel *Bathymodiolus thermophilus* colonizes and may, in some cases, displace the tubeworms (Hessler et al., 1988). Time–series observations show that this sequence can correspond to a change in the temperature and chemistry of vent fluids (Shank et al., 1998), although Hessler et al. (1988) and Mullineaux et al. (2000) document species succession under conditions of constant venting and conclude that biological interactions also play a role. Johnson et al. (1994) noted that mussels colonized tubeworm clumps and diverted the hydrothermal fluids away from the tubeworm plumes with their shells. They suggested that this reduction in the supply of hydrogen sulfide to the tubeworms' uptake organs essentially starved them and led to their death. Mullineaux et al. (2000) found that *Riftia* larvae settled onto experimental surfaces only if *Tevnia* was present, whereas *Tevnia* was able to settle alone. The colonization patterns did not correspond to gradients in the physicochemical environment, nor did they reflect larval availability, so it is possible that *Riftia* colonists were facilitated by the smaller pioneer species. The mechanism for facilitation is unknown, although it does not appear to be mediated by the tube material alone (i.e., *Riftia* does not settle preferentially in response to *Tevnia* tubes; Hunt et al., 2004) and does not appear to be obligate. Most likely, the facilitation involves the establishment of a critical microbial flora or release of a chemical settlement cue.

In situ manipulative studies of succession and predation at hydrothermal vents have investigated interactions between other vent species at the East Pacific Rise, including the predatory fish, scavenging crabs, grazing limpets, and suspension-feeding polychaetes. Using caging studies to exclude the predators/scavengers, and bait studies to investigate prey choice, Micheli et al. (2002) found that fish preyed selectively on limpets. When predators were excluded, limpet populations increased, and they excluded most sessile, tube-building, species. Mullineaux et al. (2003) used sequential colonization studies to demonstrate that early colonists at vents had an important influence on later arrivals, and in setting the structure of adult communities. When sessile, tube-building, species settled in areas with low food supply, they facilitated settlement of other sessile colonists. When mobile limpets or sessile tube builders settled in areas of high food supply, they inhibited colonization by all other species, whether mobile or sessile. Fish mediated these interactions by preying selectively on limpets above a threshold size (Sancho et al., 2005). These results demonstrate that species interactions are mediated by predators and alter the structure of communities through mechanisms that vary along gradients of vent fluid flux.

IV. BIOGEOGRAPHY AND DIVERSITY

The species composition of vent faunas varies among geographically disjunct ridges, and each ridge system can be characterized by a few distinct species. As new vents are discovered in unexplored regions, new biogeographic provinces, each with a distinctive fauna, are erected. Among these are the: East Pacific Rise, Northeastern Pacific, Mid-Atlantic Ridge, Central Indian Ocean, and Western Pacific (Fig. 9-3). The northeastern Pacific fauna shares many species and genera with the East Pacific Rise, but has lower species diversity and lacks notable groups, such as endemic decapods (Tunnicliffe et al., 1998). For instance, one species of vestimentiferan tubeworm, *Ridgeia piscesae*, occurs in the northeastern Pacific, in contrast to the three related species found on the East Pacific Rise. Tunnicliffe (1988) suggests that a vicariant event (the subduction of an extensive ancestral north–south ridge by the North American plate) isolated a subsample of eastern Pacific species onto the northern fragment of the ridge. At western Pacific vents, large, symbiont-bearing gastropods dominate many communities (Hessler and Lonsdale, 1991), although bivalves and tubeworms also occur (Kojima, 2002). Vent chimneys at Atlantic vents are covered by several different species of bresiliid shrimp, except in a few sites where shrimp are absent or rare and bivalves dominate (Desbruyères et al., 2000). Indian Ocean vents have only recently been explored. Their inhabitants appear to be related to western Pacific species (Hashimoto et al., 2001; Van Dover et al., 2001), with the exception of a prominent shrimp related to species in the Atlantic.

Along a ridge, individual vents may differ in species occurrence or relative abundances, but clines in species composition or distinct faunistic boundaries have not been observed (e.g., Juniper et al., 1990; Van Dover and Hessler, 1990). These differences between neighboring vents may be the result of environmental variation (e.g. in hydrothermal fluid flux rates or water chemistry) or stochastic events (Hessler et al., 1985); or of water depth, habitat disturbance, or character of water column particulates (Desbruyères et al., 2001). In addition, composition of vent faunas may be influenced by initial colonists or may change dramatically over time (Shank et al., 1998), leading to a mosaic of patches at different stages of succession.

Diversity patterns at hydrothermal vents are difficult to interpret because different regions have been sampled with varying effort and different techniques. Nevertheless, a comparison of species numbers between biogeographic provinces gives a qualitative perspective. As a starting point, diversity at vents is considered to be low compared with most other deep-sea and shallow-water habitats (Jollivet, 1996; Juniper and Tunnicliffe, 1997). Among vents, diversity on the East Pacific Rise is high and perhaps comparable with the western Pacific, whereas diversity at the northeastern Pacific vents is reduced. The lowest number of species is found at the mid Atlantic vents. Juniper and Tunnicliffe (1997) suggested that a combination of regional age, disturbance, habitat heterogeneity, and habitat area were responsible for these observed patterns. For instance, the Atlantic basin is relatively young compared to the Pacific, and the vents are stable, widely spaced, and homogeneous with respect to habitat type. These characteristics all would contribute to low diversity relative to the Pacific, where the regional species pool has long been in existence, the vents are disturbed frequently enough to mitigate competitive exclusion, the vent habitat area is larger (i.e., vents are more densely spaced), and habitat type varies between vents.

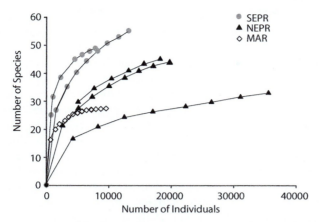

FIGURE 9-6. Comparison of diversity in the vent communities on the Mid-Atlantic Ridge (MAR) and East Pacific Rise (southern = SEPR; northern = NEPR), using sample-based species-effort curves. Each point is a mean diversity value based on 100 randomization operations (without replacement) for each curve. Redrawn from Turnipseed et al. (2003).

Recent studies have approached diversity through more quantitative means, by restricting sampling to invertebrates associated with mussel beds (Turnipseed et al., 2003; Van Dover and Trask, 2000) or vestimentiferan tubeworm clumps (Tsurumi and Tunnicliffe, 2003). Replicate sampling of known volumes or areal coverage allows these investigators to plot species–effort curves and evaluate the effect of sampling on species numbers. Turnipseed et al. (2003) found that diversity in vent mussel beds was lowest in the mid Atlantic and higher at the southern East Pacific Rise than the northern East Pacific Rise (Fig. 9-6). They suggest that the wide spacing of vents in the mid Atlantic limits species dispersal and contributes to extinction of populations at individual vent fields, leading to lowered diversity. They argue that this process is more important in reducing diversity than short vent life spans, as has been suggested by population geneticists (Craddock et al., 1995; Sibuet and Olu, 1998). In contrast, Tsurumi (2003) suggests that diversity at northeastern Pacific vents is limited by frequent disturbance and habitat transience.

V. METAPOPULATION MODELS FOR VENT FAUNAL DIVERSITY

Metapopulation modeling approaches are well suited to investigation of species diversity patterns (van Woesik, 2000; Wilson, 1992). As an example of such an approach, we address the question of why local species diversity differs among geographically separate ridge systems (e.g., eastern Pacific and mid Atlantic). Although some of these patterns have been explained in terms of evolutionary processes and vicariance events (e.g., Tunnicliffe, 1988), we want to investigate whether ecological-scale processes constitute feasible alternatives. Our ultimate objectives are to explore the hypotheses raised by vent researchers to explain diversity patterns. Are they the result of (1) disturbance, as suggested by Craddock et al. (1995); (2) habitat area or heterogeneity, as suggested by Juniper and Tunnicliffe (1997); or (3) habitat separation, as suggested by Turnipseed et al. (2003)? Or alternatively, might the differences in diversity be due to species interactions? To begin evaluating these hypotheses, we develop a model that incorporates the dynamics of vent habitat, species' colonization ability, and facilitation as an example of species' interactions. Complementary examples of the use of metapopulation approaches to address community patterns are presented in chapters on coral reefs (Mumby and Dytham) and general community structure (Karlson) in this book.

For simplicity we start with a spatially implicit model and make no effort to include the various scales of vent spacing observed in nature. As a consequence, we will not be able to study the implications of spatiotemporal correlation in vent activity. We could use a spatially explicit model to address this issue, but that is outside the scope of this study. Also for simplicity, we include facilitation, but not exclusion (competition or predation), in our evaluation of species interactions. Exclusion appears to be a fundamental part of successional sequences at vents and elsewhere, and will be a component of future modeling efforts on this topic.

For our theoretical analyses, we use metapopulation community models (or *metacommunity models*; Wilson, 1992) that describe species as collections of spatially discrete populations connected by dispersal. These models take the form of *patch-occupancy models* (Caswell and Cohen, 1991; Levins, 1970) that describe species as occupying a set of discrete habitat patches (i.e., vents). The *state* of a patch is a list of the species that are present in it. Patch-occupancy models keep track of the proportion of the patches that are in each state. An important assumption of these models is that species can disperse, with equal ease, between any two patches. This assumption allows us to write the models as (relatively) simple systems of differential equations.

Our models are different than most in two ways. First, most metapopulation and metacommunity models assume a static configuration of habitat types (but see Hanski, 1999; Nee and May, 1992). Many marine habitats, however, are known to vary over time. Alterations in hydrothermal fluid flux at deep-sea vents can cause those habitats to switch from suitable to uninhabitable (or vice versa) on timescales as short as the generation times of inhabitants. Our models explicitly include these "vent dynamics."

Second, most metacommunity models account for a very small number of species. The reason is called the *curse of dimensionality*. In a model for m species that allows any number of these species to occupy a single patch, the number of states is 2^m; the number of possible transitions is 2^{2m}. With 20 species there would be more than one million states and 10^{12} transitions. The curse is that it becomes impossible to specify the rules for these transitions even for species pools of modest size. Theorists have tried to reduce the state space in a number of ways. In some models, patches are the size of a single individual, so that only one species can occupy a patch at a time. In other models, inferior competitors are eliminated from patches immediately upon the arrival of a superior competitor. Most often, theoreticians avoid the curse by simply keeping the number of species small; one- and two-species models are the most frequently studied. Our models, however, allow for many species by introducing a hierarchy that allows us to specify transition rules in an economical way.

A. A Null Model

To begin, imagine a habitat composed of many discrete vent sites (patches) that are identical in every way, except that at any given time, some of them will be dormant and, hence, uninhabitable. Imagine that all the vents are equidistant from each other. Imagine that the active vents can support an arbitrary number of species, and that dormant vents become active and active vents become dormant at fixed rates. Finally, populate this imaginary dynamic landscape with identical species that do not interact in any way.

This imaginary vent community is, of course, highly contrived. It ignores the obvious fact that real species are different in their reproductive output, longevity, and dispersal abilities. It ignores the fact that some species do compete with each other whereas others facilitate each other's growth. It ignores the facts that real vent habitats are spatially heterogeneous, that some vents are closer to each other than others, and that vents can typically support only a limited number of species at one time. But it is exactly because of these contrivances that this imaginary vent ecosystem is so useful. Because it leaves out almost every interesting biological interaction, we can use it to derive baseline statistics against which we can compare the results of models that incorporate biological processes that we care about.

We constructed a mathematical model of the imaginary vent ecosystem just described, a so-called *null model*, using the following recipe. First, identify every species in the community with a number between 1 and m; m is then the size of the species pool. Let N be the total number of vents (both active and dormant). Let S_x be the number of dormant vents, S_i be the number of vents at which species i is present, and $\bar{S}_i$ be the number of active vents at which species i is absent. Assume that dormant vents become active at the rate R, and active vents become dormant at the rate D. Because the vent sites are assumed to be equidistant from each other, we will assume that a vent without species i will be colonized by species i at a rate that is proportional to the total number of patches that species i currently occupies. We call the constant of proportionality the *colonization rate*, C, and assume that C decreases with the distance between patches. Then, for each $i = 1, 2, \ldots, m$, we have

$$\frac{dS_x}{dT} = D(N - S_x) - RS_x \tag{1a}$$

$$\frac{d\bar{S}_i}{dT} = RS_x - \frac{C}{N}S_i\bar{S}_i - D\bar{S}_i \tag{1b}$$

$$\frac{dS_i}{dT} = \frac{C}{N}S_i\bar{S}_i - DS_i. \tag{1c}$$

In this model, and in the facilitation model that we describe in the next section, species only go extinct when an occupied vent becomes dormant. This assumption is not as bad an approximation as it might appear at first, because the timescale of vent dynamics is approximately the same as the generation time of the species that inhabit them.

For a given species, model (1) has two equilibrium states. At the first, species i is eliminated from every vent:

$$S_x = \frac{N}{(R/D)+1}, \quad \bar{S}_i = \frac{N(R/D)}{(R/D)+1}, \quad S_i = 0. \tag{2}$$

At the second equilibrium,

$$S_x = \frac{N}{(R/D)+1}, \quad \bar{S}_i = \frac{N}{(C/D)}, \quad S_i = \left[\frac{(R/D)}{(R/D)+1} - \frac{1}{(C/D)}\right]N; \tag{3}$$

species i persists if $S_i > 0$.

Note that the equilibrium levels of the states are always proportional to the total number of vents (N). This suggests that we need only keep track of the fractions,

$$s_x = \frac{S_x}{N}, \quad \bar{s}_i = \frac{\bar{S}_i}{N}, \quad s_i = \frac{S_i}{N}, \tag{4}$$

rather than the numbers S_x, $\bar{S}_i$, or S_i. Also note that while model (1) contains three parameters (D, R, C) in addition to N, these parameters always appear in the two ratios R/D and C/D. Thus, at least as far as the equilibria are concerned, the important values are activation and colonization rates relative to the dormancy rate, not the values of the rates themselves. This suggests that we should measure time in units of $1/D$, the average length of time that a vent remains active. Using

$$t = TD, \quad r = \frac{R}{D}, \quad \text{and} \quad c = \frac{C}{D}, \tag{5}$$

Chapter 9 A Metapopulation Approach to Interpreting Diversity

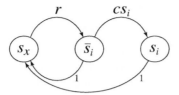

FIGURE 9-7. A graphical representation of the null model (6). Arrows indicate potential transitions among the three states: s_x (dormant), $\bar{s}_i$ (active but uninhabited by species i), and s_i (inhabited by species i). Per capita rates for each transition are indicated next to the arrows.

along with the change of variables (4) in model (1) gives the dimensionless model

$$\frac{ds_x}{dt} = 1 - s_x - rs_x \tag{6a}$$

$$\frac{d\bar{s}_i}{dt} = rs_x - cs_i\bar{s}_i - \bar{s}_i \tag{6b}$$

$$\frac{ds_i}{dt} = cs_i\bar{s}_i - s_i. \tag{6c}$$

The dynamics of this system are summarized in Figure 9-7.

Equation (6a), for ds_x/dt, is decoupled from the rest of system (6) because it represents geological rather than biological processes. Whenever $ds_x/dt = 0$, and hence $s_x = 1/(1+r)$, we say that the vent system is in "geological equilibrium." In geological equilibrium, the fraction of vents that are active is

$$f = 1 - s_x = \frac{r}{1+r}. \tag{7}$$

It is often more convenient to think of f rather than r when trying to interpret the results of our model.

When, in addition to being in geological equilibrium, $d\bar{s}_i/dt = 0$ and $ds_i/dt = 0$, we say that the system is in "biogeological equilibrium." In the dimensionless parameters, the geological equilibrium (2) is

$$s_x = 1 - f, \quad \bar{s}_i = f, \quad s_i = 0, \tag{8}$$

and the biogeological equilibrium (3) is

$$s_x = 1 - f, \tag{9a}$$

$$\bar{s}_i = \frac{1}{c} \equiv \bar{s}, \tag{9b}$$

$$s_i = f - \frac{1}{c} \equiv s \tag{9c}$$

respectively. Species will only persist in the vent system (i.e. $s > 0$) if

$$cf > 1. \tag{10}$$

The quantity cf is the expected number of new vents that would be colonized by a population at a single inhabited vent over the active period of that vent when the system is in geological equilibrium and all other active vents are uninhabited.[1] Inequality (10) can be interpreted as saying that in order for a species to survive, a single-vent population must be able to colonize at least one other vent before its vent becomes dormant.

1. Diversity Indices

To compare diversity between regions, we introduce indices of diversity at both the local (individual vent) and regional (vent system) scales. A measure of local diversity, α, is the expected number of species at a single active vent. For the null model,

$$\alpha = \frac{ms}{\bar{s}+s} = m\left(1 - \frac{1}{cf}\right). \tag{11}$$

Figure 9-8 shows how α changes as a function of c and f for a regional pool consisting of 25 species. Not surprisingly, α increases with both the amount of suitable habitat and the colonization rate.

One measure of regional diversity is simply the expected number of species that the entire vent system can support, α_{reg}. Under the null model, if one species can persist, all species can persist, so

$$\alpha_{reg} = \begin{cases} m, & \text{if } cf > 1 \\ 0, & \text{otherwise.} \end{cases} \tag{12}$$

A second regional diversity measure, H, is based on the relative proportions of the various species in the system. Let p_i be the relative proportion of species i. Then, setting $p_i \ln p_i = 0$ whenever $p_i = 0$, we define

$$H = -\sum_{i=1}^{m} p_i \ln p_i. \tag{13}$$

For a fixed species pool size, H is maximized when every species is equally likely. This is exactly the case in the null model where

$$p_i = \frac{s_i}{\sum_{i=1}^{m} s_i} = \frac{s}{ms} = \frac{1}{m}, \tag{14}$$

and hence $H = \ln m$.

[1] It is the analog of the basic reproductive number R_0 in epidemiology.

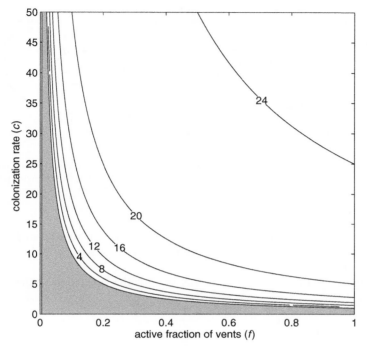

FIGURE 9-8. Contours of local diversity, as measured by the expected number of species in any patch (α), when the species pool contains $m = 25$ species and community dynamics are governed by the null model (6). In the grey area, all species are extinct.

B. Facilitation

As mentioned in Section III, succession is a well-documented process at hydrothermal vents, and facilitation appears to play a prominent role in the successional sequence. Early species in the sequence accelerate the colonization of later arrivals, and in some cases their presence may be required for later species to become established. Here, we examine how diversity indices introduced earlier in equations (11) through (13) change from their values under the null model when we introduce obligate facilitation.

To begin, we must first specify a successional hierarchy. Without loss of generality, we number the species such that species i must be present at the vent before species $i + 1$ can colonize. Next, it helps to define a new set of state variables. We keep s_x for the fraction of all vents that are dormant, but now we

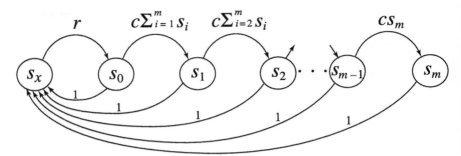

FIGURE 9-9. A graphical representation of the facilitation model (15). Arrows indicate potential transitions among the $m + 2$ states: s_x (dormant), s_0 (active but uninhabited), and s_i (inhabited by species 1 through i). Per capita rates for each transition are indicated next to the arrows.

introduce s_0 as the fraction of all vents that are active but uninhabited. A vent inhabited by species i will also have every species of lower rank. A vent with species 1 through i is in state s_i.

Incorporating this into the null model gives us the facilitation model

$$\frac{ds_x}{dt} = \sum_{j=0}^{m} s_j - rs_x \tag{15a}$$

$$\frac{ds_0}{dt} = rs_x - \left(c\sum_{j=1}^{m} s_j\right) s_0 - s_0 \tag{15b}$$

$$\frac{ds_i}{dt} = \left(c\sum_{j=i}^{m} s_j\right) s_{i-1} - \left(c\sum_{j=i+1}^{m} s_j\right) s_i - s_i, \quad \text{for } 1 \leq i \leq m-1, \tag{15c}$$

$$\frac{ds_m}{dt} = cs_m s_{m-1} - s_m. \tag{15d}$$

Here again, we have rescaled the variables as in (5), so that time is measured in average vent active periods. A diagram of all the states and transitions is shown in Figure 9-9.

The facilitation model (15) has the same geological equilibrium as the null model (c.f., Equation 8), but now there are m biogeological equilibria. For each j, $1 \leq j \leq m$, there is an equilibrium at

$$\hat{s}_x = 1 - f, \tag{16a}$$

$$\hat{s}_0 = 1/c, \tag{16b}$$

$$\hat{s}_i = \begin{cases} 1/c, & \text{for } 1 \leq i \leq j-1, \\ f - (j/c), & \text{for } i = j, \\ 0, & \text{for } i > j. \end{cases} \tag{16c}$$

For a given set of parameters, some of these equilibria have negative values. None of these is relevant. Among the rest, only the equilibrium with the largest number of species is stable. This number is determined by the values of f and c. To have $\hat{s}_k > 0$, we must have $k < fc$.

From the equilibrium values we can calculate ϕ_i, the fraction of active vents that are inhabited by species i:

$$\phi_i = \frac{1}{f}\sum_{j=i}^{m} \hat{s}_j = \begin{cases} 1-[i/(fc)], & \text{for } i \leq fc, \\ 0, & \text{for } i > fc. \end{cases} \quad (17)$$

The species frequency decreases as species rank in the hierarchy increases. This contrasts with the species distribution under the null model, which is even.

The expected number of species inhabiting a single active vent is

$$\alpha = \frac{1}{f}\sum_{i=0}^{m} i\hat{s}_i. \quad (18)$$

Contours of α as a function of f and c are shown in Figure 9-10A. As in the null model, local diversity increases with both these parameters. In Figure 9-10B, we have plotted local diversity in the facilitation model relative to its value in the null model. For small values of f and c, facilitation dramatically reduces local diversity. The reduction is less when f and c are both large. Figure 9-10B suggests that it is when suitable habitat is "sparse" (i.e. most vents are dormant [low f] and they are difficult to colonize [low c]) that we should expect facilitative interactions to have a larger effect on local diversity.

For the facilitation model, the regional species richness α_{reg} is the largest integer not exceeding either fc or m. Thus, in contrast to the null model, there is an upper limit to the number of species that the vent system can support in addition to the species pool size, and this number grows in proportion to both f and c. To calculate H, our measure of regional diversity, we use

$$p_i = \frac{\phi_i}{\sum_{i=1}^{m} \phi_i} \quad (19)$$

in formula (13). Patterns in regional diversity are similar to local diversity (Fig. 9-11). Again, it is when suitable habitat is sparse that facilitation has its largest effect.

VI. SUMMARY

Hydrothermal vent communities are apt examples of metacommunities: a group of interacting species inhabiting a set of habitat patches that are connected via

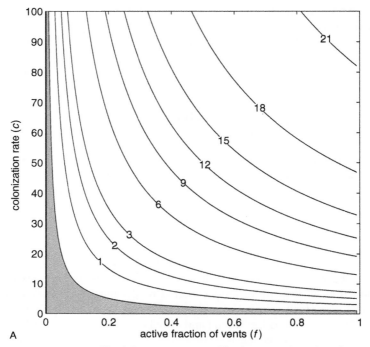

FIGURE 9-10A. Contours of local diversity, as measured by the expected number of species in any patch (α), when the species pool contains $m = 25$ species and community dynamics are governed by the facilitation model (15).

dispersal and subject to local extinction. Every year we learn more about the geology and chemistry that control the location, amount, and quality of this unique habitat. At the same time, we are compiling more information about the abundance and distribution of the species that are endemic to vent systems. As it stands, however, there are few well-documented geographic patterns in the biodiversity of vent species. Furthermore, too little is known about their life histories to justify very detailed models. In this chapter we have analyzed two simple metacommunity models to try to determine whether the hypotheses that have been advanced to describe the few patterns that have been documented are consonant with mathematical reasoning.

As it turns out, the metapopulation models we studied support one shared intuition of vent researchers about the effects of vent dynamics: all else being equal, diversity is elevated where suitable vent habitat is plentiful. As pointed out earlier, researchers have also propounded a variety of causes for the observed

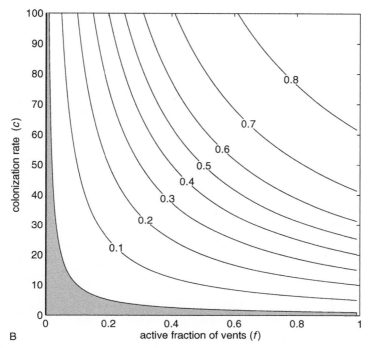

FIGURE 9-10B. Local diversity as measured relative to the null model. In the grey areas, all species are extinct.

differences in faunal diversity between vent systems, and between vent habitats and other benthic habitats. Diversity differences have been attributed to, for example, differences in the extent of barriers to dispersal (and hence to colonization), differences in the spacing of suitable habitat patches, and differences in rates of habitat disturbance. When translated into a mathematical model like the ones we have described in this chapter, these differences would appear as differences in the parameters C, R, and D. An important result of our analysis is that it is the ratios C/D and R/D that are the relevant quantities for determining diversity statistics, rather than the rates C, R, and D themselves. As a result, what may at first appear to be alternative explanations for diversity patterns may in fact be different aspects of the same explanation.

Our analysis also raises the possibility that species interactions, particularly obligate facilitative interactions, become important when habitat is sparse and difficult to colonize. These interactions may result in diversity patterns that cannot be predicted using only considerations of vent dynamics or species' colonization abilities. We are only beginning to understand how vent species

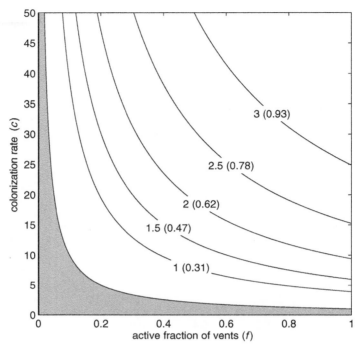

FIGURE 9-11. Contours of regional diversity, as measured by H, when the species pool contains $m = 25$ species and community dynamics are governed by the facilitation model (15). Numbers in parentheses give the ratio of H to its value under the null model. In the grey areas, all species are extinct.

actually interact. Further experiments to establish the nature of these interactions are crucial if we ever hope to determine whether these interactions actually do influence biodiversity.

There is also a need for more theoretical work. In her review of the ecology of Mid-Atlantic Ridge vents, Van Dover (1995) called for a "rigorous theoretical analysis of the consequences of [vent dynamics] . . . for comparative community structure." (277) We join her in that call, and hope that the simple null and facilitation models we have described here will provide a useful framework for further analysis. Along these lines, there are a number of elaborations of our models that we think should be addressed in future research. Among the most interesting to us are the incorporation of exclusion processes other than vent closure (e.g., competition or predation), an allowance for nonobligate facilitation, and the possibility of spatiotemporal correlation in vent dynamics. This last elaboration will require development of spatially explicit versions of our models.

REFERENCES

Berg, C.J., and Van Dover, C.L. (1987). Benthopelagic macrozooplankton communities at and near deep-sea hydrothermal vents in the eastern Pacific Ocean and the Gulf of California. *Deep-Sea Research.* **43**, 379–401.
Butterfield, D., Jonasson, I., Massoth, G., Feely, R., Roe, K., Embley, R., Holden, J., McDuff, R., Lilley, M., and Delaney, J. (1997). Seafloor eruptions and evolution of hydrothermal fluid chemistry. *Philosophical Transactions of the Royal Society of London, Series (A Math Phys Sci).* **355**, 369–386.
Caswell, H., and Cohen, J.E. (1991). Disturbance, interspecific interaction and diversity in metapopulations. *Biological Journal of the Linnean Society.* **42**, 193–218.
Chevaldonné, P., Jollivet, D., Vangriesheim, A., and Desbruyères, D. (1997). Hydrothermal-vent alvinellid polychaete dispersal in the eastern Pacific. 1. Influence of vent site distribution, bottom currents, and biological patterns. *Limnology and Oceanography.* **42**, 67–80.
Childress, J.J., and Fisher, C.R. (1992). The biology of hydrothermal vent animals: Physiology, biochemistry, and autotrophic symbioses. *Oceanography and Marine Biology Annual Review.* **30**, 61–104.
Craddock, C., Hoeh, W.R., Lutz, R.A., and Vrijenhoek, R.C. (1995). Extensive gene flow in the deep-sea hydrothermal vent mytilid *Bathymodiolus thermophilus*. *Marine Biology.* **124**, 137–146.
Craddock, C., Lutz, R.A., and Vrijenhoek, R.C. (1997). Patterns of dispersal and larval development of archaeogastropod limpets at hydrothermal vents in the eastern Pacific. *Journal of Experimental Marine Biology and Ecology.* **210**, 37–51.
Cummins, P.F., and Foreman, M.G.G. (1998). A numerical study of circulation driven by mixing over a submarine bank. *Deep-Sea Research Part I—Oceanographic Research Papers.* **45**, 745–769.
Desbruyères, D., Almeida, A., Biscoito, M., Comtet, T., Khripounoff, A., Le Bris, N., Sarradin, P.M., and Segonzac, M. (2000). A review of the distribution of hydrothermal vent communities along the northern Mid-Atlantic Ridge: Dispersal vs. environmental controls. *Hydrobiologia.* **440**, 201–216.
Desbruyères, D., Biscoito, M., Caprais, J.C., Colaco, A., Comtet, T., Crassous, P., Fouquet, Y., Khripounoff, A., LeBris, N., Olu, K., Riso, R., Sarradin, P.M., Segonzac, M., and Vangriesheim, A. (2001). Variations in deep-sea hydrothermal vent communities on the Mid-Atlantic Ridge near the Azores plateau. *Deep-Sea Research Part I—Oceanographic Research Papers.* **48**, 1325–1346.
Desbruyères, D., Chevaldonné, P., Alayse, A.M., Jollivet, D., Lallier, F.H., Jouin–Toulmond, C., Zal, F., Sarradin, P.M., Cosson, R., Caprais, J.C., Arndt, C., O'Brien, J., Guezennec, J., Hourdez, S., Riso, R., Gaill, F., Laubier, L., and Toulmond, A. (1998). Biology and ecology of the "Pompeii worm" (*Alvinella pompejana* Desbruyères and Laubier), a normal dweller of an extreme deep-sea environment: A synthesis of current knowledge and recent developments. *Deep-Sea Research Part II—Topical Studies in Oceanography.* **45**, 383–422.
Fisher, C.R., Childress, J.J., Arp, A.J., Brooks, J.M., Distel, D., Favuzzi, J.A., Macko, S.A., Newton, A., Powell, M.A., Somero, G.N., and Soto, T. (1988). Physiology, morphology, and composition of *Riftia pachyptila* at Rose Garden in 1985. *Deep-Sea Research.* **35**, 1745–1758.
Fornari, D.J., and Embley, R.W. (1995). Tectonic and volcanic controls on hydrothermal processes at the mid-ocean ridge: An overview based on near-bottom and submersible studies. In *Seafloor hydrothermal systems: physical, chemical, biological, and geological interactions*, (S.E. Humphris, R.A. Zierenberg, L.S. Mullineaux, and R.E. Thomson, eds., Geophysical Monograph Vol. 91, pp. 1–46). Washington, DC: American Geophysical Union.
France, S.C., Hessler, R.R., and Vrijenhoek, R.C. (1992). Genetic differentiation between spatially-disjunct populations of the deep-sea, hydrothermal vent-endemic amphipod *Ventiella sulfuris*. *Marine Biology.* **114**, 551–559.
Fustec, A., Desbruyères, D., and Juniper, S.K. (1987). Deep-sea hydrothermal vent communities at 13°N on the East Pacific Rise: Microdistribution and temporal variations. *Biological Oceanography.* **4**, 121–164.

Grigg, R. (1997). Benthic communities on Lo'ihi submarine volcano reflect high-disturbance environment. *Pacific Science.* **51**, 209–220.

Hanski, I. (1999). *Metapopulation ecology.* Oxford: Oxford University Press.

Hashimoto, J., Ohta, S., Gamo, T., Chiba, H., Yamaguchi, T., Tsuchida, S., Okudaira, T., Watabe, H., Yamanaka, T., and Kitazawa, M. (2001). First hydrothermal vent communities from the Indian Ocean discovered. *Zoological Science.* **18**, 717–721.

Haymon, R.M., Fornari, D.J., Edwards, M.H., Carbotte, S., Wright, W., and MacDonald, K.C. (1991). Hydrothermal vent distribution along the East Pacific Rise Crest (9°09′–54′N) and its relationship to magmatic and tectonic processes on fast-spreading mid-ocean ridges. *Earth Planet Sci Lett.* **104**, 513–534.

Haymon, R.M., Fornari, D.J., von Damm, K.L., Lilley, M.D., Perfit, M.R., Edmond, J.M., Shanks, III, W.C., Lutz, R.A., Grebmeier, J.M., Carbotte, S., Wright, D., McLaughlin, E., Smith, M., Beedle, N., and Olson, E. (1993). Volcanic eruption of the mid-ocean ridge along the East Pacific Rise crest at 9°45–52′N: Direct submersible observations of sea-floor phenomena associated with an eruption event in April, 1991. *Earth Planet Sci Lett.* **119**, 85–101.

Helfrich, K.R., and Speer, K.G. (1995). Oceanic hydrothermal circulation: Mesoscale and basin-scale flow. In *Seafloor hydrothermal systems: physical, chemical, biological, and geological interactions* (S.E. Humphris, R.A. Zierenberg, L.S. Mullineaux, and R.E. Thomson, eds., Geophysical Monograph Vol. 91, pp. 347–356). Washington, DC: American Geophysical Union.

Hessler, R.R., and Lonsdale, P.F. (1991). Biogeography of Mariana trough hydrothermal vent communities. *Deep-Sea Research.* **38**, 185–199.

Hessler, R.R., Smithey, W.M., Boudrias, M.A., Keller, C.H., Lutz, R.A., and Childress, J.J. (1988). Temporal change in megafauna at the Rose Garden hydrothermal vent (Galápagos Rift; eastern tropical Pacific). *Deep-Sea Research.* **35**, 1681–1709.

Hessler, R.R., Smithey, W.M., and Keller, C.H. (1985). Spatial and temporal variation of giant clams, tubeworms and mussels at deep-sea hydrothermal vents. *Bulletin of the Biological Society of Washington.* **6**, 465–474.

Hunt, H.L., Metaxas, A., Jennings, R.M., Halanych, K., and Mullineaux, L.S. (2004). Testing biological control of colonization by vestimentiferan tubeworms at deep-sea hydrothermal vents (East Pacific Rise, 9°50′N). *Deep-Sea Research Part I—Oceanographic Research Papers.* **51**, 225–234.

Johnson, K.S., Childress, J.J., Beehler, C.L., and Sakamoto, C.M. (1994). Biogeochemistry of hydrothermal vent mussel communities: The deep-sea analogue to the intertidal zone. *Deep-Sea Research.* **41**, 993–1011.

Jollivet, D. (1996). Specific and genetic diversity at deep-sea hydrothermal vents: An overview. *Biodiversity and Conservation.* **5**, 1619–1653.

Jollivet, D., Chevaldonné, P., and Planque, B. (1999). Hydrothermal-vent alvinellid polychaete dispersal in the eastern Pacific. 2. A metapopulation model based on habitat shifts. *Evolution.* **53**, 1128–1142.

Jollivet, D., Desbruyères, D., Bonhomme, F., and Moraga, D. (1995). Genetic differentiation of deep-sea hydrothermal vent alvinellid populations (Annelida: Polychaeta) along the East Pacific Rise. *Heredity.* **74**, 376–391.

Juniper, S.K., and Tunnicliffe, V. (1997). Crustal accretion and the hot vent ecosystem. *Philosophical Transactions Royal Society London A.* **355**, 459–474.

Juniper, S.K., Tunnicliffe, V., and Desbruyères, D. (1990). Regional-scale features of northeast Pacific, East Pacific Rise, and Gulf of Aden vent communities. In *Gorda ridge* (G.R. McMurray, ed., pp. 265–278). New York: Springer-Verlag.

Karl, D. (1995). The microbiology of deep sea hydrothermal vents. New York: CRC Press.

Kojima, S. (2002). Deep-sea chemoautosynthesis-based communities in the northwestern Pacific. *J Oceanogr.* **58**, 343–363.

Lalou, C., Reyss, J.E., Brichet, E., Arnold, M., Thompson, G., Fouquet, Y., and Rona, P.A. (1993). Geochronology of TAG and Snake Pit hydrothermal fields, Mid-Atlantic Ridge: Witness to a long and complex hydrothermal history. *Earth and Planetary Science Letters.* **97**, 113–128.

Lavelle, J.W., and Cannon, G.A. (2001). On sub-inertial oscillations trapped by the Juan de Fuca Ridge, North East Pacific. *J Geophys Res.* **106**, 31,099–31,116.

Ledwell, J.R., Montgomery, E.T., Polzin, K.L., St. Laurent, L.C., Schmitt, R.W., and Toole, J.M. (2000). Evidence for enhanced mixing over rough topography in the abyssal ocean. *Nature.* **403**, 179–182.

Levesque, C., Juniper, S.K., and Marcus, J. (2003). Food resource partitioning and competition among alvinellid polychaetes of Juan de Fuca Ridge hydrothermal vents. *Marine Ecology Progress Series.* **246**, 173–182.

Levins, R. (1970). Patch occupancy models. *Lectures on Mathematics and the Life Sciences.* **2**, 77–107.

Luther, G.W., Rozan, T.F., Taillefert, M., Nuzzio, D.B., DiMeo, C., Shank, T.M., Lutz, R.A., and Cary, S.C. (2001). Chemical speciation drives hydrothermal vent ecology. *Nature.* **410**, 813–816.

Lutz, R.A., Jablonski, D., Rhoads, D.C., and Turner, R.D. (1980). Larval dispersal of a deep-sea hydrothermal vent bivalve from the Galapagos Rift. *Marine Biology.* **57**, 127–133.

MacDonald, K.C. (1982). Mid-ocean ridges: Fine scale tectonic, volcanic and hydrothermal processes within the plate boundary zone. *A Rev Earth Planet Sciences.* **10**, 155–190.

MacDonald, K.C., Becker, K., Spiess, F.N., and Ballard, R.D. (1980). Hydrothermal heat flux of the "black smoker" vents on the East Pacific Rise. *Earth and Planetary Science Letters.* **48**, 1–7.

Marsh, A.G., Mullineaux, L.S., Young, C.M., and Manahan, D.T. (2001). Larval dispersal potential of the tubeworm *Riftia pachyptila* at deep-sea hydrothermal vents. *Nature.* **411**, 77–80.

Mercier, H., and Speer, K. (1998). Transport of bottom water in the Romanche Fracture Zone and the Chain Fracture Zone. *Journal of Physical Oceanography.* **28**, 779–790.

Micheli, F., Peterson, C.H., Mullineaux, L.S., Fisher, C., Mills, S.W., Sancho, G., Johnson, G.A., and Lenihan, H.S. (2002). Predation structures communities at deep-sea hydrothermal vents. *Ecological Monographs.* **72**, 365–382.

Mullineaux, L.S., Fisher, C.R., Peterson, C.H., and Schaeffer, S.W. (2000). Vestimentiferan tubeworm succession at hydrothermal vents: Use of biogenic cues to reduce habitat selection error? *Oecologia.* **123**, 275–284.

Mullineaux, L.S., and France, S.C. (1995). Dispersal of deep-sea hydrothermal vent fauna. In *Seafloor hydrothermal systems: Physical, chemical, biological, and geochemical interactions* (S.E. Humphris, R.A. Zierenberg, L.S. Mullineaux, R.E. Thomson, eds., pp. 408–424). Washington, DC: American Geophysical Union.

Mullineaux, L.S., Mills, S.W., Sweetman, A., Beaudreau, A.H., Hunt, H.L., Metaxas, A., and Young, C.M. (2001). Larval distributions near hydrothermal vents at 9°50′N, East Pacific Rise. Larval biology, dispersal and gene flow at hydrothermal vents: Report from the LARVE Results Symposium. Available at www.ridge2000.bio.psu.edu/wkshop_repts/LARVESympRep4.rtf.pdf.

Mullineaux, L.S., Peterson, C.H., Micheli, F., and Mills, S.W. (2003). Successional mechanism varies along a gradient in hydrothermal fluid flux at deep-sea vents. *Ecological Monographs.* **73**, 523–542.

Mullineaux, L.S., Speer, K.G., Thurnherr, A.M., Maltrud, M.E., and Vangriesheim, A. (2002). Implications of cross-axis flow for larval dispersal along mid-ocean ridges. *Cahiers de Biologie Marine.* **43**, 281–284.

Nee, S., and May, R.M. (1992). Dynamics of metapopulations: Habitat destruction and competitive coexistence. *Journal of Animal Ecology.* **61**, 37–40.

Newman, W.A. (1985). The abyssal hydrothermal vent invertebrate fauna: A glimpse of antiquity? *Bulletin of the Biological Society of Washington.* **6**, 231–242.

Rouse, G.W. (2001). A cladistic analysis of Siboglinidae Caulery, 1914 (Polychaeta, Annelida): Formerly the phyla pogonophora and vestimentifera. *Zoological Journal of the Linnean Society.* **132**, 55–80.

Sancho, G., Fisher, C.R., Mills, S., Micheli, F., Johnson, G.A., Lenihan, H.S., Peterson, C.H., and Mullineaux, L.S. (2005). Selective predation by the zoarcid fish *Thermarces cerberus* at hydrothermal vents. *Deep-Sea Research* **152**, 837–844.

Sarrazin, J., Juniper, S.K., Massoth, G., and Legendre, P. (1999). Physical and chemical factors influencing species distributions on hydrothermal sulfide edifices of the Juan de Fuca Ridge, northeast Pacific. *Marine Ecology Progress Series*. **190**, 89–112.

Shank, T.M., Fornari, D.J., von damm, K.L., Lilley, M.D., Haymon, R.M., and Lutz, R.A. (1998). Temporal and spatial patterns of biological community development at nascent deep-sea hydrothermal vents (9°50′N, East Pacific Rise). *Deep-Sea Res Pt II-Top St Oce*. **45**, 465–515.

Sibuet, M., and Olu, K. (1998). Biogeography, biodiversity and fluid dependence of deep-sea cold-seep communities at active and passive margins. *Deep-Sea Research*. **45**, 517–567.

Tsurumi, M. (2003). Diversity at hydrothermal vents. *Global Ecol Biogeogr*. **12**, 181–190.

Tsurumi, M., and Tunnicliffe, V. (2003). Tubeworm-associated communities at hydrothermal vents on the Juan de Fuca Ridge, northeast Pacific. *Deep-Sea Res*. **50**, 611–629.

Tunnicliffe, V. (1988). Biogeography and evolution of hydrothermal-vent fauna in the eastern Pacific Ocean. *Proc R Soc Lond Ser B*. **233**, 347–366.

Tunnicliffe, V. (1991). The biology of hydrothermal vents: Ecology and evolution. *Oceanogr Mar Biol A Rev*. **29**, 319–407.

Tunnicliffe, V., Embley, R.W., Holden, J.F., Butterfield, D.A., Massoth, G.J., and Juniper, S.K. (1997). Biological colonization of new hydrothermal vents following an eruption on Juan de Fuca Ridge. *Deep-Sea Res Pt I-Oceanog Res*. **44**, 1627–1644.

Tunnicliffe, V., McArthur, A., and McHugh, D. (1998). A biogeographic perspective of the deep-sea hydrothermal vent fauna. *Advances in Marine Biology*. **34**, 355–442.

Turnipseed, M., Knick, K., Lipcius, R., Dreyer, J., and Van Dover, C. (2003). Diversity in mussel beds at deep-sea hydrothermal vents and cold seeps. *Ecol Lett*. **6**, 518–523.

Tyler, P.A., and Young, C.M. (1999). Reproduction and dispersal at vents and cold seeps. *Journal of the Marine Biological Association of the United Kingdom*. **79**, 193–208.

Van Dover, C.L., Homphris, S.E., Fornari, D., Cavanaugh, C.M., Collier, R., Goffredi, S.K., Hashimoto, J., Lilley, M.D., Reysenbach, A.L., Shank, T.M., Von Damm, K.L., Banta, A., Gallant, R.M., Vrijenhoek, R.C., et al. (2001). Biogeography and ecological setting of Indian Ocean hydrothermal vents. *Science*. **294**, 818–823.

Van Dover, C.L. (1995). Ecology of Mid-Atlantic Ridge hydrothermal vents. In *Hydrothermal vents and processes* (L.M. Parson, C.L. Walker, and D.R. Dixon, eds., vol 87, pp. 257–294). Geological Society Special Publication, London: Geological Society.

Van Dover, C.L. (2000). *The ecology of deep-sea hydrothermal vents*. Princeton, NJ: Princeton University Press.

Van Dover, C.L., and Hessler, R.R. (1990). Spatial variation in faunal composition of hydrothermal vent communities on the East Pacific Rise and Galapagos spreading center. In *Gorda Ridge* (G.R. Murray, ed., pp. 253–264). New York: Springer-Verlag.

Van Dover, C.L., and Trask, J.L. (2000). Diversity at deep-sea hydrothermal vent and intertidal mussel beds. *Marine Ecology-Progress Series*. **195**, 169–178.

Van Woesik, R. (2000). Modelling processes that generate and maintain coral community diversity. *Biodiversity and Conservation*. **9**, 1219–1233.

Von Damm, K.L. (2000). Chemistry of hydrothermal vent fluids from 9°–10°N, East Pacific Rise: "Time zero," the immediate posteruptive period. *Journal of Geophysical Research Solid Earth*. **105**, 11203–11222.

Vrijenhoek, R.C. (1997). Gene flow and genetic diversity in naturally fragmented metapopulations of deep-sea hydrothermal vent animals. *Journal of Heredity*. **88**, 285–293.

Wilson, D.S. (1992). Complex interactions in metacommunities, with implications for biodiversity and higher levels of selection. *Ecology*. **73**, 1984–2000.

Won, Y., Young, C.R., Lutz, R.A., and Vrijenhoek, R.C. (2003). Dispersal barriers and isolation among deep-sea mussel populations (Mytilidae: Bathymodiolus) from eastern Pacific hydrothermal vents. *Mol Ecol*. **12**, 169–184.

PART IV

Plants and Algae

PART IV

Plants and Algae

CHAPTER 10

A Metapopulation Perspective on the Patch Dynamics of Giant Kelp in Southern California

DANIEL C. REED, BRIAN P. KINLAN, PETER T. RAIMONDI,
LIBE WASHBURN, BRIAN GAYLORD, and PATRICK T. DRAKE

I. Introduction
II. Dynamics of Giant Kelp Populations
III. Factors Affecting Colonization
   A. Life History Constraints
   B. Modes of Colonization
   C. Spore Production, Release, and Competency
   D. Postsettlement Processes
IV. Spore Dispersal
   A. Factors Affecting Colonization Distance
   B. Empirical Estimates of Spore Dispersal
   C. Modeled Estimates of Spore Dispersal
V. Connectivity Among Local Populations
VI. Summary
   References

I. INTRODUCTION

Large brown algae in the Order Laminariales are conspicuous inhabitants of shallow subtidal reefs in cool seas worldwide. This diverse group of seaweeds, known as *kelps*, consists of 27 genera that vary tremendously in size, morphology, life span, and habitat (Kain, 1979; Dayton, 1985; Estes and Steinberg, 1988). Species differ greatly even within genera, as evidenced in *Laminaria*, whose congeners include annuals and long-lived perennials inhabiting areas ranging from the tropics to the High Arctic, and from the intertidal down to depths of 70 m

(Kain, 1979). Most kelps are short in stature and extend no more than a meter or two from the bottom. They commonly occur in aggregations called *beds*, which often form a dense subsurface canopy near the seafloor. Several species, however, grow very large (as much as 45 m in length). These "giant kelps" contain gas-filled structures that allow them to produce a floating canopy that extends to the surface in water depths as great as 30 m. Beds of these giant kelps are frequently called *kelp forests*, because their vertical structure and multiple vegetation layers resemble terrestrial forests (Darwin, 1860; Foster and Schiel, 1985). Kelps are very fast growing, and kelp beds are considered to be among the most productive ecosystems in the world, comparable, for example, to tropical rain forests (Mann, 1973, 1982). The ecology of kelp stands and the diverse communities that they support have been summarized in several comprehensive reviews (e.g., North, 1971; Mann, 1982; Dayton, 1985; Schiel and Foster, 1986; Witman and Dayton, 2001).

The demographics and population dynamics of kelps are as wide ranging as the diverse morphologies and growth habits that characterize the Order. In this chapter we limit our discussions to the giant kelp *Macrocystis*—an extensively studied, ecologically important genus that is widely distributed in cool seas of the northern and southern hemispheres (Womersley, 1954). *Macrocystis* forms extensive forests off the coast of California, and a considerable amount is known about their biology and ecology in this region (see reviews by North, 1971, 1994; Foster and Schiel, 1985; and Murray and Bray, 1993). Discontinuities in hard substrate in the nearshore cause *Macrocystis* to be distributed in discrete patches of varying size that expand and contract in response to biotic and abiotic changes in the environment. The dynamic and sometimes asynchronous behavior of these patches has long been recognized, yet metapopulation theory has rarely been invoked to explain it. This is due in part to an insufficient understanding of the limits of dispersal in kelps, the conditions that promote exchange among discrete kelp patches, and the frequencies at which these conditions occur.

In this chapter we examine the patch dynamics of *Macrocystis* in southern California from a metapopulation perspective. We begin by synthesizing existing and new information pertaining to the metapopulation structure and dynamics of *Macrocystis* in the Southern California Bight. Next we review the biological and physical factors that affect colonization, and we present new empirical and theoretical estimates of spore dispersal distance for varying oceanic conditions. This information is used to estimate levels of connectivity among discrete kelp patches for different current regimes. We define connectivity as demographic exchange between patches, which in the case of giant kelp occurs primarily via the passive transport of propagules. We intentionally do not discuss mechanisms responsible for the maintenance of extant populations (e.g., short distance dispersal that results in self-replenishment). Instead, we focus on empirical and theoretical estimates of local population extinction, local population establishment (i.e., colonization), and immigration, which are considered to be "in the hearth of metapopulation ecology" (Hanski, 1999, p. 27). We emphasize studies done in southern California, the system with which we are most familiar, and we draw heavily from our own research when addressing issues pertaining to dispersal and connectivity among local populations. We conclude with a discussion of the applicability of the metapopulation concept to giant kelp, and identify future research needed to improve characterization of metapopulation dynamics in kelps and other seaweeds.

II. DYNAMICS OF GIANT KELP POPULATIONS

Local populations of *Macrocystis* fluctuate greatly in time and space in response to a complex of predictable (seasonal) and unpredictable events. Increased water motion associated with winter storms and swell is a major source of plant mortality (Dayton and Tegner, 1984; Ebeling et al., 1985; Seymour et al., 1989). The frequency and intensity of storm events vary unpredictably among years, causing erratic annual fluctuations in population size (Rosenthal et al., 1974; Foster, 1982; Dayton et al., 1992). Likewise, differences in depth and wave exposure cause rates of storm-related mortality to vary substantially among sites (Dayton et al., 1984; Graham et al., 1997; Edwards, 2004). Prolonged periods of warm, nutrient-depleted water such as those associated with El Niño southern oscillation (ENSO) events can lead to local (Zimmerman and Robertson, 1985; Reed et al., 1996) and widespread (Dayton and Tegner, 1989) kelp loss, and prevent subsequent recovery. Finally, intensive grazing (most notably by sea urchins) can eliminate entire beds (reviewed in Dayton, 1985; Foster and Schiel, 1985; and Harrold and Pearse, 1987). Conditions that promote or suppress outbreaks of sea urchin grazing are often localized in southern California, which causes asynchrony in the dynamics of local populations in this region (Ebeling et al., 1985; Harrold and Reed, 1985; Reed et al., 2000).

Population growth in *Macrocystis* is solely dependent on sexual reproduction; fragmentation, clonal growth, and other forms of vegetative reproduction do not occur. Recruitment of new plants occurs when favorable conditions of light, nutrients, and primary space coincide with periods of abundant spore supply. These factors are most likely to co-occur in the winter and spring, and depend on both chance events and the local density of adult plants (Deysher and Dean, 1986; Reed, 1990; Graham, 2000). Like many terrestrial forests, giant kelp forests have a complex vertical structure composed of several canopy layers. Competition for sunlight and space among canopy members plays an important role in regulating the recruitment of new individuals (Pearse and Hines, 1979; Dayton et al., 1984; Reed and Foster, 1984; Reed et al., 1997). Adult *Macrocystis* are the dominant competitors for light (Dayton et al., 1999) and provide the nearest source of spores for recruitment. Disturbances that alter the abundance of adult plants can interact with density-dependent processes to produce kelp forests with qualitatively different dynamics and size structures (Dean and et al., 1989; Nisbet and Bence, 1989; Burgman and Gerard, 1990; Tegner et al., 1997).

Discrete stands of giant kelp go extinct and reappear at irregular intervals. Large-scale phenomena such as El Niño events occur unpredictably in time and can produce widespread kelp loss. For example, large waves and adverse growing conditions associated with the strong El Niños of 1982 to 1983 and 1997 to 1998 eliminated *Macrocystis* from most areas of southern California and Baja California (Dayton and Tegner, 1989; Edwards, 2004). The additive effects of smaller scale but equally intense disturbances, such as those resulting from intensive sea urchin and amphipod grazing can cause giant kelp to display much higher rates of local extinction and recolonization at some sites (Ebeling et al., 1985; Tegner and Dayton, 1987).

Predictions from theoretical models also suggest that local populations of *Macrocystis* have a relatively high probability of extinction. Burgman and Gerard (1990) examined persistence in *Macrocystis* using a stage-structured population model that incorporated environmental and demographic stochasticity. Their model predicted a 60% chance that the adult density of a local population will fall to zero during a 20-year period; the occurrence of an El Niño event increased the likelihood of extinction probability to 80%.

To assess the regional-scale generality of the previously mentioned field observations and model predictions, we estimated rates of patch extinction and colonization from long-term aerial observations of giant kelp forests in southern California. Since 1958, ISP Alginates, Inc., a San Diego–based kelp harvesting company, has conducted aerial surveys of *Macrocystis* beds in southern California. During these surveys, observers use canopy area and density to gauge the biomass of kelp harvestable from the surface (0–1 m depth); visual estimates of biomass are then calibrated to actual harvested amounts. After a 10-year period of ground-truthing, aerial biomass estimation methods were standardized in January 1968. Subsequent surveys were carried out, on an approximately monthly basis, by one of two trained observers (D. Glantz, personal communication). Survey data were interpolated onto a regular monthly grid (B. Kinlan, unpublished data). Here, we use surveys conducted between January 1968 and October 2002 (418 months) from an approximately 500-km stretch of coast between Pt. Arguello and the U.S.–Mexico border (Fig. 10-1).

These surveys provide a long-term record of the presence or absence of giant kelp canopy in administrative kelp beds defined by the California Department of Fish and Game (approximately 2–20 km in along-coast extent; Fig. 10-1), with sufficient temporal resolution to identify administrative bedwide extinction and recolonization events. However, greater spatial resolution is needed to identify discrete patches of habitat that can potentially be colonized by giant kelp (i.e.,

Chapter 10 A Metapopulation Perspective on the Patch Dynamics of Giant Kelp

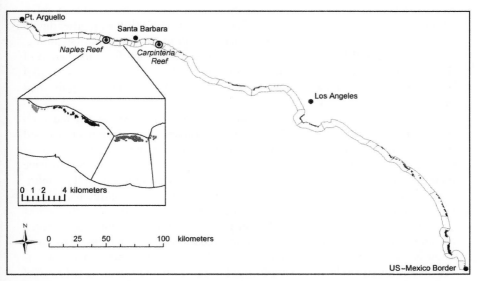

FIGURE 10-1. Map of the mainland coast of southern California, depicting giant kelp canopy detected by aerial infrared photography in 1989, 1999, and 2002 (black shading) and administrative kelp bed units (outlines) assigned by the California Department of Fish and Game. (Inset) Detail of three discrete patches of giant kelp (different shades of gray) identified according to the criteria described in Section II. These patches are separated by more than 500 m at their closest point. Note that a single administrative bed may contain multiple patches, and patches can extend across administrative bed boundaries. Patches are considered to "belong" to the bed in which the majority of surface canopy falls. (Composite map and patch classification by B. Kinlan, using unpublished data from the California Department of Fish and Game.)

firm substrates at appropriate depth; hereafter referred to simply as *patches*) as required for an analysis of metapopulation dynamics. To identify discrete patches of giant kelp habitat, we used digital maps of giant kelp canopy occurrence derived from aerial infrared imagery (~3–5 m resolution) of the California coast from Pt. Arguello to the U.S.-Mexico border for the years 1989, 1999 and 2002 (produced by California Dept. of Fish and Game). Our collective analyses of these images and the 40 + year data set collected by ISP Alginates showed that at least two of these surveys (1989, 2002) captured giant kelp canopies at their annual peak, and all three surveys were conducted in years when giant kelp biomass was near its 20-year high across the region (B. Kinlan, unpublished data). We therefore used the combined giant kelp canopy area identified by these three maps to approximate the distribution of giant kelp patches in each of the California Department of Fish and Game administrative beds (Fig. 10-1). Patches were defined as discrete areas in which the composite *Macrocystis* canopy (i.e., that estimated by overlaying images of the three aerial infrared surveys) was either contiguous or separated by gaps of less than 500 m. Comparison with other digital giant kelp canopy maps available for smaller portions of this region (spanning, in some cases, >30 years and 200 km of coast) suggests that the combination of the chosen three region-wide surveys done in 1989, 1999, and 2002 captured more than 95% of habitat patches (North et al., 1993; B. Kinlan, unpublished data).

To estimate rates of patch extinction and colonization, we considered each patch within a given California Department of Fish and Game administrative bed to be "occupied" for any month in which ISP Alginates noted the presence of surface canopy kelp in that administrative bed. We considered patches to have gone "extinct" when no surface canopy was detected within the area of the administrative bed for 6 or more consecutive months. Under most conditions, subsurface juvenile plants would grow to form a surface canopy in 6 months or less (Foster and Schiel, 1985). Note that all patches within the confines of a given administrative kelp bed are considered occupied whenever the biomass estimate exceeds zero and, conversely, all are considered extinct when no canopy is detected. Consequently, this method could over- or underestimate actual rates of extinction and colonization, depending on the degree to which patches within an administrative kelp bed fluctuate synchronously. However, because each administrative kelp bed contains only a small number of discrete patches (mean, 2.8; interquartile range, 1–4) and patch fluctuations are positively autocorrelated at small spatial scales (i.e., 1–10 km; B. Kinlan, unpublished data), the resulting biases in extinction and colonization rates should be relatively small.

At a regional scale, occupancy of the giant kelp habitat mosaic is extremely dynamic (Fig. 10-2). During the 34-year study period, the estimated fraction of patches occupied in southern California approached 100% in some months, but dipped to approximately 0% after a major El Niño event (1982–1984). In fact, for much of the time from 1982 to 1984, no surface canopy was detected in the aerial biomass surveys. We know from diver surveys that some scattered subsurface juveniles were present during this period (e.g., Dayton and Tegner, 1984; Dayton et al., 1999), but clearly they did not grow to form significant surface canopies within our 6-month "window" for defining extinction. This highlights the fact that, under very stressful environmental conditions, Macrocystis may experience suppressed recruitment and growth of juvenile stages that delays the formation of a surface canopy (Dean and Jacobsen, 1984, 1986; Kinlan et al., 2003). Delayed growth of juveniles could, in certain cases, lead to recovery of local populations even in the absence of nearby spore sources (Ladah et al., 1999), in a manner similar to the "propagule rain" effect described by Gotelli (1991).

Extinction probabilities, defined here as the monthly probability of a patch going from occupied to extinct, ranged from 0.005 to 0.292 (mean ± SD, 0.057

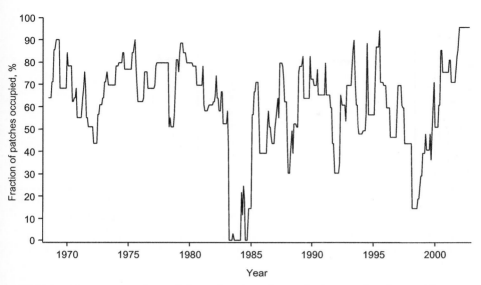

FIGURE 10-2. Percent of giant kelp patches along the mainland coast of southern California (n = 69 patches) occupied on a monthly basis from January 1968 to October 2002 as estimated from surface canopy observations made during aerial overflights.

± 0.063; Fig. 10-3A). Recolonization probabilities, defined as the monthly probability of a patch going from extinct to occupied, ranged from 0.023 to 0.200 (mean ± SD, 0.080 ± 0.040; Fig. 10-3B). These monthly rates agree with models and observations that suggest kelp forest patches are highly dynamic at the scale of months to years (e.g., the 0.8 probability of extinction in a 20-year period cited earlier corresponds to a monthly rate of just 0.0067). On average, extinction of a patch in our study region lasted from six months to four years (Fig. 10-4A), and patches remained occupied for one to five years (Fig. 10-4B). However, in certain cases extinctions lasted as little as a few months or as much as 13 years (Fig. 10-4C) and patches of kelp persisted for several months to 15 years (Fig. 10-4D).

Extinction and recolonization rates varied with patch size and patch isolation (Fig. 10-5). We used the square root of patch area as a measure of size, and the average size of surrounding patches weighted by the inverse square of distance as a measure of isolation (Thomas and Hanski, 1997). Patch isolation explained more variation in extinction (Fig. 10-5B) and recolonization (Fig. 10-5D) rates than patch size (Figs. 10-5A, C), suggesting that immigration rates are dependent on distance and source population size. The lower extinction rates in highly connected (i.e., low-isolation) patches indicate that rescue effects may play an important role in patch dynamics (Brown and Kodrik–Brown, 1977; Hanski,

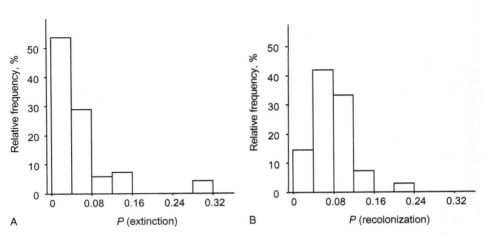

FIGURE 10-3. Monthly rates of extinction (A) and recolonization (B) of giant kelp patches along the mainland coast of southern California, based on a 34-year monthly time series of surface canopy biomass. Bin size, 0.04; ticks denote lower edges of bins.

1999). The lower colonization rates in highly isolated patches (Fig. 10-5D) indicate that immigration rates may limit recolonization of isolated patches. The statistical significance of the relatively low correlations between patch size and extinction (Fig. 10-5A) and recolonization (Fig. 10-5C) was driven primarily by the two or three largest patches. Large kelp forests may have a low chance of stochastic extinction because of their large population size. Moreover, the greater amount of suitable habitat in large kelp forests may increase the likelihood that at least some portion of the patch is recolonized. Collectively, these results confirm impressions from smaller scale studies that kelp forests are dynamic mosaics, characterized by frequent extinction and recolonization from nearby patches.

III. FACTORS AFFECTING COLONIZATION

A. LIFE HISTORY CONSTRAINTS

Basic knowledge of kelp life history is important for understanding the dynamics of local populations and the degree of connectivity among them. A characteristic feature of all kelps is that they undergo an alternation of generations between a macroscopic diploid sporophyte (a spore-producing plant) and a microscopic haploid gametophyte (a gamete-producing plant; Fritsch, 1945). Meiosis occurs in the adult sporophyte to produce male and female zoospores

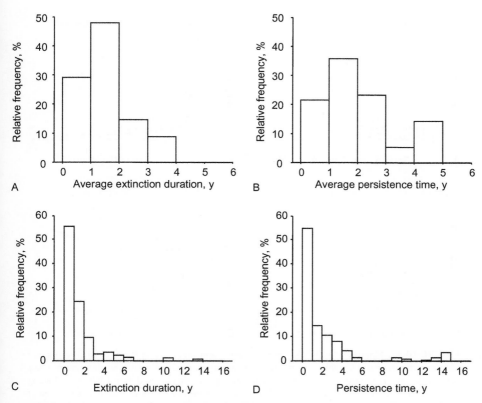

FIGURE 10-4. Duration of patch extinctions and patch persistence of giant kelp along the mainland coast of southern California based on a 34-year monthly time series of surface canopy biomass. Upper panels (A, B) show histograms of the average extinction and persistence time for 69 discrete patches. Lower panels (C, D) show the durations of all extinction and persistence intervals observed during the 34-year study, revealing extremely long and extremely short extinction and persistence intervals not reflected by the averages in (A) and (B).

that are the primary dispersive stage. After a relatively short dispersal period (i.e., hours to days; Reed et al., 1992; Gaylord et al., 2002) zoospores (hereafter referred to as *spores*) settle to the bottom and germinate into sessile, free-living, microscopic gametophytes. In contrast to most marine organisms, fertilization in kelps occurs after dispersal, when a pheromone released by the female gametophyte triggers the liberation of sperm from the male gametophyte and guides the sperm to the nonmotile egg (Müller, 1981). The distance over which the pheromone is effective in attracting sperm is believed to be less than 1 mm (Boland et al., 1983). Consequently, recruitment into the sporophyte generation

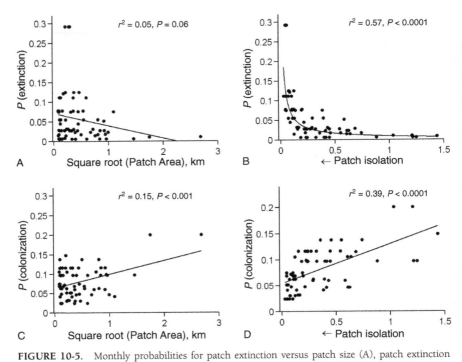

FIGURE 10-5. Monthly probabilities for patch extinction versus patch size (A), patch extinction versus patch isolation (B), patch recolonization versus patch size (C), and patch recolonization versus patch isolation (D). Data are from 69 discrete patches of giant kelp observed from 1968 to 2002 along the mainland coast of southern California. Patch size is defined as $L_i = \sqrt{A_i}$, where A_i is the area of patch i in square kilometers. Patch isolation is defined as the inverse distance-weighted average size of surrounding patches that are occupied in any given month, averaged over all $T = 418$ months, $I_j = \frac{1}{T} \cdot \sum_{t=1}^{T} \left[\left(\sum_{i \neq j} L_i \cdot \frac{1}{D_{i,j}^2} \right) \Big/ \left(\sum_{i \neq j} \frac{1}{D_{i,j}^2} \right) \right]$, where $D_{i,j}$ denotes the linear distance between occupied patches i and j. Note that low values of the patch isolation index correspond to a high degree of isolation (indicated by arrow). Lines show fitted least-squares regressions:

(A) $P(\text{extinction}) = -0.0316\sqrt{(\text{patch area})} + 0.0708$.
(B) $\ln[P(\text{extinction})] = -0.8643 \ln(\text{isolation}) - 4.7217$.
(C) $P(\text{colonization}) = 0.0349\sqrt{(\text{patch area})} + 0.0644$.
(D) $P(\text{colonization}) = 0.0760(\text{isolation}) + 0.0544$.

r^2 and P values for regressions are as shown.

is largely confined to areas of relatively dense spore settlement (e.g., >1 spore per square millimeter), where the probability of encounter between male and female gametes is sufficiently high to ensure fertilization (Reed, 1990; Reed et al., 1991). Thus, a major constraint limiting the distances over which kelps are able to colonize is the dilution of spores that accompanies their dispersal. This constraint on colonization distance decreases with increasing size of the source population (Anderson and North, 1966; Reed et al., 1997). This is because an increase in the concentration of spores at the point of release generally results in a proportional increase in the concentration of spores at a given distance from that point.

Some variants of this life history involving the production of sporophytes without fertilization (i.e., apomixis) have been described from laboratory cultures for several species of kelp, including *Macrocystis* (reviewed by Lewis, 1996). However, the development of kelp sporophytes from unfused gametes (i.e., parthenogenesis) or from gametophytes that do not produce gametes (i.e., apogamy) typically results in abnormalities, and there is little evidence that kelp sporophytes produced via apomixis are common in nature.

B. Modes of Colonization

Local emigration and immigration of *Macrocystis* occurs in one of two ways: via dispersal of microscopic spores or via the transport of large sporophytes that become dislodged and set adrift (hereafter referred to as *drifters*). The tiny biflagellated spore (approximately 6 μm in diameter) is the only motile stage in an otherwise sedentary life form (kelp sperm are also motile, but are believed to disperse no farther than a few millimeters; Müller, 1981; Boland et al., 1983). Swimming speeds of kelp spores are relatively slow (approximately 180 μm/second; C. Amsler, personal communication); however, and the distances that spores disperse are determined largely by advective currents and vertical mixing (Gaylord et al., 2002). Factors that influence these physical processes may play an important role in determining the extent of spore dispersal (Gaylord et al., 2004). Kelp itself may be particularly important in this regard because drag from the fronds acts to slow down currents that pass through the forest (Jackson and Winant, 1983; Jackson, 1998). Consequently, spores released near the center of the forest are more likely to be retained than spores released closer to the downstream edge (Graham, 2003).

Emigration via drifters occurs when wave forces rip whole plants off the bottom, and ocean currents export them out of the forest. This most frequently occurs during storms, which can remove entire populations of *Macrocystis* and transport them en masse (Rosenthal et al., 1974; Dayton and Tegner, 1984; Ebeling et al., 1985; Seymour et al., 1989; Edwards, 2004). The percentage of drifters that successfully immigrate to new reefs and establish residency has been poorly documented; however, it is likely to be quite low. Most plants set adrift during winter storms appear to end up on the beach soon after becoming detached (ZoBell, 1971; Harrold and Lisin, 1989), or are transported offshore (Kingsford, 1995; Hobday, 2000).

Successful immigration involving drifters is not necessarily contingent upon them taking up residence at a new site. Drifters have the capacity to produce and release spores during transport, and thus they have the potential to influence the colonization of neighboring reefs by providing a localized source of spores. However, it is important to recognize that a drifting plant or plant fragment constitutes a relatively small spore source, and constraints on colonization distance resulting from dilution effects are expected to be high, especially in the case when drifters are transported high in the water column. Such constraints argue that any colonization originating from spores released by drifters would occur in isolated patches only along the drift trajectories. Such localized recruitment contrasts greatly with the widespread, relatively uniform recruitment of *Macrocystis* that is typically observed after large disturbance events (Dayton and Tegner, 1984; Dayton et al., 1992; Edwards, 2004), even at sites located relatively far from the nearest source of spores (e.g., Ebeling et al., 1985; Reed et al., 2004). For drifters to account for such widespread phenomena would seemingly require two coincident factors: environmental conditions suitable for kelp recruitment, and either a constant supply or an adequate residence time of a large number of fecund drifters distributed over a substantial area. As far as we know, such conditions have never been reported. Thus, although spore dispersal from drifters may play a valuable role in occasional long-distance dispersal events that are important for biogeographical expansion and genetic exchange, they do not appear to play a major role in rapid recolonization events, which typify the patch dynamics of giant kelp in California.

The recovery of giant kelp populations destroyed by a disturbance need not be dependent on immigration if sufficient numbers of benthic microscopic stages survive the disturbance (Dayton, 1985; Kinlan et al., 2003). It has been suggested that banks of microscopic forms may function as a survival mechanism for benthic macroalgae in ways that are analogous to seed banks of terrestrial plants (Chapman, 1986; Hoffmann and Santelices, 1991). Under these circumstances, dispersal from another patch is not necessary for explaining recolonization events because local populations have the potential to be self-replenishing, even in the event of a prolonged absence of reproductive adults. Such may have been the case for *Macrocystis* near its southern limit in Baja California, after its widespread disappearance during the 1997–1998 El Niño. *Macrocystis* recolonized depopulated sites where the nearest known source of spores was more than 100 km away (Ladah et al., 1999). Recruitment occurred at least 6 months after all adult *Macrocystis* had succumbed to poor growing conditions. The source for these recruits was assumed to be benthic microscopic stages that persisted through the adverse

El Niño conditions. Such prolonged survival of microscopic kelp stages does not appear to be common in other parts of *Macrocystis'* range. Several experimental studies done in southern California indicated that microscopic stages of *Macrocystis* have a relatively short life span (typically less than a couple of weeks), and that the vast majority of recruitment originates from recently settled spores (Deysher and Dean, 1986; Reed et al., 1988; Reed, 1990; Reed et al., 1994; Reed et al., 1997). Moreover, dense recruitment of *Macrocystis* to newly constructed artificial reefs located several kilometers from the nearest spore source (Davis et al., 1982; Reed et al., 2004) provides conclusive evidence that spore immigration (whether released from attached plants on a neighboring reef, or from immigrant drifters) is a feasible means of colonization, and that the metapopulation concept is appropriate for explaining the dynamics of discrete giant kelp beds in southern California.

C. Spore Production, Release, and Competency

Macrocystis is not only one of the world's fastest growing autotrophs, but it is also one of the most fecund. Spores are produced in blades termed *sporophylls* that are located near the base of the plant. Each sporophyll may contain as many as 10 billion spores, and any given plant may produce a crop of 100 or more sporophylls at least twice per year (Reed et al., 1996; Graham, 2002). Plants generally begin producing spores during their first year after attaining a size of four to eight fronds and a wet somatic mass of 8 to 10 kg (Neushul, 1963). Spores appear to be released continuously throughout the year, with peaks occurring in early winter and late spring/early summer (Anderson and North, 1967; Reed et al., 1996). Aperiodic events such as increased water motion associated with storms may accelerate the rate of spore release. Reed et al. (1997) observed a 50% decrease in sorus area of plants immediately after a large storm. Half this decrease resulted from a reduction in sorus length resulting from sporophyll erosion, which accompanies spore release (D.C. Reed, unpublished data). Unlike the bull kelp *Nereocystis luetkeana*, which displays strong diel periodicity in spore release (Amsler and Neushul, 1989b), *Macrocystis* shows little within-day variation in rates of spore liberation (Graham, 2003). Consequently, the timing of tidal or wind-driven changes in flow that occur on a daily basis is likely to be of little importance in determining the transport of spores.

Actively swimming spores of giant kelp are typically released into the plankton within 1 m of the bottom (Gaylord et al., 2002). They can remain swimming for several days, but most stop within 24 hours, regardless of whether they reach the bottom (Reed et al., 1992). There is some evidence that *Macrocystis* spores have a short (i.e., several hours) precompetency period during which time germination is impaired, should settlement occur prematurely (Reed and Lewis,

1994). Such phenomena could serve to promote outcrossing and reduce the adverse effects of inbreeding depression, which in *Macrocystis* are quite severe (Raimondi et al., 2004).

While in the plankton, spores are able to maintain net-positive photosynthetic rates under light conditions that are typical of the subtidal environment in which they are found (Amsler and Neushul, 1991). Photosynthesis, however, is not essential for spore motility. Like many marine larvae, kelp spores contain large internal lipid reserves that serve to fuel swimming and germination (Reed et al., 1992, 1999; Brzezinski et al., 1993). Energy derived from photosynthesis allows spores to conserve their internal lipid reserves and swim for a longer period of time (Reed et al., 1992). Spores that stop swimming before contacting a surface do not immediately die, but germinate in the water column and continue to grow and develop. Germinating in the water column, however, is not without costs. Although spore motility has little effect on the distance over which kelp spores are dispersed (Gaylord et al., 2002), it may greatly enhance a spore's ability to find a high-quality microsite in which to settle. *Macrocystis* spores (and those of the palm kelp *Pterygophora californica*) exhibit chemotaxis toward nutrients that not only stimulate settlement, but also promote growth and development after germination (Amsler and Neushul, 1989a, 1990). Hence, factors that prolong the swimming stage of spore development may increase the chances of successful settlement and recruitment, because a motile spore is better able to a select a favorable microsite than a nonmotile, planktonic germling. Perhaps more important is the increased dilution of propagules that accompanies a planktonic germling that remains in the water column for extended periods, reducing its chance of finding a mate and successfully reproducing.

D. Postsettlement Processes

The production of sporophytes from gametophytes requires a complex set of biotic and abiotic conditions. Light, nutrients, temperature, and sediments need to be within critical threshold levels for gametophytes to grow and reproduce (Devinny and Volse, 1978; Lüning and Neushul, 1978; Deysher and Dean, 1984, 1986; Kinlan et al., 2003). The co-occurrence of these factors in southern California is unpredictable in space and time because of variation in oceanographic conditions (Deysher and Dean, 1986). Biological processes can also alter levels of light and nutrients to influence patterns of sporophyte recruitment. For example, unlike the large *Macrocystis* sporophyte that monopolizes light, microscopic gametophytes are poor competitors for light and space, and readily succumb to larger and/or faster growing algae (Reed and Foster, 1984; Reed 1990). Not surprisingly, recruitment of *Macrocystis* sporophytes is typically greatest on surfaces lacking other biota (Reed and Foster, 1984; Ebeling et al., 1985;

Reed et al., 1997). Sporophyte production is also influenced by strong intra- and interspecific competition of microscopic stages (Reed, 1990; Reed et al., 1991). The need for spores to settle at high densities to ensure fertilization coupled with the drastic difference in size between the gametophyte and sporophyte phases essentially guarantee that strong density-dependent mortality will occur during the production of sporophytes. Additional mortality to early life stages of kelp results from grazing invertebrates and fishes, which can scour the bottom and cause patchiness in sporophyte recruitment over a range of spatial scales (Dean et al., 1984; Harris et al., 1984; Leonard, 1994).

IV. SPORE DISPERSAL

A. FACTORS AFFECTING COLONIZATION DISTANCE

The distance that a spore is dispersed is determined by the length of time it spends in the plankton and the speed, direction, and timescales of variability of the currents that transport it while it is suspended. Suspension times are influenced by the height above the bottom that a spore is released, and its net sinking rate. Turbulence produced from waves and currents, wind-driven surface mixing, water stratification, shoreline bathymetry, and bottom roughness all interact to influence net sinking rates of spores (Gaylord et al., 2004). In the case of small particles like kelp spores, which are nearly neutrally buoyant, turbulence acts to increase sinking rates (McNair et al., 1997). Even very slight turbulence can drastically reduce the average time it takes a spore to contact the bottom. For example, the mean suspension time for a *Macrocystis* spore released 42 cm off the bottom in still water is approximately 97 hours, but is expected to be only about 9 hours under conditions of a 2-cm per second current and 0.5-m waves, at least in regions outside kelp forests above relatively smooth sand flats (Gaylord et al., 2002). Within-forest processes will attenuate current speeds and likely reduce rates of vertical mixing, extending spore suspension times to a certain extent (reviewed in Gaylord et al., 2004). The ultimate effect of a forest on overall transport distance is less clear, however, because the degree to which the counteracting effects of slower currents and reduced mixing offset one another has not been examined in any detail. Note also that the dispersal of a spore does not necessarily end upon first contact with the bottom. Turbulent shear near the seabed may resuspend spores after their initial contact and allow them to bounce along the seafloor. Such saltation of spores may occur in *Macrocystis* because spore attachment appears to be greatly reduced in even moderate flows (approximately 15 cm/second; Gaylord et al., 2002).

Spore dispersal distance is not the sole determinant of colonization distance in kelps. As mentioned earlier (see Section III.A), colonization distance also depends on the size of the spore source, due to dilution effects that accompany spore dispersal. Source size is determined by the density and fecundity of the parental population, and the degree to which adults within a population release their spores in concert (Reed et al., 1997). The synchronous release of spores during conditions that promote advection may extend the distance of colonization beyond that expected in the absence of reproductive synchrony. In the case of *Macrocystis*, increased water motion during storms may trigger large pulses of spore release and promote greater dispersal (Reed et al., 1988, 1997). The importance of storms in promoting colonization extends beyond increased dispersal, because storms also create bare space, which increases the likelihood of successful colonization (Dayton and Tegner, 1984; Ebeling et al., 1985; Reed et al., 1997). Such episodes of storm-enhanced spore dispersal and colonization may also play an important role in the dynamics of local populations if spores that arrive during storm events contribute disproportionately more to recruitment than locally produced spores released during calm conditions. This is possible in kelps, because conditions favorable for recruitment generally follow storms, which reduce competition for light and space (Cowen et al., 1982; Dayton and Tegner, 1984; Dayton et al., 1992), and promote enhanced spore settlement (Reed et al., 1988, 1997).

B. Empirical Estimates of Spore Dispersal

We know of only three published accounts of dispersal in *Macrocystis*: two in which dispersal distances were inferred from observations of the density of young sporophytes at varying distances from different sized groups of adults (Anderson and North, 1966; Reed et al., 2004), and one in which dispersal distance was estimated using the density of newly settled gametophytes at 0, 3, and 10 m from isolated adults (Reed et al., 1988). Although these studies have helped to broaden our understanding of dispersal in *Macrocystis*, limitations in their temporal and spatial resolution render them inadequate for determining the extent to which dispersal varies in time and space. Such information is needed to determine levels of connectivity among local kelp populations.

We collected data on water motion simultaneously with data on spore dispersal using two different experimental designs to obtain a more comprehensive understanding of spore dispersal in *Macrocystis* and the processes that affect it. One study involved estimating dispersal from individual adult sporophytes, whereas the other entailed estimating dispersal from an experimental population of adult sporophytes. Both experiments were done on a nearly flat area of sandy bottom at 10 m depth, near Carpinteria, California, that was at least 1 km away from the nearest *Macrocystis* patch. Having an isolated spore source is key to obtaining empirical estimates of spore dispersal because it allows one to investigate dispersal distances from the nearest known source of spores without interference from neighboring spore sources. We estimated dispersal from individual adults by transplanting three mature sporophytes 50 m apart from each other in a line perpendicular to shore. On 60 dates between January 1998 and April 1999 we recorded the densities of recently settled spores (i.e., gametophytes) on arrays of ground-glass microscope slides placed north, south, east, and west of each of the three sporophytes. Slides were positioned approximately 15 cm above the bottom on a PVC post anchored in the sand at distances of 0.5, 1, 5, and 10 m from each sporophyte. Additional slides were placed 50 m east and west of the inshore and offshore sporophytes in the array to detect any dispersal over longer distances. In the second study, the experimental population of *Macrocystis* used to examine dispersal from a larger group of fertile individuals was created by transplanting 64 adult sporophytes in a uniform array (i.e., spaced 3 m on center) to a 25 × 25-m area. Spore settlement in this experiment (hereafter referred to as the *kelp bed experiment*) was recorded on microscope slides positioned 3, 6, 12, 18, 24, 48, 72, 96, and 120 m north, south, east, and west of the edge of the sporophyte array on 29 dates between June 1999 and September 1999. Slides used to collect newly settled spores in both experiments were placed in the field for 2 to 3 days, collected, transported to nearby laboratory facilities and sampled for spore settlement as described in Reed et al. (1988).

When averaged over all trials, spore settlement decreased with distance as a negative power function in both the individual sporophyte and kelp bed experiments (Fig. 10-6). These general patterns of spore dispersal are similar to those described for *Macrocystis* in earlier studies (see references cited earlier). To explore spatial and temporal patterns of spore dispersal, we used nonlinear regression analysis to estimate dispersal as a function of distance for each trial in each of the two experiments. The equations produced from these regressions were used to calculate the x-intercept, which represents an estimate of the maximum distance that spores dispersed in a given trial. These maximum values were then used to produce frequency distributions for each experiment, showing the percentage of trials in which spore dispersal extended out to different distances (Fig. 10-7). Results indicate there was a greater range in dispersal among trials in the experiment involving individual plants than in the kelp bed experiment. Dispersal in 70% of trials involving individual plants did not exceed 16 m, whereas in 5% of trials it was estimated to be more than 2000 m. By contrast, maximum dispersal in nearly all trials from the kelp bed experiment ranged from 80 to 500 m. This difference may have been due in part to differences in the level of spatial resolution in estimates of dispersal distance, particularly in the case of the individual plant experiment, in which we projected dispersal far beyond our maximum sampling distance (50 m). Perhaps more important, trials for the individual plant experiment were done over a 15-month period that encompassed a

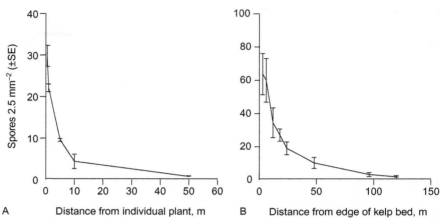

FIGURE 10-6. Spore settlement density as a function of distance from the spore source for the individual plant experiment (A) and the experimental kelp bed (B). Data are means (±1 SE) of spores averaged across all trials. Note the difference in the scale of the vertical axes of (A) and (B).

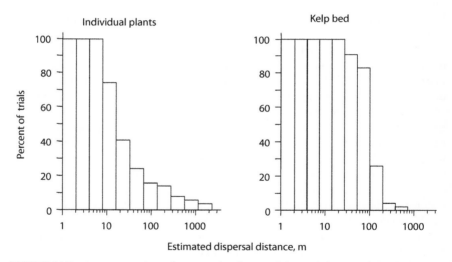

FIGURE 10-7. Inverse cumulative frequency distribution of dispersal distances for both the individual plant and kelp bed experiments. Shown are the probabilities that the maximum estimated dispersal in a trial occurred to distance (X) or more.

TABLE 10-1. Depth-averaged currents and significant wave heights for the individual plant and kelp bed experiments

Value	Currents, cm/second		Significant wave height, m	
	Individual	Kelp bed	Individual	Kelp bed
Minimum	0.095	0.067	0.125	0.160
Maximum	36.984	25.243	1.384	0.717
Mean	2.152	2.258	0.484	0.372

Mean current velocities were calculated over the duration of each experiment; minimum and maximum speeds are from hourly averages.

wider range of oceanographic conditions than the 3-month kelp bed experiment that was done during a calm summer period (Table 10-1). The two variables expected to have the most influence on spore dispersal are currents and waves (Gaylord et al., 2002), both of which were lower in the kelp bed experiment compared to the individual plant experiment.

C. MODELED ESTIMATES OF SPORE DISPERSAL

Gaylord et al. (2002) developed a physically based model for dispersal of macroalgal spores, with specific reference to *Macrocystis*. This model linked wave and current conditions to a boundary layer model of turbulence enabling prediction of profiles of vertical mixing in nearshore habitats. A random walk approach was used in conjunction with the profile of vertical mixing and the rate of spore sinking to simulate the vertical movement of spores after their release from a given height above the bottom, and to estimate the time required for them first to contact the seafloor. An estimate of the dispersal distance was obtained by multiplying the time for a spore to reach the bottom by current speed. This approach provided a rough estimate of the distances spores are transported before first contacting the bottom for fixed wave and current conditions.

Flows in nature, however, are not constant, and the general construct described cannot be expected to predict accurately the dispersal distances at specific locations or times. To address this complicating factor, we extended the approach of Gaylord et al. (2002) to account for variation in flow caused by oscillating currents, which change speed and direction over time. Although one could also incorporate variation in wave height and wave period, such measurements are not as readily available, and as shown in Gaylord et al. (2002), are likely to play a lesser role than currents in determining dispersal distances. Our approach in predicting spore dispersal therefore proceeded as follows. First, we computed the shear velocities corresponding to a range of current and wave conditions in

10 m of water that interact with a seabed characterized by a physical roughness height of 0.08 m, as in Gaylord et al. (2002). Second, we assumed a release of 1000 spores every 2 hours. Third, we tracked the position of these spores until they first contacted the bottom, using the random walk approach described earlier, but with changing flow conditions updated every 20 minutes. This included updating shear velocities that were dictated by wave and current conditions. We repeated this three-step process using current data from two 30-day periods—January 15 to February 15, 2002, and June 1 to 30, 2002—corresponding to the winter and late spring/early summer peaks in spore release exhibited by *Macrocystis* in southern California (Reed et al., 1996). To simulate better the transport velocity that sweeps suspended spores horizontally near the bottom, we incorporated a linear decrease in current speed in the lowermost 1 m of the water column. This follows from the observation that the velocity profile in the lower 10% of a boundary layer is usually logarithmic (Schlichting, 1979; Grant and Madsen, 1986), which approaches a linear gradient for small distances above the seafloor. Because waves are typically largest in the winter in southern California, we assigned a wave height of 1.0 m in winter (January–February) and 0.5 m in June, both with a 10-second period. As noted earlier, waves were held constant within each 30-day period.

We used data on current speed from two shallow reefs in the Santa Barbara Channel that experience different flow regimes to model temporally varying currents: Carpinteria Reef, located in about 12 m of water, and Naples Reef, located in about 17 m of water (Fig. 10-1). Giant kelp forests commonly occur on both reefs. Currents at both sites were measured from bottom-mounted acoustic Doppler current profilers (ADCPs) placed at the outside edge of the forests at each site as part of the ongoing Santa Barbara Coastal Long-Term Ecological Research Program. Current data from the ADCPs were averaged into 1-m bins that extended from about 2 m above the bottom up to about 1 to 2 m below the surface, depending on wave conditions. Current velocities were recorded at 2-minute intervals and were then averaged to 20-minute intervals for use in the dispersal model. Currents at the two sites flow approximately parallel to isobaths and parallel to the coastline, which runs roughly east–west (Fig. 10-1). Currents were averaged vertically and then rotated into principal axis directions for the dispersal model. The major principal axes at both sites are oriented parallel to isobaths, so hereafter this is referred to as the *alongshore direction* (positive approximately eastward). Similarly, the minor principal axis was taken to be the across-shore direction (positive onshore, or approximately northward). Alongshore currents were used in the model because they are much stronger than across-shore currents at both sites (data not shown). Furthermore, currents in shallow water tend to flow parallel to isobaths (e.g., Pedlosky, 1987), so this is likely to be the dominant direction for spore dispersal.

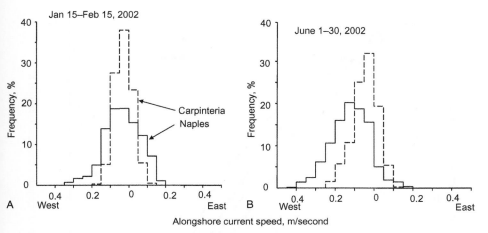

FIGURE 10-8. Histograms of alongshore (principal axis) current velocities at Naples Reef (solid lines) and Carpinteria Reef (dashed line) for two 30-day periods in 2002. Westward currents are to the left of zero and eastward currents are to the right. (A) January 15–February 15: Mean ± SD, 0.04 ± 0.10 m/s and 0.03 ± 0.05 m/s westward for Naples Reef and Carpinteria Reef, respectively. (B) June 1–30: Mean ± SD, 0.12 ± 0.10 m/s and 0.05 ± 0.07 m/s westward for Naples Reef and Carpinteria Reef, respectively.

Alongshore currents at both sites vary strongly on a wide range of timescales, but exhibit prominent tidal fluctuations. At Carpinteria Reef, tidal fluctuations are semidiurnal whereas at Naples Reef they are diurnal (data not shown). Tidal fluctuations often cause reversals of alongshore currents at both sites, particularly in winter. In summer, current reversals are less common because coastal sea level changes typically force a strong westward flow along the mainland coast of the Santa Barbara Channel (Harms and Winant, 1998).

Histograms of alongshore currents show that current speeds were, on average, two to three times faster at Naples Reef than at Carpinteria Reef, and were generally higher in June than in January/February (Fig. 10-8). Currents were, on average, two to three times faster at Naples Reef compared to Carpinteria Reef. Maximum current speeds were westward at approximately 0.4 m/sec at Naples Reef, but only approximately 0.2 m/sec at Carpinteria Reef. The histograms also show the predominance of westward flow, especially in June. Current speeds experienced by most giant kelp populations in southern California are likely to fall within the range encompassed by these two sites and time periods.

The patterns of spore dispersal predicted from our model reflect differences in current speed distributions observed between the two sites and seasons (Fig. 10-9). Of the four model runs, dispersal was predicted to be greatest for

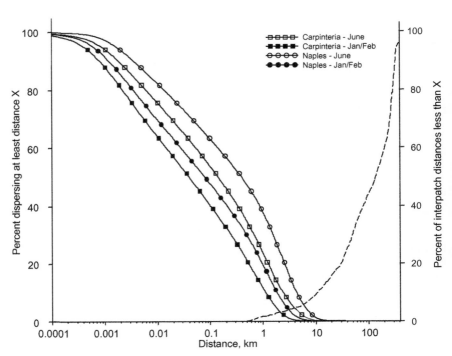

FIGURE 10-9. Dispersal potential of giant kelp spores versus distance between discrete patches. Spore dispersal distributions (solid lines) were simulated on the basis of measured currents at Carpinteria Reef (squares) and Naples Reef (circles) from June 1 to 30, 2001 (open symbols) and January 15 to February 15, 2001 (closed symbols; see text for details). The left vertical axis indicates the probability of dispersing at least as far as the distance on the horizontal axis. The distribution of distances between patches (dashed line) was calculated from distances between all pairs of occupied patches in all months of the study period (418 months). The right vertical axis indicates the percentage of distances between patches that were less than the distance on the horizontal axis.

Naples Reef during June and shortest for Carpinteria during January/February. The more persistent westward flow at both sites probably accounts for the greater dispersal distances in June; more symmetrical current speed distributions probably account for smaller dispersal distances in January/February. Median dispersal distances ranged between 40 m for Carpinteria in February and 400 m for Naples in June. Ten percent of all spores (i.e., the 90th percentile of the spore dispersal distribution) dispersed approximately 1 km at Carpinteria Reef in January/February and 4 km at Naples Reef in June before first contacting the bottom. These values are of the same general magnitude as those derived from our empirical studies discussed earlier.

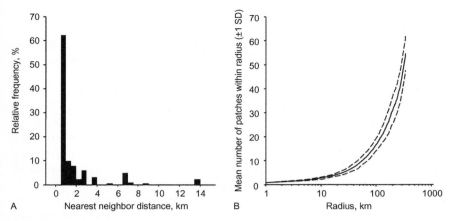

FIGURE 10-10. Spatial distribution of *Macrocystis* patches along the mainland coast of southern California. (A) Mean distance to the nearest occupied neighboring patch, calculated for all patches in the region (69 patches) averaged over the 418-month study period. (B) Mean number of occupied patches (vertical axis) within a given radius of a patch (horizontal axis), averaged over the 418-month study period. Dashed lines indicate ±1 SD.

V. CONNECTIVITY AMONG LOCAL POPULATIONS

Variability in simulated and empirical distributions of spore dispersal suggests that connectivity among discrete patches of *Macrocystis*—and hence the metapopulation dynamics of giant kelp forests—may be strongly influenced by local oceanographic conditions. To examine the connectivity of discrete patches of *Macrocystis* in southern California, we combined dispersal simulations based on the four oceanographic scenarios modeled earlier (i.e., Carpinteria Reef and Naples Reef in January/February and in June) with the 34-year monthly time series of patch extinction–recolonization described in Section II. Average distances between neighboring patches varied from 0.5 to 14 km, with 80% of all patches occurring within 2 km of another patch (Fig. 10-10A). Because each patch may have multiple neighbors, a richer description of spatial structure can be gained by considering the number of neighbors encountered as a function of distance from a patch. In southern California, patches have on average relatively few (approximately one to three) neighbors within 10 km; the number of neighboring patches increases rapidly in neighborhoods greater than 10 km in size (Fig. 10-10B).

The distribution of distances among all possible combinations of patches of giant kelp in southern California ranges from hundreds of meters to hundreds of kilometers (Fig. 10-9, dashed line). When our simulated spore dispersal profiles are overlaid on this pattern, we find that the level of potential connectivity among all patches (i.e., the probability that any two randomly selected patches in southern California are directly connected by spore dispersal) ranged from 0.37 to 1.58%, depending on the specific current regime (see overlap between the distributions of dispersal distance [solid lines] and interpatch distance [dashed line] in Fig. 10-9).

Estimates of the proportion of propagules dispersing a given distance before first contacting the bottom provide only a crude measure of the actual level of connectivity among discrete patches. More accurate measures of the degree of connectivity require information on the absolute numbers of propagules that are exchanged between neighboring patches and on their probability of survival after settlement. This is particularly important in kelps, because their spores need to settle at high densities to ensure subsequent reproduction (see Section III.A). Obtaining information on absolute estimates of spore exchange rates for *Macrocystis* is difficult for a number of reasons. The models of dispersal presented here assume spores settle at first contact with the bottom, but saltation after primary contact could substantially extend dispersal (Gaylord et al., 2002). The degree to which this occurs in nature, however, is unknown. Moreover, the probability of reproductive success for spores that have spent long periods in the water column, particularly those that have stopped swimming and/or germinated, is poorly understood, as are the chances of reproduction between different-age gametophytes. All these processes could influence the "effective" dispersal distance of kelp spores in nature, and deserve further detailed study.

As noted in Section III.A, the density of spores that settle at a given distance from their point of release is proportional to the size of their parental spore source. Our modeled estimates of connectivity assume that spore sources (i.e., the standing crop of spores in a patch) are spatially and temporally homogenous. Clearly this is not true, and our values of connectivity may be underestimated in the case of large and/or continuous spore sources, and overestimated in the case of small and/or sporadic spore sources. The effects of variation in the size of the spore source on the level of connectivity can be evaluated to some extent by examining connectivity under different threshold levels of spore dispersal. The rationale here is that a larger proportion of spores will settle at densities sufficient for fertilization when released from a large spore source compared with a smaller spore source. Other factors influencing spore source strength include the abundance and per capita fecundity of adult plants (Graham, 2003; Reed et al., 2004), the degree to which they display synchrony in spore release (Reed et al., 1997), and oceanographic conditions affecting dilution during transport (Gaylord et al., 2004). Thus, a dispersal threshold defined by the 50th percentile of the spore dispersal distribution could be viewed as representing connectivity to a relatively small/weak spore source, whereas a larger/stronger spore source might result in connectivity at distances reached by only 10% of spores (i.e., the 90th percentile of the dispersal distance distribution).

Figure 10-11 shows frequency distributions of average connectivity (i.e., number of neighboring patches connected via dispersal) for different dispersal thresholds, defined by using percentiles of the dispersal distance probability distribution. Depending on site, season, and the dispersal threshold chosen to define connectivity, an average patch of *Macrocystis* may be completely isolated or have as many as four or five connected neighbors. For example, during June conditions at Carpinteria Reef, 10% of spores (equivalent to a cutoff percentile of 90%) were estimated to disperse at least $x_c = 2.4$ km (Fig. 10-11). Under these conditions, 42% of patches would exchange spores with one other patch and 5% of patches would exchange with four other patches; 19% would not exchange spores with any other patches. Higher current speeds, such as for June conditions at Naples, would result in exchanges among more patches. That the level of connectivity was highly sensitive to the choice of the threshold dispersal rate argues that connectivity among discrete patches of giant kelp depends greatly on factors affecting the strength of the spore source (e.g., the standing crop of spores in a patch) as well as the spacing among patches. Thus, spatial and temporal patterns of adult fecundity in giant kelp may influence regional patterns of colonization in much the same manner as has been found for acroporid corals in the Great Barrier Reef (Hughes et al., 2000).

VI. SUMMARY

Our analyses of local extinctions, colonization, and immigration (via spore dispersal) suggest that the metapopulation concept is likely to prove useful in explaining the population dynamics and genetic structure of *Macrocystis*. Limitations on dispersal in giant kelp appear to prevent patches within a region from behaving as a single large population. However, although most patches within a region are not directly linked, neither are they completely isolated. The average kelp bed in southern California appears to be connected to one to three neighboring kelp beds via spore dispersal for a relatively wide range of oceanographic conditions. More important, connectivity even among nearby patches seems to be mediated by spores that travel far beyond the median dispersal distance (i.e., 75th–90th percentile; Fig. 10-11); that is, a relatively small fraction of dispersing spores accounts for a disproportionate amount of interpatch connectivity. As a result, persistence of a giant kelp metapopulation depends on the tails of the dispersal curve and cannot be predicted simply from the average or median dispersal distance. This result is consistent with theoretical predictions that (re)colonization and population spread are highly dependent on the tails of a dispersal distribution (Kot et al., 1996), but contrasts with results for stable environments in which persistence is relatively insensitive to the tails of the dispersal curve (Lockwood et al., 2002).

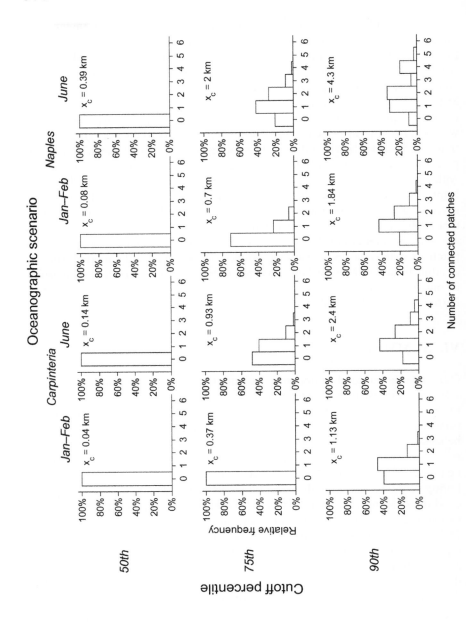

Patch size, fecundity, and proximity to neighboring patches undoubtedly exert strong influences on the level of connectivity among patches. Environmental stochasticity arising from biotic and abiotic disturbances appears to be the primary force driving extinctions and recolonizations of giant kelp populations in southern California. Local extinctions caused by recruitment failure (i.e., demographic stochasticity) to our knowledge have not been reported for giant kelp. This may be due in part to the high capacity of giant kelp forests to produce and retain large numbers of spores, which allows for self-replenishment. The relative contributions of self-seeding and spore immigration to population persistence in giant kelp is unknown and warrants further investigation.

That the vast majority of extinction events in our 34-year study period persisted for less than 2 years indicates that recruitment failure in unoccupied patches is a short-term phenomenon and the immigration of spores from neighboring patches is a common occurrence. The scenario revealed by our analyses suggests that giant kelp in southern California is a spatially structured metapopulation in which exchange occurs primarily between neighboring patches and is strongly influenced by patch size, fecundity (i.e., spore standing crop), spatial arrangement, and oceanographic conditions. Because giant kelp populations are distributed in a narrow depth range along the coastline, and because nearshore currents (the dominant mechanism for dispersal) flow primarily alongshore, the stepping-stone exchange among neighboring patches is approximately one-dimensional. Such limitations on connectivity, coupled with large geographical gradients in environmental conditions, likely account for the ecotypic variation observed for *Macrocystis* in California (Kopczak et al., 1991).

Compared with many terrestrial habitats, the aqueous medium through which kelp propagules disperse is relatively unstructured. Moreover, unlike passively dispersed kelp spores and drifters, immigrants in many terrestrial metapopulations exhibit complex behavioral interactions with the heterogenous landscape that influence connectivity between patches. However, it is important to recognize that the physical properties of the nearshore ocean are not devoid of spatial and temporal structure, and certain areas or times may be subjected to currents

◀

FIGURE 10-11. Estimated connectivity of giant kelp patches along the mainland coast of southern California for the four simulated current regimes (columns) and for three spore dispersal scenarios (rows). Connectivity is defined as the number of occupied patches within the effective dispersal distance of a source patch. Spore dispersal was simulated under the current regimes measured at two sites (Carpinteria Reef and Naples Reef) and two seasons (January 15–February 15, 2001; June 1–30, 2001). Different spore dispersal scenarios (which may correspond to variation in patch size, fecundity, degree of synchrony in spore release, and diffusive dilution) are simulated by choosing effective dispersal distances (x_c) ranging from the 50th to the 90th cutoff percentiles of the four spore dispersal profiles. Values reported are averages for the 69 patches calculated over the 418-month study period.

and waves that are more conducive to promoting connectivity than others. Indeed, our analyses of connectivity involving Naples Reef and Carpinteria Reef indicate this to be the case. Thus the nearshore habitat where kelp occurs may represent a more dynamic analog of the terrestrial concept of a landscape matrix (Weins, 1977). Perhaps even more important to the metapopulation structure and dynamics of giant kelp are the effects of a heterogenous environment on the ability of immigrants to become established, grow, and reproduce. Shallow reefs may differ greatly from each other in habitat quality as a result of differences in topography, wave exposure, sedimentation, nutrients, and other geophysical and chemical properties. Differences in colonization success, growth, and reproduction among patches of different quality may greatly affect the magnitude of "effective" connectivity between patches.

Our estimates of spore dispersal when viewed in the context of the size and distribution of discrete patches suggest that many local populations of *Macrocystis* in the Southern California Bight are "on the edge" with respect to connectivity. Relatively small increases in distances between patches could lead to substantial increases in demographic isolation. This is of particular concern because the last century has seen an increase in the average distance between reefs potentially habitable by giant kelp, due in part to anthropogenic impacts including substrate burial and reduced water clarity near major ports, warm water effluent from generating stations, and municipal sewage outfalls (Crandall, 1912; Harger, 1983; Wilson and North, 1983; Schroeter et al., 1993; Bence et al., 1996). Continued changes in the configuration of nearshore habitat suitable for giant kelp forests in this region will likely present severe challenges for conservation, persistence, and local adaptation/evolution of giant kelp populations. However, because the level of connectivity among patches is strongly dependent on the length of the "effective" dispersal tail, additional research is needed to determine the amount of dispersal that constitutes connectivity, which broadly defined includes spore dispersal, settlement, and postsettlement success.

The strong dependence of giant kelp connectivity and patch dynamics on environmental factors such as geomorphology (distribution of rocky substrate) and oceanography (wave disturbance and nutrient stress) suggests that the metapopulation structure of this species is likely to differ substantially among regions. For example, populations along stretches of the coast of Baja California, Mexico, are far more isolated than any of the patches we studied, whereas central California kelp populations occur in a near-continuous band throughout much of the region. A better understanding of regional variation in metapopulation structure may help to explain observations of extreme variation in persistence and dynamics of this species across its global range (reviewed in North, 1994).

Finally, we note that because our approximations of patch extinction and recolonization suffer from methodological limitations that could lead to either over- or underestimates (see Section II), actual connectivity may be somewhat different from that suggested by our analyses. For example, our modeling did not incorporate cross-shore flows or dilution resulting from lateral mixing—two poorly understood processes that probably limit effective dispersal distance and thus reduce connectivity among patches. A detailed examination of these and other potential limitations of our analyses is beyond the scope of this chapter. Clearly, further research is needed to characterize extinction and colonization at large scales using high-resolution mapping, as well as to quantify more accurately the effects of spore saltation; delayed recruitment; variation in the number, spacing, and fecundity of plants within a patch; and various physical processes (e.g., variability in waves and turbulence) on the effective colonization distance of giant kelp. Studies of population genetics may prove useful in this regard. The estimates of connectivity presented here provide a platform for future studies on the metapopulation ecology of giant kelp and other seaweeds.

VII. ACKNOWLEDGMENT

The authors are grateful to the National Science Foundation for the continued support of their investigations of the mechanisms that promote colonization in kelp following local extinctions. They thank N. Morgan for assistance in analyzing current data and M. Anghera for his tireless efforts in the field. D. Glantz kindly provided access to ISP Alginates, Inc., historical records of giant kelp abundance. J. Kum, M. Johnson, N. Wright, and M. Merrifield facilitated access to California Department of Fish and Game kelp surveys. B. Turner provided technical assistance in development of the composite kelp map. Funding for the preparation of this chapter was supplied by the National Science Foundation (grant nos. OCE99-82105 and OCE02-41447), and the Fannie and John Hertz Foundation. This is contribution number 199 from PISCO, the Partnership for Interdisciplinary Studies of Coastal Oceans, funded primarily by the Gordon and Betty Moore Foundation and the David and Lucile Packard Foundation.

REFERENCES

Amsler, C.D., and Neushul, M. (1989a). Chemotactic effects of nutrients on spores of the kelps *Macrocystis pyrifera* and *Pterygophora californica*. *Mar. Biol.* **102**, 557–564.
Amsler, C.D., and Neushul, M. (1989b). Diel periodicity of spore release from the kelp *Nereocystis luetkeana* (Mertens) Postels et Ruprecht. *J. Exp. Mar. Biol. Ecol.* **134**, 117–127.
Amsler, C.D., and Neushul, M. (1990). Nutrient stimulation of spore settlement in the kelps *Macrocystis pyrifera* and *Pterygophora californica*. *Mar. Biol.* **107**, 297–304.
Amsler, C.A., and Neushul, M. (1991). Photosynthetic physiology and chemical composition of spores of the kelps *Macrocystis pyrifera*, *Nereocystis luetkeana*, *Laminaria farlowii*, and *Pterygophora californica* (Phaeophyceae). *J. Phycol.* **27**, 26–34.
Anderson, E.K., and North, W.J. (1966). In situ studies of spore production and dispersal in the giant kelp *Macrocystis pyrifera*. *Proc. International Seaweed Symp.* **5**, 73–86.
Anderson, E.K., and North, W.J. (1967). Zoospore release rates in giant kelp *Macrocystis*. *Bull. South. Calif. Acad. Sci.* **66**, 223–232.

Bence, J.R., Stewart–Oaten, A., and Schroeter, S.C. (1996). Estimating the size of an effect from a before-after-control-impact paired series design: The predictive approach applied to a power plant study. In *Detecting ecological impacts: Concepts and applications in coastal habitats* (R.J. Schmitt and C.W. Osenberg, eds., pp. 133–151). San Diego: Academic Press.

Boland, W., Marner, F.J., Jaenicke, L., Müller, D.G., and Folster, E. (1983). Comparative receptor study in gamete chemotaxis of the seaweeds *Ectocarpus siliculosus* and *Cutleria multifida*: An approach to interspecific communication of algal gametes. *Eur. J. Biochem.* **134**, 97–103.

Brown, J.H., and Kodric–Brown, A. (1977). Turnover rates in insular biogeography: Effect of immigration on extinction. *Ecology.* **58**, 445–449.

Brzezinski, M., Reed, D.C., and Amsler, C.D. (1993). Neutral lipids as major storage products in zoospores of the giant kelp, *Macrocystis pyrifera*. *J. Phycol.* **29**, 16–23.

Burgman, M.A., and Gerard, V.A. (1990). A stage-structured, stochastic population model for the giant kelp *Macrocystis pyrifera*. *Mar. Biol.* **105**, 15–23.

Chapman, A.R.O. (1986). Population and community ecology of seaweeds. In *Advances in marine biology* (J.H.S. Baxter and A.J. Southwood, eds., pp. 1–161). London: Academic Press.

Cowen, R.C., Agegian, C.R., and Foster, M.F. (1982). The maintenance of community structure in a central California giant kelp forest. *J. Exp. Mar. Biol. Ecol.* **64**, 189–201.

Crandall, W.C. (1912). The kelps of the southern California coast: Fertilizer resources. In *U.S. 62nd Congress, 2nd Senate Session* (pp. 209–213). Doc. 190. Appendix N.

Darwin, C. (1860). *The voyage of the* Beagle. Garden City, NY: Anchor Books, Doubleday and Co.

Davis, N., VanBlaricom, G.R., and Dayton, P.K. (1982). Man-made structures on marine sediments: Effects on adjacent benthic communities. *Mar. Biol.* **70**, 295–303.

Dayton, P.K. (1985). The ecology of kelp communities. *Annu. Rev. Ecol. Syst.* **16**, 215–245.

Dayton, P.K., Currie, V., Gerrodette, T., Keller, B., Rosenthal, R., and Van Tresca, D. (1984). Patch dynamics and stability of some southern California kelp communities. *Ecol. Monogr.* **54**, 253–289.

Dayton, P.K., and Tegner, M.J. (1984). Catastrophic storms, El Niño and patch stability in a southern California kelp community. *Science.* **224**, 283–285.

Dayton, P.K., and Tegner, M.J. (1989). Bottoms beneath troubled waters: Benthic impacts of the 1982–1984 El Niño in the temperate zone. In *Global ecological consequences of the 1982–83 El Niño–southern oscillation* (P.W. Glynn, ed., pp. 433–472). Oceanographic series no. 52. Amsterdam: Elsevier.

Dayton, P.K., Tegner, M.J., Edwards, P.B., and Riser K.L. (1999). Temporal and spatial scales of kelp demography: The role of oceanographic climate. *Ecol. Mongr.* **69**, 219–250.

Dayton, P.K., Tegner, M.J., Parnell, P.E., and Edwards, P.B. (1992). Temporal and spatial patterns of disturbance and recovery in a kelp forest community. *Ecol. Monogr.* **62**, 421–445.

Dean, T.A., Schroeter, S.C., and Dixon, J. (1984). Effects of grazing by two species of sea urchins (*Strongylocentrotus franciscanus* and *Lytechinus anamesus*) on the recruitment and survival of two species of kelp (*Macrocystis pyrifera* and *Pterygophora californica*). *Mar. Biol.* **78**, 301–313.

Dean, T.A., and Jacobsen, F.R. (1984). Growth of juvenile *Macrocystis pyrifera* (Laminariales) in relation to environmental factors. *Mar. Biol.* **83**, 301–311.

Dean, T.A., and Jacobsen, F.R. (1986). Nutrient-limited growth of juvenile kelp, *Macrocystis pyrifera* during the 1982–1984 "El Niño" in southern California. *Mar. Biol.* **90**, 597–601.

Dean, T.A., Thies, K., and Lagos, S.L. (1989). Survival of juvenile giant kelp: The effects of demographic factors, competitors and grazers. *Ecology.* **70**, 483–495.

Devinny, J.S., and Volse, L.A. (1978). Effects of sediments on the development of *Macrocystis pyrifera* gametophytes. *Mar. Biol.* **48**, 343–348.

Deysher, L., and Dean, T.A. (1984). Critical irradiance levels and the interactive effects of quantum irradiance and quantum dose on gametogenesis in the giant kelp, *Macrocystis pyrifera*. *J. Phycol.* **20**, 520–524.

Deysher, L., and Dean, T.A. (1986). *In situ* recruitment of the giant kelp, *Macrocystis pyrifera*: Effects of physical factors. *J. Exp. Mar. Biol. Ecol.* **103**, 41–63.
Ebeling, A.W., Laur D.R., and Rowley, R.J. (1985). Severe storm disturbances and the reversal of community structure in a southern California kelp forest. *Mar. Biol.* **84**, 287–294.
Edwards, M.S. (2004). Estimating scale-dependency in disturbance impacts: El Niños and giant kelp forests in the northeast Pacific. *Oecologia*, **138**, 436–447.
Estes, J.A., and Steinberg, P.D. (1988). Predation, herbivory, and kelp evolution. *Paleobiology.* **14**, 19–36.
Foster, M.S. (1982). The regulation of macroalgal associations in kelp forests. In *Synthetic and degradative processes in marine macrophytes* (L. Srivastava, ed., pp. 185–205). Berlin: Walter de Gruyter.
Foster, M.S., and Schiel, D.R. (1985). *The ecology of giant kelp forests in California: A community profile.* Biological report 85(7.2). Slidell, LA: United States Fish and Wildlife Service.
Fritsch, F.E. (1945). *The structure and reproduction of the algae* (vol. II). London: Cambridge University Press.
Gaylord, B., Reed, D.C., Raimondi, P.T., Washburn, L., and McLean, L. (2002). A physically-based model of macroalgal spore dispersal in the wave and current-dominated nearshore. *Ecology.* **83**, 1239–1251.
Gaylord, B., Reed, D.C., Washburn, L., and Raimondi, P.T. (2004). Physical–biological coupling in spore dispersal of kelp forest macroalgae. *J. Mar. Systems* **49**, 19–39.
Gotelli, N.J. (1991). Metapopulation models: The rescue effect, the propagule rain and the core satellite hypothesis. *American Naturalist.* **138**, 768–776.
Graham, M.H. (2000). Planktonic patterns and processes in the giant kelp *Macrocystis pyrifera*. PhD dissertation. San Diego: University of California.
Graham, M.H. (2002). Prolonged reproductive consequences of short-term biomass loss in seaweeds. *Mar. Biol.* **140**, 901–911.
Graham, M.H. (2003). Coupling propagule output to supply at the edge and the interior of a giant kelp forest. *Ecology.* **84**, 1250–1264.
Graham, M.H., Harrold, C., Lisin, S., Light, K., Watanabe, J.M., and Foster, M.S. (1997). Population dynamics of giant kelp *Macrocystis pyrifera* along a wave exposure gradient. *Mar. Ecol. Prog. Ser.* **148**, 269–279.
Grant, W.D., and Madsen, O.S. (1986). The continental-shelf bottom boundary layer. *Annu. Rev. Fluid Mech.* **18**, 265–305.
Hanski, I. (1999). *Metapopulation ecology.* New York: Oxford University Press.
Harger, B. (1983). A historical overview of kelp in southern California. In *The effects of waste disposal on kelp communities* (W. Bascom, ed., pp. 70–83). Southern California Coastal Water Research Project, Long Beach. Long Beach, CA: Southern California Coastal Water Research Project.
Harms, S., and Winant, C.D. (1998). Characteristic patterns of the circulation in the Santa Barbara Channel. *J. Geophys. Res.* **130**, 3041–3065.
Harris, L.G., Ebeling, A.W., Laur, D.R., and Rowley, R.J. (1984). Community recovery after storm damage: A case of facilitation in primary succession. *Science.* **224**, 1336–1338.
Harrold, C., and Lisin, S. (1989). Radio-tracking rafts of giant kelp: Local production and regional transport. *J. Exp. Ecol. Mar. Biol.* **130**, 237–251.
Harrold, C., and Pearse, J.S. (1987). The ecological role of echinoderms in kelp forests. In *Echinoderm studies* (M. Jangoux and J.M. Lawrence, eds., vol. 2, pp. 137–233). Rotterdam: A.A. Balkema.
Harrold, C., and Reed, D.C. (1985). Food availability, sea urchin grazing and kelp forest community structure. *Ecology.* **63**, 547–560.
Hobday, A.J. (2000). Abundance and dispersal of drifting kelp *Macrocystis pyrifera* rafts in the Southern California Bight. *Mar. Ecol. Prog. Ser.* **195**, 101–116.
Hoffmann, A.J., and Santelices, B. (1991). Banks of algal microscopic forms: Hypotheses on their functioning and comparisons with seed banks. *Mar. Ecol. Prog. Ser.* **79**, 185–194.

Hughes, T.P., Baird, A.H., Dinsdale, E.A., Moltschaniwskyi, N.A., Pratchett, M.S., Tanner, J.E., and Willis, B.L. (2000). Supply-side ecology works both ways: The link between benthic adults, fecundity, and larval recruits. *Ecology.* **81**, 2241–2249.
Jackson, G.A. (1998). Currents in the high drag environment of a coastal kelp stand off California. *Cont. Shelf Res.* **17**, 1913–1928.
Jackson, G.A., and Winant, C.D. (1983). Effects of a kelp forest on coastal currents. *Cont. Shelf. Res.* **2**, 75–80.
Kain, J.S. (1979). A view of the genus *Laminaria*. *Oceanogr. Mar. Biol. Annu. Rev.* **17**, 101–161.
Kingsford, M.J. (1995). Drift algae: A contribution to near-shore habitat complexity in the pelagic environment and an attractant for fish. *Mar. Ecol. Prog. Ser.* **116**, 297–301.
Kinlan, B.P., Graham, M.H., Sala, E., and Dayton, P.K. (2003). Arrested development of giant kelp (*Macrocystis pyrifera*, Phaeophyceae) embryonic sporophytes: A mechanism for delayed recruitment in perennial kelps? *J. Phycol.* **39**, 47–57.
Kopczak, C.D., Zimmerman, R.C., and Kremer, J.N. (1991). Variation in nitrogen physiology and growth among geographically isolated populations of the giant kelp *Macrocystis pyrifera*—Phaeophyta. *J. Phycol.* **27**, 149–158.
Kot, M., Lewis, M.A., and van den Driessche, P. (1996). Dispersal data and the spread of invading organisms. *Ecology.* **77**, 2027–2042.
Ladah, L.B., Zertuche-Gonzalez, J.A., and Hernandez-Carmona, G. (1999). Giant kelp (*Macrocystis pyrifera*, Phaeophyceae) recruitment near its southern limit in Baja California and mass disappearance during ENSO 1997–1998. *J. Phycol.* **35**, 1106–1112.
Leonard, G.H. (1994). Effect of the bat star *Asterina miniata* (Brandt) on recruitment of the giant kelp *Macrocystis pyrifera* C. Agardh. *J. Exp. Mar. Biol. Ecol.* **179**, 81–98.
Lewis, R.J. (1996). Chromosomes of the brown algae. *Phycologia.* **35**, 19–40.
Lockwood, D.R., Hastings, A., and Botsford, L.W. (2002). The effects of dispersal patterns on marine reserves: Does the tail wag the dog? *Theo. Pop. Biol.* **61**, 297–309.
Lüning, K., and Neushul, M. (1978). Light and temperature demands for growth and reproduction of laminarian gametophytes in southern and central California. *Mar. Biol.* **45**, 297–309.
Mann, K.H. (1973). Seaweeds: Their productivity and strategy for growth. *Science.* **182**, 975–981.
Mann, K.H. (1982). *Ecology of coastal waters.* Berkeley: University of California Press.
McNair, J.N., Newbold, J.D., and Hart, D.D. (1997). Turbulent transport of suspended particles and dispersing benthic organisms: How long to hit bottom? *J. Theor. Biol.* **188**, 29–52.
Müller, D.G. (1981). Sexuality and sexual attraction. In *The biology of seaweeds* (C.S. Lobban and M.J. Wynne, eds., pp. 661–674). Oxford: Blackwell Scientific Publications.
Murray, S.N., and Bray, R.N. (1993). Benthic macrophytes. In *Ecology of the Southern California Bight* (M.D. Dailey, D.J. Reish, and J.W. Anderson, eds., pp. 304–368). Berkeley: University of California Press.
Neushul, M. (1963). Studies on giant kelp, *Macrocystis*. II. Reproduction. *Am. J. Bot.* **50**, 354–359.
Nisbet, R.M., and Bence, J.R. (1989). Alternative dynamic regimes for canopy forming kelp: A variant on density vague population regulation. *Am. Nat.* **134**, 377–408.
North, W.J. (1971). *The biology of giant kelp beds* (Macrocystis) *in California.* Lehre: Verlag Von J. Cramer.
North, W.J. (1994). Review of *Macrocystis* biology. In *Biology of economic algae* (I. Akatsuka, ed., pp. 447–527). The Hague, Netherlands: SPB Academic Publishing.
North, W.J., James, D.E., and Jones, L.G. (1993) History of kelp beds (*Marcocystis*) in Orange and San Diego counties, California. *Hydrobiologia.* **260/261**, 277–283.
Pearse, J.S., and Hines, A.H. (1979). Expansion of a central California kelp forest following the mass mortality of sea urchins. *Mar. Biol.* **51**, 83–91.
Pedlosky, J. (1987). *Geophysical fluid dynamics* (2nd ed.). New York: Springer-Verlag.

Raimondi, P.T., Reed, D.C., Gaylord, B., and Washburn, L. (2004). Effects of self-fertilization in the giant kelp, *Macrocystis pyrifera*. *Ecology*. **85**, 3267–3276.
Reed, D.C. (1990). The effects of variable settlement and early competition on patterns of kelp recruitment. *Ecology*. **71**, 776–787.
Reed, D.C., Amsler, C.D., and Ebeling, A.W. (1992). Dispersal in kelps: Factors affecting spore swimming and competency. *Ecology*. **73**, 1577–1585.
Reed, D.C., Anderson, T.W., Ebeling, A.W., and Anghera, M. (1997). The role of reproductive synchrony in the colonization potential of kelp. *Ecology*. **78**, 2443–2457.
Reed, D.C., Brzezinski, M.A., Coury, D.A., Graham, W.M., and Petty, R.L. (1999). Neutral lipids in macroalgal spores and their role in swimming. *Mar. Biol.* **133**, 737–744.
Reed, D.C., Ebeling, A.W., Anderson, T.W., and Anghera, M. (1996). Differential reproductive responses to fluctuating resources in two seaweeds with different reproductive strategies. *Ecology*. **77**, 300–316.
Reed, D.C., and Foster, M.S. (1984). The effects of canopy shading on algal recruitment and growth in a giant kelp (*Macrocystis pyrifera*) forest. *Ecology*. **65**, 937–948.
Reed, D.C., Laur, D.R., and Ebeling, A.W. (1988). Variation in algal dispersal and recruitment: The importance of episodic events. *Ecol. Monogr.* **58**, 321–335.
Reed, D.C., and Lewis, R.J. (1994). Effects of an oil and gas production effluent on the colonization potential of giant kelp (*Macrocystis pyrifera*) zoospores. *Mar. Biol.* **119**, 277–283.
Reed, D.C., Lewis, R.J., and Anghera, M. (1994). Effects of an open coast oil production outfall on patterns of giant kelp (*Macrocystis pyrifera*) recruitment. *Mar. Biol.* **120**, 26–31.
Reed, D.C., Neushul, M., and Ebeling, A.W. (1991). Role of settlement density on gametophyte growth and reproduction in the kelps *Pterygophora californica* and *Macrocystis pyrifera* (Phaeophyceae). *J. Phycol.* **27**, 361–366.
Reed, D.C., Raimondi, P.T., Carr, M.H., and Goldwasser, L. (2000). The role of dispersal and disturbance in determining spatial heterogeneity in sedentary organisms. *Ecology*. **81**, 2011–2026.
Reed, D.C., Schroeter, S.C., and Raimondi, P.T. (2004). Spore supply and habitat availability as sources of recruitment limitation in giant kelp. *J. Phycol.* **40**, 275–284.
Rosenthal, R.J., Clarke, W.D., and Dayton, P.K. (1974). Ecology and natural history of a stand of giant kelp *Macrocystis pyrifera*, off Del Mar, California. *Fish. Bull.* **72**, 670–684.
Schiel, D.R., and Foster, M.S. (1986). The structure of subtidal algal stands in temperate waters. *Oceanogr. Mar. Biol. Annu. Rev.* **24**, 265–307.
Schlichting, H. (1979). *Boundary-layer theory* (7th ed.). New York: McGraw-Hill.
Schroeter, S.C., Dixon, J.D., Kastendiek, J.D., Smith R.O., and Bence, J.R. (1993). Detecting the ecological effects of environmental impacts: A case study of kelp forest invertebrates. *Ecol. Appl.* **3**, 331–350.
Seymour, R., Tegner, M.J., Dayton, P.K., and Parnell, P.E. (1989). Storm wave induced mortality of giant kelp *Macrocystis pyrifera* in southern California. *Estuar. Coast. Shelf Sci.* **28**, 277–292.
Tegner, M.J., and Dayton, P.K. (1987) El Niño effects on southern California kelp forest communities. *Adv. Ecol. Res.* **17**, 243–279.
Tegner, M.J., Dayton, P.K., Edwards, P.B., Riser, K.L., et al. (1997). Large-scale, low frequency oceanographic effects on kelp forest succession: A tale of two cohorts. *Mar. Ecol. Prog. Ser.* **146**, 117–134.
Thomas, C.D., and Hanski, I. (1997). Butterfly metapopulations. In *Metapopulation dynamics: Ecology, genetics and evolution* (I. Hanski and M. Gilpin, eds., pp. 359–386). San Diego: Academic Press.
Wiens, J.A. (1997). Metapopulation dynamics and landscape ecology. In *Metapopulation dynamics: Ecology, genetics and evolution* (I. Hanski and M. Gilpin, eds., pp. 43–62). San Diego: Academic Press.
Wilson, K.H., and North, W.J. (1983). A review of kelp bed management in southern California. *J. World Maricult. Soc.* **14**, 347–359.

Witman, J.D., and Dayton, P.K. (2001) Rocky subtidal communities. In *Marine community ecology* (M.D. Bertness, S.D. Gaines, and M.E. Hay, eds., pp. 339–360). Sunderland, MA: Sinauer Associates.

Wormersley, H.B.S. (1954). The species of *Macrocystis* with special reference to those on southern Australia coasts. *Univ. Calif. Publ. Bot.* **27**, 109–132.

Zimmerman, R.C., and Robertson, D.L. (1985). Effects of the 1983 El Niño on growth of giant kelp *Macrocystis pyrifera* at Santa Catalina Island. *Limnol. Oceanogr.* **30**, 1298–1302.

ZoBell, C.E. (1971). Drift seaweeds on San Diego county beaches. In *The biology of giant kelp beds* (Macrocystis) *in California* (W.J. North, ed., pp. 269–314). Lehre: Verlag Von J. Cramer.

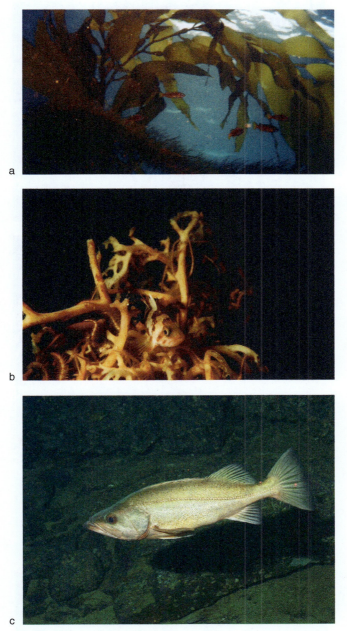

FIGURE 3-4. Typical rocky reef fishes of the Northeast Pacific. a. newly recruited juvenile rockfish sheltering in kelp canopy (J. Hyde). b. benthic juvenile calico rockfish sheltering in a kelp holdfast (J. Hyde). c. bocaccio rockfish, (J. Butler). d. yelloweye rockfish, at rear, and vermilion rockfish, in front, (J. Butler). e. lingcod, (J. Butler). f. gravid cowcod, (J. Butler).

FIGURE 3-4. *Continued*

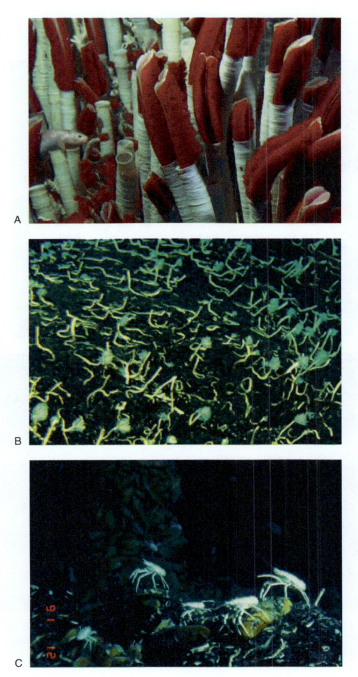

FIGURE 9-2. Hydrothermal vent fauna on East Pacific Rise: (A) thicket of vestimentiferan tubeworms providing habitat for bythograeid crabs and fish; (B) field of suspension-feeding serpulid polychaetes; (C) vent mussels (with yellowish periostracum covering shell) and galatheid crabs.

FIGURE 11-2. (A) Patch of *Halophila decipiens* from the West Florida Shelf. Scale bar = 1 cm. (B) Edge of *H. decipiens* patch from West Florida Shelf showing sediments with marked ripples. Photos courtesy of M.O. Hall.

CHAPTER 11

Seagrasses and the Metapopulation Concept: Developing a Regional Approach to the Study of Extinction, Colonization, and Dispersal

SUSAN S. BELL

I. Introduction
II. Seagrass Reproduction
III. Patches: Colonization and Extinction
IV. Examples of Potential Seagrass Metapopulations
   A. *Halophila decipiens* on the West Florida Shelf: A Local Population That Exhibits Patch Extinction and Regional Recruitment
   B. *Phyllospadix scouleri* on Exposed Pacific Shores: Regional Patterns of Suitable Sites Exist among a Matrix of Unsuitable Sites; Patch Extinction Documented
   C. *Halophila johnsonii* in Southeastern Florida: A Well-defined Regional Population with Limited Dispersal between Patches; Extinction of Patches Documented
V. The Metapopulation Model and Seagrass Populations: A Useful Concept?
   A. Collection of Good Information on Spatial Organization of Seagrasses
   B. Collection of Genetic Information to Help Analyze Spatial Structure of a Population
   C. Seed Dormancy: A Special Problem for Plant Populations

VI. Summary
VII. Acknowledgments
References

I. INTRODUCTION

Metapopulation theory is largely based on a dynamic aspect of population maintenance that embraces a regional perspective (Levins, 1970). Generally the equilibrium, or strictest, view embraces a balance between periods of extinction and colonization, with migration between patches. Although metapopulation theory has focused mainly on animal populations (e.g., Hanski and Gilpin, 1997) it has also expanded our thinking about dynamics of plant populations from a local to a regional scale, and forced a consideration of the interconnectedness of local populations. Specifically, although local populations (arbitrarily defined for each study by researchers) may go extinct, a set of local populations (= metapopulation) may ultimately persist because of the influx of dispersers from other populations (see Hanski and Simberloff, 1997, for terminology).

In a review, Husband and Barrett (1996) point out that, in contrast to animal populations, very little attention has been directed at plant metapopulations. In the terrestrial literature most attention has been focused on annual and short-lived perennials inhabiting ephemeral environments. Few studies have expanded the detailed demography to a regional scale (i.e., a large area that encompasses a set of local populations) and only a handful has examined the relationship between demography at the local scale and the overall regional characteristics of the population. In general, although there are extensive studies on the population biology on a number of terrestrial plants used as models by population biologists, the literature on plant metapopulations is extremely limited.

There are a number of complicating issues of applying metapopulation theory to plant populations. Husband and Barrett (1996) state that even the mere measurement of extinction is not straightforward for plants, especially when plants display strong seed dormancy. Clearly seed dormancy may change the probability of a patch being "occupied" at a given time and the rates of colonization—both integral measures of metapopulation models. It is also difficult to determine what should be done with the belowground component of a plant population and this has not been resolved. Likewise plants may not be homogeneously distributed across a landscape and many plants have, for the most part, localized dispersal. Thus, nonuniform spatial patterns may be the norm for many plant populations. Finally, although dispersal in animals may be observable in many cases, following dispersal of plants remains an extremely difficult task.

In a recent paper, Freckleton and Watkinson (2002), building upon the work of Husband and Barrett (1996), further examined the concept of plant dynamics at the regional scale and metapopulation theory. They again called into question whether the concept of a metapopulation was one that was applicable to plant populations because of widespread spatial heterogeneity of habitat, limited dispersal, and the presence of localized interactions among plants. This view was shared by Bullock et al. (2002), who also suggested that metapopulations may not be common in plant populations. Freckleton and Watkinson (2002) also argued, however, that consideration of metapopulation models was in fact a useful exercise, because it forced plant ecologists to expand their view of plant populations to the larger regional level. Metapopulation models may be difficult to apply at this time and much discussion has been generated on this topic (see Ehrlen and Ericksson, 2003). But Freckleton and Watkinson (2002) proposed that the simultaneous monitoring of local and regional level of population dynamics of plants may provide an impetus for some new thinking. This is the tone adopted for the following discussion on seagrasses as metapopulations.

The previous reviews of plant metapopulations evolved from studies in terrestrial settings, typically ignoring the plant populations from marine habitats. One such assemblage that merits scrutiny is the seagrasses, a group representing about 60 species of marine flowering plants, often dominating coastal habitats worldwide. The possible application of metapopulation theory to seagrasses is a logical inquiry given that recent evidence has been presented indicating that these taxa do in fact display spatially discrete or patchy distributions in a variety of locations. Furthermore, the beginnings of a regional landscape approach to seagrass have emerged (e.g., Robbins and Bell, 1994; Kendrick et al., 1999; Frederiksen et al., 2004). In addition, quantitative information on seagrass dispersal, as well as seagrass dynamics including colonization and extinction, is starting to accumulate. Thus, in this chapter I explore whether seagrass populations have characteristics that make a metapopulation approach a useful one for investigation.

In the following discussions I review information about the three important components of metapopulation theory—dispersal, colonization, and extinction—with special emphasis on a patch dynamic context. Additionally I provide an overview of reproductive biology of seagrasses as it relates to dispersal capabilities and discuss selected examples of possible seagrass populations that have somewhat unique features, making them potential candidates for further evaluation from a metapopulation perspective. Finally I provide suggestions for future avenues of targeted research on the topic of seagrass metapopulations and a regional perspective.

II. SEAGRASS REPRODUCTION

Key parameters of interest in the study of metapopulations are dispersal rates and distances. Thus it is instructive to review what is generally known about these topics for seagrasses. Seagrasses reportedly display both asexual and sexual reproduction (Hemminga and Duarte, 2000). Asexual reproduction via clonal growth is thought to be the primary mechanism by which seagrasses obtain space and thus expand patches. The growth rate of seagrass patches is not very well documented at the present time, although some seagrass horizontal rhizomes are reported to expand by meters per year (e.g., Inglis, 2000b; Rollon et al., 2001). However the available data suggest that clonal growth varies under different abiotic conditions (Jensen and Bell, 2001; Hemminga and Duarte, 2002), so that features of patch expansion rates are likely to be species specific and vary temporally within a location. Asexual reproduction through drifting and subsequent reestablishment of fragments of seagrasses has also been proposed (e.g., Ewanchuk and Williams, 1996; Harwell and Orth, 2002; Campbell, 2003) although this is generally considered to be a rare event.

With few possible exceptions (e.g., Kaldy and Dunton, 1999; Inglis, 2000b; Lacap et al., 2002), studies of clonal seagrasses suggest that local persistence and long-distance dispersal are often achieved by asexual spreading via rhizome elongation or vegetative fragments rather than by sexual recruitment (Terrados, 1993; Marba and Duarte, 1998; Rasheed, 1999; Kenworthy, 2000; Rollon et al., 2001). Although some seagrass genera produce buoyant fruits that disperse relatively long distances, or disperse seeds in rafting flowering shoots (e.g., Harwell and Orth, 2002), most seeds lack dispersal-enhancing characteristics and fall rapidly through the water column once they are released from the flower or fruit (see Orth et al., 1994; Orth et al., 2004; Ruckelshaus, 1996). Notably, some seagrass species (e.g., *Cymodocea nodosa, Halodule wrightii,* and *Halophila decipiens*) release their seeds directly into the sediments, thereby forming seed banks, which in turn limit seed dispersal (Inglis, 2000a,b).

For some seagrass species, flowering and fruit/seed production is well documented, and seed dispersal is thought to be a possible mechanism by which new patches are colonized. There are but limited estimates of seed dispersal distances, some from direct observations or experiments (e.g., Orth et al., 1994) or often calculated as possible distances based upon the longevity of a seed, its rate of fall to the sediment, and some range of current velocity (e.g., Lacap et al., 2002) and an excellent review is available in Orth et al. (2004). It is also possible that seagrass seeds can be moved along with sediments under conditions of bedload transport (discussed later). Storm events are thought to be the major mechanisms by which these long-distance events might occur (e.g., Nakaoka, 2001; Lacap et al., 2002; Reusch, 2002). Interestingly, Kendall (2004) suggests that hurricanes every 10 to 50 years in the Caribbean may exhibit a "storm stimulus" whereby pollination and seed dispersal may occur. Finally, the importance of seed dispersal and recruitment to patch colonization was highlighted by Peterken and Conacher (1997), who reported that sexual reproduction was a major mechanism by which *Zostera capricorni* colonized patches in Australia. Rapid recovery of patches by seedlings was recorded, as was an abundant seed supply in the sediment. The authors, in fact, suggested that the high seed density and recruitment of seedlings found at their study site may help maintain other populations of this species within the larger bay area, although no data to verify this suggestion are currently available.

The documentation of isolated plant populations located long distances away from existing ones is thought to provide additional support for the idea of seagrass dispersal by fragments (e.g., Harwell and Orth, 2001). But isolated populations may be the result of other processes as well. Rollon et al. (2001) reported the occurrence of an apparently isolated seagrass bed composed of *Thalassia hemphrichii*, a species widely distributed throughout the Philippines. This species reproduces sexually and researchers found 30% of all shoots to produce flowers. Seed production was estimated to be 260 seeds/m^2. The interesting question is how a seagrass bed that is more than 500 km away from where the known closest seagrass beds are located originated. The authors argued that long-distance dispersal of seeds was unlikely given that seeds have the propensity to settle to the bottom when released. Another possibility would be drifting fruits, but the seeds have no dormancy over the time period that would be required to travel from the closest site to remote island location (given average current speeds), and seeds would not be viable once they reached the island site. The authors therefore suggest that the seeds may have been transported by human intervention, a dispersal mechanism that may be closely associated with shipping activities.

III. PATCHES: COLONIZATION AND EXTINCTION

The theory of metapopulations has, as a necessary condition, that populations live in identifiable, discrete patches. Likewise, metapopulation models embrace the concept that these patches represent a continuous area of space with all the necessary resources for the persistence of a local population, and these patches lie within a matrix of mostly unsuitable habitat (see Hanski and Simberloff, 1997). Thus, determining whether seagrasses exhibit such a patchy spatial structure provides a first step in evaluating the usefulness of a metapopulation approach.

The seagrass literature has a number of examples of studies that have embraced a patch dynamics approach of seagrasses across a variety of species, although admittedly this approach is relatively new. One of the first studies investigating the existence and fate of patches was conducted by Olesen and Sand-Jensen (1994), who examined the patch dynamics of *Zostera marina* in a semienclosed embayment in northern Denmark. Their study recorded that some patches displayed a high mortality rate in their system. Vidondo et al. (1997) also found that many *Cymodocea nodosa* patches in their Mediterranean study site had a high rate of patch mortality. They also reported patch colonization in their system, with the colonization rate exceeding the patch extinction rate (Vidondo et al., 1997). Other reports of the fates of patches are found in a variety of studies (e.g., Bell et al., 1999; Ramage and Schiel, 1999; Robbins and Bell, 2000; Jensen and Bell, 2001).

These studies reiterate that formation of seagrass patches may be a result of a variety of processes encompassing colonization and extinction. In marine as well as other systems, disturbance has been identified as a major factor underlying spatial heterogeneity of plant populations by often changing the substrate, the physical environmental, and/or resource availability (Farina, 1998). This has been invoked to explain the variation in seagrass distribution as well (Marba and Duarte, 1995; Bell et al., 1999). Physical disturbance, whether biological (Townsend and Fonseca, 1998) and/or physical (tidal currents, waves, and sedimentation [Fonseca and Bell, 1998; Robbins and Bell, 2000]; anoxia [Campbell, 2003]; ice scouring [Robertson and Mann, 1984]), may shape the spatial patterns of seagrass landscapes by causing gaps within seagrass beds (Bell et al., 1999) and by influencing the lateral growth behavior of the seagrass itself (Marba and Duarte, 1995; Townsend and Fonseca, 1998; Fonseca et al., 2002). Disturbance appears to mold seagrass spatial signature, especially if seagrasses cannot quickly reoccupy the disturbed area of a seagrass bed, and provides a possible explanation for patch persistence/extinction.

In some locations seagrass patch extinction may be linked to a pathogenic stage of a disease caused by protists of the genus *Labyrinthula*. Seagrass blades harboring *Labyrinthula* may display extensive lesion coverage and high incidence of infection (Burdick et al., 1993). Ralph and Short (2002) recently showed that *Labyrinthula* spp. is a primary pathogen of the seagrass *Z. marina*, and that infection by this saprophytic "slime mold" may increase the mortality rate of seagrass shoots. Infection of *Z. marina* by *L. zosterae* is thought to have caused the "wasting disease" of the 1930s and loss of seagrass in the United States and eastern Europe (deJonge and deJonge, 1992), and is suspected to be a primary cause in the decline of some seagrass patches more recently (Short et al., 1986, 1988; Burdick et al., 1993; see also Bowles and Bell, in press).

Algal blooms of rhizophytic macroalgae (genus *Caulerpa*) may also impact seagrass patch persistence. Such blooms have become common events in selected geographic locations (e.g., *C. taxifolia* in the Mediterranean; Meinesz and Hesse, 1991). This rapidly growing species can spread by fragmentation or stolon elongation and appears to outcompete other vegetation in the shallow-water habitats. In fact, the macroalgae may displace seagrass by overgrowing seagrass blades or invading bare areas more quickly than seagrass (Ceccherelli et al., 2000). Additionally the extensive canopy cover produced by *Caulerpa* may lead to anoxia of underlying sediments, thereby damaging seagrass rhizomes. Therefore, by a variety of mechanisms, *Caulerpa* can change the patch structure of seagrass landscapes.

Patterns of seagrass spatial persistence are also influenced by animal activities, with evidence suggesting that fauna may be responsible for modifying/disrupting seagrass patch structure. For example, stingray disturbance may maintain the spatial structure of these *Z. marina* patches (Townsend and Fonseca, 1998). In addition, herbivores of seagrasses such as urchins, turtles, birds, and marine mammals, may also contribute to patch structure by feeding on plants. In some cases patches may be destroyed, as evidenced by some recent examples of extensive destruction of wide swaths of seagrass beds by large aggregations of sea urchins (Rose et al., 1999; Alcoverro and Mariani, 2002). Dugongs, which feed on both above- and belowground components of plants, are reported to leave feeding trails and thus enhance patch formation within some seagrass beds, especially those of *Halophila ovalis* (Nakaoka and Aioi, 1999; Masini et al., 2001). These herbivore activities are highly variable in space and time, but in selected cases herbivore grazing events may remove significant amounts of seagrass and alter extinction rates of patches.

In addition to biological agents, with activities that may contribute to the spatial arrangement of the seagrass patches, physical factors such as sediment movement may cause patch mortality, and physical energy may influence the large-scale patterns of vegetation structure within seagrass landscapes. For example, Marba and Duarte (1995) and Marba et al. (1994) reported that the seagrass (*C. nodosa*) off northeastern Spain was arranged into a large number of patches, and this was strongly correlated to the inshore migration of sand waves. Physical setting modifying seagrass (mainly *H. wrightii*) landscapes was suggested by Robbins and Bell (2000), who argued that hydrodynamic factors were mainly responsible for observed seagrass distributions. Blowouts have been previously discussed as events contributing to patch loss (Patriquin, 1975) and recently have been reported in Australian seagrass beds (Kirkman and Kirkman, 2000). Strong physical forces on seagrasses may also affect long-term patterns of seagrass patch dynamics by influencing sexual reproduction. Inglis (2000b) noted that the formation of persistent seed banks was reduced in seagrass habitats in Australia that experienced strong water flow, and suggested that large-scale patterns of seed distributions may reflect historical patterns of seed production and disturbance. A general discussion of the relationship between hydrodynamic regime and seagrass spatial structure over regional scales is available in Fonseca and Bell (1998) and Fonseca et al. (2002).

Although information on patch extinction is beginning to be assembled, detecting patch origins/colonization is more problematic and often is gleaned from a serendipitous event. Hammerstrom et al. (2006) documented the existence of a seed bank for *H. decipiens* in offshore seagrass beds, which serves as the dominant mode of new seagrass patches. Robbins and Bell (2000) reported the appearance of new patches of *H. wrightii* over a 2-year period. Asexual growth was strongly implicated as the mechanism underlying patch growth in this study, because seeds were not produced by plants at these, or nearby, sites.

The transport of viable fragments of seagrass shoots over large distances remains the suggested mode of seagrass recruitment into unvegetated areas in some locations. Harwell and Orth (2001) provided support for this theory by showing that reproductive shoots of *Z. marina* were present in 70% of polychaete (*Diopatra*) tube structures examined in the Chesapeake Bay. The authors suggested that perhaps significant worm presence can be a catalyst for the colonization or recovery of seagrasses. However, the relative contribution of dislodged and transported seagrass shoots to the establishment of new seagrass patches remains understudied for most species (see also Harwell and Orth, 2002). Campbell (2003) reported a case of successful fragment establishment, with 33% of fragments of the seagrass *Posidonia australis* that recruited to an area successfully producing new rhizomes. In contrast, fragments of *P. coriacea* showed no similar response. Albeit rare, yet successful, establishment of fragments could lead to patch persistence, and documentation of such occurrences is a missing element in the study of patch dynamics.

Thus, formation and extinction of new seagrass patches may arise from a number of biotic and abiotic processes that may vary with respect to both the temporal and spatial scale over which they operate and their impact on different species (Fig. 11-1 and Table 11-1). Moreover, the potential for connectivity between patches as a function of a dispersal mechanism may also vary spatially and temporally, and such mechanisms are summarized in Table 11-1. As is evident from the summary in Table 11-1, there are numerous mechanisms maintaining connectivity of populations across seagrass taxa, and these vary with respect to both the distance over which they are thought to be effective and their frequency of occurrence. However, a major challenge remains to gather more quantitative information on dispersal distances, especially those involving infrequent events that occur over large spatial scales.

IV. EXAMPLES OF POTENTIAL SEAGRASS METAPOPULATIONS

True metapopulations for any group aligned with the classical Levins model may be difficult to identify (Harrison and Taylor, 1997). A more relaxed definition of a metapopulation may reflect regional assemblages of populations of seagrasses with asynchronous extinction of patches, but with migration allowing connectivity among patches (Harrison and Taylor, 1997; see also Kritzer and Sale, 2003). However, although neither of these definitions of a metapopulation has been directly applied to seagrass populations, some species have ecological attributes that may make them logical candidates for future examination. In the following

Chapter 11 Seagrasses and the Metapopulation Concept

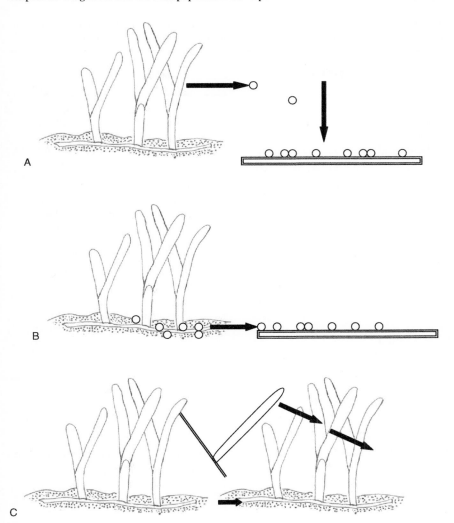

FIGURE 11-1. Major mechanisms of colonization of new seagrass patches (right) from regional sources (left). Arrows indicate direction of propagule dispersal; circles indicate seeds. (A) Seed dispersal through the water column and settlement into a new patch. (B) Seed dispersal through bedload transport. (C) Asexual colonization of a new patch by fragmented seagrass moving via the water column and then settling into sediments or by rhizome extension through the sediment. See Table 11-1 for the spatial scale and temporal frequency of these and other mechanisms for connectivity of seagrass populations.

TABLE 11-1. Comparison of Mechanisms by Which Seagrass Patches May be Connected and the Suggested Spatial Scale and Temporal Frequency Over Which They Operate

Mechanism	Spatial Scale	Temporal Scale or Frequency	Examples	References
Rhizome extension	Meters	Common	*Halodule wrightii*	Jensen and Bell, 2001
Sea grass fragment	Meters–?	Rare	*Zostera marina*	Ewanchuk and Williams, 1996; Reusch, 2002
Bedload transport of seeds	<10–>100 m	Infrequent to frequent	*Halodule uninervis, Zostera marina, Halophila decipiens*	Orth et al, 1994; Inglis, 2000a; Bell et al. (submitted)
Seed dispersal	Meters, 100 km	Frequent	*Syringodium filiforme, Enhalus acoroides, Thalassia hemprichii*	Lacap et al., 2002; Kendall, 2004
Human transport	10–100 km	n.a.	*Thalassia hemprichii, Halophila johnsoni*	Rollon et al., 2001; Kenworthy (unpublished)
Animal transport	10–100 km??	n.a.	*Zostera marina*	Thayer et al., 1984

Examples of sea grass species are provided. See Figure 11–1 for additional illustrations.
n.a. = no information available.

sections, three species of seagrasses are discussed with respect to one or more critical components of metapopulation theory: colonization, extinction, and dispersal.

A. *HALOPHILA DECIPIENS* ON THE WEST FLORIDA SHELF: A LOCAL POPULATION THAT EXHIBITS PATCH EXTINCTION AND REGIONAL RECRUITMENT

Halophila decipiens is the only tropical circumglobal seagrass, living subtidally at approximately 6.0- to 30-m depths (Kenworthy, 2000). The species is has been recorded in unconsolidated sediments across the eastern Gulf of Mexico (the West Florida Shelf), distributed widely and attaining highest densities in this geographic region. *H. decipiens* is characterized by small blades (2–3 cm in length) only two cells thick and rapid leaf–pair turnover (days), and displays a strong seasonal pattern of growth (June–October) controlled by a combination of light and temperature (Dawes et al., 1989). The seagrass releases small seeds (approximately 0.5 mm in diameter) into the sediment, thus contributing to the formation of a seed bank. After the growing season roots, rhizomes and blades deteriorate. Unlike other seagrass genera, a perennial, extensive rhizome system does not form, and reappearance of beds is from seed banks and perhaps some, extremely rare, overwintering vegetative fragments.

Based on sequential underwater video and spatial mapping of the underwater habitats on the West Florida Shelf, Bell et al. (submitted) noted that the patterns of seagrass and hard-bottom cover were altered dramatically over a 1-year period. Specifically, they recorded that sand and *H. decipiens* appeared at a portion of a site that had been hard bottom during the previous year. Because virtually all *H. decipiens* shoots die at the end of the growing season and seeds were found within the sandy sediments, Bell et al. (submitted) argued that this must have been a result of seeds being transported into the location along with the sand. Thus, secondary seed dispersal appears to explain the patterns of seagrass distribution on the West Florida Shelf, with some seeds moving at least 150 m over a year (Chambers and McMahon, 1994). Although seeds released by this seagrass may also fall near parent plants, movement en masse of the seagrass seed reservoir appears to be an important component of dispersal. The mechanisms responsible for transporting the seed bank in this coastal area are linked to physical disturbance/transport of the unconsolidated sediments within which seeds were found (Fig. 11-2), with hurricanes strongly implicated as the physical process moving the sediments (see also Phillips et al., 1990).

Thus over large spatial (or regional) scales, patches of suitable habitat may be represented by sandy sediments that have a heterogeneous distribution, and hard-bottom areas would be considered unsuitable habitats on the West Florida Shelf. The relative proportion of suitable to unsuitable sites, often of interest in discussions of metapopulations, is unknown at this time, however. Seagrass beds here are almost exclusively dependent on dispersal by seeds, and dispersal occurs mainly via sediment movement. Patches of seagrass on the West Florida Shelf may go extinct if sediments are moved away from, or no seeds move into, the area. Finally, the redistribution of sediment and the seed bank possibly from a hurricane appears to mold the spatial signature of the resulting seagrass landscape in this offshore area.

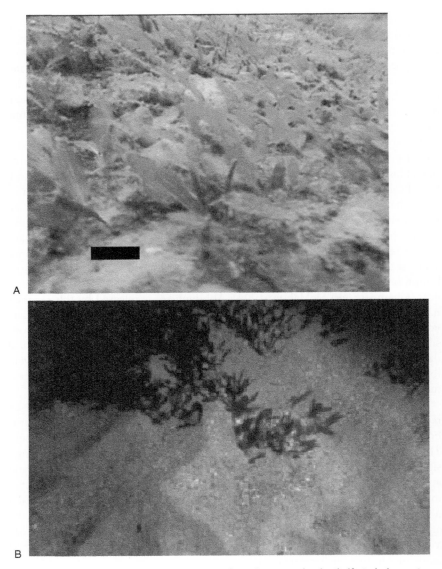

FIGURE 11-2. (A) Patch of *Halophila decipiens* from the West Florida Shelf. Scale bar = 1 cm. (B) Edge of *H. decipiens* patch from West Florida Shelf showing sediments with marked ripples. Photos courtesy of M.O. Hall. (See color plate)

B. *Phyllospadix scouleri* on Exposed Pacific Shores: Regional Patterns of Suitable Sites Exist Among a Matrix of Unsuitable Sites; Patch Extinction Documented

Phyllospadix scouleri inhabits the Pacific coast of North America with distributions limited to hard substrate in rocky intertidal and shallow subtidal areas. *P. scouleri* is a perennial and is dioecious, producing both male and female flowers. Barbed seeds are released by this species, which subsequently disperse in the water column. Fruits are negatively buoyant and drift along the bottom until they locate a suitable host site for germination. After finding a suitable attachment site, the plant undergoes vegetative growth and patches are maintained by asexual growth. Information on patch growth of a related species, *P. torreyi*, suggests that the species grows slowly, with single rhizomes extending but 1 m in 2 years (Stewart, 1989).

The heterogeneous distribution of *P. scouleri* exemplifies a scenario in which suitable sites within the intertidal/shallow subtidal can be well identified. This species requires branched algae for successful seed establishment and will not settle on its own species (Turner, 1983). Seeds attach to the algal structure, germinate, and roots are eventually produced that then attach the plant to rocks. Thus, if any areas lack or have a low amount of branched algae, recruitment of *Phyllospadix* will not occur and, unlike many other seagrass species, unsuitable conditions can be relatively easy to identify.

Within the hard substrate, patches go extinct via disturbance from waves, desiccation, and sand burial (Turner, 1983). Local extinction of patches from a number of mortality sources is known to occur, but the spatial extent of extinctions and source of new colonists are not currently available. Ultimately the colonization of new patches would require recruits arriving from seed dispersal. In this system, large-scale spatial distribution and patch structure would ultimately depend upon the availability and spatial distribution of branched algae. By examining large-scale spatial structure and patch dynamics of this seagrass, along with measuring seed dispersal, an assessment of the interconnectedness of a regional assemblage of local patches is possible.

C. *Halophila johnsonii* in Southeastern Florida: A Well-defined Regional Population with Limited Dispersal between Patches; Extinction of Patches Documented

Halophila johnsonii, measuring 0.5 to 2.5 cm in length, has been reported to inhabit only the southeastern coast of Florida between 25°45′N and 27°50′N latitude. Genetic analyses suggest that the plant is most closely related to a congener in the Indo-Pacific (Jewitt–Smith et al., 1997). The species has been charted in intertidal and shallow subtidal areas to a depth of 3 m, in both clear and turbid waters. This species is rare, comprising an average of 2% cover over its distributional range. *H. johnsonii* is known to exist in discrete patches, with dimensions ranging from $1\,cm^2$ to $300\,m^2$, and yearly fluctuations in seagrass patches have been reported (Morris et al., 2001). Patches are thought to go extinct from sediment instability.

New patches of *H. johnsonii* are produced by asexual reproduction. The species produces flowers in the northern part of its range, but only female flowers have been recorded (Jewitt–Smith et al., 1997). In addition, because no seeds have been found to date, seed dormancy does not appear to be an issue for this population. Plant fragments have been collected from the coastal drift line and floating in the water, both of which are capable of producing new plants. Moreover, plant fragments may be transported by boating activities, which are extensive in the area (J. Kenworthy, personal communication).

Patches of *H. johnsonii* may display extinctions and colonization, with asexual reproduction maintaining the population. The species is being considered for placement on the endangered species list because of its restricted range and limited dispersal abilities. The spatial extent of patch extinction and colonization remains unknown, however, and the small size and low frequency of occurrence of this plant will require special sampling strategies to document accurately the persistence of patches in this setting. Hanski and Simberloff (1997) suggest that local breeding populations that are distinct and small are very useful for a metapopulation approach, and thus *H. johnsonii* may be a logical candidate for further investigation.

V. THE METAPOPULATION MODEL AND SEAGRASS POPULATIONS: A USEFUL CONCEPT?

Grimm et al. (2003) presented an overview of the metapopulation concept and asked if it was applicable to marine populations in general. In their discussion, Grimm et al. (2003) commented that fragmented seagrass habitats might serve as a basis for metapopulations of animals, but they did not address the usefulness of the metapopulation model for marine plants. However, two of their suggestions for improvements in a regional approach to marine populations, and hence consideration of the metapopulation concept, serve as excellent foci for future investigations on seagrasses. Likewise, review of the information outlined in this chapter reveals that some critical information on seagrass populations and their spatial structure is currently lacking and accordingly represent areas for further exploration in the process of evaluating the usefulness of metapopulation models. I address these suggestions for future research and information gaps in the following sections.

A. Collection of Good Information on Spatial Organization of Seagrasses

The application of metapopulation theory requires that an adequate representation of the spatial structure of a population and data on the incidence of patch extinction and colonization across regional spatial scales be obtained. For seagrasses, this would require large-scale (e.g., landscape) monitoring of seagrass distribution over a time period designed to detect patch extinctions and colonization. Asynchronous extinction of patches is suggested to be an integral part of metapopulation models, and comparison of the timing of extinction across patches is thus desirable. In reality, however, the lack of data on spatial structure of seagrasses is a major obstacle hindering the evaluation of seagrasses as metapopulations. To date there are but a handful of studies that directly investigate seagrass landscape ecology (e.g., Robbins and Bell, 2000), although many address issues that help interpret patch dynamics. One problem lies with obtaining adequate spatial representation of seagrasses, especially over a regional-level landscape level. Although the collection of data on underwater landscapes still poses some formidable problems (Robbins and Bell, 1994), advances within remote sensing and the increasing use of GIS for analyses, seagrass landscape data should soon become more accessible to researchers.

Another problem lies in the temporal extent of seagrass studies. The time period of observation should be aligned with extinctions, which may range in duration from occurring frequently (within years) or at low frequency (e.g., hurricanes) spanning decades (Table 11-1). If even infrequent events become important in the maintenance of metapopulations, then extensive spatial information on seagrasses may be requisite over long time periods to document the dynamics of both local and regional populations. Some historical records of seagrass spatial distributions do exist, and quantification of patch turnover in these systems (e.g., Frederiksen et al., 2004) may serve as a starting point for application of the metapopulation concept. The warning of Dupre and Ehrlen (2002) is especially appropriate for interpretation of such spatial representations, however. They noted that an isolated patch of a selected species may not be the result of successful dispersal, and colonization but may represent a remnant of a formerly continuous distribution. Use of chronological sequences of seagrass coverage along with estimates of seagrass seed dormancy, albeit poorly known at this time, could provide critical information to distinguish between these alternatives.

B. Collection of Genetic Information to Help Analyze Spatial Structure of a Population

There is an emerging set of studies on seagrasses that can be used as a basis for developing investigations on genetic structure of regional populations. The studies conducted to date provide a variety of results. For example, recent studies on the Caribbean seagrass *Thalassia testudinum* have reported a high degree of genotypic similarity among samples collected across large geographical distances (Schlueter and Guttman, 1998; Waycott and Barnes, 2001), and dispersal of vegetative fragments has been invoked as a possible explanation for such patterns. In contrast, Fain et al. (1994) reported that geographically separated populations of the same species (*Z. marina*) were found to be genetically distinct. Similarly, Procaccini and Mazzella (1998) reported genetic differentiation between populations separated by large geographical distances. For other seagrass species, dispersal of propagules on fruit-bearing shoots has been proposed as a major factor influencing gene flow among populations, and movement of propagules underlies discussions of seagrasses as metapopulations linked via frequent medium to long-distance gene flow events (Reusch, 2002).

Overall, the information on genetic differences among local and regionally connected seagrass populations is beginning to be assembled. However, results appear to be highly dependent on the method used to assess genotypic variability (e.g., see Waycott, 1998). But as Sork et al. (1999) note, interest in metapopulation and regional-level dispersal of plants has fostered the development of new approaches to estimate gene flow based upon combined information on dispersal distances and patterns of patch distribution over large spatial scales. Thus, critical examination of metapopulation theory will encourage that a landscape perspective on demography and genetics be developed (e.g., Husband and Barnett, 1996). Although this approach has been initiated in terrestrial settings, similar approaches await application in seagrass systems.

C. Seed Dormancy: A Special Problem for Plant Populations

Seed dormancy certainly has major implications for application of metapopulation models in seagrasses (see Freckleton and Watkinson, 2002, for more details). For some seagrasses, a persistent seed bank is found in sediments. Thus, although the aboveground structure of plants may disappear from a patch, the population may still remain viable in that patch because of a seed bank. Thus seed dormancy plays a complicating role in what is considered extinction of patches. For metapopulation theory to operate, it is essential that the patch that goes extinct be colonized by propagules from a distant area. Thus it will be necessary not only to know rates of colonization and extinction of patches for these species of seagrasses, but also to identify the source of new colonists for the patches. As Freckleton and Watkinson (2002) stated when considering metapopulation models, we are now concerned not only with dispersal in space but also dispersal in time (via the seed bank). For seagrasses, data on seed dormancy are sparse and in need of attention (see Ortho et al., 2000).

VI. SUMMARY

Orth et al. (1994) were among the first to use the term *metapopulation* in reference to seagrasses. They argued that *Z. marina* seeds were capable of only very limited dispersal under normal conditions, and that reliance on surviving patches to recolonize "sink" populations of this species at even moderate distances was probably not feasible. In contrast, Inglis (2000b) argued that measures for protection of seagrass populations must consider both local and metapopulation dynamics, given that local recruitment was often influenced by regional abundance of seagrasses and dispersal patterns. He also suggested that some populations were a major source of all propagules. Such discordance among studies is not unexpected. It is highly probable that the possibility of metapopulation structure operating in seagrasses will, in fact, vary among species and even within species from different locations, given the wide variety of biological and ecological characteristics outlined in this chapter. As a next step it will be necessary for seagrass studies to adopt a regional or landscape perspective of patch colonization, extinction, and dispersal, setting the stage for more detailed assessment of metapopulation biology and a more detailed understanding of the persistence of seagrass populations on the local scale. In fact, the documentation of extinction and colonization of seagrass local populations makes these systems somewhat uniquely capable of application of metapopulation theory, in contrast to other marine taxa for which the models do not seem to apply strictly or for which extinction and colonization are difficult to assess (see Kritzer and Sale, 2003). Collection of more detailed demographic information on seagrasses, and improved descriptions of seagrass spatial structure, are required to evaluate better whether a regional view of local persistence of seagrasses is appropriate and informative.

VII. ACKNOWLEDGMENTS

Comments by the editors and two anonymous reviewers helped focus the ideas in this chapter. This work was supported in part by a grant from the National Science Foundation (OCE0337052) to F. Thomas and S. Bell.

REFERENCES

Alcoverro, T., and Mariani, S. (2002). Effects of sea urchin grazing on seagrass (*Thalassodendron ciliatum*) beds of a Kenyan lagoon. *Mar. Ecol. Prog. Ser.* **226**, 255.

Bell, S.S., Robbins, B.D., and Jensen, S.L. (1999). Gap dynamics in a seagrass landscape. *Ecosystems.* **2**, 493.

Bowles, J.W., and Bell, S.S. (2004). Simulated herbivory and the dynamics of disease in *Thalassia testudinum*. *Mar. Ecol. Prog. Ser.*

Bullock, J.M., Moy, I.L., Pywell, R.F., Coulson, S.J., Nolan, A.M., and Caswell, H. (2002). Plant dispersal and colonization processes at local and landscape scales. In *Dispersal ecology* (J.M. Bullock, R. Kenward, and R. Hails, eds., pp. 279–302). Malden, MA: Blackwell Scientific.

Burdick, D.M., Short, F.W., and Wolf, J. (1993). An index to assess and monitor the progression of wasting disease in eelgrass *Zostera marina*. *Mar. Ecol. Prog. Ser.* **94**, 83.

Campbell, M.L. (2003). Recruitment and colonization of vegetative fragments of *Posidonia australis* and *Posidonia coriacea*. *Aquat. Bot.* **76**, 175.

Ceccherelli, G., Piazzi, L., and Cinelli, F. (2000). Response of the non-indigenous *Caulerpa racemosa* (Forsskal) J. Agardh to the native seagrass *Posidonia oceanica* (L.) Delile: Effect of density of shoots and orientation of edges of meadows. *J. Exp. Mar. Biol. Ecol.* **243**, 227.

Chambers, J.C., and McMahon, J.A. (1994). A day in the life of a seed: Movements and fates of seeds and their implications for natural and managed systems. *Ann. Rev. Ecol. Syst.* **25**, 263.

Dawes, C.J., Loban, C.S., and Tomasko, D.A. (1989). A comparison of the physiological ecology of the seagrasses *Halophila decipiens* Gatenfeld and *H. johnsonii* Eiseman from Florida. *Aquat. Bot.* **33**, 149.

deJonge, V.N., and deJonge, D.J. (1992). Role of tide, light and fisheries in the decline of *Zostera marina* in the Dutch Wadden Sea. *Neth. Instit. Sea Res. Publ. Ser.* **20**, 161.

Dupre, C., and Ehrlen, J. (2002). Habitat configuration, species traits and plant distributions. *J. Ecol.* **90**, 796.

Ehrlen, J., and Ericksson, O. (2003). Large scale spatial dynamics of plants: A response to Freckleton & Watkinson. *J. Ecol.* **91**, 316.

Ewanchuk, P.J., and Williams, S.L. (1996). Survival and re-establishment of vegetative fragments of eelgrass (*Zostera marina*). *Can. J. Bot.* **74**, 1584.

Fain, S.R., DeTomaso, A., and Alberte, R.S. (1994). Characterization of disjunct populations of *Zostera marina* (eelgrass) from California: Genetic differences resolved by restriction–fragment length polymorphism. *Mar. Biol.* **112**, 683.

Farina, A. (1998). *Principles and methods in landscape ecology.* New York: Chapman and Hall.

Fonseca, M.S., and Bell, S.S. (1998). Influence of physical setting on seagrass landscapes near Beaufort, North Carolina, U.S.A. *Mar. Ecol. Prog. Ser.* **171**, 109.

Fonseca, M.S., Whitfield, P.E., Kelly, N.M., and Bell, S.S. (2002). Modeling seagrass landscape pattern and associated ecological attributes. *Ecol. Appl.* **12**, 218–237.

Freckleton, R.P., and Watkinson, A.R. (2002). Large scale spatial dynamics of plants: Metapopulations, regional ensembles, and patchy populations. *J. Ecol.* **90**, 419.

Frederiksen, M., Krause-Jensen, D., Holmer, M., and Sund Laursen, J. (2004). Long-term changes in area distribution of eelgrass (*Zostera marina*) in Danish coastal waters. *Aquat. Bot.* **78**, 167.

Grimm, V., Reise, K., and Strasser, M. (2003). Marine metapopulations: A useful concept? *Helgol. Mar. Res.* **56**, 222.

Hammerstrom, K., Kenworthy, W.J., Fonseca, M.S., Whitfield, P.E., and Willis, K.L. (2006). Seedbank biomass and productivity of *Halophila decipiens*, a deep water seagrass on the West Florida continental shelf. *Aquat. Bot.* (in press).

Hanski, I.A., and Gilpin, M.E., eds. (1997). *Metapopulation biology: Ecology, genetics and evolution*. San Diego: Academic Press.
Hanski, I.A., and Simberloff, D. (1997). The metapopulation approach, its history, conceptual domain and application to conservation. In *Metapopulation biology: Ecology, genetics and evolution* (I.A. Hanski and M.E. Gilpin, eds., pp. 5–26). New York: Academic Press.
Harrison, S., and Taylor, A.D. (1997). Empirical evidence for metapopulation dynamics. In *Metapopulation biology: Ecology, genetics and evolution* (I.A. Hanski and M.E. Gilpin, eds., pp. 27–42) New York: Academic Press.
Harwell, M.C., and Orth, R.J. (2001). Influence of a tube-dwelling polychaete on the dispersal of fragmented reproductive shoots of eelgrass. *Aquat. Bot.* **70**, 1.
Harwell, M.C., and Orth, R.J. (2002). Long distance dispersal potential in a marine macrophyte. *Ecology.* **83**, 3319.
Hemminga, M.A., and Duarte, C.M. (2000). *Seagrass ecology.* Cambridge, UK: Cambridge University Press.
Husband, B.C., and Barrett, S.C.H. (1996). A metapopulation perspective in plant and population biology. *J. Ecol.* **84**, 461.
Inglis, G.J. (2000a). Disturbance-related heterogeneity in the seed banks of a marine angiosperm. *J. Ecol.* **88**, 88.
Inglis, G.J. (2000b). Variation in the recruitment behaviour of seagrass seeds: Implications for population dynamics and resource management. *Pac. Conserv. Biol.* **5**, 251.
Jensen, S.L., and Bell, S.S. (2001). Seagrass growth and patch dynamics: Cross-scale morphological plasticity. *Plant Ecol.* **155**, 201.
Jewitt-Smith, J., McMillan, C., Kenworthy, W.J., and Bird, K. (1997). Flowering and genetic banding patterns of *Halophila johnsonii* and conspecifics. *Aquat. Bot.* **59**, 323.
Kaldy, J.E., and Dunton, K.H. (1999). Ontogenetic photosynthetic changes, dispersal and survival of *Thalassia testudinum* (turtle grass) seedlings in a sub-tropical lagoon. *J. Exp. Mar. Biol. Ecol.* **240**, 193.
Kendall, M.S., Battista, T., and Hillis-Starr, Z. (2004). Longterm expansion of a deep *Syringodium filiforme* meadow in St. Croix, U.S. Virgin Islands: the potential role of hurricanes in the dispersal of seeds. *Aquat. Bot.* **78**, 15.
Kendrick, G.A., Eckersley, J., and Walker, D.I. (1999). Landscape-scale changes in seagrass distribution over time: A case study from Success Bank, Western Australia. *Aquat. Bot.* **65**, 293.
Kenworthy, W.J. (2000). The role of sexual reproduction in maintaining populations of *Halophila decipiens*: Implications for the biodiversity and conservation of tropical seagrass ecosystems. *Pac. Conserv. Biol.* **5**, 260.
Kirkman, H., and Kirkman, J. (2000). Long-term seagrass monitoring near Perth, Western Australia. *Aquat. Bot.* **67**, 319.
Kritzer, J.P., and Sale, P.F. (2003). Metapopulation ecology in the sea: From Levins' model to marine ecology and fisheries science. *Fish Fisher.* **4**, 1.
Lacap, C.D.A., Vermaat, J.E., Rollon, R.N., and Nacorda, H.M. (2002). Propagule dispersal of the SE Asian seagrasses *Enhalus acoroides* and *Thalassia hemprichii*. *Mar. Ecol. Prog. Ser.* **235**, 75.
Levins, R. (1970). Extinction. In *Some mathematical questions in biology* (M. Gerstenhabu, ed., pp. 77–107) Providence, RI: American Mathematical Society.
Marba, N., Cebrian, J., Enriquez, S., and Duarte, C.M. (1994). Migration of large-scale subaqueous bedforms measured with seagrasses (*Cymodocea nodosa*) as tracers, *Limnol. Oceanogr.* **39**, 126.
Marba, N., and Duarte, C.M. (1995). Coupling of seagrass (*Cymodocea nodosa*) patch dynamics to subaqueous dune migration. *J. Ecol.* **83**, 381.
Marba, N., and Duarte, C.M. (1998). Rhizome elongation and seagrass clonal growth. *Mar. Ecol. Prog. Ser.* **174**, 269.

Masini, R.J., Anderson, P.K., and McComb, A.J. (2001). A *Halodule*-dominated community in a subtropical embayment: Physical environment, productivity, biomass and impact of dugong grazing. *Aquat. Bot.* **71**, 179.

Meinesz, A., and Hesse, B. (1991). Introduction of the tropical alga *Caulerpa taxifolia* and its invasion of the northwest Mediterranean. *Oceanol. Acta.* **14**, 415.

Morris, L.J., Hall, L.M., and Virnstein, R.W. (2001). *Field guide for fixed transect monitoring in the Indian River Lagoon*. Palatka, FL: St. Johns River Water Management District.

Nakaoka, M. (2001). Small-scale variation in a benthic community at an intertidal flat in Thailand: Effects of spatial heterogeneity of seagrass vegetation. *Benthos Res.* **56**, 63.

Nakaoka, M., and Aioi, K. (1999). Growth of seagrass *Halophila ovalis* at dugong trails compared to existing within-patch variation in a Thailand intertidal flat. *Mar. Ecol. Prog. Ser.* **184**, 97.

Olesen, B., and Sand-Jensen, B. (1994). Patch dynamics of eelgrass *Zostera marina*. *Mar. Ecol. Prog. Ser.* **106**, 147.

Orth, R.J., Harwell, M.C., Bailey, M.E., Bartholomew, A., Jawad, J.T, Lombana, A.V., Moore, K.A., Rhode, J.M., and Woods, H.E. (2000). A review of issues in seagrass seed dormancy and germination: Implications for conservation and restoration. *Mar. Ecol. Prog. Ser.* **200**, 277.

Orth, R.J., Harwell, M.C., and Inglis, G.I. (2004). Ecology of seagrass seeds and seagrass dispersal processes: Emerging paradigms. In *Seagrass biology: A treatise* (T. Larkum, R. Orth, and C. Duarte, eds.), pp. 111–134. New York: Kluwer.

Orth, R.J., Luckenbach, M., and Moore, K.A. (1994). Seed dispersal in a marine macrophyte: Implications for colonization and restoration. *Ecology.* **75**, 1927.

Patriquin, D.G. (1975). "Migration" of blowouts in seagrass beds at Barbados and Carriacou, West Indies, and its ecological and geological implications. *Aquat. Bot.* **1**, 163.

Peterken, C.J., and Conacher, C.A. (1997). Seed germination recolonisation of *Zostera capricorni* after grazing by dugongs. *Aquat. Bot.* **59**, 333.

Phillips, N.W., Gettleson, D.A., and Spring, K.D. (1990). Benthic biological studies of the southwest Florida Shelf. *Amer. Zool.* **30**, 65.

Procaccini, C., and Mazzella, L. (1998). Population genetic structure and gene flow in the seagrass *Posidonia oceanica* assessed using microsatellite analysis. *Mar. Ecol. Prog. Ser.* **169**, 133.

Ralph, P.J., and Short, F.T. (2002). Impact of the wasting disease pathogen, *Labyrinthula zosterae*, on the photobiology of eelgrass *Zostera marina*. *Mar. Ecol. Prog. Ser.* **226**, 265.

Ramage, D.L., and Schiel, D.R. (1999). Patch dynamics and response to disturbance of the seagrass *Zostera novazelandica* on intertidal platforms in southern New Zealand. *Mar. Ecol. Prog. Ser.* **189**, 275.

Rasheed, M.A. (1999). Recovery of experimentally created gaps within a tropical *Zostera capricorni* (Aschers) seagrass meadow, Queensland Australia. *J. Exp. Mar. Biol. Ecol.* **235**, 183.

Reusch, T.B. (2002). Microsatellites reveal high population connectivity in eelgrass (*Zostera marina*) in two contrasting coastal areas. *Limnol. Oceanogr.* **47**, 78.

Robbins, B.D., and Bell, S.S. (1994). Seagrass landscapes: A terrestrial approach to the marine subtidal environment. *Trends Ecol. Evol.* **9**, 301.

Robbins, B.D., and Bell, S.S. (2000). Dynamics of a subtidal seagrass landscape: Seasonal and annual change in relation to water depth. *Ecology.* **81**, 1193.

Robertson, A.I., and Mann, K.H. (1984). Disturbance by ice and life-history adaptations of the seagrass *Zostera marina*. *Mar. Biol.* **80**, 131.

Rollon, R.N., Cayabyab, N.M., and Fortes, M.D. (2001). Vegetative dynamics and sexual reproduction of monospecific *Thalassia hemprichii* meadows in the Kalayaan Island Group. *Aquat. Bot.* **71**, 239.

Rose, C.D., Sharp, W.C., Kenworthy, W.J., Hunt, J.H., Lyons, W.G., Prager, E.J., Valentine, J.F., Hall, M.O., Whitfield, P.E., and Fourqurean, J.W. (1999). Overgrazing of a large seagrass bed by the sea urchin *Lytechinus variegatus* in outer Florida Bay. *Mar. Ecol. Prog. Ser.* **190**, 212.

Ruckelshaus, M. (1996). Estimation of genetic neighborhood parameters from pollen and seed dispersal in the marine angiosperm *Zostera marina*. *Evolution.* **50**, 856.

Schlueter, M.A., and Guttman, S.I. (1998). Gene flow and genetic diversity of turtle grass, *Thalassia testudinum*, Banks ex König, in the lower Florida Keys. *Aquat. Bot.* **61**, 147.

Short, F.T., Ibelings, B.W., and den Hartog, C. (1988). Comparison of a current eelgrass wasting disease to the wasting disease in the 1930s. *Aquat. Bot.* **30**, 295.

Short, F.T., Mathieson, A.C., and Nelson, J.I. (1986). Recurrence of the eelgrass wasting disease at the border of New Hampshire and Maine, U.S.A. *Mar. Ecol. Prog. Ser.* **29**, 88.

Sork, V.L., Nason, J., Campbell, D.R., and Fernandez, J.F. (1999). Landscape approaches to historical and contemporary gene flow in plants. *Trends Ecol. Evol.* **14**, 219.

Stewart, J.G. (1989). Maintenance of a balanced shifting boundary between the seagrass *Phyllospadix* and algal turf. *Aquat. Bot.* **33**, 223.

Terrados, J. (1993). Sexual reproduction and seed banks of *Cymodocea nodosa* (Ucria) Ascherson meadows on the southeast Mediterranean coast of Spain. *Aquat. Bot.* **46**, 293.

Thayer, G.W., Bjorndal, K.A., Ogden, J.C., Williams, S.L., and Zieman, J.C. (1984). Role of herbivores in seagrass communities. *Estuaries.* **7**, 371.

Townsend, E.C., and Fonseca, M.S. (1998). Bioturbation as a potential mechanism influencing spatial heterogeneity of North Carolina seagrass beds. *Mar. Ecol. Prog. Ser.* **169**, 123.

Turner, T. (1983). Facilitation is a successional mechanism in a rocky intertidal community. *Amer. Natur.* **121**, 729.

Vidondo, B., Duarte, C.M., Middelboe, A.L., Stefansen, K., Lützen, T., and Nielsen, S.L. (1997). Dynamics of a landscape mosaic: Size and age distributions, growth and demography of seagrass *Cymodocea nodosa* patches. *Mar. Ecol. Prog. Ser.* **158**, 131.

Waycott, M. (1998). Genetic variation its assessment and implications for conservation of seagrasses. *Mol. Ecol.* **7**, 793.

Waycott, M., and Barnes, P.A.G. (2001). AFLP diversity within and between populations of the Caribbean seagrass *Thalassia testudinum* Hydrocharitaceae. *Mar. Biol.* **136**, 1021.

PART V

Perspectives

CHAPTER 12

Conservation Dynamics of Marine Metapopulations with Dispersing Larvae

LOUIS W. BOTSFORD and ALAN HASTINGS

I. Introduction
II. Single Population Persistence
III. Metapopulation Persistence
   A. Consequences for the Distribution of Meroplanktonic Species
   B. Consequences for the Success of Marine Reserves
IV. Role of Variability
V. Discussion
VI. Summary
   References

I. INTRODUCTION

The distribution of marine populations over space is important to the extent that most are not well mixed (i.e., each individual does not have an equal probability of encountering each other individual). Their dynamics therefore depend on movement, and are complicated to a point that is not easy to unravel with current analytical tools and understanding. There are two primary modes of movement in marine populations: dispersal during a larval phase in early life, and directed, migratory or random local movement of juveniles and adults throughout later life. The complexities of dynamics arise from the interaction of the structure of single populations (i.e., the age or size structure of the population) and movement of individuals of various ages and sizes across the spatial structure of the population. We currently have a reasonably good understanding of the dynamics of single populations. Even with age structure, stochasticity, and density dependence, we know what causes these populations to persist, have more or less variability, be stable, and increase or decrease. However, we do not have a similar understanding of populations with the further complications of spatial structure and movement, especially if there is spatial heterogeneity. Thus, the study of populations broken up into interconnected, well-mixed subpopulations (i.e., marine metapopulations) is currently of great interest.

In marine systems there is an interest in whether explicit consideration of the metapopulation structure changes the characteristics of population dynamics. For example, will metapopulation structure lead to greater or less variability than in single populations? How does the metapopulation structure change persistence, stability, and rates of increase? These questions necessarily take on a spatial aspect, so in addition there is a new characteristic of interest: how abundance covaries over space. Note that we are using a broader definition of metapopulation than has sometimes been used in terrestrial systems to analyze the role of local extinctions and colonizations (Kritzer and Sale, 2004). Specifically, our definition of a metapopulation is simply a number of well-mixed subpopulations connected by more restricted movement among them. In particular, for this chapter we examine the very common case in marine systems of populations with relatively sedentary adults and connection through larval dispersal only.

Broadening our approach to management problems in marine conservation dynamics to include explicit consideration of metapopulation structure increases the level of difficulty, but may be necessary to deal with real-world complexities. For example, including spatial structure in fisheries management requires we account for spatial structure and movement, and there may be further issues such as heterogeneity in fishing pressure and habitat. The question then naturally arises of whether explicitly including consideration of these issues in management would be an improvement over models and concepts based on considering only averages over space (e.g., Fogarty, 1998). Will including spatial aspects lead to better predictions? What are the extant uncertainties? Also, explicitly spatial management is under increasing consideration, in the form of marine reserves, MPAs, and rotating spatial harvest (e.g., Gerber et al., 2003; Sale et al., 2005). Do these strategies increase sustainability and reduce uncertainty as some promise? What level of understanding is necessary for spatially explicit models to achieve better results than nonspatial ones?

Because connectivity of subpopulations is clearly an essential element of metapopulation dynamics, another topical issue regarding management of marine metapopulations is the implications of increasing efforts to understand movement among marine populations. For example, various empirical efforts have indicated that larval dispersal distances may be shorter than previously believed (Todd, 1998; Swearer et al., 1999; Warner et al., 2000; Cowen et al., 2004). How would the implied changes in our view of population connectivity affect fishery management and alter the design and resultant effect of marine reserves in terms of yield and uncertainty?

Here we focus on one aspect: addressing questions related to *persistence*—the issue of primary concern in a number of conservation applications, from fisheries (which must at least be sustainable) to the preservation of biodiversity. Using several different definitions, we explore the question of what makes spatially distributed marine populations persist. We discuss implications of the results for spatially specific management in marine conservation, primarily oriented toward the design of marine reserves. The approach could also be useful in the protection of habitat and recovery of endangered species. We address issues associated with larval dispersal only, hence our results are most applicable to species with sedentary adults such as invertebrates and reef-dwelling fishes (see Sale and Kritzer, 2003, for a comparative review of recent progress in metapopulation aspects for these two groups). We focus mainly on deterministic approaches here, although we discuss the role of variability.

II. SINGLE POPULATION PERSISTENCE

During the past couple of millennia, humankind has developed an understanding of what makes single, well-mixed (i.e., nonspatial) populations continue to persist and increase or decrease. Demographic tables date back to the Roman Republic and Empire (Hutchinson 1978), and their consequence for rates of increase was understood by the 18th century with the development of Euler's relationship. Modern ecology texts (e.g., Hastings, 1997) describe the fact that single, linear (i.e., not density-dependent) populations grow geometrically, and one simple condition for population increase is that lifetime reproduction, R_0, exceed one,

$$R_0 \geq 1. \tag{1}$$

This condition expresses the condition for persistence (i.e., increasing from near-zero abundance, in terms of the question of whether each individual replaces itself). Lifetime reproduction, R_0, is related to population growth rate, λ, in that their values will always both be either greater or less than 1.0. R_0 simply lacks the timescale of a growth rate. This replacement notion has been useful in formulating human demographic concepts such as how to achieve zero population growth, but its appreciation in the marine area is limited by the fact that it arises from age-structured models that lack both density dependence and environmental variability.

Explicit consideration of the effects of a random environment on single linear populations does not change their characteristic geometric growth; they merely grow more slowly with greater environmental variance, at least as indicated by the mode of their distribution of abundance (Tuljapurkhar, 1982; Tuljapurkhar and Orzack, 1980; Lande and Orzack, 1988). The logarithm of total abundance at time t of an age-structured population with randomly varying survivals and reproductive terms (independent over time) is normally distributed with a mean that grows linearly at a rate $\ln(\bar{\lambda}) - \sigma^2$, where $\bar{\lambda}$ is the growth rate of the population with each parameter set to its mean value, and σ^2 depends on the covariance of the environmental variability among population parameters, and the sensitivities of λ to changes in those parameters (see Tuljapurkhar, 1990, or Caswell, 2001, for details; see Cisneros et al., 1997, for an example). Thus the geometric nature of the population growth is not changed, but persistence can only be described in terms of probabilities of persistence. The values of λ required for a specific probability of persistence will increase with increasing variability. Associated with these values of λ will be a specific value of R_0 for any specific population. Thus variability does not change the nature of the condition for persistence, it merely raises the required R_0.

Description of the effects of density dependence on persistence of age-structured populations arose out of investigations of the (local) stability of single populations, with density-dependent recruitment beginning in the 1970s (e.g., Botsford and Wickham, 1978; Roughgarden et al., 1985; see Botsford, 1997, for references). Each involved an expression for equilibrium, most of which allow interpretation of the condition for persistence. Because this chapter is about marine populations, we refer to the graphical interpretation from Sissenwine and Shepherd (1987), which is commonly used in fisheries management (Fig. 12-1). The equilibrium of a single population with density-dependent recruitment is determined by the intersection of the nonlinear recruit–egg dependence, and a straight line through the origin with slope 1/LEP, where LEP is lifetime egg production. Taking fishing as an example with a clear, age-specific effect on mortality, as fishing increases, the age structure is increasingly truncated and LEP declines. The equilibrium moves to the left until the point at which 1/LEP equals the slope of the egg–recruit relationship. Beyond that point the population will no longer persist. This condition is similar to the condition $R_0 = 1$ in the linear models. Because the condition for persistence occurs at the linear part of the recruit–egg dependence, we can assume that to a first approximation, small amounts of environmental variability will simply lower the threshold value of R_0 with increasing variance.

This simplified, assumption-laden model provides a reasonable means of understanding and estimating the persistence of single populations in the marine environment. Persistence declines as a fishery (or other anthropogenic influence) changes the age structure and decreases LEP, the ability of individuals in the population to replace themselves. LEP is used in the management of marine fisheries as a reference point in the precautionary approach to marine fisheries (FAO, 1995) and in the management of some U.S. fisheries. However, its use is limited by uncertainty, both in the estimated values of current LEP and in the threshold

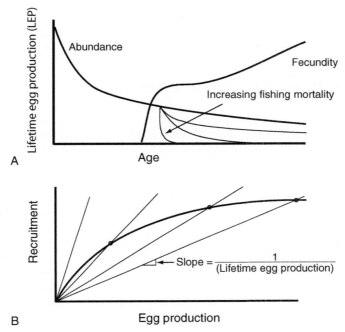

FIGURE 12-1. Changes in the age structure of a single, nonspatial population as a result of fishing, and the consequent changes in the equilibrium level of a density-dependent, age-structured population. (A) As the age structure becomes more truncated, lifetime egg production, the sum over age of the product of fecundity and survival, declines with increasing fishing mortality. (B) When lifetime egg production declines, the slope of the line whose intersection with the egg-to-recruit relationship determines equilibrium becomes steeper, moving the equilibrium to lower levels, and eventually to zero.

value indicated by the slope of the egg–recruit relationship. To address the latter to some degree, fishery scientists currently describe the threshold value in terms of a common fraction of natural, unfished LEP required for marine populations to persist (FLEP, where F stands for the word *fraction*; this is also termed *spawning production ratio* [or SPR] in the fisheries literature), with possible phylogenetic variation in that fraction (Goodyear, 1993; Mace and Sissenwine, 1993; Myers et al., 1999). Thus, a fished population is presumed to persist if

$$LEP_F/LEP \geq FLEP_{min} \qquad (2)$$

where LEP_F is defined as the lifetime reproduction of the fished population, and LEP is the lifetime egg production of the unfished, natural population. Values near $FLEP_{min} = 0.35$ have been used both as target and limit reference points on the basis of both empirical (Mace and Sissenwine, 1993) and (somewhat) theoretical (Clark, 1990) calculations. This value appears to have been too low in some cases (Clark, 2002; Ralston, 2002).

III. METAPOPULATION PERSISTENCE

Analysis of single populations provides a clear view of the requirements for sustainable individual populations, and results identify a key uncertainty in conservation management of a single population (i.e., in the value of $FLEP_{min}$), but they provide few clues regarding the persistence and management of metapopulations. Unfortunately, we do not have a simple, interpretable model or expression for persistence of interconnected populations distributed over space. We can, of course, determine persistence for specific parameter values through simulation, and can also apply multiple-population linear models to determine lack of stability about the origin (e.g., Armsworth, 2002), but understanding metapopulation persistence as clearly as single population persistence is not yet possible. When we go from the single population replacement path to multiple populations interconnected by larval dispersal (not to mention adult movement), we do not have the same level of intuition regarding what makes a population persist.

However, we can gain some level of understanding of population persistence in marine metapopulations by further interpretation of some of the results obtained thus far in the spatial design of marine reserves. Botsford et al. (2001) formulated the problem of persistence of a marine metapopulation in its simplest form—a periodic system of reserves along an infinite coastline with fishing between reserves causing complete removal. The species being protected was assumed to have sedentary adults and a larval dispersal pattern that declined exponentially over space with a specific mean dispersal distance. Dispersal was followed by a nonlinear recruit–settlement relationship such as the one shown in Figure 12-1, with settlement replacing egg production on the x-axis. They took two different approaches to the question of how population persistence depends on the spatial configurations of marine reserves. One approach was based upon existing theory, with persistence defined as lack of local stability about zero abundance (Van Kirk and Lewis, 1997); the other was an ad hoc approach based on a hockey stick stock–recruitment model and adequate point-by-point replacement. These approaches led to similar results: that populations would persist when a quantity involving the dispersal pattern and the spatial configuration of marine reserves was greater than the same minimum value of FLEP ($FLEP_{min}$) referred to earlier regarding the management of fisheries on presumed single, well-mixed populations (Equation 2). These results could be depicted in terms of the combinations of fraction of coastline in reserves and normalized size of the reserves (i.e., [reserve width]/[dispersal distance]), with $FLEP_{min}$ assumed to be

Chapter 12 Conservation Dynamics of Marine Metapopulations

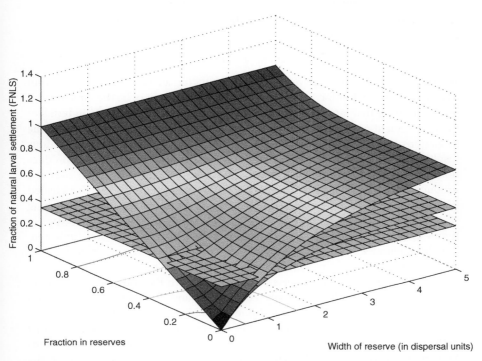

FIGURE 12-2. Metapopulation persistence as determined by the spatial structure of marine reserves along a heavily fished (assumed complete removal) coastline. Two different approaches to determining persistence (the curved surfaces)—an ad hoc approach and an application of Van Kirk and Lewis (1997)—show the values of fraction of coastline in reserves and size of the reserves (relative to the mean dispersal distance of the species) for which replacement through spatial connectivity (i.e., the fraction of natural lifetime settlement (FNLS) occurring for each configuration) exceeds the minimum required for persistence (here assumed to be 0.35, the plane).

0.35 (Fig. 12-2). Following the right-hand axis we see that a population can persist with just a single reserve if the alongshore dimension of the reserve is as large as the mean dispersal distance. Following the left-hand axis, we see that regardless of how small the reserves are, the population can persist if a specific fraction of the coastline is in marine reserves. That fraction was the same minimum FLEP required when managing the age structure of single fished populations (shown here to be 0.35). The latter result (from the left-hand axis) is referred to as a *network effect*, because persistence depends on a number of reserves acting together.

The key to understanding this condition intuitively is the interpretation of the quantity involving the dispersal pattern and the spatial configuration of marine reserves that was required to be greater than $FLEP_{min}$. For the theoretical approach, it can be written as

$$LEPp_{rr} > FLEP_{min} \qquad (3)$$

where p_{rr} is the fraction of larvae released from within a reserve that settles within a reserve. In this definition, the destination of a larva need not be the same reserve as the origin of that larva. Thus, this relationship can be understood in terms of the replacement concept used in the management of single populations, with the limitations resulting from an abbreviated age structure replaced by the limitations resulting from diminished possible areas for successful settlement and reproduction (Fig. 12-3). In an unfished, natural population, LEP is the sum over the product of the natural age structure and fecundity at age (Fig. 12-3A). For a fished population, LEP is diminished because the age structure is truncated by fishing (Fig. 12-3B). For a system of reserves and "scorched earth" fishing (i.e., complete removal, as used in this model), the age structure in reserves is not truncated, but the replacement level is reduced by the diminished area available for successful settlement and reproduction (Fig. 12-3C).

This view of fishing and reserves is consistent with the conventional ecological view of sustainability with self-recruitment, and the required relationship between reserve size and dispersal distance (Skellam, 1951). But, more important, it also aids understanding of the less commonly understood potential network effect of reserves by clarifying the reason why persistence of populations in marine reserves requires the same fraction of the coastline be in reserves as the value of LEP required for persistence in a fished population without reserves. The concept of replacement over space through dispersal is also referred to in Armsworth's (2002) results of analysis of a metapopulation model with density-dependent subpopulations.

Extensions of this theoretical approach to persistence are consistent with and build upon this result for the simplest case. Adding the possibility of fishing at a specified fishing mortality rate, rather than having complete removal, reduces the normalized size of a reserve required for persistence, as well as the fraction of the coastline required (Lockwood et al., 2002). Replacing the assumption of an infinite coastline with a finite coastline and allowing the loss of larvae outside its boundaries also changes conditions in the expected way. This extension adds a second type of viable habitat to the condition for persistence, but the expression is still in terms of two values of LEP—one for the fished population and one for the unfished—and probabilities of originating in one habitat type and settling successfully in another for all habitat pairs (i.e., reserve to reserve, fished to fished, reserve to fished, and fished to reserve). Lockwood et al. (2002) also showed that the persistence results in Botsford et al. (2001) did not depend on the shape of the dispersal pattern, as long as it was centered at the origin. When larval dispersal patterns include substantial advection (i.e., are not

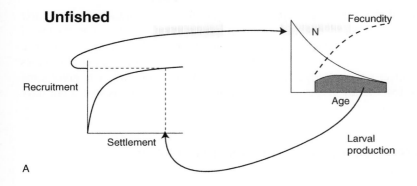
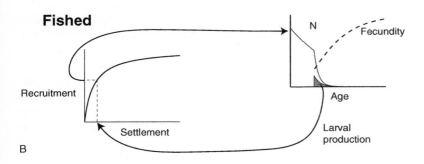
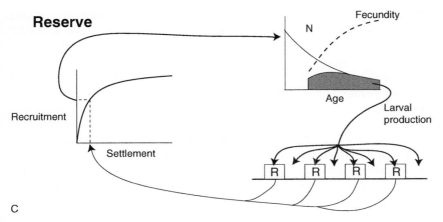

FIGURE 12-3. (A–C) Effects on replacement through truncation of age structure or fragmentation of settlement habitat by fishing and marine reserves. Individual replacement through reproduction, and therefore population persistence, can be changed from (A) the natural situation by either (B) a truncated age structure through fishing a single, nonspatial population, or (C) by having a limited amount of area in reserves, with the remainder very heavily fished.

symmetrical about the origin), persistence diminishes substantially (Gaylord and Gaines, 2000; Botsford et al., 2001). In response to calls for variable spacing of marine reserves (e.g., Palumbi, 2003) to account for differences in dispersal distances, Kaplan and Botsford (2005) showed that random spacing of reserves did not affect population persistence, except in rare cases when reserves were clumped together by chance in a situation in which populations would not have persisted with periodic reserves.

A. Consequences for the Distribution of Meroplanktonic Species

Although these studies were initiated in the context of marine reserves, as they continue to develop to address a greater number of habitat types and random spacing, they will have increasingly useful implications for the distribution of natural and fished meroplanktonic populations without reserves. Instead of the distribution of reserve and nonreserve habitat, we could consider distributions of acceptable and unacceptable habitat types along a coastline (e.g., rock outcroppings, sandy bottom). They suggest (1) that such populations can persist on isolated areas larger than the scale of dispersal or (2) that species dispersing longer distances than the size of habitat patches can persist if the fraction of coastline in reserves is greater than LEP_{min}.

From a conservation standpoint, the reduction in total area of adult habitat for meroplanktonic species is a concern (e.g., the characterization of essential fish habitat in the United States). As the area of suitable adult habitat declines, the results indicate that species will be able to persist either with single large patches on the scale of dispersal distances, or with smaller patches that cover a specific fraction of the coastline (35%?). As important is the fact that the results involve a concept that links the spatial coverage to the harvest limits (i.e., that the fraction of coastline required for persistence is related to the minimum lifetime egg production required for persistence in a conventionally fished population).

These suggestions are of course an abstraction of very complex systems with substantial outstanding uncertainties. The major uncertainties are, unfortunately, the key aspects: dispersal patterns and the fraction of natural lifetime egg production required for a population to persist. Nonetheless, the results provide some guidance regarding justification for preserving a specific fraction of the natural habitat, at a time when such proposals are coming under increasing scrutiny.

B. Consequences for the Success of Marine Reserves

There are many ongoing efforts to implement marine reserves, but the spatial design of these reserves focuses on choosing a combination of locations that includes the desired habitats, communities and species (e.g., Leslie et al., 2003), whereas the conditions for persistence are seldom accounted for (we know of no example). It is therefore of interest to examine the consequences of implementing a single reserve on a coastline on which the species of interest is heavily fished. The question of the ultimate improvement in population persistence brought about by implementing a marine reserve is typically not addressed, because adding a reserve is presumed to be an improvement on the existing conditions affecting population persistence. However, that may not ultimately be the case.

Persistence results obtained thus far indicate that if the fishing mortality rate outside reserves remains the same, persistence will increase for species with mean (alongshore) dispersal distances up to a bit larger than the alongshore dimension of the reserve (e.g., see the normalized reserve size axis in Fig. 12-2, and Lockwood et al., 2002). If, on the other hand, the available fishing effort remains the same but is redistributed over the lesser remaining area (see Smith and Wilen, 2003, and Halpern et al., 2005, for analyses of shifts in fishing effort), the maximum mean dispersal distance of species protected will be lower. Furthermore, if effort increases with the passage of time, as it generally has in most fisheries, the dispersal distances protected will be even smaller. These likely increases in fishing effort outside reserves suggest that when projecting future benefits of marine reserves, some attention should be paid to specifying the conditions under which the populations they purport to protect will be sustained.

IV. ROLE OF VARIABILITY

The dynamics of variability in spatially distributed populations is more complex than in single populations because of the opportunities for transfer of individuals between subpopulations, and it is not as well understood. Early results regarding persistence of multiple unconnected populations with varying degrees of correlation among environments are relevant as background, and different approaches to other definitions of metapopulations have been taken. We focus here on progress with the definition of marine metapopulations given earlier in this chapter for which several important results have been developed. In particular, results suggest that with variability included in models of the type discussed here, not only does connectivity play a critical role, but the issue of the "connectivity" within the variability (i.e., the spatial correlation of variability) also becomes important.

It is clear that a metapopulation will have a longer time to extinction than a single population. But beyond this obvious conclusion, the role played by connectivity, and the time to extinction of single subpopulations, especially in a variable environment, is less easy to intuit. Hill et al. (2002) investigated the relative importance of connectivity within metapopulations and correlations in environmental variability in affecting extinction times in simple models. (Their analysis is in the context of salmon populations in which connectivity is the result of straying of spawning adults away from natal streams, but the results are directly applicable to connectivity through larval dispersal.) Straying (connectivity) has the largest effect on increasing extinction time if correlation is zero, but the effect of an increase in correlation on extinction time is substantial even for high correlations. For low environmental correlations, even very small straying rates (0.02 per generation) effectively eliminate the possibility of extinction in two-patch metapopulations over ecologically relevant timescales. This work also has the general conclusion of emphasizing the importance of the contrast between connectivity and environmental correlation.

V. DISCUSSION

The results obtained thus far regarding conservation management of spatial population dynamics of marine populations lead in the direction of an interpretation of persistence that is consistent with existing, single population tools based on the replacement concept, and extend it to replacement over space in addition to age. Our thinking regarding the impact of human intrusions on marine metapopulations, and our plans for mitigating them, need to be cast in terms of how they affect overall population replacement at each point, in terms of the potential for dispersal to and from other habitats as well as the potential contribution to population growth at each point (i.e., local LEP or R_0). The simplest case examined was in a reserve context and involved only two types of habitat: very good and useless (unfished and complete removal by fishing). This case obviously has some implications for the simplest kind of nonreserve problem: how population persistence depends on the spatial distribution of suitable adult habitat and the larval dispersal patterns of the species. Including a third habitat type, as in Lockwood et al. (2002) enables a broader class of characterizations and applications, and further expansion of the number of habitat types seems warranted.

The concept that the fraction of space available is related to population growth rate and persistence in this way may not be immediately acceptable to some. Although the idea that greater egg production per individual leads to higher persistence and growth rate, the companion idea that the same holds for space available for settlement and consequent reproduction is not as palatable to some (Sale et al., 2005, Box 1).

This is also a shift from the common view of reserves being placed in a system of sources and sinks, with the question frequently being whether they should be placed in sources and sinks (e.g., Allison et al., 1998; Tuck and Possingham, 2000; Morgan and Botsford, 2001; Armsworth, 2002). The question shifts from one of how many larvae are being produced at a location, to one that also includes how many are returning? Sustainability, at least, depends on replacement paths through dispersal, not just dispersal from the reserve.

These musings suggest that, at the least, we have the basis for suggesting fruitful future avenues for development and research needs. The tie to the use of a replacement notion in fisheries illuminates the fact that there is a fundamental uncertainty in conservation management of populations whether managed spatially (e.g., as in marine reserves) or with conventional limits on catch and size. That uncertainty is in the slope of the recruit–egg relationship at the origin, or alternatively $FLEP_{min}$, the fraction of lifetime egg production required for population persistence (Botsford and Parma, 2005). This is a fundamental uncertainty in fishery management: How hard can we fish a population before it collapses? And the developing theory adds a spatial component to it. There is just as much of a need to improve our knowledge of this factor for the spatial management of fisheries, including the design of marine reserves, as there is for the sake of conventional fishery management. We need a better understanding of how this parameter depends on phylogeny, type of marine system (e.g., upwelling vs. nonupwelling shelf), and other factors. This could be done both empirically (e.g., Myers et al., 1999) or possibly through extension of life history theory.

For spatial management, estimation of the current value of p_{rr} in $FLEPp_{rr}$ on the left-hand side of Equation 3 requires knowledge of the species' dispersal pattern, a second area in which improvement is needed (e.g., Cowen et al., 2004). Knowledge of dispersal pattern will be needed to estimate the values of other similar terms describing the transfer between habitat types, as models become more realistic. It has been shown that, unlike other problems in population ecology, metapopulation persistence does not depend critically on the shape of the tails of the dispersal kernel (e.g., Lockwood et al., 2002) However, advective dispersal (i.e., nonsymmetrical dispersal kernels) does affect persistence (Botsford et al., 2001; Gaylord and Gaines, 2000). Even though the current level of uncertainty regarding dispersal patterns is high, it seems worthwhile to assess reserve design in terms of the dependence on the spatial scale of dispersal and the presence/absence of strong alongshore flows.

Here we have used a common definition of persistence, but other related definitions would provide additional information. It is important to realize that the common definition of persistence as a lack of stability about zero abundance has limitations. For example, it does not yield information on where subpopulations will persist and what their relative abundances will be. Also, it can be satisfied in a somewhat trivial way by the existence of a small persistent piece of the total potential population range. The ad hoc approach to determining persistence in Botsford et al. (2001) can be applied point by point, and other approaches may be possible.

The study of the effects of random environments on marine metapopulations described here (i.e., Hill et al., 2002) identifies essential mechanisms and the role of correlation over space, but further development of the effect of redundancy provided by multiple subpopulations on persistence in a random environment is warranted (e.g., Allison et al., 2003). This idea is vital for understanding the dynamics of marine metapopulations, and essentially is in the spirit of the rescue effect (Hanski and Gaggiotti, 2004). In managed systems, there can be multiple scales and concepts of connectivity: the role of environmental correlation, the movement of adults, the movement of larvae, and the movement of fishing effort. Reconciling and understanding the different spatial and temporal scales inherent in these different forces is a real challenge for the future. We have focused here on populations with sedentary adults. There is a need for analysis of how juvenile and adult movements change these results, as they surely will. Important examples from fished species include Smedbol and Wroblewski (2002) and McQuinn (1997).

The results described here for sustainability of species dispersing specific distances also need to be applied to the actual multispecies assemblages in potential reserves. The group of species projected to persist within reserve may not turn out to be those that actually persist, because the differing level of connectivity among species dispersing different distances may change the outcome of competitive links among species in the community. There is a growing awareness that trophic interactions should be accounted for in projecting the benefits of reserves. For example, a recent meta-analysis of empirical reports of the effects of marine reserves showed that reserves result in increasing levels of top trophic levels and that as much as a third of species, presumably at lower trophic levels, appeared to be reduced to lower abundance (Micheli et al., 2004).

In addition to needs for further research, there is as great a need for the application of metapopulation dynamics to problems in marine conservation. Although the title of this chapter suggests that we are describing the dynamic relationships used in actual management for marine conservation, (1) very few fisheries are managed on a spatially explicit basis, and (2) even in the recently burgeoning area of the most explicit spatial management—marine reserves—few planning and design efforts explicitly consider population persistence. Although advocates for marine reserves frequently fault fishery managers for failing to provide for sustainability of fished stocks, they themselves seldom describe the sustainability in reserves. An expeditious way of including connectivity has been suggested by Briers (2002), but more development is needed.

When examining a developing scientific approach, it is usually informative to compare that development with other precedents. In the case of marine metapopulations, it is somewhat surprising that the many developments in the general area of metapopulations have been independent of the particular model appropriate for marine species. Rather than models with continuous spatial distributions of abundance, with dispersal in the larval stage subject to strong physical forcing, and problems in which complete extinction is not the central practical issue, metapopulation models in ecology have tended to be discrete in space (with patches), many are occupancy rather than abundance based, and extinction and recolonization are frequent (e.g., Hixon et al., 2002; Hanski and Gaggiotti, 2004; Kritzer and Sale, 2004). Although they have not dealt with marine populations, there are some interesting observations. In the analysis of the persistence problem examined here, space does become discrete, but in terms of habitat types rather than physically distinct habitats (following Van Kirk and Lewis, 1997). Also, the fact that the important factors end up being combinations of LEP and dispersal terms is reminiscent of models as far back as Skellam's (1951) characterization of population persistence as a function of rate of exponential increase and diffusivity. It was also a means chosen by Lande (1987) to expand on Levins' (1970) original model to include spatial consideration. He characterized the demographic potential of populations in terms of a combination of R_0 and a dispersal parameter.

Results thus far shed some light on current trends in research regarding components of marine metapopulations. For example, the general trend of increasing evidence for local retention of larvae and shorter dispersal distances implies that subpopulations can be managed more independently. Because shorter distance dispersers require smaller reserves for persistence, this is encouraging news for those involved in reserve implementation; however, it also means that benefits outside reserves resulting from spillover (e.g., greater fish catch) will not extend as far (Hastings and Botsford, 2003; Botsford et al., 2004). Another recent topical issue is the concern that large females being absent from fished populations severely impacts persistence, and needs to be addressed by implementation of marine reserves (Berkeley et al., 2004). The emerging point of view described here provides a means of examining the effects of both age distribution and spatial distribution in a common context.

Because of their distribution over space, metapopulation descriptions provide the opportunity for understanding a spatially variable dynamic interaction between fishing and abundance that has serious consequences for conservation. *Serial depletion* is the frequently occurring process of sequentially fishing down subpopulations within a metapopulation beginning with the most valuable and working toward the least (e.g., Orensanz et al., 1998). If space is not specifically accounted for in the assessment of such a fishery, serial depletion creates the problem that apparent abundance, as determined from catch-per-unit effort, essentially does not decline perceptibly until the last population has been fished (Prince and Hilborn, 1998). In particular, including spatial aspects is important for understanding the idea that the initial targets of fisheries may not represent a random choice among the individuals in a population, and therefore the population consequences may be more or less severe than a nonspatial model would predict.

In this chapter we have focused on sustainability, a population characteristic of great topical interest because of current concerns for marine conservation. However, progress has been recently made in the understanding of other characteristics of metapopulation behavior. Population stability and the potential for cyclical behavior is of interest in the marine environment (e.g., blue crab [Tang, 1985], acorn barnacles [Roughgarden et al., 1985], and Dungeness crab [Botsford and Hobbs, 1995; also see Fogarty and Botsford, this volume]). In metapopulations, consideration of this phenomenon raises the additional characteristic of synchrony of cyclical behavior over space. The essential questions then are how the metapopulation structure affects stability and the likelihood of population cycles, and how synchronous those cycles would be. We do not cover this topic here, but refer the reader to a brief description of recent results showing the increasing breadth of the dispersal pattern can lead to synchrony of cyclical behavior over distances on the order of the dispersal distance, in alternating bands of synchronous and antisynchronous behavior (see the description by Fogarty and Botsford, this volume).

VI. SUMMARY

Management and planning for marine conservation will likely benefit from explicit consideration of the metapopulation nature of marine populations. Understanding of the associated metapopulation dynamics is in its infancy, and more effort is needed. It does not appear to be a focus of metapopulation efforts in ecology (Hanski and Gaggiotti, 2004). The concept of individual replacement has been useful in single population marine management (i.e., in fisheries), and is supported by basic theory of age-structured models with environmental variability and density dependence. Extension of the replacement concept to include connectivity over space, in addition to the degree of "replacement connectivity" in the age structure of a population, is an emerging aspect of work on metapopulation persistence, and it appears that it can provide valuable intuition for managers of marine conservation.

REFERENCES

Allison, G.W., Gaines, S.D., Lubchenco, J., and Possingham, H.P. (2003). Ensuring persistence of marine reserves: Catastrophes require adopting an insurance factor. *Ecological Applications.* **13**, S8–S24.

Chapter 12 Conservation Dynamics of Marine Metapopulations

Allison, G.W., Lubchenco, J., and Carrr, M.H. (1998). Marine reserves are necessary but not sufficient for marine conservation. *Ecological Applications.* **8**, S79–S92.

Armsworth, P.R. (2002). Recruitment limitation, population regulation, and larval connectivity in reef fish metapopulations. *Ecology.* **83**, 1092–1104.

Berkeley, S.A., Chapman, C., Sogard, S.M. (2004). Maternal age as a determinant of larval growth and survival in marine fish. *Ecology.* **85**, 1258–1264.

Botsford, L.W. (1997). Dynamics of populations with density-dependent recruitment and age structure. In *Structured Population Models in Marine, Terrestrial, and Freshwater Systems* (S. Tuljapurkar and H. Caswell, eds.). New York: Chapman and Hall, 371–408.

Botsford, L.W., Hastings, A., and Gaines, S.D. (2001). Dependence of sustainability on the configuration of marine reserves and larval dispersal distance. *Ecology Letters.* **4**, 144–150.

Botsford, L.W., and Hobbs, R.C. (1995). Recent advances in the understanding of cyclic behavior of Dungeness crab (*Cancer magister*) populations. *ICES Marine Sciences Symposium.* **199**, 157–166.

Botsford, L.W., Kaplan, D.M., and Hastings, A. (2004). Sustainability and yield in marine reserve policy. *American Fisheries Society Symposium.* **42**, 75–86.

Botsford, L.W., and Parma, A.M. (2005). Uncertainty in marine management. In *Marine conservation biology* (E. Norse and L. Crowder, eds., pp. 375–392). Covelo, CA: Island Press.

Botsford, L.W., and Wickham, D.E. (1978). Behavior of age-specific, density-dependent models and the northern California Dungeness crab (*Cancer magister*) fishery. *Journal of the Fisheries Research Board of Canada.* **35**, 833–843.

Briers, R.A. (2002). Incorporating connectivity into reserve selection procedures. *Biological Conservation.* **103**, 77–83.

Caswell, H. (2001). *Matrix population models: Construction, analysis and interpretation.* Sunderland, MA: Sinauer Associates.

Cisneros–Mata, M.A., Botsford, L.W., and Quinn, J.F. (1997). Projecting viability of *Totoaba macdonaldi*, a population with unknown age-dependent variability. *Ecological Applications.* **7**, 968–980.

Clark, W.G. (1990). Groundfish exploitation rates based on life history parameters. *Canadian Journal of Fisheries and Aquatic Science.* **48**, 734–750.

Clark, W.G. (2002). F35% revisited ten years later. *North American Journal of Fisheries Management.* **22**, 251–257.

Cowen, R.K., Gawarkiewiccz, G., Pineda, J., Thorrold, S., and Werner, F. (2004). *Population connectivity in marine systems. Report of a workshop to develop science recommendations for the National Science Foundation, November 2002.* Durango, CO., Washington, DC: National Science Foundation.

FAO. (1995). *Code of conduct for responsible fisheries.* 41 pp. Rome: FAO, United Nations (Available on www.fao.org).

Fogarty, M.J. (1998) Implications of larval dispersal and directed migration in American lobster stocks: Spatial structure and resilience. *Can. Spec. Publ. Fish. Aquat. Sci.* **125**, 273–283.

Gaylord, B., and Gaines, S.D. (2000). Temperature or transport? Range limits in marine species mediated solely by flow. *American Naturalist.* **155**, 769–789.

Gerber, L.R., Andelman, S.J., Botsford, L.W., Gaines, S.D., Hastings, A., Palumbi, S.R., and Possingham, H.P. (2003). Population models for marine reserve design: A retrospective and prospective synthesis. *Ecological Applications.* **13**, S47–S64.

Goodyear, C.P. (1993). Spawning stock biomass per recruit in fisheries management: Foundation and current use. In *Risk evaluation and biological reference points for fisheries management* (S.J. Smith, J.J. Hunt, and D. Rivard, eds.). *Can. Spec. Publ. Fish. Aquat. Sci.* **120**, 67–81.

Halpern, B.S., Gaines, S.D., and Warner, R.R. (2005). Confounding effects of the export of production and the displacement of fishing effort from marine reserves. *Ecol. Appl.* **14**, 1248–1256.

Hanski, I., and Gaggiotti, O.E. (2004). *Ecology, genetics, and evolution of metapopulations.* Amsterdam: Elsevier.

Hastings, A. (1997). *Population biology: Concepts and models.* New York: Springer-Verlag.
Hastings, A., and Botsford, L.W. (2003). Are marine reserves for fisheries and biodiversity compatible? *Ecological Applications.* **13**, S65–S70.
Hill, M.F., Hastings, A., and Botsford, L.W. (2002). The effects of small dispersal rates on extinction times in structured metapopulation models. *American Naturalist.* **160**, 389–402.
Hixon, M.A., Pacala, S.W., and Sandin, S.A. (2002). Population regulation: Historical context and contemporary challenges of open vs. closed systems. *Ecology.* **83**, 1490–1508.
Hutchinson, G.E. (1978). *An introduction to population ecology.* New Haven, CT: Yale University Press.
Kaplan, D.M., and Botsford, L.W. (2005). Effects of variability in spacing of coastal marine reserves on fisheries yield and sustainability. *Can. J. Fish. Aquat. Sci.* **62**, 905–912.
Kritzer, J.P., and Sale, P.F. (2004). Metapopulation ecology in the sea: From Levins' model to marine ecology and fisheries science. *Fish and Fisheries.* **5**, 131–140.
Lande, R. (1987). Extinction thresholds in demographic models of territorial populations. *American Naturalist.* **130**, 624–635.
Lande, R., and Orzack, S.H. (1988). Extinction dynamics of age-structured populations in a fluctuating environment. *Proceedings of the National Academy of Science USA.* **85**, 7418–7421.
Leslie, H., Ruckelshaus, M., Ball, I.R., Andelman, S., and Posssingham, H.P. (2003). Using siting algorithms in the design of marine reserves. *Ecological Applications.* **13**, S185–S198.
Levins, R. (1970). Extinction. In *Some Mathematical Problems in Biology* (M. Desternhaber, ed., pp. 77–107). Providence, RI: American Mathematical Society.
Lockwood, D.R. (2002). The effects of larval dispersal and spatial heterogeneity on the design of marine reserves. Ph.D Thesis, University of California–Davis, 96pp.
Lockwood, D.L., Hastings, A., and Botsford, L.W. (2002). The effects of dispersal patterns on marine reserves: Does the tail wag the dog? *Theoretical Population Biology.* **61**, 297–309.
Mace, P.M., and Sissenwine, M.P. (1993). How much spawning per recruit is enough? In *Risk evaluation and biological reference points for fisheries management* (S.J. Smith, J.J. Hunt, and D. Rivard, eds.). *Can. Spec. Pub. Fish. Aquat. Sci.* **120**, 101–118.
McQuinn, I.H. (1997). Metapopulations and the Atlantic herring. *Reviews in Fish Biology and Fisheries.* **7**, 297–329.
Micheli, F., Halpern, B.S., Botsford, L.W., and Warner, R.R. (2004). Trajectories and correlates of community change in no-take marine reserves. *Ecological Applications.* **14**, 1709–1723.
Morgan, L.E., and Botsford, L.W. (2001). Managing with reserves: Modeling uncertainty in larval dispersal for a sea urchin fishery. In *Proceedings of the Symposium on Spatial Processes and Management of Marine Populations* (pp. 667–684). Fairbanks, AK: University of Alaska Sea Grant College Program.
Myers, R.A., Bowen, K.G., and Barrowman, N.J. (1999). Maximum reproductive rate of fish at low population sizes. *Can. J. Fish. Aquat. Sci.* **56**, 2404–2419.
Orensanz, J.M., Armstrong, J., Armstrong, D., and Hilborn, R. (1998). Crustacean resources are vulnerable to serial depletion: The multifaceted decline of crab and shrimp fisheries in the Greater Gulf of Alaska. *Reviews in Fish Biology and Fisheries.* **8**, 117–176.
Palumbi, S.R. (2003). *Marine reserves: a tool for ecosystem management and conservation.* Arlington, VA: Pew Oceans Commission.
Prince, J., and Hilborn, R. (1998). Concentration profiles and invertebrate fisheries management. In *Proceedings of the North Pacific Symposium on Invertebrate Stock Assessment and Management* (G.S. Jamieson and A. Campbell, eds.). *Can. Spec. Publ. Fish. Aquat. Sci.* **125**, 187–196.
Ralston, S. (2002). West coast groundfish policy. *North American Journal of Fisheries Management.* **22**, 249–250.
Roughgarden, J., Iwasa, Y., and Baxter, C. (1985). Demographic theory for an open marine population with space-limited recruitment. *Ecology.* **66**, 54–67.

Sale, P.F., Cowen, R.K., Danilowicz, B.S., Jones, G.P., Kritzer, J.P., Lindeman, K.C., Planes, S., Polunin, N.V.C., Russ, G.R., Sadovy, Y.J., and Steneck, R.S. (2005). Critical science gaps impede use of no-take fishery reserves. *Trends Ecol. Evo.* **20**, 74–80.

Sale, P.F., and Kritzer, J.P. (2003). Determining the extent and spatial scale of population connectivity: Decapods and coral reef fishes compared. *Fisheries Research.* **65**, 153–172.

Sissenwine, M.P., and Shepherd, J.G. (1987). An alternative perspective on recruitment overfishing and biological reference points. *Can. J. Fish. Aquat. Sci.* **44**, 913–918.

Skellam, J.G. (1951). Random dispersal in theoretical populations. *Biometrika.* **38**, 196–218.

Smedbol, R.K., and Wroblewski, J.W. (2002). Metapopulation theory and northern cod population structure: Interdependency of subpopulations in recovery of a groundfish population. *Fisheries Research.* **55**, 161–174.

Smith, M.D., and Wilen, J.E. (2003). Economic impacts of marine reserves: The importance of spatial behavior. *Journal of Environmental Economics and Management.* **46**, 183–206.

Swearer, S.E., Caselle, J.E., Lea, D.W., and Warner, R.R. (1999). Larval retention and recruitment in an island population of a coral-reef fish. *Nature.* 402, 799–802.

Tang, Q. (1985). Modification of the Ricker stock recruitment model to account for environmentally induced variation in recruitment with particular reference to the blue crab fishery in Chesapeake Bay. *Fish. Res.* **3**, 13–21.

Todd, C.D. (1998). Larval supply and recruitment of benthic invertebrates: Do larvae always disperse as much as we believe? *Hydrobiologia.* **375/376**, 1–21.

Tuck, G.N., and Possingham, H.P. (2000). Marine protected areas for spatially structured exploited stocks. *Marine Ecology Progress Series.* **192**, 89–101.

Tuljapurkar, S.D. (1982). Population dynamics in variable environments. II. Correlated environments, sensitivity analysis and dynamics. *Theoretical Population Biology.* **21**, 114–140.

Tuljapurkar, S.D. (1990). Population dynamics in variable environments. *Lecture Notes in Biomathematics.* **85**, 1–154.

Tuljapurkar, S.D., and Orzack, S.H. (1980). Population dynamics in variable environments. I. Long run growth rates and extinction. *Theoretical Population Biology.* **18**, 314–342.

Van Kirk, R.W., and Lewis, M.A. (1997). Integrodifference models for persistence in fragmented habitats. *Bulletin of Mathematical Biology.* **59**, 107–137.

Warner, R.R., and Swearer, S.E., and Caselle, J.E. (2000). Larval accumulation and retention: Implications for the design of marine reserves and essential fish habitat. *Bulletin of Marine Science.* **66**, 821–830.

CHAPTER **13**

Genetic Approaches to Understanding Marine Metapopulation Dynamics

MICHAEL E. HELLBERG

I. Introduction
   A. Subdivision in Marine Populations
   B. The Island Model and Its Limitations
II. Delineating Populations
III. Inferring Patterns of Connectivity
IV. Inferring Nonequilibrium Population Dynamics
   A. Population Extinction and Recolonization
   B. Source–Sink Relationships
   C. Mixing in the Plankton and the Genetics of Larval Cohorts
V. Summary
VI. Acknowledgment
   References

I. INTRODUCTION

Dispersal by pelagic larvae provides a means of linking distant marine populations both demographically and genetically. The genes that such larvae carry from their natal population and transplant elsewhere can reveal these links. The interpretation of genetic data that allows for such inferences, however, is laden with assumptions and caveats, many of which do not always make the translation from the population geneticists who gather it to other biologists who would seek to utilize it. Furthermore, although many genetic analyses of marine organisms continue to rest on untenable assumptions of equilibrium (discussed later), newly emerging approaches that can provide insights into the dynamic processes that characterize metapopulations have been little used. In this chapter, I briefly outline how genetic approaches can help define three fundamental aspects of marine metapopulation dynamics: what a population is, how populations are connected, and what impact population extinction–recolonization events (and other nonequilibrium dynamics) have.

A. Subdivision in Marine Populations

Even thousands of kilometers of open ocean may be insufficient to subdivide species with long-lived feeding larvae into subpopulations (e.g., Lessios et al., 2001b; Bowen et al., 2001). However, ongoing larval connection over such broad scales may be exceptional (reviewed in Swearer et al., 2002). Instead, many marine populations, even some with larvae that spend weeks in the plankton, show a substantial degree of self-recruitment (Jones et al., 1999; Swearer et al., 1999). If such local larval retention is great enough that populations are self-replacing, and such demographic isolation is sustained for many generations, then subpopulations will become genetically differentiated. Such differentiation has been detected in marine species with pelagic larvae at surprisingly modest spatial scales (approximately 500–1000 km; e.g., Barber et al., 2002; Taylor and Hellberg, 2003; see Hellberg et al., 2002).

The subdivision of species into self-recruiting populations has important evolutionary and ecological consequences. Local recruitment facilitates local adaptation, because alleles favorable in a particular setting can be selected for without having to compromise for their lower fitness in different environments. Over time, these variants may become coadapted with other locally favored alleles, resulting in coadapted gene complexes that do poorly when mixed with migrant genotypes (see Burton et al., 1999, for a marine example; see Barton and Whitlock, 1997, for theoretical background). Genetic subdivision also has practical consequences for determining sustainable levels of harvesting (Shaklee et al., 1999), evaluating the safety of translocation programs (Johnson, 2000), and designing marine reserves (Palumbi, 2003).

B. The Island Model and Its Limitations

The simplest way to model population subdivision is using the island model (Fig. 13-1). Under the island model, discrete subpopulations containing a uniform number (N_e) of diploid hermaphrodites are connected by migration. Generations are discrete, and the migration rate (m) is constant each generation. Migrants are drawn from across all subpopulations with equal probability. This model has supplied a framework for theoretical work in ecology and evolution, as well as the basis for inferring the number of migrants between populations ($N_e m$) based upon levels of genetic subdivision, measured by the relative genetic variance among compared to within populations (F_{ST}; see Neigel, 2002).

Chapter 13 Genetic Approaches to Understanding Marine Metapopulation Dynamics   433

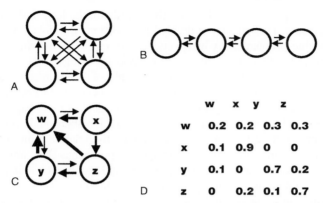

FIGURE 13-1. (A–D). Schematic representations of the island model (A), the stepping stone model (B), and asymmetrical migration among populations (C). (D) A migration matrix for migration rates is illustrated in (C). Recipient populations are listed in rows; source populations, in columns. Values along the diagonal indicate the proportion of each generation retained locally.

No natural population, however, strictly conforms to the assumptions of the island model. Different populations vary in size, sometimes by several orders of magnitude (Sinclair, 1988). N_e refers explicitly to the effective population size, the number of individuals contributing genetically to the next generation, and thus determines the strength of genetic drift. Mating systems (Nunney, 1993) and fluctuations in population size (Vucetich et al., 1997), among other factors, can result in N_e being several orders of magnitude lower than census population size (Hedgecock, 1994; Turner et al., 2002). Dispersal between populations may vary across different habitats within a range (Johnson and Black, 1991) or may be strongly asymmetrical (Hoskin, 2000; Wares et al., 2001). Both population size and dispersal rates fluctuate over time. In the extreme, these fluctuations can cause some populations to go extinct. The genetic consequences of such population extinctions hinge on the size and origin of subsequent colonizing groups. In addition, population dynamics and patterns of connectivity that hold at one spatial scale may be completely different at others (Hellberg, 1995).

Departures from the assumptions of the island model occur within marine populations to an extent determined by the biological attributes and the environmental circumstances of particular species and populations. The central question for evolutionary geneticists is which of these departures from the island model is significantly impact the partitioning of genetic variation among populations. These patterns of genetic variation, however, can also be turned on their head and used as tools to infer demographic processes, and in some cases even to estimate the parameters of metapopulation models. Here I outline which processes can be inferred using genetic data, and provide examples of where such approaches have provided insights into the population biology and evolution of marine organisms. This chapter is divided into sections that focus on a particular demographic problem for which genetic data can offer insights. In each of these sections, I briefly outline the logical support for using genetic data to make a particular inference, outline the power and pitfalls of the approach, and provide specific examples of when genetic approaches have been used to make the inference in question in marine organisms. The overall goal of this chapter is not to review genetic subdivision in marine species (see Grosberg and Cunningham, 2001; Hellberg et al., 2002), nor is it to evaluate the relative utility of different genetic markers (Sunnucks, 2000) or analytical approaches (Hey and Machado, 2003) for addressing questions in population biology. Rather, I hope that by framing the evolutionary genetics of marine metapopulations from the perspective of specific questions about demography, the reader will get a sense of when a genetic approach is warranted and how genetic data can contribute to a fuller understanding of the dynamics of marine populations.

II. DELINEATING POPULATIONS

It is perhaps a hallmark of ecology and evolutionary biology that we choose to study subjects ("communities," "species") whose very definitions we cannot agree upon. Certainly "population" falls into this category, with its varying definitions reflecting whatever it is that the user wants to know about. For ecologists and fisheries managers, the question may boil down to how large a region must be considered for stock–recruitment dynamics to make sense. If 95% of recruits originate within, say, 10 km of a population, then the few additional recruits that wander in from potentially great distances are inconsequential. Evolutionary biologists, on the other hand, want to know the scale at which populations are sufficiently isolated so that adaptive differences or new species can evolve. In this case, even a few immigrants from afar per generation may suffice to keep populations genetically homogeneous, even if their demographics are largely independent. Thus, both evolutionary questions and genetic answers are quite sensitive to low levels of dispersal. As a result, evolutionary questions are usually cast at larger and longer scales than ecological ones. Confusion can result when ecologists and evolutionary biologists interpret each others' results, especially when spatial scales are referred to as small or large, but distances are not stated explicitly. As a rule, an ecological population may be no larger than an evolutionary one, and is often much smaller.

In theory, genetic markers are exquisitely sensitive to the movement of genes between populations. Just one migrant per generation between populations is sufficient to swamp significant differentiation by genetic drift (Slatkin, 1987). When the enormous reproductive output of many individual marine organisms is considered, along with the high census population sizes at many locales, getting just one larva every generation (on average) to travel between populations and subsequently reproduce seems a modest requirement, especially when larvae often seem designed to sustain themselves for long periods in the plankton (Pechenik, 1999). Indeed, allozyme (Winans, 1980; Nishida and Lucas, 1988) and mtDNA surveys (Lessios et al., 1998) that have found little sign of genetic differentiation even at transoceanic spatial scales support the idea that some marine populations are demographically open.

The great sensitivity of genetic markers to detect migration in some ways undermines the general application of genetic analyses to delineate populations. Certainly populations that are, for all practical purposes, closed demographically can still exchange sufficient gene flow to stem genetic divergence, as envisioned by Sinclair's (1988) member–vagrant hypothesis. This sensitivity of genetic markers, however, makes the inferences drawn from those cases in which the data speak clearly for closed populations all the more powerful.

In the extreme, genetic markers can distinguish cryptic species (Knowlton, 1993), which by definition do not exchange migrants and thus constitute reciprocally closed populations. In the simplest case, geographically widespread nominal species are revealed to be composed of multiple geographically distinct cryptic species (e.g., deep sea fish [Miya and Nishida, 1998], foraminifers [de Vargas et al., 1999], and scyphozoans [Dawson and Jacobs, 2001]). More critical to understanding population dynamics are those instances in which genetic data have helped to distinguish co-occurring cryptic species (corals [Knowlton et al., 1992], tunicates [Tarjuelo et al., 2001]; and brittlestars [Sponer and Roy, 2002]). Notably, such co-occurring cryptic species often exhibit nonrandom associations with depth, so that genetic surveys of species with broad microhabitat distributions should note where sampled individuals had been living.

Phylogeographic breaks occur when genealogical lineages within a species exhibit abrupt geographical change. Like geographic cryptic species (drawing a distinction between these categories can be difficult), the demographic inference is straightforward: Such populations are not exchanging migrants and have not done so for some time. More subtle are cases in which genetic breaks are not geographically abrupt (or at least available sampling cannot allow for such an inference), but fixed genetic differences occur between more distant populations. The classic marine example for this comes from the tide pool copepod *Tigriopus californicus*, which exhibits deep mitochondrial and nuclear sequence differentiation among Californian populations that can still interbreed in the laboratory (Burton, 1998; Burton et al., 1999). Jiang et al. (1995) used variation in mitochondrial repeats to infer similar genetic breaks (and thus persistent demographic isolation) between populations of abalone living on the eastern coast of Taiwan. Some recent work has also revealed deep genetic splits between tropical island populations, both in the Pacific (stomatopods [Barber et al., 2002], grouper [Rhodes et al.,

2003], and cowries [Paulay and Meyer, 2002]) and the Caribbean (gobies [Taylor and Hellberg, 2003]). In this last case, populations as close as 1000 km (Curaçao and Barbados) were reciprocally monophyletic, which, when combined with a molecular clock calibration and a coalescent estimation technique (Kuhner et al., 1998), suggests these populations have not exchanged migrants that went on to reproduce for more than 75,000 years—a long time considering the 3-week larval duration of the reef fish studied.

At the lowest end of the differentiation spectrum are instances when genetic differentiation may be subtle, but this result is nonetheless surprising given the apparent opportunities for interpopulation exchange. Again, as for cryptic species, the most enlightening cases involve populations occurring in different habitats at the same sites (the gastropod *Littorina saxatilis* [Johannesson et al., 1993], the marine angiosperm *Zostera marina* [Ruckelshaus, 1998] and the coral *Favia fragum* [Carlon and Budd, 2002]). Ruzzante et al. (1996b) found that microsatellite allele frequencies distinguished populations of cod that overwintered inshore from populations that overwintered in warmer waters offshore. The genetic distinctions drawn by these presumably neutral markers were correlated with biomolecules of functional importance: Levels of antifreeze proteins in the blood were higher for cod from inshore populations than from offshore populations. The consistency of temporal differences can also enhance the power to infer closure from subtle genetic differences. Lewis and Thorpe (1994) found small differences among populations of the scallop *Aequipecten opercularis*, but these differences held across year classes (recognized by annual shell rings), substantiating their conclusion that populations were largely closed.

Often, however, one finds little genetic differentiation between populations. In many cases this is because levels of gene flow are high—a situation that makes accurate genetic estimation of the number of migrants between populations very difficult (Waples, 1998; Neigel, 2002). In other instances, however, independent approaches (including different sets of genetic markers) suggest that the genetically similar populations are not demographically connected. So why aren't such populations genetically distinguishable?

Certainly selection acting directly on the genetic markers used (or loci closely linked to them) is one possibility. For example, mitochondrial (Reeb and Avise, 1990) and nuclear restriction fragment length polymorphism (RFLP) (Karl and Avise, 1992) data revealed a sharp phylogeographic break at Cape Canaveral in Florida within oysters. No such discontinuity was detected by a previous allozyme survey (Buroker, 1983), suggesting that stabilizing selection on the allozymes had masked real isolation by homogenizing allele frequencies. The switch away from allozyme-based genetic surveys (which require amino acid variation) to sequence-based variation (most of which is silent and thus presumably selectively neutral) may sometimes remedy this problem. However, two recent studies on cod suggest that caution is still warranted. Strong stabilizing selection on just one (GM798, later identified as pantophysin, *Pan*I) of 10 nuclear RFLP markers was sufficient to mask the strength of a correlation between gene flow and distance (Pogson et al., 2001). Subsequent sequencing of *Pan*I alleles suggested that populations from coastal Norway and Russia, and from the open Barents Sea were closed to a greater degree than previously thought (Pogson and Fevolden, 2003).

Another explanation for genetic homogeneity despite population isolation is the time required for genetic differences to evolve. New mutations must increase in frequency to detectable levels. Alternatively, existing variation must be altered by genetic drift, a slow process in all but the smallest populations. Thus, if an island population goes extinct, is recolonized, and then persists in total demographic isolation, it will take many generations before the closed state of the island becomes genetically apparent. For F_{ST}, the summary statistic most commonly used to summarize genetic subdivision and calculate levels of gene flow, the time (in generations) required to go just halfway to such an equilibrium between migration and genetic drift after a change in levels of gene flow is $\ln2/(2m + 1/2N_e)$ (Crow and Aoki, 1984). Thus, low levels of migration and small population sizes can slow the approach to drift–migration equilibrium to a degree that it may never be reached. For example, in looking at gene flow between populations of *Balanophyllia elegans*, a broad-ranging solitary coral with crawling larvae, Hellberg (1994) calculated that it would require more than 40,000 years for estimates to go halfway to their equilibrium values. This is far more time than that between climatic changes that must alter this species' geographic range (see also Jollivet et al., 1999, and Skold et al., 2003).

F_{ST}-based approaches necessarily assume equilibrium, and so are likely to mislead when this assumption is violated. Coalescent models, which focus on gene genealogies (Hudson, 1990), are less sensitive to assumptions of equilibrium. In fact, coalescent methods have been developed to determine whether populations are experiencing low levels of ongoing gene flow or are demographically isolated (Nielsen and Slatkin, 2000; Nielsen and Wakeley, 2001). These new methods have already found wide application in understanding the history of genetic exchange among marine populations, being used to infer connectivity and isolation between populations of bivalves at hydrothermal vents (Won et al., 2003), to estimate rates of introgression between hybridizing coral species (Vollmer and Palumbi, 2002), and to determine whether a transoceanic snail introduction was human mediated (Wares et al., 2002).

Still, coalescent approaches work over timescales longer than those of concern to most ecologists and conservation biologists (about N_e generations). New analytical methods (Pritchard et al., 2000; Falush et al., 2003), however, can recognize groups of individuals that have been isolated for far shorter periods, as few as eight generations. These analyses take advantage of the high levels of polymorphism at some genetic markers (microsatellites and amplified fragment length polymorphisms [AFLPs]). Multilocus genotypes are clustered to minimize linkage disequilibrium (the nonrandom association of alleles at different loci) using a Bayesian approach implemented by the program *structure*. This approach has proved exquisitely sensitive. Chicken breeds isolated for as few as 20 generations were grouped with 95% accuracy (Rosenberg et al., 2001). In addition, populations are not assumed to be at drift–gene flow equilibrium nor must they be designated a priori; both the number of clusters and their membership can be determined within the framework of the analysis. This approach can thus reveal genetic breaks analogous to those detected by phylogeographic sequence analyses, but which separate populations isolated for far shorter periods of time. For example, Baums et al. (Baums et al., 2005), examining subdivision in the elkhorn coral *Acropora palmata*, found a genetic break between populations east and west of the Mona Passage (between Puerto Rico and Hispañola), a line coincident with a previously recognized biogeographic breakpoint (Starck and Colin, 1978; Taylor and Hellberg, 2003).

In summary, the size of populations defined by strong genetic differentiation sets an upper bound on the scale of dispersal over shorter ecological time frames. More subtle genetic differentiation must consider the complications of selection and historical disequilibria on genetic markers (especially when population sizes are large), but even small genetic differences can provide insights to population boundaries when combined with other biological, environmental, and geographical data. Emerging analyses based on multilocus genotyping may finally begin to reconcile the ecological and genetic definitions of "population."

III. INFERRING PATTERNS OF CONNECTIVITY

The island model assumes that dispersal is equally likely among all populations within a species. Certainly this assumption is violated in the sea; dispersal usually occurs more often between neighboring populations than distant ones, and prevailing currents probably result in strong asymmetries to propagule exchange. Analyses using genetic data to infer more complicated patterns of connectivity fall into two categories: those that evaluate *overall* patterns of gene flow and those that estimate complete migration matrices, with separate estimates of migration for all possible pairs of *specific* populations in both directions.

The simplest step from the island model toward realism is the steppingstone model (Fig. 13-1), in which all genetic exchange takes place between immediately neighboring populations. Genetic similarity between populations in a steppingstone scenario should diminish with distance. Specifically, Slatkin (1993) showed that the slope of $\log(\hat{M})$ (denoting estimated gene flow for pairs of populations) versus $\log(\text{distance})$ will be -1 for steppingstones arrayed in a one-dimensional (say, along a linear coastline) and -0.5 for a two-dimensional steppingstone system (e.g., island populations within an open ocean).

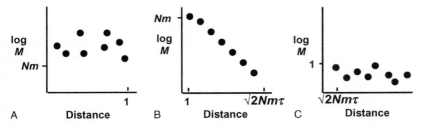

FIGURE 13-2. The relationship between allele frequency-based estimates of gene flow (M), time to equilibrium, and spatial scale. Assume all populations are of size N, with migration rate m between neighboring, linearly distributed populations that are one distance unit apart. τ generations have passed since a geographic radiation that spread genetically similar migrants among all populations. (A) Island model gene flow at small spatial scales. Subpopulations separated by one unit of distance or less will experience high levels of gene flow among all subpopulations. (B) Stepping stone gene flow at intermediate spatial scales. Populations separated by more than one unit of distance, but less than $\sqrt{2Nm}\,\tau$ units will have had time for gene flow and drift to equilibrate, resulting in an inverse relationship between M and distance of separation. (C) Nonequilibrium at large spatial scales. For populations separated by large distances ($>\sqrt{2Nm}\,\tau$ units), gene flow and drift will not have had time to equilibrate and drift will dominate, resulting in no strong relationship between M and distance. See Slatkin (1993) for details.

Patterns of genetic differentiation consistent with the steppingstone model are prevalent in taxa with limited larval dispersal capabilities. For example, gene flow and distance are inversely correlated among populations of *B. elegans* from a 50-km stretch of northern California coast. Both the slope and correlation coefficient of this relationship fall within the range expected for a one-dimensional stepping stone at equilibrium given the number and heterozygosity of allozyme markers used (Hellberg, 1995). However, this equilibrial pattern does not extrapolate to larger spatial scales (Fig. 13-2). Over most of its range (3000 km), the slope and strength of the correlation between gene flow and distance in *B. elegans* lies below the equilibrium stepping stone expectations (Hellberg, 1994, 1995). This apparent paradox has a simple explanation: disequilibrium between gene flow and genetic drift (Fig. 13-2). Because the time to equilibrium decreases with increasing levels of gene flow (Crow and Aoki, 1984), steppingstone populations that are close together (and thus exchange more migrants) will establish an equilibrial relationship between inferred levels of gene flow and distance more rapidly than distant populations (Slatkin, 1993; Hellberg, 1995; Hutchison and Templeton, 1999; see Fig. 13-2). This switch from a relationship consistent with a steppingstone at modest spatial scales to patterns owing more to historical associations at larger spatial scales has been seen in a number of marine animals (Lavery et al., 1995; Planes and Fauvelot, 2002).

Given the problems of estimating gene flow in high-dispersal species, one might expect that determining patterns of exchange in such organisms would also prove problematic. Certainly species with very high dispersal potential (e.g., echinoids) may display patterns attributable to metapopulation dynamics only at the largest possible spatial scales (e.g., the tropical Pacific; Lessios et al., 2003). Nevertheless, patterns consistent with the steppingstone model have been detected in some high-dispersal species (e.g., cod; Pogson et al., 2001). Furthermore, a significant relationship between gene flow and distance is unlikely to result from noise, and thus can boost confidence in inferences drawn from subtle genetic differences (Palumbi, 2003). Again, though, selection deserves consideration, because neighboring populations likely inhabit similar selective regimes, which could shape variation at multiple loci in parallel, even in the face of gene flow.

The stepping stone model adds the realism of limited dispersal to the island model, but still makes the assumptions that equally distant populations will exchange identical numbers of migrants and that equal numbers of migrants move between populations in both directions. These seem unlikely to hold strictly in oceans full of complex currents with velocities that may outstrip the swimming speeds of many larvae. Still, ascertaining the true patterns of connectivity among populations remains perhaps the greatest goal for marine molecular ecologists, especially because these patterns have tremendous implications for conservation and management (Roberts, 1997).

Migration matrices (Fig. 13-1) cannot be reliably obtained using pairwise estimates of gene flow derived from F_{ST} for a couple reasons. First, these offer no way of dealing with asymmetrical gene flow between populations. More importantly, the variance of F-statistics is high as a result of both high variance among loci (Nei and Maruyama, 1975; Robertson, 1975) and the inverse function used to calculate gene flow ($N_e m$) from F_{ST}. Thus, although F_{ST}-based estimates of gene flow can be useful for quantifying genetic exchange among populations conforming to the island model, or for inferring an overall pattern of gene flow that takes place primarily between neighbors, estimates for a particular pair of populations will be unreliable.

Recently, maximum likelihood analyses have been developed that simultaneously estimate the migration matrix and the effective population sizes of sampled populations (Beerli and Felsenstein, 1999, 2001) or population divergence times (Nielsen and Wakely, 2001). Although the potential of these approaches to estimate full migration matrices is great, to date they have been used on marine species primarily to infer differences between past and present population sizes (Turner et al., 2002; Roman and Palumbi, 2003). Given the number of parameters estimated in these analyses, it is not surprising that the confidence intervals around the estimates they produce can be broad. One way to address this is by gathering genetic data from multiple independent genetic markers. Alternatively, estimates from existing data can be improved by tailoring system-specific analyses that incorporate genetic, geographic, and demographic data (e.g., Gaggiotti et al., 2002). Coalescent models have also been used to address an unrealistic assumption of models of steppingstone dispersal: that there are no edge effects. Results show that not all neighbors are equal; on average, shared ancestry dates back further in time for individuals from populations in the center of the range than for those in populations near the ends of the range (Hey, 1991; Wilkins and Wakeley, 2002).

Even before such sophisticated analyses are broadly applied to marine populations, however, one surprising conclusion now seems inevitable: patterns of genetic differentiation often owe little to present-day current patterns (Benzie and Williams, 1997; Palumbi et al., 1997; O'Foighil and Jozefowicz, 1999; reviewed in Benzie, 1999). This may be the result of variation in flow patterns over time and the ability of larvae to control their dispersal (Leis, 2002). Currents and other present-day barriers that correlate with strong genetic breaks are probably holding apart lineages that differentiated long ago, rather than marking the site of primary divergence (Rocha et al., 2002; Bilodeau et al., in preparation), much in the manner of how the position of hybrid zones are determined (Barton and Hewitt, 1985). Certainly the data are already sufficient to warn against predictions of long-term connectivity between marine populations based on present-day surface currents.

Shorter term isolation may be another matter, however, and again multilocus genotyping may provide the bridge between population genetics and ecology. Because individuals can be clustered without reference to where they were collected, recent migrants can be identified and probabilities of origin assigned to potential source populations (Cornuet et al., 1999; Wilson and Rannala, 2003). This allows for the estimation of true migration matrices, estimates of migration that are not scaled to N_e. Borrell et al. (2004) used this approach to show most shrimp along the southern coast of Cuba were retained in their natal populations. Baums et al. (2005) also inferred high levels of local recruitment in *A. palmata*. Furthermore, they found that migration rates were asymmetrical and, in contrast to the patterns noted previously, agreed with prevailing currents during this species' breeding season.

IV. INFERRING NONEQUILIBRIUM POPULATION DYNAMICS

In the two previous sections, the tacit assumption (nonequilibrium caveat aside) has been that levels of genetic exchange between populations remain at constant levels for generations. Nature is less constant and predicable than this. For many marine species, years may pass with little or no larval settlement before a successful cohort establishes and dominants demographically, with profound ecological effects (Underwood and Keough, 2001; Witman et al., 2003). In habitats with frequent and severe disturbance patterns, populations may even go extinct before colonists arrive. Studying such rare events directly can be difficult. Years of monitoring may not catch settlement pulses, or momentous changes may receive attention only after the possibility of determining previous conditions has passed (e.g., Russ et al., 1996).

The genetic signature of some types of nonequilibrium population dynamics are very clear. For example, recent population bottlenecks (Luikart et al., 1998; Garza and Williamson, 2001) can be inferred with confidence, and have been detected in several marine taxa (Knight et al., 1987; Hoelzel et al., 1993). Rapid population expansions also create a characteristic genetic signature (Rogers and Harpending, 1992), used by Lessios et al. (2001a) to show that populations of the Caribbean urchin *Diadema antillarum* had exploded long before humans could have altered its ecosystem.

Other situations are more complex. Fluctuations in both population size and migration rate over time enhance genetic variance among populations, and geographic variation in these parameters has an even larger effect (Whitlock, 1992). These complications mean that, unlike the case for subdivided populations at equilibrium, no single value for gene flow can be estimated (Pannell and Charlesworth, 2000); more parameters must be estimated simultaneously. Still, genetic data, especially when combined with temporal sampling strategies, can often reveal patterns of extinction–recolonization, population source–sink relationships, and the extent to which larvae from different parents and populations are mixed in the plankton.

A. Population Extinction and Recolonization

The genetic effects of recolonization via a propagule model can be seen in its simplest form when a single colonization event follows a broad extinction. Temperate species inhabiting broad latitudinal ranges have been pushed north and south by fluctuating climatic conditions throughout the Pleistocene (Roy et al., 1996). During warming cycles, species can extend their ranges poleward to reestablish populations that perished during the previous cooling bout. If the colonists effecting this range expansion are few and drawn from a single source population (most probably those at the leading edge of the range), then genetic variation within the recolonized range will be markedly lower than elsewhere (see Austerlitz et al., 1997, for a more detailed mathematical treatment). For example, genetic diversity within populations of a gastropod with limited larval dispersal (*Acanthinucella spirata*) to the north of Point Conception (which probably went extinct during the Wisconsonian glaciation) was a small subset of that found in refugial populations to the south (Hellberg et al., 2001). Similar patterns of poleward genetic poverty have been found in other coastal temperate species (corals [Hellberg, 1994], snails [Marko, 1998], and teleosts [Bernardi, 2000]), as well as oceanic copepods (Bucklin and Wiebe, 1998) and mackerel (Nesbo et al.,

2000). Such colonization bottlenecks are also evident at far smaller spatial scales when populations become established on newly available substrate (Ayre, 1984; Johannesson and Warmoes, 1990).

The situation becomes more complex and less intuitive in a metapopulation setting, in which populations turn over asynchronously. One way of sorting out the dynamics of extinction–recolonization processes under such circumstances is by sampling populations repeatedly. Such repeated sampling can occasionally catch populations that have recently been recolonized, as evidenced by rapid changes at multiple characters set against a background of temporal stability among other populations (see isopod examples in Lessios et al., 1994, and Piertney and Carvalho, 1995).

The impacts of population turnover can also be examined using F_{ST}. Before considering how F_{ST} is changed by population turnover, note that F_{ST} is affected by levels of variation within as well as between populations (Charlesworth, 1998). This is apparent in Nei's (1973) definition of G_{ST} (conceived as a multiallelic measure of F_{ST}): $G_{ST} = (f_0 - \bar{f})/(1 - \bar{f})$, where f_0 and $\bar{f}$ are the probabilities that two alleles sampled from the same population or the entire species respectively are the same. Thus, increasing homozygosity within populations (f_0, for randomly mating diploid populations) will increase F_{ST}. Stated differently based on Slatkin's (1991) coalescent definition, F_{ST} will increase as the time to common ancestry within populations decreases relative to that among populations.

Wade and McCauley (1988) were the first to frame the genetic effects of extinction and recolonization on population differentiation in terms of F_{ST} (see Pannell and Charlesworth, 2000, for more on the development of genetic models of metapopulations). Specifically, they found that although genetic diversity within individual populations or within a species as a whole always decreased when population extinction was added to an island model, among-population differentiation measured by F_{ST} could either increase or decrease depending upon (1) the size of the propagule of colonists that recolonizes a population relative to equilibrium levels of gene flow, and (2) where the colonists came from. If colonists were all drawn from a single source population (the propagule pool model), they likely harbor less variation than an average population at equilibrium; hence, the effect of population extinctions is to drive up interpopulation differentiation and F_{ST}. More precisely, F_{ST} will increase if K (the size of the colonizing propagule) $< 2N_e m$ (twice the number of migrants between populations per generation) + 1/2. On the other hand, if colonists are drawn randomly from the entire metapopulation (the migrant pool model) and the number of colonists is large relative to normal gene flow, then recently founded populations will be well mixed compared with other populations, and this extra homogenization furnished by population turnover will decrease overall population subdivision. Broadly speaking then, the effects of extinction and recolonization on genetic differentiation depend on whether colonization increases or decreases the genetic variance among populations.

Repeated samplings can also be used to track changes in F_{ST} over time. Under the propagule pool model (single source), the low levels of variation present just after recolonization should increase over time, with F_{ST} decreasing along the way. In contrast, under the migrant pool (colonists drawn from the full range), variability within populations should diminish as time since colonization goes on, with F_{ST} increasing. Note that these changes in F_{ST} over time are driven primarily by changes in f_0 (population homozygosity), which are then reflected in levels of differentiation among populations.

It would seem reasonable to expect propagule model conditions to hold when dispersal capability is limited relative to the scale of population separation, whereas broad mixing at a given spatial scale should lean toward the migrant pool model. Such expectations, however, cannot substitute for genetic data, even when substantiated by direct observations of recolonization. Dybdahl (1994) gathered data on extinction rates and number of colonists in *Tigriopus californicus* directly, but his genetic data (demonstrating that older tide pool populations were more differentiated than younger ones) turned out to contradict predictions based on these ecological data.

Another way to assess the impact of population turnover is to compare levels of differentiation among habitats with different extinction probabilities. Ruckelshaus (1998) found less subdivision among subtidal populations of eelgrass than high (and more extinction prone) intertidal ones, with mid-intertidal populations falling in between. Thus, population differentiation was inversely correlated with extinction probability, consistent with population turnover enhancing differentiation.

Alternatively, recent theory supports using measures other than subdivision as a means of recognizing population turnover. For instance, extinction–recolonization and ongoing migration produce different frequency distributions of polymorphic nucleotide sites, allowing the relative strengths of these processes to be determined (Wakely and Aliacar, 2001). Population turnover leads to characteristic patterns of within-population allele diversity (Tajima's D) and gene tree shape as well (Pannell, 2003). These approaches are not, however, without their complications and caveats: Changes in the number of populations within a species can confound efforts to distinguish migration and population turnover (Wakeley and Aliacar, 2001), and a large number of polymorphic sites (and thus the isolation and sequencing of many genetic loci) are required for statistical power (Pannell, 2003).

B. Source–Sink Relationships

Once population turnover has been established, either genetically or by more direct means, genetic data can be used to assess components of a specific metapopulation model. The stronger the independent information on the system and its dynamics, the better the ability to frame genetically refutable hypotheses. In the simplest case, genetic data can refute a proposed qualitative model for population dynamics. For example, Jolly et al. (2003) knew that populations of the polychaete *Pectinaria koreni* inhabited isolated patches of fine muddy sands along the French coast of the English Channel, and that some populations were transient whereas others were persistent. Given this species' pelagic larval duration (2 weeks) and short life span (15–18 months), they reasoned that recruits from persistent populations probably maintained demographic sink populations. However, genetic data revealed significant differentiation both between sites and across years, a pattern inconsistent with a metapopulation structure, in which sink populations are maintained by a homogeneous rain of well-mixed and broadly dispersed larvae.

Gaggiotti et al. (2002) used a Bayesian analysis of microsatellite data to obtain quantitative estimates of the relative contributions of several established gray seal colonies (potential sources) to the founding of three new colonies (sinks, at least initially). Genetic differentiation among colonies was subtle, and confidence intervals around their estimates were broad. However, in combination with data on colony reproductive output, they were able to conclude that both geographical proximity and colony productivity played a role in determining the relative contributions of different population sources. Such situation-specific approaches hold great promise (Beerli and Felsenstein, 2001; Wilson and Rannala, 2003).

C. Mixing in the Plankton and the Genetics of Larval Cohorts

For practical reasons, marine ecologists must often treat the pelagic larvae as actors in a black box. It is just too difficult to follow such tiny creatures in such a large environment over extended periods of time. The promise of being able to follow larvae indirectly using genetic markers is the primary attraction of molecular approaches for many marine biologists. It is ironic, then, that until recently so few efforts have been made to learn about the secrets of larvae by directly analyzing their genes. When these efforts have been taken, they have been able to eliminate some alternative hypotheses about the origins of recruiting cohorts, but have also exposed startling and confusing patterns of variation that suggest major events occurring during pelagic larval dispersal still escape our understanding.

A recurring question has been the degree to which settling cohorts are composed of larvae that have remained together since their origin in a single population or even from a single mother. Avise and Shapiro (1986) used the numbers, frequencies, and distributions of allozyme alleles and genotypes to conclude that recently settled aggregations of a serranid reef fish were no more closely related to each other than random. Subsequent studies using more variable microsatellite markers have likewise found fish larvae to be well mixed (Herbinger et al., 1997). Planes et al. (2002) also found that kinship within a settling cohort of *Naso unicornis* (Acathuridae) was not especially high. However, using otoliths to estimate spawning date, they found that larvae of the same age were highly related. Thus, although larvae that settle synchronously may mix individuals from different spawning sites and dates, individual sibships may recruit together— a surprising result given the approximately 75-day pelagic larval durations of *N. unicornis*.

These recent data provide insight into the mechanisms underlying a genetic pattern consistently observed in marine species with planktonic larvae: chaotic patchiness. This pattern was first described by Johnson and Black (1984). They sampled settling limpet larvae multiple times over 2 years, comparing genetic variation among these cohorts with geographic patterns in established adults as well as with meteorological data. They found that temporal genetic variation between different recruitment pulses equaled the geographic subdivision among adult populations. Subsequent work has repeatedly found this pattern of temporal variation among larval cohorts that equals or exceeds geographic variation in adult populations (Watts et al., 1990; Edmands et al., 1996; Ruzzante et al., 1996a; Toonen, 2001). This pattern can also be detected in horizontal studies when cohorts are distinguishable, as when backdated using otoliths (Lenfant and Planes, 2002), although the potential for postsettlement selection arises (Toonen, 2001). These studies establish that, just as pulses of larval settlement can have long-lasting impacts on demography, so too does the unique history of these pulses impact genetic variation.

The mechanism underlying this temporal variation has yet to be firmly established. One possibility is that larvae originate in different populations over the course of time (Kordos and Burton, 1993). However, the low geographic variation seen in adults of the species exhibiting chaotic patchiness suggest this is not a general explanation. Furthermore, Johnson and Black (1984) were able to test this hypothesis explicitly because adult populations varied clinally at one allozyme locus and current patterns changed throughout the year. The pattern of temporal genetic differences in limpet recruits that they found conflicted with the predictions of the different populations of origin hypothesis.

The remaining hypotheses generate great controversy because they posit fundamentally different views of marine demography, specifically of effective population sizes. Johnson and Black (1984) found a strong correlation between air temperature and the frequency of a particular allozyme allele, consistent with a role for selection reshaping variation within larval cohorts during their pelagic dispersal. Selection should be most efficient when effective population sizes are large, which seems reasonable based upon census population sizes. Alternatively, the "sweepstakes" hypothesis (Hedgecock, 1994) argues that pelagic larval dispersal is a highly risky business, with massive mortality and failure likely at every stage from spawning to recruitment. As a result, most adults do not contribute reproductively to the next generation. However, fecundity in these same species is so high that output from just a few pairings can suffice to sustain populations. As a result, and despite high census population counts, these species in fact have low effective population sizes, with temporal variation among recruits resulting from genetic drift. This view has been bolstered by studies finding temporal variation among adults (Hedgecock et al., 1992) and larvae (Li and Hedgecock, 1998) in populations known to be closed and by genetic analysis indicating a low $N_e:N$ ratio (Turner et al., 2002), as well as by the high kinships among identically aged settlers found by Planes et al. (2002). Recent work (Flowers et al., 2002), however, found no genetic evidence for the large variance in female reproductive success predicted by the sweepstakes hypothesis.

Alternative hypotheses for explaining chaotic genetic patchiness need not be mutually exclusive, however. Data from the porcelain crab (*Petrolisthes cinctipes*) are consistent with the action of all three of the primary hypotheses acting synergistically (Toonen, 2001). Recruits in each year could be assigned to different sites (consistent with the varying origin hypothesis), and the number of adults contributing to the larval pool in any given season was a small subset of the census population (consistent with the sweepstakes hypothesis). By following individual cohorts of recruits through time, Toonen (2001) also showed that allele frequencies changed nonrandomly among sites, consistent with local adaptation to sites. Thus for *P. cinctipes*, variation in the source of larval recruits, sweepstakes reproductive success, and natural selection all combine to contribute to a chaotic pattern of population genetic structure.

V. SUMMARY

Genetic markers can contribute to the recognition and understanding of metapopulations in the sea in a number of ways. Much work has gone into measuring subdivision among populations, detecting phylogeographic breaks, and exposing cryptic species (see Section II), all of which has helped to define the spatial scale of populations for different species. More geographically explicit work (see Section III) has confirmed the expected relationship between gene flow and distance in several marine species, but also pointed out how this relationship may only hold at intermediate spatial scales, with smaller scales approximating a true island model and larger ones in nonequilibria dominated by historical events. The absolute scales of these transitions between different models of connectivity vary among species. Finally, the population turnover characteristic of strictly defined metapopulations should be detectable by genetic means (see Section IV.A), but supporting demographic data and temporal sampling regimes greatly strengthen any conclusions that can be drawn. The genetic impact of population turnover hinges on whether colonists are small groups of related individuals or large groups drawn from a larger geographically mixed pool (although it is not clear how often marine species fit the classic metapopulation structure assumed by the models making these predictions). Although this may seem predictable based on larval dispersal capability, one of the few marine studies to address this issue directly (Dybdahl, 1994) found genetic results that contradicted ecological predictions. Genetic studies of settling larvae collected at different times (see Section IV.C) have revealed an even more vexing problem: Variation among temporal samples of larvae often exceeds that seen among geographic samples of adults. Competing explanations for this pattern (selection and sweepstakes reproductive success) predict wildly different effective population sizes, which in turn have consequences for the interpretation of all other population genetic data.

Progress toward a better understanding of marine population dynamics will come from both new analyses and new types of data. New likelihood and Bayesian analyses based on the coalescent (Beerli and Felsenstein, 2001), some of which take advantage of temporal change (Wang and Whitlock, 2003), were designed without assumptions of equilibrium and promise to estimate effective population sizes and migration matrices simultaneously. Such analyses, like all those mentioned earlier, make inferences about population demography over long periods of time (on the order of N generations or longer). Other new techniques utilize linkage disequilibria between different loci to infer migration patterns in the recent past (Cornuet et al., 1999; Pritchard et al., 2000; Wilson and Rannala, 2003). The application of such analyses to marine species may help to reconcile apparent conflicts between patterns of connectivity predicted by present-day currents and those inferred from genetic data.

All these new analytical approaches are data hungry, and increasingly the data they will analyze will come from nuclear gene sequences (Brumfield et al., 2003; Zhang and Hewitt, 2003). Studies of nuclear sequence data will allow for the same types of coalescent analyses developed for mtDNA sequences, but with the statistical power that comes from multiple independent markers. These new data will certainly present analytical challenges, but many of them will turn out to be opportunities as well. For example, recombination rates between variable sites adds an analytical complication not generally present in nonrecombining mtDNA. However, finding linkage disequilibria between sites within nuclear genes, as seen in *rag1* between different Caribbean populations of the goby *Elacatinus horsti* (Taylor and Hellberg, 2006), provides strong evidence of long-term isolation given that rates of recombination are usually faster than those of mutation. Still, for the foreseeable future, the quantity of highly heterozygous markers needed to utilize powerful new multilocus genotyping approaches are beyond the sequencing capabilities of most marine genetic laboratories, and will come instead from microsatellites and AFLPs.

Ultimately, the greatest progress in understanding the dynamics of marine populations will come from coupling new genetic approaches with other sources of data. Even sophisticated genetic analyses can have a difficult time detecting certain population dynamics without additional information (e.g., source–sink; Beerli and Felsenstein, 2001), and some new analyses depend on intensive temporal sampling (Wang and Whitlock, 2003). Chronological markers laid down by otoliths in fish (Lenfant and Planes, 2002) and shell bands in molluscs (Lewis and Thorpe, 1994) offer special opportunities for temporal sampling. Combining judicious choice of focal study species, careful survey data, multiple genetic markers, and emerging analyses should reveal more of the hidden workings of marine populations.

VI. ACKNOWLEDGMENT

The author thanks Mark Dybdahl, Rob Toonen, John Wares, anonymous reviewers, and the editors for insightful comments. Special thanks go to Yaisel Borrell for helping the author to see the power of multilocus genotyping.

REFERENCES

Austerlitz, F., Jung–Muller, B., Godelle, B., and Gouyon, P.-H. (1997). Evolution of coalescence times, genetic diversity and structure during colonization. *Theoretical Population Biology.* **51**, 148–164.

Avise, J.C., and Shapiro, D.Y. (1986). Evaluating kinship of newly settled juveniles within social groups of the coral reef fish *Anthias squamipinnis. Evolution.* **40**, 1051–1059.

Ayre, D.J. (1984). The effects of sexual and asexual reproduction on geographic variation in the sea anemone *Actinia tenebrosa. Oecologia.* **62**, 222–229.

Barber, P.H., Palumbi, S.R., Erdmann, M.V., and Moosa, M.K. (2002). Sharp genetic breaks among populations of *Haptosquilla pulchella* (Stomatopoda) indicates limits to larval transport: Patterns, causes, and consequences. *Molecular Ecology.* **11**, 659–674.

Barton, N., and Hewitt, G.M. (1985). Analysis of hybrid zones. *Annual Review of Ecology and Systematics.* **16**, 113–148.

Barton, N.H., and Whitlock, M.C. (1997). The evolution of metapopulations. In *Metapopulation biology* (I. Hanski and M.E. Gilpin, eds., pp. 183–210). San Diego: Academic.

Baums, I.B., Miller, M.W., and Hellberg, M.E. (2005). Regionally isolated populations of an Imperiled Caribbean coral, *Acropora palmata. Molecular Ecology.* **14**, 1377–1390.

Beerli, P., and Felsenstein, J. (1999). Maximum-likelihood estimation of migration rates and effective population numbers in two populations using a coalescent approach. *Genetics.* **152**, 763–773.

Beerli, P., and Felsenstein, J. (2001). Maximum likelihood estimation of a migration matrix and effective population sizes using a coalescent approach. *Proceedings of the National Academy of Sciences USA.* **98**, 4563–4568.

Benzie, J.A.H. (1999). Genetic structure of coral reef organisms: Ghosts of dispersal past. *American Zoologist.* **39**, 131–145.

Benzie, J.A.H., and Williams, S.T. (1997). Genetic structure of giant clam (*Tridacna maxima*) populations in the West Pacific is not consistent with dispersal by present-day ocean currents. *Evolution*. **51**, 768–783.
Bernardi, G. (2000). Barriers to gene flow in *Embiotoca jacksoni*, a marine fish lacking a pelagic larval stage. *Evolution*. **54**, 226–237.
Borrell, Y., Espinosa, G., Romo, J., Blanco, G., Vazquez, E., and Sanchez, J.A. (2004). DNA microsatellite variability and genetic differentiation among natural populations of the Cuban white shrimp *Litopenaeus schmitti*. *Marine Biology*. **144**, 327–333.
Bowen, B.W., Bass, A.L., Rocha, L.A., Grant, W.S., and Robertson, D.R. (2001). Phylogeography of the trumpetfishes (*Aulostomus*): Ring species complex on a global scale? *Evolution*. **55**, 1029–1039.
Brumfield, R.T., Beerli, P., Nickerson, D.A., and Edwards, S.V. (2003). The utility of single nucleotide polymorphisms in inferences of population history. *Trends in Ecology and Evolution*. **18**, 249–256.
Bucklin, A., and Wiebe, P.H. (1998). Low mitochondrial diversity and small effective population sizes of the copepods *Calanus finmarchicus* and *Nannocalanus minor*: Possible impact of climatic variation during recent glaciation. *Journal of Heredity*. **89**, 383–392.
Buroker, N.E. (1983). Population genetics of the American oyster *Crassostrea virginica* along the Atlantic coast and the Gulf of Mexico. *Marine Biology*. **75**, 99–112.
Burton, R.S. (1998). Intraspecific phylogeography across the Point Conception biogeographic boundary. *Evolution*. **52**, 734–745.
Burton, R.S., Rawson, P.D., and Edmands, S. (1999). Genetic architecture of physiological phenotypes: Empirical evidence for coadapted gene complexes. *American Zoologist*. **39**, 451–462.
Carlon, D.B., and Budd, A.F. (2002). Incipient speciation across a depth gradient in a scleractinian coral? *Evolution*. **56**, 2227–2242.
Charlesworth, B. (1998). Measures of divergence between populations and the effects of forces that reduce variability. *Molecular Biology and Evolution*. **15**, 538–543.
Cornuet, J.-M., Piry, S., Luikart, G., Estoup, A., and Solignac, M. (1999). New methods employing multilocus genotypes to select or exclude populations as origins of individuals. *Genetics*. **153**, 1989–2000.
Crow, J.F., and Aoki, K. (1984). Group selection for a polygenic behavioral trait: Estimating the degree of population subdivision. *Proceedings of the National Academy of Sciences USA*. **81**, 6073–6077.
Dawson, M.N., and Jacobs, D.K. (2001). Molecular evidence for cryptic species of *Aurelia aurita* (Cnidaria, Scyphozoa). *Biological Bulletin*. **200**, 92–96.
de Vargas, C., Norris, R., Zaninetti, L., Gibb, S.W., and Pawlowski, J. (1999). Molecular evidence of cryptic speciation in planktonic foraminifers and their relation to oceanic provinces. *Proceedings of the National Academy of Sciences USA*. **96**, 2864–2868.
Dybdahl, M.F. (1994). Extinction, recolonization, and the genetic structure of tidepool copepod populations. *Evolutionary Ecology*. **8**, 113–124.
Edmands, S., Moberg, P.E., and Burton, R.S. (1996). Allozyme and mitochondrial DNA evidence of population subdivision in the purple sea urchin *Strongylocentrotus purpuratus*. *Marine Biology*. **126**, 443–450.
Falush, D., Stephens, M., and Pritchard, J.K. (2003). Inference of population structure using multilocus genotype data: Linked loci and correlated allele frequencies. *Genetics*. **164**, 1567–1587.
Flowers, J.M., Schroeter, S.C., and Burton, R.S. (2002). The recruitment sweepstakes has many winners: Genetic evidence from the sea urchin *Strongylocentrotus purpuratus*. *Evolution*. **56**, 1445–1453.
Gaggiotti, O.E., Jones, F., Lee, W.M., Amos, W., Harwood, J., and Nichols, R.A. (2002). Patterns of colonization in a metapopulation of grey seals. *Nature*. **416**, 424–427.
Garza, J.C., and Williamson, E.G. (2001). Detection of reduction in population size using data from microsatellite loci. *Molecular Ecology*. **10**, 305–318.

Grosberg, R.K., and Cunningham, C.W. (2001). Genetic structure in the sea: From populations to communities. In *Marine community ecology* (M.D. Bertness, S.D. Gaines, and M.E. Hay, eds., pp. 61–84). Sunderland, MA: Sinauer Associates.

Hedgecock, D. (1994). Does variance in reproductive success limit effective population size of marine organisms? In *Genetics and evolution of aquatic organisms* (A. Beaumont, ed., pp. 122–134). London: Chapman & Hall.

Hedgecock, D., Chow, V., and Waples, R.S. (1992). Effective population numbers of shellfish broodstocks estimated from temporal variance in allelic frequencies. *Aquaculture.* **108**, 215–232.

Hellberg, M.E. (1994). Relationships between inferred levels of gene flow and geographic distance in a philopatric coral, *Balanophyllia elegans. Evolution.* **48**, 1829–1854.

Hellberg, M.E. (1995). Stepping-stone gene flow in the solitary coral *Balanophyllia elegans*: Equilibrium and nonequilibrium at different spatial scales. *Marine Biology.* **123**, 573–581.

Hellberg, M.E., Balch, D.P., and Roy, K. (2001). Climate-driven range expansion and morphological evolution in a marine gastropod. *Science.* **292**, 1707–1710.

Hellberg, M.E., Burton, R.S., Neigel, J.E., and Palumbi, S.R. (2002). Genetic assessment of connectivity among marine populations. *Bulletin of Marine Science.* **70**, 273–290.

Herbinger, C.M., Doyle, R.W., Taggart, C.T., Lochmann, S.E., Brooker, A.L., Wright, J.M., and Cook, D. (1997). Family relationships and effective population size in a natural cohort of Atlantic cod (*Gadus morhua*) larvae. *Canadian Journal of Fisheries and Aquatic Sciences.* **54(suppl. 1)**, 11–18.

Hey, J. (1991). A multi-dimensional coalescent process applied to multi-allelic selection models and migration models. *Theoretical Population Biology.* **39**, 30–48.

Hey, J., and Machado, C.A. (2003). The study of structured populations: New hope for a difficult and divided science. *Nature Reviews Genetics.* **4**, 535–543.

Hoelzel, A.R., Halley, J., O'Brien, S.J., Campagna, C., Arnbom, T., Leboeuf, B., Ralls, K., and Dover, G.A. (1993). Elephant seal genetic variation and the use of simulation models to investigate historical population bottlenecks. *Journal of Heredity.* **84**, 443–449.

Hoskin, M.G. (2000). Effects of the East Australian Current on the genetic structure of a direct developing muricid snail (*Bedeva hanleyi*): Variability within and among local populations. *Biological Journal of the Linnean Society.* **69**, 245–262.

Hudson, R.R. (1990). Gene genealogies and the coalescent process. In *Oxford surveys in evolutionary biology 7* (D.J. Futuyma and J. Antonovics, eds., pp. 1–44). Oxford: Oxford University Press.

Hutchison, D.W., and Templeton, A.R. (1999). Correlation of pairwise genetic and geographic distance measures: Inferring the relative influences of gene flow and drift on the distribution of genetic variability. *Evolution.* **53**, 1898–1914.

Jiang, L., Wu, W.L., and Huang, P.C. (1995). The mitochondrial DNA of Taiwan abalone *Haliotis diversicolor* Reeve, 1846 (Gastropoda: Archeogastropoda: Haliotidae). *Molecular Marine Biology and Biotechnology.* **4**, 353–354.

Johannesson, K., Johannesson, B., and Rolán–Alvarez, E. (1993). Morphological differentiation and genetic cohesiveness over a micro-environmental gradient in the marine snail *Littorina saxatilis. Evolution.* **47**, 1770–1787.

Johannesson, K., and Warmoes, T. (1990). Rapid colonization of Belgian breakwaters by the direct developer, *Littorina saxatilis* (Olivi) (Prosobranchia, Mollusca). *Hydrobiologia.* **193**, 99–108.

Johnson, M.S. (2000). Measuring and interpreting genetic structure to minimize the genetic risks of translocations. *Aquaculture Research.* **31**, 133–143.

Johnson, M.S., and Black, R. (1984). Pattern beneath chaos: The effect of recruitment on genetic patchiness in an intertidal limpet. *Evolution.* **38**, 1371–1383.

Johnson, M.S., and Black, R. (1991). Genetic subdivision of the intertidal snail *Bembicium vittatum* (Gastropoda: Littorinidae) varies with habitat in the Houtman Abrolhos Islands, Western Australia. *Heredity.* **67**, 205–213.

Jollivet, D., Chevaldonné, P., and Planque, B. (1999). Hydrothermal-vent alvinellid polychaete dispersal in the Eastern Pacific. 2. A metapopulation model based on habitat shifts. *Evolution.* **53**, 1128–1142.

Jolly, M.T., Viard, F., Weinmayr, G., Gentil, F., Thiebaut, E., and Jollivet, D. (2003). Does the genetic structure of *Pectinaria koreni* (Polychaeta: Pectinariidae) conform to a source–sink metapopulation model at the scale of the Baie de Seine? *Helgoland Marine Research.* **56**, 238–246.

Jones, G.P., Milicich, M.J., Emslie, M.J., and Lunow, C. (1999). Self-recruitment in a coral reef fish population. *Nature.* **402**, 802–804.

Karl, S.A., and Avise, J.C. (1992). Balancing selection on allozyme loci in oysters: Implications from nuclear RFLPs. *Science.* **256**, 100–102.

Knight, A.J., Hughes, R.N., and Ward, R.D. (1987). A striking example of example of the founder effect in the mollusc *Littorina saxatilis. Biological Journal of the Linnean Society.* **32**, 417–426.

Knowlton, N. (1993). Sibling species in the sea. *Annual Review of Ecology and Systematics.* **24**, 189–216.

Knowlton, N., Weil, E., Weigt, L.A., and Guzman, H.M. (1992). Sibling species in *Montastrea annularis,* coral bleaching, and the coral climate record. *Science.* **255**, 330–333.

Kordos, L.M., and Burton, R.S. (1993). Genetic differentiation of Texas Gulf Coast populations of the blue crab *Callinectes sapidus. Marine Biology.* **117**, 227–233.

Kuhner, M.K., Yamato, J., and Felsenstein, J. (1998). Maximum likelihood estimation of population growth rates based on the coalescent. *Genetics.* **149**, 429–434.

Lavery, S., Moritz, C., and Fielder, D.R. (1995). Changing patterns of population structure and gene flow at different spatial scales in *Birgus latro* (the coconut crab). *Heredity.* **74**, 531–541.

Leis, J.M. (2002). Pacific coral-reef fishes: The implications of behaviour and ecology of larvae for biodiversity and conservation, and a reassessment of the open population paradigm. *Environmental Biology of Fishes.* **65**, 199–208.

Lenfant, P., and Planes, S. (2002). Temporal genetic changes between cohorts in a natural population of a marine fish, *Diplodus sargus. Biological Journal of the Linnean Society.* **76**, 9–20.

Lessios, H.A., Garrido, M.J., and Kessing, B.D. (2001a). Demographic history of *Diadema antillarum,* a keystone herbivore on Caribbean reefs. *Proceedings of the Royal Society of London B.* **268**, 2347–2353.

Lessios, H.A., Kane, J., and Robertson, D.R. (2003). Phylogeography of the pantropical sea urchin *Tripneustes*: Contrasting patterns of population structure between oceans. *Evolution.* **57**, 2026–2036.

Lessios, H.A., Kessing, B.D., and Robertson, D.R. (1998). Massive gene flow across the world's most potent marine biogeographic barrier. *Proceedings of the Royal Society of London B.* **265**, 583–588.

Lessios, H.A., Kessing, B.D., and Pearse, J.S. (2001b). Population structure and speciation in tropical seas: Global phylogeography of the sea urchin *Diadema. Evolution.* **55**, 955–975.

Lessios, H.A., Weinberg, J.R., and Starczak, V.R. (1994). Temporal variation in populations of the marine isopod *Excirolana*: How stable are gene frequencies and morphology? *Evolution.* **48**, 549–563.

Lewis, R.I., and Thorpe, J.P. (1994). Temporal stability of gene frequencies within genetically heterogeneous populations of the queen scallop *Aequipecten (Chlamys) opercularis. Marine Biology.* **121**, 117–126.

Li, G., and Hedgecock, D. (1998). Genetic heterogeneity, detected by PCR SSCP, among samples of larval Pacific oysters (*Crassostrea gigas*) supports the hypothesis of large variance in reproductive success. *Canadian Journal of Fisheries and Aquatic Sciences.* **55**, 1025–1033.

Luikart, G., Allendorf, F.W., Cornuet, J.-M., and Sherwin, W.B. (1998). Distortion of allele frequency distributions provides a test for recent population bottlenecks. *Journal of Heredity.* **89**, 238–247.

Marko, P.B. (1998). Historical allopatry and the biogeography of speciation in the prosobranch snail genus *Nucella. Evolution.* **52**, 757–774.

Miya, N., and Nishida, M. (1998). Speciation in the open ocean. *Nature.* **389**, 803–804.

Nei, M. (1973). Analysis of gene diversity in subdivided populations. *Proceedings of the National Academy of Sciences USA.* **70**, 3321–3323.
Nei, M., and Maruyama, T. (1975). Lewontin–Krakauer test for neutral genes. *Genetics.* **80**, 395.
Neigel, J.E. (2002). Is F_{ST} obsolete? *Conservation Genetics.* **3**, 167–173.
Nesbo, C.L., Rueness, E.K., Iversen, S.A., Skagen, D.W., and Jakobsen, K.S. (2000). Phylogeography and population history of Atlantic mackerel (*Scomber scombrus* L.): A genealogical approach reveals genetic structuring among the eastern Atlantic stocks. *Proceedings of the Royal Society of London B.* **267**, 281–292.
Nielsen, R., and Slatkin, M. (2000). Likelihood analysis of ongoing gene flow and historical association. *Evolution.* **54**, 44–50.
Nielsen, R., and Wakeley, J. (2001). Distinguishing migration from isolation: A Markov chain approach. *Genetics.* **158**, 885–896.
Nishida, M., and Lucas, J.S. (1988). Genetic differences between geographic populations of the crown-of-thorns starfish throughout the Pacific region. *Marine Biology.* **98**, 359–368.
Nunney, L. (1993). The influence of mating system and overlapping generations on effective population size. *Evolution.* **47**, 1329–1341.
O'Foighil, D., and Jozefowicz, C.J. (1999). Amphi-Atlantic phylogeography of direct-developing lineages of *Lasaea*, a genus of brooding bivalves. *Marine Biology.* **135**, 115–122.
Palumbi, S.R. (2003). Population genetics, demographic connectivity, and the design of marine reserves. *Ecological Applications.* **13**, S146–S158.
Palumbi, S.R., Grabowsky, G., Duda, T., Geyer, L., and Tachino, N. (1997). Speciation and population genetic structure in tropical Pacific sea urchins. *Evolution.* **51**, 1506–1517.
Pannell, J.R. (2003). Coalescence in a metapopulation with recurrent extinction and recolonization. *Evolution.* **57**, 949–961.
Pannell, J.R., and Charlesworth, B. (2000). Effects of metapopulation processes on measures of genetic diversity. *Proceedings of the Royal Society of London B.* **355**, 1851–1864.
Paulay, G., and Meyer, C. (2002). Diversification in the tropical Pacific: Comparisons between marine and terrestrial systems and the importance of founder speciation. *Integrative and Comparative Biology.* **42**, 922–934.
Pechenik, J.A. (1999). On the advantages and disadvantages of larval stages in benthic marine invertebrate life cycles. *Marine Ecology Progress Series.* **177**, 269–297.
Piertney, S.B., and Carvalho, G.R. (1995). Microgeographic genetic differentiation in then intertidal isopod *Jaera albifrons* Leach. II. Temporal variation in allele frequencies. *Journal of Experimental Marine Biology and Ecology.* **188**, 277–288.
Planes, S., and Fauvelot, C. (2002). Isolation by distance and vicariance drive genetic structure of a coral reef fish in the Pacific Ocean. *Evolution.* **56**, 378–399.
Planes, S., Lecaillon, G., Lenfant, P., and Meekan, M. (2002). Genetic and demographic variation in new recruits of *Naso unicornis*. *Journal of Fish Biology.* **61**, 1033–1049.
Pogson, G.H., and Fevolden, S.E. (2003). Natural selection and the genetic differentiation of coastal and Arctic populations of the Atlantic cod in northern Norway: A test involving nucleotide sequence variation at the pantophysin (PanI) locus. *Molecular Ecology.* **12**, 63–74.
Pogson, G.H., Taggart, C.T., Mesa, K.A., and Boutilier, R.G. (2001). Isolation by distance in the Atlantic cod, *Gadus morhua*, at large and small geographic scales. *Evolution.* **55**, 131–146.
Pritchard, J.K., Stephens, M., and Donnelly, P. (2000). Inference of population structure using multilocus genotype data. *Genetics.* **155**, 945–959.
Reeb, C.A., and Avise, J.C. (1990). A genetic discontinuity in a continuously distributed species: Mitochondrial DNA in the American oyster, *Crassostrea virginica*. *Genetics.* **124**, 397–406.
Rhodes, K.L., Lewis, R.I., Chapman, R.W., and Sadovy, Y. (2003). Genetic structure of camouflage grouper, *Epinephelus polyphekadion* (Pisces: Serranidae), in the western central Pacific. *Marine Biology.* **142**, 771–776.

Roberts, C.M. (1997). Connectivity and management of Caribbean coral reefs. *Science.* **278**, 1454–1457.

Robertson, A. (1975). Remarks on the Lewontin–Krakauer test. *Genetics.* **80**, 396.

Rocha, L.A., Bass, A.L., Robertson, D.R., and Bowen, B.W. (2002). Adult habitat preferences, larval dispersal, and the comparative phylogeography of three Atlantic surgeonfishes (Teleostei: Acanthuridae). *Molecular Ecology.* **11**, 243–252.

Rogers, A.R., and Harpending, H. (1992). Population growth makes waves in the distribution of pairwise genetic differences. *Molecular Biology and Evolution.* **9**, 552–569.

Roman, J., and Palumbi, S.R. (2003). Whales before whaling in the North Atlantic. *Science.* **301**, 508–510.

Rosenberg, N.A., et al. (2001). Empirical evaluation of genetic clustering methods using multilocus genotypes from 20 chicken breeds. *Genetics.* **159**, 699–713.

Roy, K., Valentine, J.W., Jablonski, D., and Kidwell, S.M. (1996). Scales of climatic variability and time averaging in Pleistocene biotas: Implications for ecology and evolution. *Trends in Ecology and Evolution.* **11**, 458–463.

Ruckelshaus, M. (1998). Spatial scale of genetic structure and an indirect estimate of gene flow in eelgrass, *Zostera marina*. *Evolution.* **52**, 330–343.

Russ, G.R., Lou, D.C., and Ferreira, B.P. (1996). Temporal tracking of a strong cohort in the population of a coral reef fish, the coral trout, *Plectropomus leopardus* (Serranidae: Epinephelinae), in the central Great Barrier Reef, Australia. *Canadian Journal of Fisheries and Aquatic Sciences.* **53**, 2745–2751.

Ruzzante, D.E., Taggart, C.T., and Cook, D. (1996a). Spatial and temporal variation in the genetic composition of a larval cod (*Gadus morhua*) aggregation: Cohort contribution and genetic stability. *Canadian Journal of Fisheries and Aquatic Sciences.* **53**, 2695–2705.

Ruzzante, D.E., Taggart, C.T., Cook, D., and Goddard, S. (1996b). Genetic differentiation between inshore and offshore Atlantic cod (*Gadus morhua*) off Newfoundland: Microsatellite DNA variation and antifreeze level. *Canadian Journal of Fisheries and Aquatic Sciences.* **53**, 634–645.

Shaklee, J.B., Beacham, T.D., Seeb, L., and White, B.A. (1999). Managing fisheries using genetic data: Case studies from four species of Pacific salmon. *Fish. Res.* **43**, 45–78.

Sinclair, M. (1988). *Marine populations.* Seattle: University of Washington Press.

Slatkin, M. (1987). Gene flow and the geographic structure of natural populations. *Science.* **236**, 787–792.

Slatkin, M. (1991). Inbreeding coefficients and coalescence times. *Genetical Research.* **58**, 167–175.

Slatkin, M. (1993). Isolation by distance in equilibrium and non-equilibrium populations. *Evolution.* **47**, 264–279.

Skold, M., Wing, S.R., and Mladenov, P.V. (2003). Genetic subdivision of a sea star with high dispersal capability in relation to physical barriers in a fjordic seascape. *Marine Ecology Progress Series.* **250**, 163–174.

Sponer, R., and Roy, M.S. (2002). Phylogeographic analysis of the brooding brittle star *Amphipholis squamata* (Echinodermata) along the coast of New Zealand reveals high cryptic genetic variation and cryptic dispersal potential. *Evolution.* **56**, 1954–1967.

Starck, W.A., II, and Colin, P.L. (1978). *Gramma linki*: A new species of grammid fish from the tropical western Atlantic. *Bulletin of Marine Science.* **28**, 146–152.

Sunnucks, P. (2000). Efficient genetic markers for population biology. *Trends in Ecology and Evolution.* **15**, 199–203.

Swearer, S.E., Caselle, J.E., Lea, D.W., and Warner, R.R. (1999). Larval retention and recruitment in an island population of a coral-reef fish. *Nature.* **402**, 799–802.

Swearer, S.E., Thorrold, S.R., Shima, J.S., Hellberg, M.E., Jones, G.P., Robertson, D.R., Morgan, S.G., Selkoe, K.A., Ruiz, G.M., and Warner, R.R. (2002). Evidence for self-recruitment in benthic marine populations. *Bulletin of Marine Science.* **70**, 251–271.

Chapter 13 Genetic Approaches to Understanding Marine Metapopulation Dynamics

Tarjuelo, I., Posada, D., Crandall, K.A., Pascual, M., and Turon, X. (2001). Cryptic species of *Clavelina* (Ascidiacea) in two different habitats: Harbours and rocky littoral zones in the northwestern Mediterranean. *Marine Biology.* **139**, 455–462.

Taylor, M.S., and Hellberg, M.E. (2003). Genetic evidence for local retention of pelagic larvae in a Caribbean reef fish. *Science.* **299**, 107–109.

Taylor, M.S., and Hellberg, M.E. (2006). Comparative phylogeography of a genus of coral reef fishes: biogeographical and genetic concordance in the Caribbean. *Molecular Ecology.* In press.

Toonen, R.J. (2001). Genetic analysis of recruitment and dispersal patterns in the porcelain shore crab, *Petrolisthes cinctipes*. PhD dissertation. Davis, CA: Center for Population Biology, University of California.

Turner, T.F., Wares, J.P., and Gold, J.R. (2002). Genetic effective size is three orders of magnitude smaller than adult census size in an abundant, estuarine-dependent marine fish (*Sciaenops ocellatus*). *Genetics.* **162**, 1329–1339.

Underwood, A.J., and Keough, M.J. (2001). Supply-side ecology: The nature and consequences of variations in recruitment of intertidal organisms. In *Marine community ecology* (M.D. Bertness, S.D. Gaines, and M.E. Hay, eds., pp. 183–200). Sunderland, MA: Sinauer Associates.

Vucetich, J.A., Waite T.A., and Nunney, L. (1997). Fluctuating population size and the ratio of effective to census population size. *Evolution.* **51**, 2017–2021.

Vollmer, S.V., and Palumbi, S.R. (2002). Hybridization and the evolution of reef coral diversity. *Science.* **296**, 2023–2025.

Wade, M.J., and McCauley, D.E. (1988). Extinction and recolonization: Their effects on the genetic differentiation of local populations. *Evolution.* **42**, 995–1005.

Wakeley, J., and Aliacar, N. (2001). Gene genealogies in a metapopulation. *Genetics.* **159**, 893–905.

Wang, J., and Whitlock, M.C. (2003). Estimating effective population size and migration rates from genetic samples over space and time. *Genetics.* **163**, 429–446.

Waples, R.S. (1998). Separating the wheat from the chaff: Patterns of genetic differentiation in high gene flow species. *Journal of Heredity.* **89**, 438–450.

Wares, J.P., Gaines, S.D., and Cunningham, C.W. (2001). A comparative study of asymmetric migration events across a marine biogeographic boundary. *Evolution.* **55**, 295–306.

Wares, J.P., Goldwater, D.S., Kong, B.Y., and Cunningham, C.W. (2002). Refuting a controversial case of a human-mediated marine species introduction. *Ecology Letters.* **5**, 577–584.

Watts, R.J., Johnson, M.S., and Black, R. (1990). Effects of recruitment on genetic patchiness in the urchin *Echinometra mathaei* in western Australia. *Marine Biology.* **105**, 145–152.

Whitlock, M.C. (1992). Temporal fluctuations in demographic parameters and the genetic variance among populations. *Evolution.* **46**, 608–615.

Wilkins, J.F., and Wakeley, J. (2002). The coalescent in a continuous, finite, linear population. *Genetics.* **161**, 873–888.

Wilson, G.A., and Rannala, B. (2003). Bayesian inference of recent migration rates using multilocus genotypes. *Genetics.* **163**, 1177–1191.

Winans, G.A. (1980). Geographic variation in the milkfish *Chanos chanos*. I. Biochemical evidence. *Evolution.* **34**, 558–574.

Witman, J., Genovese, S.J., Bruno, J.F., McLaughlin, J.W., and Pavlin, B.I. (2003). Massive prey recruitment and the control of rocky subtidal communities on large spatial scales. *Ecological Monographs.* **73**, 441–462.

Won, Y., Young, C.R., Lutz R.A., and Vrijenhoek, R.C. (2003). Dispersal barriers and isolation among deep-sea mussel populations (Mytilidae: *Bathymodiolus*) from eastern Pacific hydrothermal vents. *Molecular Ecology.* **12**, 169–184.

Zhang, D.-X., and Hewitt, G.M. (2003). Nuclear DNA analysis in genetic studies of populations: Practice, problems and prospects. *Molecular Ecology.* **12**, 563–584.

CHAPTER 14

Metapopulation Dynamics and Community Ecology of Marine Systems

RONALD H. KARLSON

I. Introduction
   A. Scale of Dispersal
   B. Dispersal and Population Dynamics
   C. Relevance to Marine Metacommunities
II. Metacommunities and Species–Area Relationships
   A. Background
   B. Relevance to Marine Metacommunities
   C. Regional-Scale Differentiation
   D. Summary
III. Metacommunities and Local–Regional Species Richness Relationships
   A. Background
   B. Relevance to Marine Metacommunities
   C. Effects of Transport Processes and Relative Island Position
   D. Summary
IV. Metacommunities and Relative Species Abundance Patterns
   A. Background
   B. Metacommunities and Relative Species Abundance Patterns
   C. Relevance to Marine Metacommunities
   D. Summary
V. Summary
   References

I. INTRODUCTION

Sound science depends on theory for the development of the conceptual framework for empirical investigations and explanatory models of real systems. In a recent volume on dispersal (Clobert et al., 2001), several of the contributing authors expressed words of caution as to how theoretical models have been used in studies of gene flow, mechanisms of dispersal, dispersal behavior, and the evolution of dispersal. Like so many areas of science, the theoretical literature on dispersal has rapidly expanded well beyond empirical tests of theory (e.g., the literature on deterministic chaos in population ecology). Consequently, there are few solid empirical studies permitting us to discriminate among different theoretical explanations for dispersal phenomena. One reason for this problem is the difficulty in studying dispersal directly in most organisms. For population geneticists, the study of gene flow has traditionally used indirect methods and there are very few studies in which these have been corroborated by direct estimation procedures (Barton, 2001).

As we apply the notion of the metapopulation to marine systems, we must consider dispersal as the fundamental process linking populations. Yet dispersal has different meanings in different contexts (Clobert et al., 2001). We must take care to identify the units of dispersal (genes, individuals, or species) and the spatiotemporal scale at which dispersal operates. In population genetics, it is recommended that gene flow and genetic differentiation be studied with standard genetic methods using nearby local populations (Barton, 2001; Ross, 2001; Rousset, 2001). If one uses populations separated by large distances with low levels of gene flow, the assumption of equilibrium in the degree of genetic differentiation is likely to be violated and model predictions are strongly biased. Thus the population ecologist interested in the structure and dynamics of metapopulations is advised to use appropriate models matching the spatiotemporal scales spanned by individuals dispersing among populations.

For the community ecologist and biogeographer, it is the dispersal of species that is of interest (e.g., Hubbell, 2001). Although the former may focus on how dispersal influences the dynamics of interacting members of ecological communities (see Roughgarden, 1989; Paine, 1994), the latter typically focuses on the effects of dispersal on biogeographical assemblages over very large spatiotemporal scales (e.g., consideration of oceanic gyres and major geographic barriers to dispersal; as in Connolly et al., 2003). The words of caution regarding matching the scale of gene flow with appropriately scaled models apply here as well. Because one can refer to the dispersal of species at these different spatiotemporal scales using the same terms, the intended meaning can be ambiguous and easily misunderstood (Rosen, 1988).

In the sections that follow I briefly discuss some general fundamentals of dispersal and the dynamics of metapopulations, and then consider their relevance to marine community ecology. In particular, I am interested in those aspects of metapopulation biology that may be applicable to the study of real marine communities. Empirical studies of marine communities linked by dispersal in the marine environment are not common. Consequently, it is appropriate at this early stage to take care in formulating appropriate predictions from the growing body of metacommunity theory. I emphasize the importance of matching the geographic scale of dispersal from the perspective of the selected organisms and the putatively linked habitats that might comprise a marine metacommunity. After the introductory sections in the following pages, I consider three quantitative relationships or patterns that have received considerable attention in the general community ecology literature (i.e., species–area relationships, local–regional species richness relationships, and patterns of relative species abundances). Although these attributes transcend some disciplinary boundaries as one shifts scales from local communities to metacommunities and even larger biogeographical assemblages, my primary focus is on how these relationships might be influenced by marine metacommunities at what has been called the mesoscale (Ricklefs and Schluter, 1993). For related contributions in the literature that discuss these same relationships in quite general terms, I refer the reader to Holt (1993) and Mouquet and Loreau (2003). The latter paper was published as I began revisions of this chapter. It elegantly illustrates the timeliness of the question I pose here, particularly with respect to local–regional species richness relationships and patterns of relative species abundances (see Sections III and IV respectively).

A metacommunity has been defined two different ways, depending on whether one wishes to consider multiple trophic levels or only a single group (assemblage) of trophically similar species. The first defines it as a set of local communities, each occurring at different locations and linked to one another by the dispersal of one or more of the constituent species (Gilpin and Hanski, 1991; Holt, 1997). I qualify this definition by adding that the dispersal of these species must occur at spatiotemporal scales that approximately match the dynamical changes in local community structure (species richness, diversity, evenness, composition, trophic relationships, etc.). Typically, dispersal over evolutionary or biogeographical scales (describing radiations or range extensions respectively) does not match the scale of fluctuations in community structure over ecological time. As emphasized in this chapter, a large mismatch in the scale of dispersal, scale of linked habitats, and the scale of fluctuations in community structure can lead to large discrepancies between observed and predicted metacommunity patterns.

In the second definition, "the *metacommunity* consists of all trophically similar individuals and species in a regional collection of local communities" (Hubbell, 2001). This definition explicitly excludes trophic structure as an attribute of a metacommunity, but it includes the regional scale comprising the source of all immigrating species into local assemblages as posited by competition theory (e.g., Wilson, 1992), the theory of island biogeography (MacArthur and Wilson, 1967), and Hubbell's new unified theory of biodiversity and biogeography (see further discussion in Section IV). Given these different perspectives, another word of caution is warranted when attempting to generate predictions from metacommunity theory. Purely biogeographical studies at macroecological scales use the body of theory relating to assemblages of similar species. Mesoecological studies explicitly dealing with multiple trophic levels (food webs, trophic cascades, keystone species, etc.) use theory emphasizing these trophic relationships. In studies focused on the structure of a single trophic level in the presence or absence of other trophic levels, one may need to blend these two approaches.

A. SCALE OF DISPERSAL

In this section, I briefly consider how the notion of scale applies to ecologically relevant dispersal phenomena. Some phenomena may result directly from evolved morphological adaptations favoring increased dispersal. For example, phenotypically variable dispersal traits promote escape from high densities in planthoppers (Denno et al., 1991; Denno and Roderick, 1992) and escape from local mortality in freshwater bryozoans (Callaghan and Karlson, 2002). Likewise, stoloniferous growth forms in plants and sessile invertebrates promote escape from what Harper (1985) called *resource depletion zones*. These dispersal phenomena relate to local-scale densities, disturbances, and competition. In contrast, other phenomena may be secondary consequences of these local-scale factors and not directly related to the evolution of dispersal traits. For example, the literature on long-distance dispersal by shallow-water, benthic marine invertebrates includes evidence for dispersal across major marine barriers (e.g., Scheltema, 1988) and even genetic exchange across oceans (e.g., Scheltema, 1971). Although some of this evidence may very well document historically important dispersal (e.g., Benzie, 1999), its relevance to the dynamics of populations or communities on ecological spatiotemporal scales is not clear. Strong linkage between the scale of dispersal characterizing particular taxa and dynamics at ecological and biogeographical scales should be made explicitly (e.g., Mora et al., 2003).

The dispersal capabilities of an organism across a fragmented landscape (or seascape) dictates the scale at which one should study metapopulations. Consider the lesson from the now classic studies of the bay checkerspot butterfly in the San Francisco Bay region, where serpentine outcrops provide suitable grassland habitat for plants used for food and oviposition. Ehrlich (1961) concluded that this butterfly is "extraordinarily sedentary" even in the absence of obvious physical barriers to dispersal at Jasper Ridge. In fact, it was later determined that the Jasper Ridge site supported three distinct populations each exhibiting different long-term patterns of population fluctuations (Ehrlich et al., 1975). To the south at Morgan Hill, a set of 27 habitat patches supported a butterfly metapopulation with source–sink dynamics and the characteristic pattern of local colonization and extinction events (Harrison, 1989; Harrison et al., 1988). The transient dynamics of these local populations as they "wink in and out" resulted in many empty patches; the equilibrium frequency of patch occupancy at Morgan Hill was only 30 to 41% (Harrison et al., 1988). Despite the fact that dispersal is required to maintain metapopulations, this butterfly is a "poor disperser and colonist" (Harrison, 1989). Greater dispersal capabilities would increase the frequency of occupied patches and reduce the independent dynamics of local populations to form a highly connected set of populations with more synchronous dynamics. Poorer dispersal abilities would increase the degree of isolation, asynchrony, and independence of local populations. Thus the very existence of a metapopulation depends on an organism's dispersal capabilities relative to the spatial arrangement and size of suitable habitats and the dispersal barriers among them.

The importance of the relative scale between an organism's dispersal ability and the spatial arrangement of habitat has been made numerous times in both empirical and theoretical studies (Hanski, 1999; Hubbell, 2001). This relationship strongly influences the absolute scale at which population dynamics should be studied. In marine populations, the appropriate absolute scales for such studies vary enormously in terms of the range in space and time covered during an organism's life time. Raffaelli et al. (1994) highlighted the positive correlation between body size and these spatiotemporal life-time scales for whales, fish, benthic invertebrates, zooplankton, phytoplankton, and bacteria. In addition, they placed metapopulation and community dynamics at correspondingly larger spatiotemporal scales than these life-time scales. Yet these considerations only partly determine the appropriate scales. Very small organisms (less than 1–2 mm in length) often are globally dispersed (Godfray and Lawton, 2001; Finlay, 2002) by wind and water at spatiotemporal scales exceeding those traversed by the largest marine mammals. Nearly ubiquitous dispersal is the rule for such microbial species. Thus small size does not restrict the scale at which metapopulations might operate. Furthermore, the distribution patterns of many larger organisms can be strongly influenced by dispersal abilities unrelated to body size. For example, Mayr (1982) contrasted the limited dispersal capabilities of terrestrial mammals, freshwater fishes, and earthworms compared with those of "freshwater plankton, ballooning spiders, birds, and some groups of insects." Likewise, Jackson (1986) contrasted extremely short-distance versus extremely long-distance larval dispersal among major groups of marine epibenthic invertebrates of comparable size. Very similar organisms can disperse over very different spatial scales.

In a recent study of the dispersal capabilities of labrid and pomacentrid fishes in the Indo-Pacific, Mora et al. (2003) linked the pelagic larval duration for 211 species to both community and biogeographic attributes. In these two families, the mean values for pelagic larval duration varied between 19 days and 57 days in the open ocean for 63 Indo-Pacific locations. They found a negative correlation between the regional number of species and distance from the center of high biodiversity in Indonesia and the Philippines. With an increase in this distance, the mean pelagic larval duration increased for both families as the number of species in local communities decreased by more than an order of magnitude. Mora et al. (2003) attributed these patterns to variable dispersal abilities and significant dispersal limitation in a large percentage of this fish fauna. One might also infer that the spatial extent of these assemblages would also increase with distance from the biodiversity center as the mean pelagic larval duration increases. However, how this measure of dispersal ability scales with the local dynamics of these assemblages remains uncertain.

B. Dispersal and Population Dynamics

In this short section I begin by noting some early population studies that either ignored or incorporated dispersal as an important population process. Both deterministic and stochastic approaches to modeling interpopulation dispersal have been used. Later studies introducing the notions of population sources and sinks, clonal metapopulation dynamics, and recruitment limitation are then mentioned to illustrate the parallel lines of thought on how dispersal in space and time can structure local populations. There is considerable overlap in the ideas developed in these three population-level areas and much of it may be relevant to the study of marine metacommunities.

1. Background

Historically, most simple studies of population dynamics have emphasized birth and death rather than dispersal processes, and have contrasted unbounded exponential growth with density-dependent logistic growth of populations (Hutchinson, 1978; May, 1981a; Gaines and Lafferty, 1995). Such treatments of the subject have generally been directed at dynamic fluctuations in population size in terms of the intrinsic growth rate, the supply of local resources, local densities, and biological interactions. Alternative approaches emphasizing interpopulation dispersal (immigration and emigration) also have a long history. These are especially applicable, if not fundamental, to studies of metapopulations. Among the most notable early contributions are the classic works by Andrewartha and Birch (1954), Huffaker (1958), and den Boer (1968). These authors emphasized dispersal ability and colonization–extinction processes as key factors influencing dynamic fluctuations in local population size.

Stochastic approaches to the study of how birth, immigration, death, and emigration (BIDE) influence population growth (partially reviewed in Cohen, 1969) date back at least to 1939 (Bharucha–Reid, 1960). This body of literature highlights the strongly probabilistic nature of dispersal. For many organisms, the notions of mean dispersal or random dispersal by diffusion are not adequate representations of dispersal in the real world. Instead, the entire spectrum of dispersal outcomes as depicted in a probability distribution is a more appropriate representation of reality. This is especially true in the sea, where contrasting phenomena like widespread dispersal and philopatry are both common (Jackson, 1986). Application of stochastic approaches to the study of metapopulations have used multiple methods and have been directed at fluctuations in variables representing environmental factors, population sizes, rates of population births and deaths, and rates of colonization and extinction (Gilpin and Hanski, 1991; Hanski and Gilpin, 1997; Hanski, 1999). Less common are approaches examining stochastic variation in the number and fate of dispersed offspring. In the following two sections, examples of approaches to the study of dispersal and population dynamics in a metapopulation context are briefly presented.

2. Population Sources and Sinks

Using a simplified deterministic BIDE model (based on earlier stochastic models), Pulliam (1988) conducted an algebraic analysis of one-way, habitat-specific migration from a source population (with a net reproductive surplus) to a sink population (with a net reproductive deficit). One of the key predictions from this model is that surplus individuals born in the source habitat could maintain a sink population at equilibrium, thus generating a realized niche larger than the fundamental niche. At the extreme in which the surplus in the source habitat is large and the deficit in the sink habitat is small, a large proportion of the individuals are predicted to occur in the sink habitat. This contribution was extremely timely when published and it has been very influential. It put dispersal and identification of population sources and sinks on center stage. This notion of a dynamic source–sink relationship has been extensively applied to metapopulations in general (Hanski and Gilpin, 1997; Hanski, 1999; Mouquet and Loreau, 2003) and will be addressed again in Section IV.

3. Clonal Metapopulation Dynamics

Karlson and Taylor (1992) used a stochastic model of dimorphic dispersal to generate a wide range of optimal strategies for clonal population growth, a tradition established by Hamilton and May (1977) in the theoretical literature on dispersal. This clonal habit is a life history attribute of many successful marine taxa dominating benthic environments worldwide (Jackson, 1977). The model included significant risks of local, catastrophic mortality to nondispersed offspring as well as the additional mortality incurred by other offspring during dispersal. The success of a particular clonal lineage was evaluated using the probability distribution for offspring reaching adulthood, the generating function for this distribution, and the probability of extinction for an entire clone. Karlson and Taylor (1992) specifically noted the relevance of this probabilistic approach in studies of dispersal and the dynamics of metapopulations, but to my knowledge this approach has yet to be used in metacommunity studies. Empirical investigations using this general perspective have been conducted on metapopulations of bryozoans, a common clonal taxon in both marine and freshwater systems (Okamura, 1997; Karlson, 2002b; Okamura and Freeland, 2002; Okamura et al., 2002). This body of work includes examples of both spatial and temporal dispersal because some clonal organisms produce banks of asexually produced propagules analogous to seed banks in plants.

4. Recruitment Limitation

The relevance of population sources and sinks as well as clonal metapopulation dynamics to the study of metapopulations has not been widely recognized in the marine literature because of differences in terminology and the general emphasis on nonclonal metazoans, oceanographic transport processes ("supply-side ecology"), and recruitment limitation. Although some studies of marine communities have emphasized physical and biological processes controlling the larval supply to a particular location (Underwood and Denley, 1984; Roughgarden et al., 1987, 1988), processes operating within a community as well as those operating within the water column jointly influence community structure. In areas exposed to high larval settlement rates, the former may predominate. Physical transport processes and larval mortality may be more important in areas exposed to lower settlement rates. Thus the relative importance of these processes is a direct consequence of the supply of larvae reaching the adult habitat. This emphasis has been most strongly supported from research conducted in habitats dominated by solitary metazoans.

There are multiple examples, using such solitary organisms, in which local population size is restricted by the rate of recruitment (e.g., Doherty, 1983; Gaines and Roughgarden, 1985, 1987; Sutherland, 1987; Karlson and Levitan, 1990). This rate is influenced by a composite set of processes controlling larval supply and juvenile survivorship, and the net effect on population size is called *recruitment limitation*. When the size of a recruitment-limited population depends on the influx of larvae from exogenous sources, the population is a sink (sensu Pulliam, 1988). Because many clonal organisms are also capable of recruiting to local populations by a variety of local, asexual mechanisms, recruitment limitation is less likely (Karlson, 2002b), as is the potential for important source–sink relationships. The dynamics of invertebrate populations living in soft sediments also appear to be controlled more by local processes than by recruitment limitation (Ólafsson et al., 1994).

Caley et al. (1996) actually stated that the question regarding whether population size is recruitment limited "is misguided," because in most cases "recruitment must influence local population density to some extent." Furthermore, they concluded that most marine invertebrate and fish populations are likely to be influenced by both density-dependent mortality and recruitment limitation. Thus a mix of exogenous factors controlling interpopulation dispersal and local factors controlling success within a habitat are likely to be important determinants of population size. Comparable conclusions regarding the dual importance of exogenous and endogenous factors would appear to be supported by recent evaluations of the population dynamics of the bay checkerspot butterfly (McLaughlin et al., 2002) and coral reef fishes (Mora and Sale, 2002). Yet recruitment limitation may still be relevant if one uses a metapopulation context for studying populations (see Hixon et al., 2002). A focus on the connectivity among habitats, the identification of source and sink populations, the persistence of sink populations, or the rescue of extinct local populations by migration (Hanski, 1999) may require consideration of this question.

C. Relevance to Marine Metacommunities

A marine, multispecies metacommunity must surely exhibit some of the same attributes noted earlier for metapopulations. The dynamics of the metacommunity should be primarily controlled by dispersal among patchily distributed habitats and habitat-specific colonization–extinction processes. Thus dispersal abilities and barriers to dispersal should be important. The structure of the local component communities should be influenced by the size and spatial arrangement of the set of suitable habitats comprising the larger metacommunity. Paine (1994) illustrated a hypothetical marine metacommunity that included trophic structure within habitat patches, larval exchanges among them, and even one source–sink relationship (Fig. 14-1). The level of complexity in patch interconnections and the detail within patches one considers for a metacommunity may be related to the question of interest and the modeling tools used by a particular investigator. For example, information regarding the abundance of an important keystone species in a community may be necessary to include because such species strongly control species membership in local communities. Likewise if the exclusion of species from local communities is strongly controlled by competitors or predators, one may need to include trophic structure and/or competitive relationships among species. A range of modeling approaches has been used successfully at the population level (see Case, 2000; Hanski, 1999) and these have

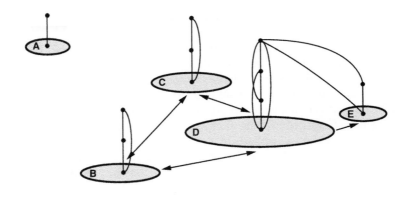

← ———————— **Unspecified spatial scale** ————————→

FIGURE 14-1. Trophic structure and the linkage of habitat patches by larval exchanges. Bidirectional larval exchanges are indicated for a source group of three patches (B, C, and D). Trophic structure is indicated by interconnected nodes and the complexity of this structure varies with patch size. A single consumer–resource interaction occurs in the small isolated patch (A), whereas the largest patch (D) has the most trophic levels and the highest incidence of omnivory. Another small patch (E) is a net sink for larvae from the largest patch (from Paine, 1994).

begun to be extended to consider metacommunities. Here I ask a more general question without regard to any particular empirical system or modeling approach. I consider some common attributes of ecological communities and ask how using a metacommunity approach might influence them. In other words, how might the existence of a metacommunity influence fundamental species–area relationships, local–regional species richness relationships, and relative species abundance patterns?

II. METACOMMUNITIES AND SPECIES–AREA RELATIONSHIPS

A. BACKGROUND

Ecologists have long recognized the quantitative relationship between the number of species in an area and the size of the area (Arrhenius, 1921). Typically these relationships are restricted to single taxa in keeping with island biogeography theory (see Section I). Although there are good reasons for there to be considerable curvilinearity in the species–area relationship (Rosenzweig, 1995), it has been most commonly depicted as a linear relationship between the logarithms

Chapter 14 Metapopulation Dynamics and Community Ecology of Marine Systems

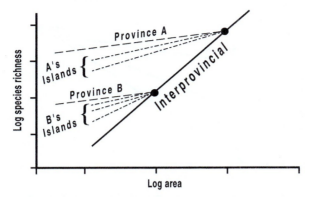

FIGURE 14-2. Species–area relationships at three spatial scales. The logarithms of species richness and area are plotted for islands within provinces A and B, large mainland source areas within provinces A and B, and interprovincial areas. Small islands differ in number of species depending on their degree of isolation from the mainland source. Typical slopes (z-values) are 0.25 to 0.45 for islands, 0.15 for large mainland areas, and 0.9 for interprovincial areas (from Rosenzweig, 1995). Reprinted with the permission of Cambridge University Press.

for species richness and area. The slope of this line is the exponent describing the power relationship between the two variables. Rosenzweig (1995) recognized four different species–area relationships at different spatial scales: (1) those representing data from a single biota within "tiny" areas (this relationship is typically convex upward rather than linear as a result of well-known sampling biases [Williams, 1964]), (2) those representing data from a single biota in larger areas from a mainland, (3) those representing data from individual islands within an archipelago, and (4) those representing data collected at the largest spatial scale from multiple biotic provinces with different evolutionary histories. Although Rosenzweig (1995) admitted that mainland and island designations are sometimes arbitrary, the essential notion is that species occurring on islands "originate entirely by immigration from outside the region (archipelago)." Thus mainland sources are implied in the expected species–area relationships for island archipelagos.

Rosenzweig (1995) illustrated the basic species–area relationships across intraprovincial and interprovincial spatial scales in Figure 14-2. The lowest typical slope (0.15) characterizes intraprovincial relationships for large mainland sources where colonization rates are high and habitat diversity is strongly correlated with area. The representation in Figure 14-2 shows only a single curve for each of two provinces, so one can infer that there is little or no intraprovincial differentiation and no gradients in regional species richness. Such regional differentiation would result in similar-size small areas with different numbers of

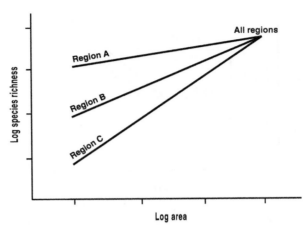

FIGURE 14-3. Hypothetical intraprovincial species–area relationships for a mainland with three differentially rich regions (A, B, and C). Small areas differ in number of species depending on location and the corresponding regional species richness.

species (Fig. 14-3). In Figure 14-2, typically steeper slopes (0.25–0.45) characterize species–area relationships for islands reflecting their degree of isolation from sources (a distance effect), effects of island size and habitat heterogeneity, and the balance between colonization and extinction rates. Because two or three curves are depicted for each of two provinces, it is recognized that similar-size islands may differ in terms of their degree of isolation. The steepest typical slopes (0.9) for species–area relationships reflect the mixing of biotas among provinces with different evolutionary histories as one considers very large areas. Thus one can anticipate considerable variation in the slope of a species–area relationship as a result of spatial scale, regional differences within provinces, the difference between mainlands and islands in terms of the source of species, the distance between islands and their mainland sources, and the number of provinces being considered.

Among the four species–area relationships considered by Rosenzweig (1995), the one that most closely matches the metacommunity is that for islands within archipelagos at intraprovincial scales. In Figure 14-2, the steepest intraprovincial species–area slopes occur in archipelagos where small islands are most isolated from mainland sources of species. At equilibrium, these islands are expected to have the lowest number of species. The least steep slopes occur in archipelagos where small islands are closest to mainland sources. Thus this distance effect is a fundamental component of the colonization dynamics of these islands.

Chapter 14 Metapopulation Dynamics and Community Ecology of Marine Systems

B. Relevance to Marine Metacommunities

Would a metacommunity approach to species–area relationships modify these expectations? In keeping with the theory of island biogeography (MacArthur and Wilson, 1967), Rosenzweig (1995) considered the influence of isolation on colonization rates and island size on extinction rates and habitat heterogeneity to generate a variable set of species–area relationships for islands within an archipelago (Fig. 14-2). In the following, I use the terms *island* and *archipelago* to refer in general to patches of marine benthic habitat as a modification of the terrestrial context used by Rosenzweig. In a marine context, such patches may occur around terrestrial islands or they may occur without such exposed land on ocean bottoms (continental shelves, slopes, or the deep sea).

For a marine metacommunity occurring in an archipelago of islands (e.g., the habitat patches depicted in Fig. 14-1), one would need to consider the additional influence of sources within the archipelago, the relative positions of the patches, the transport processes controlling interpatch dispersal, dispersal ability of the target organisms, and regional differentiation among archipelagos within the province. Thus a species–area relationship may be influenced by a prevailing current system with strong directional flow (e.g., the Kuroshio Current) or by more variable coastal currents and upwelling controlled by the strength and direction of the wind. Small islands in downstream locations would be expected to have higher colonization rates than comparable islands located upstream (e.g., Roberts 1997). Island position relative to prevailing currents is predicted to contribute to the degree of isolation and variation in the slope of the species–area curve. Likewise, small islands located near larger islands in an archipelago would have higher colonization rates than more isolated islands of comparable size (as in the rescue effect). These positional effects on colonization rates are independent of the area effect depicted in traditional species–area curves. The key expectation is that island position relative to mainland and island sources in an archipelago contributes to significant variation in colonization rates and thus influences the expected number of species on an island. As noted by Holt (1993), the expected response of local species richness can either increase (as a result of enrichment from regional sources as discussed in Section III or as a result of the positive effects of some biological interactions) or decrease (as a result of species exclusions caused by competitors or predators).

These predictions regarding how colonization rates may vary with island position relative to other islands and the prevailing transport processes are likely to be overly simplistic, because they do not incorporate species-specific variation in reproductive and larval biology. The expectations are based purely on the assumption that dispersal occurs at spatiotemporal scales that match the putative positional effects on islands. Consequently, they are likely to require modification.

The influence of broadcast spawning versus brooding in members of the target assemblage should be considered. Mechanisms promoting larval retention, low larval exchange rates among islands (Cowen et al., 2000), and nonrandom habitat selection by larvae (Jinks et al., 2002) are also likely to contribute to variation in species colonization rates and thus influence the expected number of species on islands.

C. Regional-Scale Differentiation

Additional variation in species–area relationships for islands can be attributed to the enormous spatial scale encompassed by some marine biotic provinces and large regional (intraprovincial) differences in numbers of species. The expected number of species occurring on large islands as depicted in Figure 14-2 converges on the intraprovincial species–area relationship for the mainland source. This is because, by definition, the maximum species richness for any area within an archipelago cannot exceed that for the mainland source. In the simplest case, a single archipelago might be located off the coast of a very large mainland. However, multiple oceanic archipelagos located at variable distances from continental mainlands are exposed to significant regional differences in numbers of species. Consider the equatorial regions of the Indian and Pacific Oceans, where multiple archipelagos occur across thousands of kilometers. For some taxa (e.g., corals), much of this enormous area is a single faunal province with very low levels of endemism (Hughes et al., 2002). However, there are strong regional gradients in species richness (Veron, 1995; Karlson, 1999). The regional species pools for corals vary enormously between the central Indo-Pacific with 581 named species (Veron, 2000) and extreme marginal areas of the province with far fewer species (Hughes et al., 2002). Consequently, the number of species expected on islands of a given size should reflect the size of the regional species pool rather than the total richness of the entire province. With this perspective, one can depict multiple, intraprovincial species–area relationships for archipelagos that differ in terms of isolation from the "mainland" source (the richest center of biodiversity). Each species–area relationship would converge on the regional species richness for the archipelago rather than the total number of species in the province (Fig. 14-4). Thus the number of species occurring on islands of comparable size within a province can be influenced by the size of the regional species pool (or the position of the archipelago along a regional species richness gradient) as well as the relative position within any given archipelago.

The primary differences between the intraprovincial species–area relationships depicted for islands in Figures 14-2 and 14-4 reflect differences in the time required for equilibrium to occur. Figure 14-4 provides intraprovincial expectations based on short ecological timescales for active dispersal, colonization, and

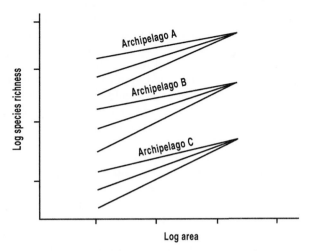

FIGURE 14-4. Hypothetical intraprovincial species–area relationships for islands within three different archipelagos (A, B, and C) distributed over a regional gradient in species richness. Archipelagos differ as a result of the degree of isolation from the "mainland" source. Small areas within an archipelago differ in number of species depending on their location relative to other islands within the archipelago. The most isolated archipelagos have the fewest expected number of species based on the regional (rather than provincial) species richness.

local extinction processes. Figure 14-2 is based on expectations over evolutionary timescales for which speciation, range extensions, and regional extinctions can influence the numbers of species. Given that millions of years may be needed for some insular faunas to reach equilibrium (e.g., Welter–Schultes and Williams, 1999), the latter predictions may be strongly biased (as argued in Section I). Although an inclusive theory is desirable (as is evident in MacArthur and Wilson, 1967; Rosenzweig, 1995; Hubbell, 2001), I emphasize the different predictions that can arise when there is a large discrepancy in the spatiotemporal scales of ecological and evolutionary processes. The consideration of regional-scale differentiation within a province may provide more appropriate expectations of species–area relationships, especially within more isolated archipelagos.

D. Summary

Conventional species–area relationships and the theory behind them highlight the importance of the spatial scale of an analysis, interprovincial differences in evolutionary history, sample area, habitat heterogeneity, island size, and distance of islands from mainland sources of species. The marine metacommunity perspective (as addressed earlier) adds the following to this mix: regional differentiation within provinces, oceanic transport processes, local transport processes within archipelagos, dispersal abilities, and the relative positions of islands in archipelagos. These factors are predicted to contribute to variation in colonization rates and thus influence the expected number of species within local communities. The expected number of species should also vary with the intensity of biological interactions with competitors and predators, whose presence in local communities is controlled by the same mix of transport processes and positional effects.

In light of recent publications highlighting important variation in species–area relationships resulting from several other factors (noted below), testing predictions will require careful planning of sampling regimes and the use of controlled experiments (Godfray and Lawton, 2001) to discriminate among alternative explanations. This is imperative if we are to understand the causal mechanisms controlling the structure of marine metacommunities. Here are but a few additional factors specifically influencing species–area relationships: sea surface temperature and productivity (Chown et al., 1998; Roy et al., 1998), nonequilibrial historical effects (Welter–Schultes and Williams, 1999), sampling effort and the probability of species detection (Cam et al., 2002), the relative abundances and spatial distributions of species (He and Legendre, 2002), spatial distributions of habitats (Seabloom et al., 2002), and scale-dependent persistence of species (Plotkin et al., 2000). In addition, taxon-specific variation resulting from biological traits like population density, population size (Frank and Shackell, 2001), habitat specialization, presence of resting stages in complex life cycles (Ricklefs and Lovette, 1999), and body size (Finlay et al., 1998; Azovsky, 2002; Matter et al., 2002) are also known to influence species–area relationships.

III. METACOMMUNITIES AND LOCAL–REGIONAL SPECIES RICHNESS RELATIONSHIPS

A. BACKGROUND

Examination of the relationship between the number of species occurring within habitats and the number of species in a set of large regions dates back to Terborgh and Faaborg (1980). They studied bird species richness on Caribbean islands within two distinct habitats and reported a curvilinear local–regional species richness relationship suggesting an upper limit to local richness. This observation has generally been interpreted as evidence that local habitats are saturated with species (a notion attributed to Elton, 1933). At the time, it was in keeping with conventional ecological theory (MacArthur, 1965) that the number of species occurring within habitats, especially in the tropics, is controlled by the

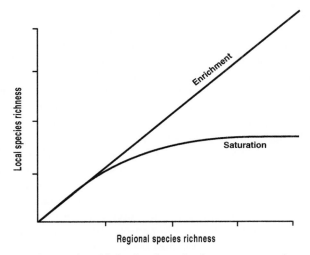

FIGURE 14-5. Two theoretical models for the relationship between species richness within a single local habitat and species richness characterizing a large region. In the enrichment model, local richness is proportional to regional richness. In the saturation model, local richness reaches an upper limit and becomes independent of regional richness in speciose regions. (From Cornell and Lawton, 1992.)

local environment (physical factors and biological interactions). In contrast, the number of species in a region is controlled by evolutionary history and regional-scale physical processes. At these two widely divergent scales, the saturation model (Fig. 14-5) predicts that within-habitat species richness becomes independent of regional species richness in speciose regions. Based on inadequacies in the data and the consideration of alternative interpretations and other measures of community structure, Wiens (1989) questioned the validity of the claim of saturation by Terborgh and Faaborg (1980). More recently, three different general reviews on this subject (Cornell and Karlson, 1997; Srivastava, 1999; Hillebrand and Blenckner, 2002) lend considerable support to an alternative model in which local, within-habitat species richness is linearly related to regional species richness (Fig. 14-5).

This alternative model (called *pool enrichment* by Cornell [1985], *regional enrichment* by Ricklefs [1987], and *proportional sampling* by Cornell and Lawton [1992]) emphasizes the influence of evolutionary history and large-scale physical phenomena on local species richness within habitats. However, this is not to say that local phenomena are not important. Both local- and regional-scale processes jointly influence local richness patterns (Cornell and Karlson, 1996; Karlson, 1999; Lawton, 2000; Karlson and Cornell, 2002). When the slope of the local–regional species richness relationship is close to one, all or nearly all species in the regional pool are represented at most localities, species membership at the habitat scale is not limited, and there are few or no unoccupied habitats. Less steep slopes may be indicative of stronger mechanisms partially limiting species membership at the habitat scale, less within-habitat spatial heterogeneity (e.g., less heterogeneity resulting from patch disturbances and fewer co-occurring successional stages), and more unoccupied habitats. Current efforts are being directed at understanding how local and regional processes operate together to generate observed species richness patterns (see Huston, 1999; Karlson and Cornell, 2002; and Mouquet and Loreau, 2003).

Some of the recent literature on local–regional species richness relationships has erroneously emphasized the saturation model (e.g., Findley and Findley, 2001), whereas some has highlighted the alternative model but used inappropriate scales for local or regional richness data (see Hillebrand and Blenckner, 2002). The former problem emerges as a consequence of undersampling at the local level (Caley and Schluter, 1997; Karlson, 2002a). The latter introduces species–area effects either by mixing multiple habitats at the local level (Srivastava, 1999), by mixing regional biota from different provinces at the larger level (Hugueny et al., 1997), or by introducing spatial autocorrelation when the two scales are closely matched (Bartha and Ittzés, 2001; Hillebrand and Blenckner, 2002). Rosenzweig and Ziv (1999) actually referred to plots of local versus regional species richness as the "echo pattern of species diversity." Such plots can be a direct consequence of species–area relationships (see Section II) by merely sampling species richness at two spatial scales (Bartha and Ittzés, 2001). However, the local scale used for generating species–area relationships is typically quite large (Rosenzweig, 1995). The use of large localities in this context obviates one's ability to infer anything about within-habitat diversity patterns. Thus it is clear that the spatial scales selected for the purpose of evaluating the saturation and enrichment models are critical (Karlson and Cornell, 2002).

B. Relevance to Marine Metacommunities

The existence of metacommunities at the regional scale is a fundamental premise for regional enrichment and linear local–regional species richness relationships. Species dispersal and the colonization of local assemblages within habitats is linked to the historical and geographical factors controlling the size and composition of regional pools of species. These pools comprise the metacommunities across differentially rich regions. Such regional-scale differentiation of metacommunities at intraprovincial scales (see Section II.C) is implicit as one examines variation in local species richness.

In the ocean, the structure of local communities is influenced by multiple control mechanisms contributing to either enrichment or species exclusions.

Chapter 14 Metapopulation Dynamics and Community Ecology of Marine Systems

These mechanisms operate across an enormous range of scales and are highly variable in their relative importance from place to place. For example, the well-studied intertidal communities of the northeast Pacific Ocean can be influenced primarily by transport mechanisms in the water column in some places (e.g., Roughgarden et al., 1987, 1988) or by biological interactions in others (see Paine, 1994). In a general context, Roughgarden (1989) contrasted these two extreme perspectives in terms of unlimited and limited membership. Unlimited membership corresponds approximately with the notion of regional enrichment, because membership within local assemblages is dictated primarily by transport and colonization processes. Limited membership results from exclusionary processes (competition, predation, or the physical environment) that can limit local species richness, as in the saturation model, or modify local richness in more open communities depending on the degree of recruitment or dispersal limitation and the species being so constrained (Section I.B).

For regionally enriched communities, Cornell and Karlson (1997) and Cornell (1999) have pointed out that the average proportion of species from a regional species pool found within local communities is merely the slope of the local–regional species richness relationship. This slope represents the average proportion (across all species) of local sites in a region occupied by a species. In this context, Hugueny and Cornell (2000) used a patch-occupancy model to illustrate how local species richness can be influenced by processes operating at the metacommunity scale. In that patch-occupancy models have been a major tool used in studying metapopulations, they suggested that "the slope of the local–regional richness relationship and patch-occupancy processes are different expressions of the same phenomenon" in metacommunities (Hugueny and Cornell, 2000). Because one fundamental prediction of metapopulation models is that "not all suitable habitat is occupied all of the time" (Hanski, 1999), one should expect low to intermediate slopes for a corresponding local–regional species richness relationship. The metacommunity example provided by Hugueny and Cornell (2000) yielded a slope of 0.536 for cynipid gall-wasp assemblages. This observation closely matched the predictions of the patch-occupancy model they used, thus supporting the linkage between local–regional species richness relationships and the patch dynamics of metacommunities. A much higher slope for the local–regional species richness relationship would not have been consistent with this metacommunity perspective (see Mouquet and Loreau, 2003, for a very different interpretation).

A comparable marine example comes from a study of coral assemblages across a 10,000-km biodiversity gradient in the Indo-Pacific (Karlson et al., 2004). Using a hierarchical sampling design across five large regions (Indonesia, Papua New Guinea, the Solomon Islands, American Samoa, and the Society Islands), we documented significant regional differences in both the local and regional species richness of corals occurring on coral reef flats, crests, and slopes. Despite these differences, the local–regional species richness relationships in each habitat were linear with a slope of 0.27. This low value for the slope indicates that many coral species in each region were not present in most local communities. The fact that the relationships were linear supports the notion of regional enrichment (rather than the concepts of saturation or limited membership) and would appear to be consistent with metacommunity patch dynamic models emphasizing colonization and local extinction processes.

C. Effects of Transport Processes and Relative Island Position

We can also consider how local–regional richness relationships are influenced by oceanic transport processes and positional effects comparable to those noted earlier for species–area relationships. When transport processes strongly limit the supply of larvae to a particular habitat, some species may become severely recruitment limited. The impact this has on local diversity will depend on which species are being limited (e.g., competitive dominants or keystone species). Because biological interactions can result in enhanced or reduced local species richness (Holt, 1993), the net effect of limited larval supplies on local–regional species richness relationships may be quite complicated and difficult to demonstrate in the field. However, such effects are most likely to be related to island position relative to oceanic currents. They should be strongest in upstream locations where larval supplies are predicted to be most limited. Likewise, the relative position of an island within an archipelago may generally enhance or reduce the supply of larvae and contribute to significant variation in local–regional species richness relationships. To test for these effects, one might contrast local samples from isolated islands with those from archipelagos (or clusters of islands within archipelagos), samples from upstream and downstream locations, or both. The most extreme differences ought to emerge from contrasts between the most isolated, transport-limited locations versus the least isolated locations with the largest larval surpluses.

These positional effects on local species richness are also likely to be modified by a variety of other variables (as noted earlier for species–area relationships in Section II.D). Among the most important of these are habitat area (e.g., Bellwood and Hughes, 2001), productivity (e.g., Fraser and Currie, 1996), and the nonequilibrial effects of disturbances and history on ecological communities (see Cornell and Karlson, 2000). Investigators should carefully evaluate these along with the issues of scale and taxon-specific variation in such things as larval biology (e.g., dispersal, behavior, predation, settlement), adult interactions (e.g., neighborhood competition, intraspecific aggregation), and other factors potentially contributing to spatial heterogeneity within ecological communities.

D. Summary

A fundamental premise for examining local–regional species richness relationships is that local ecological assemblages are embedded in differentially rich regional metacommunities. The literature on this subject strongly supports the notion of regional enrichment thus linking local communities with historical and biogeographical processes controlling the size and composition of regional species pools. Marine metacommunities are predicted to exhibit significant variation in local species richness as a result of oceanic and more local transport processes within archipelagos as well as the relative positions of islands in archipelagos. The largest effects should be evident in contrasts between more isolated, transport-limited locations and less isolated locations. Because local species richness is also influenced by the local environment and biological interactions, the nature of these effects on local–regional richness relationships are predicted to depend on which species are being limited, how they interact with other members of local assemblages, and where such assemblages occur along regional richness gradients.

IV. METACOMMUNITIES AND RELATIVE SPECIES ABUNDANCE PATTERNS

A. Background

Since the seminal papers by Preston (1948) and Fisher et al. (1943), there has been a lively debate in the ecological literature over the theoretical basis for predicting relative species abundance patterns. The disparate models predicting these patterns invoke lognormal, logarithmic series, geometric series, or brokenstick distributions based largely on different simplifying assumptions regarding the nature of interspecific competition (May, 1975, 1981b; Hayek and Buzas, 1997). All the models predict that ecological assemblages will be numerically dominated by a few common species. They differ in terms of the predicted relative frequency of rare and moderately abundant species. The extremes are predicted for the logarithmic and geometric series models. The former predicts that there are many more rare species than common species, whereas the latter predicts that rare and common species will be equally represented by the same number of species in these assemblages. The lognormal model predicts that there is a mode to the relative species abundance distribution signifying the prevalence of moderately abundant species.

B. METACOMMUNITIES AND RELATIVE SPECIES ABUNDANCE PATTERNS

Here I briefly consider two metacommunity models that predict relative species abundance patterns. The first is part of a broader body of theory linking biogeography and biodiversity (Hubbell, 2001). The second is more narrowly defined, but specifically relevant to this topic (Mouquet and Loreau, 2003). Both approaches are useful and are viewed as complementary contributions to the study of metacommunities in general. In addition, both contributions specifically note applications to marine metacommunities.

Hubbell (2001) provided a new theoretical basis for studying ecological and biogeographic assemblages of species. Many theoretical treatments of this subject have focused primarily on the use of species presence/absence data (and the proportion of occupied patches) to establish patterns of species richness across a wide range of spatiotemporal scales. Hubbell (2001) explicitly added the relative abundances of species as a key to unifying the theories of biodiversity and biogeography (more specifically, the theories of relative species abundances [MacArthur, 1957, 1960, 1972] and island biogeography [MacArthur and Wilson, 1963, 1967]). By defining the numbers of individuals of each species and the numbers of species in assemblages of trophically similar species, one can create traditional relative abundance distributions for these assemblages (see Section IV.A). Hubbell (2001) also added a birth–death process, dispersal, and speciation to generate a dynamic theory governing fluctuations in relative abundances at local and landscape scales. The birth–death process assumes that each death in a local community is replaced by a birth from the local community or by an immigrant drawn at random from a large pool of individuals representing the metacommunity. The total number of species in the metacommunity is determined by the balance between regional extinction and speciation.

The first of two key assumptions to this dynamic theory is that the probability of birth, death, and immigration are all equal on a per-capita basis. Thus the theory generates expectations based purely on relative abundances rather than species-specific differences. Common species in the metacommunity successfully immigrate into local communities more often than less common species because of their relative abundances in the metacommunity, not species-specific dispersal capabilities. Likewise, local births are distributed among species based on their local relative abundances, not fecundity differences. Rare species become extinct at local and metacommunity scales more often than common species. This assumption of equality thus restricts the predictions of the theory to neutral processes. These processes generate null expectations for testing empirical data for nonneutral patterns.

The second key assumption of the theory is that the dynamics of these assemblages are a zero-sum game. Each death is replaced exactly by a single birth or immigrant so the size of the community in terms of numbers of individuals is fixed; the landscape is saturated with individuals so that any increase in the abundance of one species is offset by a decrease in another species. Species may come and go, because speciation (two versions of this process are considered) and extinctions can occur, but community size cannot change. The number of species at any particular time is thus subject to stochastic drift according to this equilibrial process.

According to the neutral theory, the equilibrial number of species and their relative abundances in the metacommunity are controlled by what is called "a fundamental biodiversity number, θ" (Hubbell 2001). Simply stated, $\theta = 2J_m\nu$, where J_m is the number of individuals in the metacommunity and ν is the probability of a speciation event per birth in the metacommunity. "θ is asymptotically identical to Fisher's α"—a measure of species diversity from the logarithmic series distribution (Fisher et al., 1943) and a distributional feature of the metacommunity in the neutral theory (Volkov et al., 2003). This relative abundance distribution is characterized by a long tail of very rare species. In fact, in progressively larger samples from communities matching the logarithmic series distribution, the number of species represented by only one individual approaches the value of α, the largest class of species in the relative abundance distribution.

In the neutral theory, the probability that a death in a local community is replaced by an immigrant from the metacommunity is defined by the variable m. When $m = 1$, no species are dispersal limited, there are no local births, and local communities are dominated by the more common species in the metacommunity. Successful immigration by any species is controlled purely by its relative abundance in the metacommunity, the size of the local community, and number of deaths providing opportunities to immigrants. Thus one prediction of the neutral theory is that rare endemic species in the metacommunity can only persist in relatively small local populations when there is some degree of dispersal limitation, and local births of these rare species are permitted.

There are a large number of other specific neutral predictions generated by this body of theory. Virtually all of these are relevant to the dynamics of metacommunities, because this is fundamental to Hubbell's theory. These include but are not restricted to the following five types of predictions: (1) relative species abundance patterns (based on varying the degree of dispersal limitation as depicted in Fig. 14-6), (2) species–area relationships (because the number of individuals is fixed, density varies linearly with area, and numbers of species per unit area can be explicitly predicted), (3) mean extinction times for species in local communities (given the relative species abundance), (4) variation in local com-

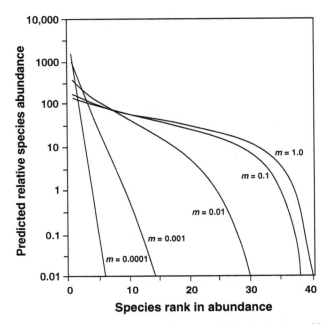

FIGURE 14-6. Predicted relative species abundance patterns for local island assemblages under different levels of dispersal limitation in accordance with the neutral theory. In this example, $\theta = 10$ and there are 1600 individuals in each local island sample. Under no dispersal limitation, each death is replaced by an immigrant from the metacommunity ($m = 1.0$). The effects of dispersal limitation on the relative abundances of species are predicted for cases in which every 10 to 10,000 deaths are replaced by a single immigrant ($m = 0.1$–0.0001). Local communities become increasingly dominated by fewer species as they become more limited by dispersal. From Hubbell, 2001. Copyright © 2001 by Princeton University Press. Reprinted by permission of Princeton University Press.

munity similarity with distance, and (5) abundance-based evolutionary extinction times. The predicted relative species abundance patterns from this theory span the full range of possible distributions (see Section IV.B), thus resolving "the conflict between Fisher and Preston" (Hubbell, 2001). The logarithmic series distribution (Fisher et al., 1943) characterizes the metacommunity, whereas local communities exhibit a range of possible distributions depending on their size and the degree of dispersal limitation.

Hubbell (2001) emphasized the remarkable fact that some of the predictions of the neutral theory closely match observed patterns exhibited by a broad range of closed-canopy forest assemblages. However, some types of assemblages clearly do not fit the predictions and some are thought to violate the assumptions of the theory. For example, an open-canopy forest assemblage in India experiences so much disturbance as a result of fire and elephants that "the population dynamics of individual tree species are to a large extent independent of one another, and are not constrained to follow zero-sum dynamics." Hubbell (2001) acknowledged that many other communities may also violate this assumption because of severe disturbances and because "all limiting resources are not, in fact, consumed." Hubbell's neutral theory represents an equilibrial set of expectations that may not apply to a broad range of ecological assemblages. On the other hand, perhaps a stochastic nonequilibrial version of the neutral theory could be devised without the assumption of zero-sum dynamics and the absolute tradeoff between local births and immigration from the metacommunity. This might be achieved using random variables to signify the species-specific numbers of surviving individuals, local births, and successful immigrants. Perhaps, then, a generating function for the probability distribution describing the number of species in local communities could provide a basis for analytical solutions for the neutral theory. Currently, we are restricted to numerical solutions (as in Volkov et al., 2003) or simulations (Hubbell, 2001).

Recently, Mouquet and Loreau (2003) published results from a "source–sink metacommunity" model that provided an alternative niche-based approach to the neutral theory. Regional niche differentiation and dispersal among a set of local communities promoted species coexistence in this model. Specifically, they modeled the performance of multiple species in multiple local communities under conditions with species-specific variation in the rates of reproduction and mortality. In keeping with lottery competition models, these rates were varied "such that each species was the best competitor in one community." Dispersal ability, on the other hand, did not vary among species. Immigration into local communities was modeled by distributing a fixed proportion of all the offspring of each species equally among all nonnatal communities. This proportion represented the amount of dispersal among local communities—a simple measure of connectivity. The results from this model support the idea that intermediate levels of dispersal can promote high species diversity and a range of relative species abundance distributions matching the lognormal, logarithmic series, and broken-stick distributions in local communities. When the dispersal proportion was either very high or very low, the relative species abundance distributions of local communities matched the geometric series, and local richness was quite low. As one would expect, regional coexistence (β diversity) was promoted at low levels of connectivity, but not at intermediate to high levels.

These predictions are based on the presence of strong competitive interactions and regional niche differentiation, yet they can be viewed to be "complementary" to Hubbell's neutral theory (Mouquet and Loreau, 2003). This comment may seem counterintuitive because the models are either niche based or dispersal based, but it emphasizes the point that empirical systems should be evaluated by careful examination of the underlying processes generating observable patterns in ecological communities. Hubbell (2001) emphasized deviations from neutral predictions as a basis for focused empirical study. Thus one might explore the mechanisms promoting the persistence of rare endemics in local communities, because such species are not predicted to do so, except under dispersal limitation. Mouquet and Loreau (2003) might emphasize examining empirical processes controlling patterns that are predicted to differ between niche- and dispersal-based approaches. For example, they noted the predicted differences in the relative species abundance distributions in the absence of dispersal limitation (a geometric series in their model vs. a logarithmic series distribution in the neutral theory). This difference justifies focused empirical studies on the nature and scale of competitive processes in local communities as well as on dispersal processes operating within the metacommunity. These are the fundamental processes resulting in these very different predictions.

C. Relevance to Marine Metacommunities

Both of these contributions are likely to be relevant to marine metacommunities. Because both approaches predict responses in the relative species abundance distributions to varying degrees of dispersal limitation, these distributions are also likely to be influenced by the transport processes and positional effects for islands within archipelagos noted previously (Sections II.B and III.C). Mouquet and Loreau (2003) explicitly noted the relevance of mobile dispersal stages in the life cycles of sessile marine species as well as many territorial coral reef fishes. This biological feature highlights the importance of regional-scale processes controlling dispersal and local-scale processes controlling success in local communities. Mouquet and Loreau (2003) recognized that future studies of these marine metacommunities are likely to require consideration of species-specific variation in dispersal abilities (as in Mora et al., 2003) and "species-specific perception of scales." That is to say that the dichotomy between local and regional scales in metacommunity models is overly simplistic. Real species disperse at different scales, so the appropriate scales associated with metacommunity processes are likely to vary from place to place and with species composition.

Likewise, the neutral theory from Hubbell (2001) has specific applications to marine metacommunities. In an earlier presentation of this theory, Hubbell (1997) considered its relevance to marine communities like coral reefs and marine intertidal zones, where sessile organisms predominate. In particular, he noted similarities between the local communities of closed-canopy forests and these marine communities where severe competition for light and space may make them consistent with the assumption of zero-sum dynamics in the neutral theory. Hubbell (1997) also recognized the vast biogeographic scales these marine organisms span and speculated that the structure and dynamics of such marine metacommunities may very well be determined more by dispersal than by niche differences among species. The unified neutral theory represents a first step in reconciling these divergent perspectives in that it clearly provides specific testable predictions that can be falsified empirically. Perhaps the difficulties in falsifying dispersal-based predictions were overemphasized in Hubbell (1997) and we can begin to discriminate between niche- and dispersal-based patterns in ecological communities. Our ultimate goal is to integrate these perspectives into a truly unified theory.

D. SUMMARY

Hubbell's neutral theory predicts that the degree of skewing of the relative species abundance distribution for a local community should be strongly influenced by the degree of dispersal limitation (Hubbell, 2001; Volkov et al., 2003). In the absence of dispersal limitation, this distribution should reflect the logarithmic series distribution typical of the metacommunity. Strong dispersal limitation is predicted to favor the abundance of rare species in local communities according to the neutral theory. The metacommunity model by Mouquet and Loreau (2003) also predicts considerable variation in relative species abundance patterns in response to dispersal among local communities (connectivity). In the absence of dispersal limitation, a geometric series distribution is predicted because high connectivity reduces regional differentiation, and membership in local communities is dominated by competition. Efforts to resolve these different predictions empirically should focus on species-specific variation in competitive and dispersal ability, and the scales at which competitive and dispersal processes operate. Relative species abundance distributions can vary dramatically across scales (Connolly et al., 2005).

V. SUMMARY

In this chapter, I consider how marine communities may be influenced by metacommunity processes operating at the mesoscale. Of fundamental importance are issues of scale, species composition, trophic diversity, dispersal abilities, dispersal barriers, transport mechanisms, and island positions within archipelagos and across differentially rich regions. It is critical that community ecologists use predictive models which match the scale of dispersal of the selected organisms. Given the variety of available models and the large difference between biogeographical models employing single trophic levels and mesoscale models emphasizing trophic relationships, the process of model selection should include careful evaluation of these alternative approaches. In some cases, a blend of these approaches may be appropriate. This position is supported by assessments from population studies emphasizing both external factors controlling interpopulation dispersal and local, within-habitat factors controlling variation in population size.

Specific consideration of three fundamental community attributes in a metacommunity context leads to a number of predictions. Species-area relationships at intra-provincial scales are predicted to vary substantially due to regional differences in species richness, transport processes, dispersal abilities, and island positions. All of these factors are external to local habitats and their primary influence is on colonization rates. Fluctuations in local species richness are also predicted due to interactions with competitors and predators. This same set of factors is predicted to influence local-regional species richness relationships with the largest contrasts being due to the degree of isolation and transport limitation. Likewise, relative species abundance patterns are predicted to vary based on scale, biological interactions, and the degree of dispersal limitation.

REFERENCES

Andrewartha, H.G., and Birch, L.C. (1954). *The distribution and abundance of animals.* Chicago: University of Chicago Press.

Arrhenius, O. (1921). Species and area. *J. Ecol.* **9**, 95–99.

Azovsky, A.I. (2002). Size-dependent species–area relationships in benthos: Is the world more diverse for microbes? *Ecography.* **25**, 273–282.

Bartha, S., and Ittzés, P. (2001). Local richness-species pool ratio: A consequence of the species–area relationship. *Folia Geo.* **36**, 9–23.

Barton, N.H. (2001). The evolutionary consequences of gene flow and local adaptation: Future approaches. In *Dispersal* (J. Clobert, E. Danchin, A.A. Dhondt, and J.D. Nichols, eds., pp. 330–340). Oxford: Oxford University Press.

Bellwood, D.R., and Hughes, T.P. (2001). Regional-scale assembly rules and biodiversity of coral reefs. *Science.* **292**, 1532–1534.

Benzie, J.A.H. (1999). Genetic structure of coral reef organisms: Ghosts of dispersal past. *Am. Zool.* **39**, 131–145.

Bharucha–Reid, A.T. (1960). *Elements of the theory of Markov processes and their applications.* New York: McGraw-Hill.

Caley, M.J., Carr, M.H., Hixon, M.A., Hughes, T.P., Jones, G.P., and Menge, B.A. (1996). Recruitment and the local dynamics of open marine populations. *Ann. Rev. of Ecol. and System.* **27**, 477–500.

Caley, M.J., and Schluter, D. (1997). The relationship between local and regional diversity. *Ecology.* **78**, 70–80.

Callaghan, T.P., and Karlson, R.H. (2002). Summer dormancy as a refuge from mortality in the freshwater bryozoan *Plumatella emarginata. Oecologia.* **132**, 51–59.

Cam, E., Nichols, J.D., Hines, J.E., Sauer, J.R., Alpizar–Jara, R., and Flather, C.H. (2002). Disentangling sampling and ecological explanations underlying species–area relationships. *Ecology.* **83**, 1118–1130.

Case, T.J. (2000). *An illustrated guide to theoretical ecology.* Oxford: Oxford University Press.

Chown, S.L., Gremmen, N.J.M., and Gaston, K.J. (1998). Ecological biogeography of southern ocean islands: species–area relationships, human impacts, and conservation. *Am. Nat.* **152**, 562–575.

Clobert, J., Danchin, E., Dhondt, A.A., and Nichols, J.D., eds. (2001). *Dispersal.* Oxford: Oxford University Press.

Cohen, J.E. (1969). Natural primate troops and a stochastic population model. *Am. Nat.* **103**, 455–477.
Connolly, S.R., Bellwood, D.R., and Hughes, T.P. (2003). Indo-Pacific biodiversity of coral reefs: Deviations from a mid-domain model. *Ecology.* **84**, 2178–2190.
Connolly, S.R., Hughes, T.P., Bellwood, D.R., and Karlson, R.H. (2005). Community structure of corals and reef fishes at multiple scales. *Science.* **309**, 1363–1365.
Cornell, H.V. (1985). Local and regional richness of cynipine gall wasps on California oaks. *Ecology.* **66**, 1247–1260.
Cornell, H.V. (1999). Unsaturation and regional influences on species richness in ecological communities: A review of the evidence. *Ecoscience.* **6**, 303–315.
Cornell, H.V., and Karlson, R.H. (1996). Species richness of reef-building corals determined by local and regional processes. *J. Anim. Ecol.* **65**, 233–241.
Cornell, H.V., and Karlson, R.H. (1997). Local and regional processes as controls of species richness. In *Spatial ecology. The role of space in population dynamics and interspecific interactions: Monographs in population biology* (D. Tilman and P. Karieva, eds., vol. 30, pp. 250–268). Princeton, Princeton University Press.
Cornell, H.V., and Karlson, R.H. (2000). Coral species richness: Ecological versus biogeographical influences. *Coral Reefs.* **19**, 37–49.
Cornell, H.V., and Lawton, J.H. (1992). Species interactions, local and regional processes, and limits to the richness of ecological communities: A theoretical perspective. *J. Anim. Ecol.* **61**, 1–12.
Cowen, R.K., Lwiza, K.M.M., Sponaugle, S., Paris, C.B., and Olson, D.B. (2000). Connectivity of marine populations: Open or closed? *Science.* **287**, 857–859.
den Boer, P.J. (1968). Spreading of risk and stabilization of animal number. *Acta Biotheoretica.* **18**, 165–194.
Denno, R.F., and Roderick, G.K. (1992). Density-related dispersal in planthoppers: Effects of interspecific crowding. *Ecology.* **73**, 1323–1334.
Denno, R.F., Roderick, G.K., Olmstead, K.L., and Döbel, H.G. (1991). Density-related migration in planthoppers (Homoptera: Delphacidae): The role of habitat persistence. *Am. Nat.* **138**, 1513–1541.
Doherty, P.J. (1983). Tropical territorial damselfishes: Is density limited by aggression or recruitment? *Ecology.* **64**, 176–190.
Ehrlich, P.R. (1961). Intrinsic barriers to dispersal in checkerspot butterfly. *Science.* **134**, 108–109.
Ehrlich, P.R., White, R.R., Singer, M.C., McKechnie, S.W., and Gilbert, L.E. (1975). Checkerspot butterflies: A historical perspective. *Science.* **188**, 221–228.
Elton, C. (1933). *The ecology of animals.* London: Methuen.
Findley, J.S., and Findley, M.T. (2001). Global, regional, and local patterns in species richness and abundance of butterflyfishes. *Ecol. Monogr.* **71**, 69–91.
Finlay, B.J. (2002). Global dispersal of free-living microbial eukaryote species. *Science.* **296**, 1061–1063.
Finlay, B.J., Esteban, G.F., and Fenchel, T. (1998). Protozoan diversity: Converging estimates of the global number of free-living ciliate species. *Protist.* **149**, 29–37.
Fisher, R.A., Corbett, A.S., and Williams, C.B. (1943). The relation between the number of species and the number of individuals in a random sample of an animal population. *J. Anim. Ecol.* **12**, 42–58.
Frank, K.T., and Shackell, N.L. (2001). Area-dependent patterns of finfish diversity in a large marine ecosystem. *Can. J. Fish. Aquat. Sci.* **58**, 1703–1707.
Fraser, R.H., and Currie, D.J. (1996). The species richness–energy hypothesis in a system where historical factors are thought to prevail: Coral reefs. *Am. Nat.* **148**, 138–159.
Gaines, S.D., and Lafferty, K.D. (1995). Modeling the dynamics of marine species: The importance of incorporating larval dispersal. In *Ecology of marine invertebrate larvae* (L. McEdward, ed., pp. 389–412). Boca Raton, FL: CRC Press.

Gaines, S.D., and Roughgarden, J. (1985). Larval settlement rate: A leading determinant of structure in an ecological community of the marine intertidal zone. *Proc. Nat. Acad. Sci. USA.* **82**, 3707–3711.
Gaines, S.D., and Roughgarden, J. (1987). Fish in offshore kelp forests affect recruitment to intertidal barnacle populations. *Science.* **235**, 479–481.
Gilpin, M., and Hanski, I., eds. (1991). *Metapopulation dynamics: Empirical and theoretical investigations.* London: Academic Press.
Godfray, H.C.J., and Lawton, J.H. (2001). Scale and species numbers. *Trends in Ecol. & Evol.* **16**, 400–404.
Hamilton, W.D., and May, R.M. (1977). Dispersal in stable habitats. *Nature.* **269**, 578–581.
Hanski, I. (1999). *Metapopulation ecology. Oxford series in ecology and evolution.* Oxford: Oxford University Press.
Hanski, I.A., and Gilpin, M.E., eds. (1997). *Metapopulation biology: Ecology, genetics, and evolution.* San Diego: Academic Press.
Harper, J.L. (1985). Modules, branches, and the capture of resources. In *Population biology and evolution of clonal organisms* (J.B.C. Jackson, L.W. Buss, and R.E. Cook, eds., pp. 1–33). New Haven, CT: Yale University Press.
Harrison, S. (1989). Long-distance dispersal and colonization in the bay checkerspot butterfly, *Euphydryas editha bayensis. Ecology.* **70**, 1236–1243.
Harrison, S., Murphy, D.D., and Ehrlich, P.R. (1988). Distribution of the bay checkerspot butterfly, *Euphydryas editha bayensis:* Evidence for a metapopulation model. *Am. Nat.* **132**, 360–382.
Hayek, L.-A., and Buzas, M.A. (1997). *Surveying natural populations.* New York: Columbia University Press.
He, F.L., and Legendre, P. (2002). Species diversity patterns derived from species–area models. *Ecology.* **83**, 1185–1198.
Hillebrand, H., and Blenckner, T. (2002). Regional and local impact on species diversity: From pattern to processes. *Oecologia.* **132**, 479–491.
Hixon, M.A., Pacala, S.W., and Sandin, S.S. (2002). Population regulation: Historical context and contemporary challenges of open vs. closed systems. *Ecology.* **83**, 1490–1508.
Holt, R.D. (1993). Ecology at the mesoscale: The influence of regional processes on local communities. In *Species diversity in ecological communities: Historical and geographical perspectives* (R.E. Ricklefs and D. Schluter, eds., pp. 77–88). Chicago: Chicago University Press.
Holt, R.D. (1997). From metapopulation dynamics to community structure. Some consequences of spatial heterogeneity. In *Metapopulation biology: Ecology, genetics, and evolution* (I.A. Hanski and M.E. Gilpin, eds., pp. 149–164). San Diego: Academic Press.
Hubbell, S.P. (1997). A unified theory of biogeography and relative species abundance and its application to tropical rain forests and coral reefs. *Coral Reefs.* **16**, S9–S21.
Hubbell, S.P. (2001). *The unified neutral theory of biodiversity and biogeography: Monographs in population biology* (vol. 32). Princeton: Princeton University Press.
Huffaker, C.B. (1958). Experimental studies on predation: Dispersion factors and predator–prey oscillations. *Hilgardia.* **27**, 343–383.
Hughes, T.P., Bellwood, D.R., and Connolly, S.R. (2002). Biodiversity hotspots, centres of endemicity, and the conservation of coral reefs. *Ecol. Lett.* **5**, 775–784.
Hugueny, B., and Cornell, H.V. (2000). Predicting the relationship between local and regional species richness from a patch occupancy dynamics model. *J. Anim. Ecol.* **69**, 194–200.
Hugueny, B., Tito de Morais, L., Mérona, B., and Ponton, D. (1997). The relationship between local and regional species richness: Comparing biotas with different evolutionary histories. *Oikos.* **80**, 583–587.
Huston, M.A. (1999). Local processes and regional patterns: Appropriate scales for understanding variation in the diversity of plants and animals. *Oikos.* **86**, 393–401.

Hutchinson, G.E. (1978). *An introduction to population ecology.* New Haven, CT: Yale University Press.
Jackson, J.B.C. (1977). Competition on marine hard substrata: The adaptive significance of solitary and colonial strategies. *Am. Nat.* **111**, 743–767.
Jackson, J.B.C. (1986). Modes of dispersal of clonal benthic invertebrates: Consequences for species distributions and genetic structure of local populations. *Bull. Mar. Sci.* **39**, 588–606.
Jinks, R.N., Markley, T.L., Taylor, E.E., Perovich, G., Dittel, A.I., Epifanio, C.E., and Cronin, T.W. (2002). Adaptive visual metamorphosis in a deep-sea hydrothermal vent crab. *Nature.* **420**, 68–70.
Karlson, R.H. (1999). *Dynamics of coral communities.* Dordrecht: Kluwer Academic Publishers.
Karlson, R.H. (2002a). Global, regional, and local patterns in species richness and abundance of butterflyfishes: Comment. *Ecology.* **83**, 583–585.
Karlson, R.H. (2002b). Population processes in modular benthic invertebrates. In *Reproductive biology of invertebrates* (R.N. Hughes, ed., vol. 11, pp. 255–282). New Delhi: Oxford & IBH Publishing.
Karlson, R.H., and Cornell, H.V. (2002). Species richness of coral assemblages: Detecting regional influences at local spatial scales. *Ecology.* **83**, 452–463.
Karlson, R.H., Cornell, H.V., and Hughes T.P. (2004). Coral communities are regionally enriched along an oceanic biodiversity gradient. *Nature* **429**, 867-870.
Karlson, R.H., and Levitan, D.R. (1990). Recruitment–limitation in open populations of *Diadema antillarum*: An evaluation. *Oecologia.* **82**, 40–44.
Karlson, R.H., and Taylor, H.M. (1992). Mixed dispersal strategies and clonal spreading of risk: Predictions from a branching process model. *Theor. Popul. Biol.* **42**, 218–233.
Lawton, J.H. (2000). *Community ecology in a changing world* (vol. 11). Oldendorf/Luhe: Ecology Institute.
MacArthur, R.H. (1957). On the relative abundance of bird species. *Proc. Nat. Acad. Sci. USA.* **43**, 293–295.
MacArthur, R.H. (1960). On the relative abundance of species. *Am. Nat.* **94**, 25–36.
MacArthur, R.H. (1965). Patterns of species diversity. *Biol. Rev.* **40**, 510–533.
MacArthur, R.H. (1972). *Geographical ecology: Patterns in the distribution of species.* New York: Harper & Row.
MacArthur, R.H., and Wilson, E.O. (1963). An equilibrium theory of insular zoogeography. *Evolution.* **17**, 373–387.
MacArthur, R.H., and Wilson, E.O. (1967). *The theory of island biogeography.* Princeton: Princeton University Press.
Matter, S.F., Hanski, I., and Gyllenberg, M. (2002). A test of the metapopulation model of the species–area relationship. *J. Biogeogr.* **29**, 977–983.
May, R.M. (1975). Patterns in species abundance and diversity. In *Ecology and evolution of communities* (M.L. Cody and J.M. Diamond, eds., pp. 81–120). Cambridge: Belknap Press of Harvard University Press.
May, R.M. (1981a). Models for single populations. In *Theoretical ecology: Principles and applications* (R.M. May, ed., 2nd ed., pp. 5–29). Sunderland, MA: Sinauer Associates.
May, R.M. (1981b). Patterns in species abundance and diversity. In *Theoretical ecology: Principles and applications* (R.M. May, ed., 2nd ed., pp. 197–227). Sunderland, MA: Sinauer Associates.
Mayr, E. (1982). *The growth of biological thought: Diversity, evolution, and inheritance.* Cambridge: The Belknap Press of Harvard University Press.
McLaughlin, J.F., Hellman, J.J., Boggs, C.L., and Ehrlich, P.R. (2002). The route to extinction: Population dynamics of a threatened butterfly. *Oecologia.* **132**, 538–548.
Mora, C., Chittaro, P.M., Sale, P.F., Kritzer, J.P., and Ludsin, S.A. (2003). Patterns and processes in reef fish diversity. *Nature.* **421**, 933–936.
Mora, C., and Sale, P.F. (2002). Are populations of coral reef fish open or closed? *Trends in Ecol. & Evol.* **17**, 422–428.

Mouquet, N., and Loreau, M. (2003). Community patterns in source–sink metacommunities. *Am. Nat.* 162, 544–557.
Okamura, B. (1997). The ecology of subdivided populations of a clonal freshwater bryozoan in southern England. *Archiv für Hydrobiologie.* 141, 13–34.
Okamura, B., and Freeland, J.R. (2002). Gene flow and the evolutionary ecology of passively dispersing aquatic invertebrates. In *Dispersal ecology* (J.M. Bullock, R.E. Kenwood, and R.S. Hails, eds., pp. 194–216). Oxford: Blackwell Publishing.
Okamura, B., Freeland, J.R., and Hatton-Ellis, T. (2002). Clones and metapopulations. In *Reproductive biology of invertebrates* (R.N. Hughes, ed., vol. 11, pp. 283–312). New Delhi: Oxford & IBH Publishing.
Ólafsson, E.B., Peterson, C.H., and Ambrose, Jr., W.G. (1994). Does recruitment limitation structure populations and communities of macro-invertebrates in marine soft sediments: The relative significance of pre- and post-settlement processes. *Oceanogr. Mar. Biol. Ann. Rev.* 32, 65–109.
Paine, R.T. (1994). *Marine rocky shores and community ecology: An experimentalist's perspective* (vol. 4). Oldendorf/Luhe: Excellence in Ecology, Ecology Institute.
Plotkin, J.B., Potts, M.D., Yu, D.W., Bunyavejchewin, S., Condit, R., Foster, R., Hubbell, S., LaFrankie, J., Manokaran, N., Seng, L.H., Sukumar, R., Nowak, M.A., and Ashton, P.S. (2000). Predicting species diversity in tropical forests. *Proc. Nat. Acad. Sci. USA.* 97, 10850–10854.
Preston, F.W. (1948). The commonness, and rarity, of species. *Ecology.* 29, 254–283.
Pulliam, H.R. (1988). Sources, sinks, and population regulation. *Am. Nat.* 132, 652–661.
Raffaelli, D.G., Hildrew, A.G., and Giller, P.S. (1994). Scale, pattern and process in aquatic systems: Concluding remarks. In *Aquatic ecology: Scale, pattern and process* (P.S. Giller, A.G. Hildrew, and D.G. Raffaelli, eds., pp. 601–606). Oxford: Blackwell Scientific Publications.
Ricklefs, R.E. (1987). Community diversity: Relative roles of local and regional processes. *Science.* 235, 167–171.
Ricklefs, R.E., and Lovette, I.J. (1999). The roles of island area per se and habitat diversity in the species–area relationships of four Lesser Antillean faunal groups. *J. Anim. Ecol.* 68, 1142–1160.
Ricklefs, R.E., and Schluter, D., eds. (1993). *Species diversity in ecological communities: Historical and geographical perspectives.* Chicago: Chicago University Press.
Roberts, C.M. (1997). Connectivity and management of Caribbean coral reefs. *Science.* 278, 1454–1457.
Rosen, B.R. (1988). Progress, problems and patterns in the biogeography of reef corals and other tropical marine organisms. *Helgoländer Meeresuntersuchungen.* 42, 269–301.
Rosenzweig, M.L. (1995). *Species diversity in space and time.* Cambridge: Cambridge University Press.
Rosenzweig, M.L., and Ziv, Y. (1999). The echo pattern of species diversity: Pattern and processes. *Ecography.* 22, 614–628.
Ross, K.G. (2001). How to measure dispersal: The genetic approach the example of fire ants. In *Dispersal* (J. Clobert, E. Danchin, A.A. Dhondt, and J.D. Nichols, eds., pp. 29–42). Oxford: Oxford University Press.
Roughgarden, J. (1989). The structure and assembly of communities. In *Perspectives in ecological theory* (J. Roughgarden, R.M. May, and S.A. Levin, eds., pp. 203–226). Princeton: Princeton University Press.
Roughgarden, J., Gaines, S.D., and Pacala, S.W. (1987). Supply side ecology: The role of physical transport processes. In *Organization of communities past and present* (J.H.R. Gee and P.S. Giller, eds., pp. 491–518). Oxford: Blackwell Scientific Publications.
Roughgarden, J., Gaines, S., and Possingham, H. (1988). Recruitment dynamics in complex life cycles. *Science.* 241, 1460–1466.
Rousset, F. (2001). Genetic approaches to estimating dispersal rates. In *Dispersal* (J. Clobert, E. Danchin, A.A. Dhondt, and J.D. Nichols, eds., pp. 18–28). Oxford: Oxford University Press.

Roy, K., Jablonski, D., Valentine, J.W., and Rosenber G. (1998). Marine latitudinal diversity gradients: Tests of causal hypotheses. *Proc. Nat. Acad. Sci. USA.* **95**, 3699–3702.

Scheltema, R.S. (1971). Larval dispersal as a means of genetic exchange between geographically separated populations of shallow-water benthic marine gastropods. *Biol. Bull.* **140**, 284–322.

Scheltema, R.S. (1988). Initial evidence for the transport of teleplanic larvae of benthic invertebrates across the East Pacific Barrier. *Biol. Bull.* **174**, 145–152.

Seabloom, E.W., Dobson, A.P., and Stoms, D.M. (2002). Extinction rates under nonrandom patterns of habitat loss. *Proc. Nat. Acad. Sci. USA.* **99**, 11229–11234.

Srivastava, D.S. (1999). Using local–regional richness plots to test for species saturation: Pitfalls and potentials. *J. Anim. Ecol.* **68**, 1–16.

Sutherland, J.P. (1987). Recruitment limitation in a tropical intertidal barnacle: *Tetraclita panamensis* (Pilsbry) on the Pacific coast of Costa Rica. *J. Exp. Mar. Biol. Ecol.* **113**, 267–282.

Terborgh, J.W., and Faaborg, J. (1980). Saturation of bird communities in the West Indies. *Am. Nat.* **116**, 178–195.

Underwood, A.J. and Denley, E.J. (1984). Paradigms, explanations, and generalizations in models for the structure of intertidal communities on rocky shores. In *Ecological communities: Conceptual issues and the evidence* (D.R. Strong, Jr., D. Simberloff, L.G. Abele, and A.B. Thistle, eds., pp. 151–180). Princeton: Princeton University Press.

Veron, J.E.N. (1995). *Corals in space and time: The biogeography and evolution of the scleractinia*. Ithaca, NY: Cornell University Press.

Veron, J.E.N. (2000). *Corals of the World*. Townsville: Australian Institute of Marine Science.

Volkov, I., Banavar, J.R., Hubbell, S.P., and Maritan, A. (2003). Neutral theory and relative species abundance in ecology. *Nature.* **424**, 1035–1037.

Welter-Schultes, F.W., and Williams, M.R. (1999). History, island area and habitat availability determine land snail species richness of Aegean islands. *J. Biogeogr.* **26**, 239–249.

Wiens, J.A. (1989). *The ecology of bird communities* (vol. I). Cambridge: Cambridge University Press.

Williams, C.B. (1964). *Patterns in the balance of nature*. London: Academic Press.

Wilson, D.S. (1992). Complex interaction in metacommunities, with implications for biodiversity and higher levels of selection. *Ecology.* **73**, 1984–2000.

CHAPTER 15

Metapopulation Ecology and Marine Conservation

LARRY B. CROWDER and WILL F. FIGUEIRA

I. Introduction
II. Sources, Sinks, and Metapopulation Dynamics
III. Case Studies
   A. Coral Reef Fishes
   B. Caribbean Spiny Lobster (*Panulirus pargus*)
   C. Red Sea Urchin (*Strongylocentrotus franciscanus*)
   D. Loggerhead Sea Turtles (*Caretta caretta*)
IV. Summary
   References

I. INTRODUCTION

Coastal ecosystems are influenced by multiple, often interacting, factors including pollution, coastal development, destruction and fragmentation of habitats, invasive species, climate change, and fishing. Although all these factors can have dramatic effects, fisheries probably account for the largest and oldest impact on marine ecosystems (Jackson et al., 2001). We've long known that fisheries dramatically reduce the biomass of target species in heavily exploited regions of the ocean (see Gulland, 1988, for a review; Pauly and Maclean, 2003). Researchers have recently raised concerns about the multiple impacts of industrialized fishing on target stocks, bycatch of long-lived species, and damage to critical habitat (Dayton et al., 1995; Pauly, 1995; Pauly et al., 1998; Watling and Norse, 1998; Pauly and Maclean, 2003). The impact on large predatory fishes now appears to be global—a recent estimate suggests that the biomass of large predatory fishes now is at least 90% lower than preindustrial levels (Myers and Worm, 2003, 2004). Nearly 80% of the observed declines occurred during the first 15 years of exploitation, often before stock assessments could be completed or fishery independent surveys begun. Many nontarget species including sharks and other elasmobranchs (Casey and Myers, 1998; Baum et al., 2003; Baum and Myers, 2004) and endangered sea turtles have undergone similar steep declines (Spotila et al., 2000; Kamezaki, 2003; Limpus and Limpus, 2003).

Concerns about the direct and indirect effects of fishing led to a United Nations resolution on restoring fisheries and marine ecosystems (United Nations, 2002). Recent calls for ecosystem-based management from both the Pew Oceans Commission (2003) and the U.S. Commission on Ocean Policy (2004) seek to place fished (and overfished) populations into a larger ecosystem context. In fact, much of the current interest in population dynamics and metapopulation dynamics in marine conservation emerges from concerns about recovering and sustaining fished populations, mitigating habitat damage resulting from fishing gear, and reducing impacts of human activities on protected and endangered species. Population dynamics modeling has a long and venerable history in fisheries, but most fisheries models focus on predicting sustainable take levels based on fishery-dependent data rather than modeling the entire life history of the species. Although the data that enter these models comes from a fished region, they generally are not spatially explicit and they do not deal explicitly with the linkages between environmental variation and recruitment variability in these organisms. Furthermore, reports of cumulative catch rates can mask subpopulation depletion (or serial overfishing) as fishers move on to another region after overfishing the current fishing grounds. In other words, the dynamics of spatially distributed metapopulations are often submerged by data aggregation and simplistic model structures.

Marine conservationists are increasingly interested in place-based approaches to managing the impacts of fisheries and other human activities on marine ecosystems (Norse et al., 2005). Approaches range from *MPAs* (simply defined as portions of ocean and coastal habitats set aside for some sort of special protection) to *marine reserves* that imply no extractive use within the protected area. Of course, protection may be required along migratory routes for some organisms to ensure their safe passage among habitat patches. Fisheries managers have long used the term *time–area closure* to refer to regions where fisheries targeting a particular species and/or using a particular fishing gear or method are prevented, at least for a period of time. Permanently protected areas have a reasonably long history in terrestrial ecosystems, but are still fairly novel in marine systems. Only a small fraction of 1% of coastal ecosystems are fully protected in marine reserves (Roberts and Hawkins, 2000). Although reserves cannot protect marine ecosystems from all possible threats, place-based approaches have substantial promise if stakeholders can agree upon appropriate objectives for management of marine ecosystems.

No-take marine reserves are an emerging and potentially powerful tool in the conservation of marine biodiversity (Lubchenco et al., 2003, and references therein). The field evidence is captivating—upon establishment of a marine reserve, one often observes localized augmentation of both the abundance and body size of previously overexploited and sparsely distributed populations characterized by small individuals (Roberts et al., 2001). But removing one source of mortality from one or a few locations in the domain of a population guarantees neither enhancement nor future persistence of a population or metapopulation as a whole (Allison et al., 2003; Lubchenco et al., 2003). Marine reserves are highly likely to serve as an effective tool in the conservation of biodiversity when used judiciously with complementary management tools, under appropriate circumstances (Lubchenco et al., 2003), and with a thorough understanding of the potential limitations (Botsford et al., 2003).

Although the influence of metapopulation and source–sink dynamics upon the efficacy of marine reserves has been widely recognized (Crowder et al., 2000; Lipcius et al., 2001; Botsford et al., 2003; Lipcius et al., 2005), scientists have not yet fully developed the theory of marine metapopulations. Nor have they established the strong empirical evidence for how metapopulation dynamics influence reserve design to benefit fisheries.

The two major postulated benefits of marine reserves to fishery stocks include (1) enhancement of the spawning stock, which subsequently magnifies recruitment of larvae, postlarvae, and juveniles to reserve and non-reserve areas; and (2) export of biomass to nonreserve areas when the exploitable segment of the stock in a reserve emigrates to nonreserve areas (Roberts and Polunin, 1991, 1993). Larval supply to downstream reefs is difficult to document directly, although recent research suggests that local to regional physics will play a large role in whether larvae are transported long distances or retained near their source (Roberts, 1997; Cowen et al., 2000; Palumbi, 2003). The benefit of exported biomass is primarily local (Roberts and Polunin, 1991, 1993; Russ and Alcala, 1996; Roberts et al., 2001; Russ and Alcala, 2004), and may increase yield-per-recruit in the fishery, depending upon exploitation rates and demographic characteristics (Hastings and Botsford, 1999; Sladek Nowlis and Roberts, 1999; Botsford et al., 2003).

Variability in recruitment drives population fluctuations in marine organisms, and understanding the linkages in space and time between recruitment and environmental variation is critical to managing many marine organisms. Bottlenecks to sustaining populations may occur in time or space, so to understand and model the entire life history of fishes or other marine organisms, researchers now resort to spatially explicit, temporally dynamic approaches. Habitat patches may be fixed (e.g. coral reefs, seamounts) or dynamic (fronts, eddies); organisms may require protection within these habitats or as they migrate (or are transported) between habitat patches. Fishing and other human activities happen at particular places and times as do activities of fished populations and protected species. So understanding how population dynamics of these organisms play out in a dynamic, patchy habitat is critical to good conservation decision making.

II. SOURCES, SINKS, AND METAPOPULATION DYNAMICS

The abundance and distribution of populations within a landscape are influenced by the types, quality, and spatial arrangement of habitat patches (Pulliam, 1988; Pulliam and Danielson, 1991; Pulliam et al., 1992; Hanski, 1994; Lipcius et al., 1997; Fogarty, 1998; Crowder et al., 2000). Similarly, the dynamics of marine species depend upon various spatial scales (Menge and Olson, 1990; Doherty and Fowler, 1994), such that many marine populations with dispersive stages should be viewed as metapopulations with interconnected subpopulations (Roughgarden et al., 1985, 1988; Botsford et al., 1994, 1998). We use the term *metapopulation* broadly to mean any group of local populations that exchange individuals. Although exchange among subpopulations could be extensive enough to convert marine metapopulations into large but spatially subdivided single populations, it seems reasonable (and conservative) to assume that many marine populations could function as real metapopulations, at least until data are garnered to prove otherwise.

Dispersive stages in species with complex life cycles (sensu Roughgarden et al., 1988) break the connection at the local scale between reproduction and recruitment, so connectivity among subpopulations is an emergent and critical property of the system (Doherty and Fowler, 1994). Obviously models that integrate ocean physics with dispersal behavior of larvae are critical to characterizing the dynamic linkages among habitat patches (Cowen et al., 2000; Morgan and Botsford, 2001; Swearer et al., 2002).

Marine species are also likely to be characterized by "source" and "sink" habitats (Lipcius et al., 1997; Fogarty, 1998; Crowder et al., 2000; Tuck and Possingham, 2000; Lipcius et al., 2001; Figueira, 2002). A fraction of the individuals may regularly occur in "sink" habitats, where the output of juveniles or adults to the spawning stock is insufficient to balance mortality. In contrast, a segment of the population may occur in "source" habitats, where the output of individuals to the spawning stock is sufficient to maintain populations in source and sink habitats (Pulliam, 1988). We distinguish between our use of the definitions of source–sink dynamics dependent on local habitat quality and its effect on demographic rates (Pulliam, 1988), and that of other authors where sources and sinks appear to pertain to origins and destinations respectively of dispersive stages (Roberts, 1997, 1998; Cowen et al., 2000).

Although a variety of metapopulation "types" have been identified (Lipcius et al., 2005) the spatially explicit nature of demography and connectivity in the life history of marine organisms makes the use of metapopulation source–sink theory critical to the goals of conservation in marine ecosystems. There is little doubt that some species may not exhibit source–sink dynamics, and source–sink structure of subpopulations can vary among species and over time. Such matters can complicate the analysis, but fundamentally the processes will still be based upon

Chapter 15 Metapopulation Ecology and Marine Conservation

the same theories. Although we are in the early stages of developing and extending these theories to marine systems, previous modeling approaches that ignored spatial interactions are inadequate to understand and manage these marine systems.

III. CASE STUDIES

In the following sections we use four case studies to illustrate the usefulness of the metapopulation approach to the goals of conservation in marine systems. The first three cases—coral reef fishes, spiny lobsters, and red sea urchins—involve life histories in which dispersal is in the larval stage. In this case, we need to understand the linkages between larval transport, environmental variation, and recruitment of larvae at particular times and places. Without a detailed understanding of these relationships, conservation decisions regarding the placement of reserves or other protections are likely to go seriously wrong. The fourth case study involves loggerhead sea turtles that nest in the southeast United States, but range as posthatchlings throughout the North Atlantic. After a decade or so they return to the continental shelf of the United States. Adults migrate from feeding grounds to breeding and nesting areas. Recent genetic information suggests that these populations are much more intricately structured than we have previously assumed.

A. Coral Reef Fishes

Despite the myriad of body forms and life histories adopted by fishes inhabiting the coral reef environments of the world, they tend very generally to share two fundamental properties: (1) Juveniles and adults reside in relatively discrete population units that are largely reproductively isolated from one other because of very low levels of migration, and (2) the early life history of most species is characterized by the production of larvae that develop over the course of a pelagic existence and ultimately return to the benthos for the juvenile and adult stages. Such highly structured habitat and disparate life history stages present serious challenges to conservation and management of coral reef fishes. Traditional fisheries methods rely on relationships between the size of the spawning stock and the subsequent number of new individuals added to the population. Unfortunately, the confined nature of adult habitat relative to the potentially wide-ranging and dynamic nature of the pelagic larval habitat tends more often than not to decouple these stock–recruit relationships.

The theory of metapopulations, however, offers a framework in which to build anew our understanding of the population dynamics of these systems. Although individual subpopulations may be demographically open at the local scale (Mora and Sale, 2002), the system of subpopulations will inevitably be closed at a larger, more regional scale. Although no system will ever be completely closed, there will certainly be a point at which larval flux across boundaries becomes unimportant to the dynamics of the metapopulation relative to the connectivity within the system. Viewed at this larger scale, we again have some hope of understanding the fluctuations in abundance and distribution of organisms over time.

Because of the directional nature of oceanographically forced dispersal in these systems, coral reef fish metapopulations have tended to be described as source–sink systems. This approach was exemplified by Roberts (1997, 1998) in describing connectivity patterns for reefs in the greater Caribbean basin based on large-scale average current data. In general, studies of coral reef fish source–sink dynamics are difficult as a result of the broad spatial scales involved as well as the challenges associated with establishing connectivity patterns. Because of this, there are currently no comprehensive field studies of the subject. However, modeling studies have looked closely at habitat, connectivity, and source–sink structure of coral reef fish.

James et al. (2002) studied connectivity patterns and metapopulation dynamics in the Cairns section of the Great Barrier Reef, Australia. Using a simulation model they analyzed connectivity patterns, identified areas of high local retention, and evaluated system persistence as a function of individual populations. Their simulations meshed an individually based dispersal model with a five-stage patch-level population model. Dispersal was driven by a fine-scale, two-dimensional depth-integrated flow model and appropriate larval behavior and biology modeled after the ambon damselfish (*Pomacentrus amboinensis*).

Using 20 years of seasonal flow data they demonstrated that the majority of larval dispersal connections between reefs in the system are relatively strong and occurred over short distances (ca. 1° of latitude), but there are some weaker connections over much longer distances (ca. 5° of latitude) that have the potential to have strong systemwide effects. They also saw that self-recruitment, defined as the proportion of recruits settling to a reef that came from that reef, tended to be rare (80% of reefs had less than 9% self-recruitment) and connectivity levels were quite variable from year to year (CV values ranged from 0.36–4.5 with a mode of about 0.75). Interestingly, one of the most strongly self-recruiting reefs, Lizard Island, was also the site where Jones et al. (1999) demonstrated very high levels of self-recruitment (15–60%) for the damselfish, *Pomacentrus amboinensis*, using mark-recapture techniques.

The metapopulation simulations done by James et al. (2002) showed quite clearly that the reefs with the strongest self-recruitment were also key reefs for the long-term viability of the metapopulation. Their simulations showed dramatic and continued reduction in overall metapopulation size (adult density) when these high self-recruiting reefs were removed from the system compared with removing reefs either by size (large ones) or at random (Fig. 15-1).

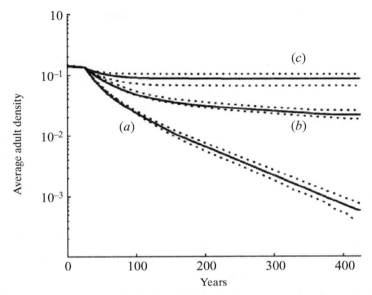

FIGURE 15-1. Great Barrier Reef, Australia, metapopulation simulation study of James et al. (2002; Fig, 5) showing the average number of adults per reef in response to the removal of populations of the 20 strongest self-seeding reefs (a), the 20 largest reefs (b), and randomly selected reefs (c). The solid line is the average for multiple replicates; the broken line represents ±1 SD. Reprinted with permission.

Although the study made no attempt to include the effects of fishing or protection from fishing in certain areas, it very clearly demonstrates the usefulness of the metapopulation approach toward that end. Using this approach James et al. (2002) were able to identify reef patches in the systems that served as key nodes, or perhaps even sources. The protection of these reef patches would be expected to be crucial to the long-term viability of the entire metapopulation.

However, determination of an individual patch's contribution to the metapopulation (its identity as either a source or sink) is not simply dependent upon dispersal and connectivity, but also upon its demographic characteristics (Figueira, 2002). Modeling studies (Figueira, 2002) have demonstrated that although the relative importance of each factor is context dependent, patch contribution is very often more sensitive to demographic characteristics than to those associated with dispersal ability. Because the demographic character of a patch is often tightly associated with the overall quality of patch habitat (Figueira, 2002), complete treatments of source–sink metapopulations require acknowledgment of habitat quality as well as connectivity.

Figueira's (2002) approach to evaluating source–sink metapopulation structure of the Florida Keys reef system did exactly that. This study also used individual-based larval movement between patches and local patch stage-structured growth. Dispersal in the simulations was driven by four different realizations of typical flow states for the Florida Keys reef system. Two of the states represented average conditions under summer and fall wind conditions whereas the other two represented conditions under average fall winds and the presence of mesoscale recirculation events known as the Tortugas and Pourtales gyres respectively. The use of different flow states allowed for evaluation of the temporal stability of resulting connectivity and source–sink patterns. In addition, local demographics were specific to one of two distinct reef habitat types (platform margin reefs and patch/mid-channel reefs) with parameter values assigned based on a 2-year study of the bicolor damselfish *Stegastes partitus* (Figueira, 2002).

The source–sink identity of individual patches was assessed in this study using a variant of the rate of population increase parameter (lambda) known as the contribution lambda (λ_c). λ_c is essentially the difference between deaths occurring within a patch and births that occur anywhere within the metapopulation as a result of production from that patch. When λ_c is greater than one, the patch functions as a source for the metapopulation whereas a λ_c value of less than one indicates that the patch is a sink. We have previously demonstrated the importance of using this metric (vs. the more traditional patch growth parameter, λ) for studies of marine source–sink metapopulations (Figueira, 2002).

The results of this study indicated that including information about habitat-specific demographics could dramatically alter the source–sink structure of patches within the system. Patches identified as sources when averaged demographic rates were used could become very strong sinks when habitat-specific, and therefore theoretically more accurate, demographic rates were used (Figs. 15-2 and 15-3). For all four flow states, between 15% and 30% of all patches changed identity from either source to sink or sink to source, with the majority of those changing from source to sink. The study also demonstrated that although the overall variability in source–sink identity for a patch from one flow state to the next could be quite large (CV, 0–40; mode, about 3), there were many patches with an identity that remained fairly constant as either source or sink under all flow states.

Conservation and management of coral reef fish is increasingly turning to spatial closures such as marine reserves. Although marine reserves seem to function quite well to protect biodiversity, their record at enhancement of surrounding fisheries has been poor (Crowder et al., 2000). There is some evidence of adult spillover into surrounding areas (see Halpern, 2002, for a review), but none concerning the effects of the assumed increases in larval output from a protected area. Modeling studies have indicated that such increases should occur over the long term (for non-coral reef fishes; Sladek Nowlis and Roberts, 1999; Tuck and Possingham, 2000), but that specific outcomes can be strongly influenced by the reserve configuration relative to average larval dispersal distances (Botsford,

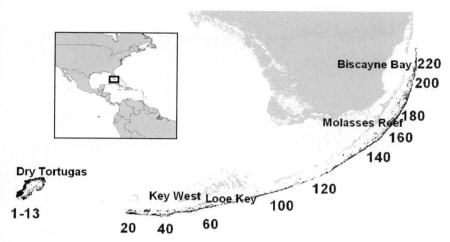

FIGURE 15-2. Model domain for metapopulation study of Florida Keys coral reef fish. Gray areas are land; black areas are reef. Numbers indicate approximate location of reefs as identified in the study. These are the reef ID numbers used in the x-axis of Figure 15-3. From Figueira, 2002, Figure 4.4.

2001). In reality, such things as connectivity and even habitat-specific demographics can be challenging if not impossible to measure with certainty.

Spatially explicit approaches such as those described earlier, which incorporate realistic dispersal as well as habitat-driven demographics, offer insight into the processes at play in these systems as well. In so doing they allow us to focus our efforts where they are needed most to conserve and manage coral reef fish more effectively. The studies described demonstrate methodology that can be used to identify reefs or areas that are strong nodes or population sources within the metapopulation, even in the face of temporal variability in patterns of connectivity or habitat quality. Such areas would be ideal targets for reserve placement. The conservation of fishes on coral reefs is just beginning to and will continue to rely on these spatially explicit metapopulation approaches.

B. CARIBBEAN SPINY LOBSTER (*PANULIRUS PARGUS*)

P. pargus is widely distributed within the Caribbean basin and serves as the basis for a large and valuable commercial fishery throughout the region. It is characterized by a complicated five-stage life history in which individuals inhabit both benthic and pelagic habitats (Stockhausen et al., 2000). Initially, gravid adult

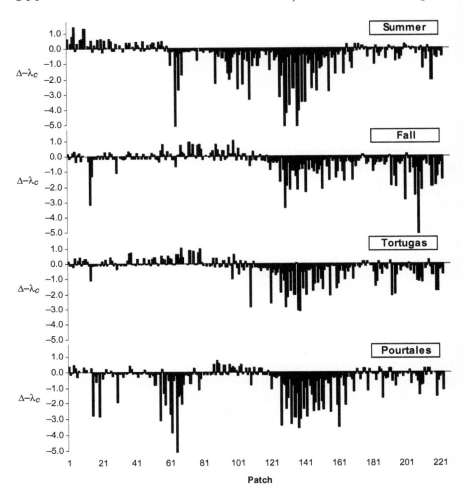

FIGURE 15-3. Magnitude of difference in patch λ_c values (explained in text) resulting from model runs using uniform patch demographic values versus using habitat-specific values from the Florida Keys metapopulation study of Figueira (2002, Fig. 4.16). Positive values indicate that including habitat-specific demographics resulted in an increase in λ_c; negative values indicate a decrease. The x-axis indicates patch number as assigned beginning with number 1 in the Dry Tortugas and increasing moving up the Keys to number 228 off Biscayne Bay in the northeast (see Fig. 15-2 for map).

females release fertilized eggs from the seaward edge of fringe reefs. The phyllosome larva that hatch from these eggs shortly thereafter develop through a series of approximately 11 larval stages during the next 4 to 9 months, during which they grow from about 1 to 12 mm in carapace length (CL). The end of this stage is marked by the metamorphosis into the puerulus postlarvae. During this stage, individuals are transparent and do not feed. They have well-developed swimming abilities and eventually migrate to shallow coastal waters during new moon nighttime tides and settle in algal clumps or among mangrove roots. After several days, the postlarvae begin to acquire pigmentation and metamorphose into the first juvenile benthic instar (Stockhausen et al., 2000).

As juveniles, *P. pargus* can be found in different habitat types as they grow progressively larger. Initially juveniles are found primarily associated with algal clumps, but as they grow to about 15 mm CL and then on to about 45 mm CL, they transition to habitats with larger refuge sizes such as sponges, octocorals, and crevices, and eventually begin to forage openly in sand and seagrass habitat, although often associating in aggregations. As juveniles approach sexual maturity they begin to migrate to deeper offshore reefs, although they remain highly gregarious (Stockhausen et al., 2000).

The strong habitat associations of the benthic stages combined with the extreme dispersal potential afforded by the long duration of the pelagic stage makes understanding population dynamics at the local scale nearly impossible. This problem is especially evident in the Florida Keys, where the presence of the late-stage larvae along the northern edge of the northeasterly flowing Florida current suggests an upstream Caribbean source of larvae to the Keys (Yeung and Lee, 2002). Although wind-driven onshore/countercurrents and mesoscale gyre recirculations may provide mechanisms by which locally produced larvae can be retained within the Florida Keys system (Yeung and Lee, 2002), the exogenous input clearly plays a significant role in the population dynamics of the system.

Effective conservation and management in a system such as this requires attention to the population at larger spatial scales—that of the metapopulation—and to the connectivity that exists within that larger system. Such ocean basin-level research has yet to be attempted for the case of the Florida Keys, but the spiny lobster population of Exuma Sound in the Bahamas has been the focus of a series of studies aimed at understanding connectivity among isolated habitats as well as the metapopulation ramifications of such connectivity. Exuma Sound is a very deep (>1000 m) and rather large (200 × 75 km) basin in the central Bahamas that is largely closed to mixing with the surrounding shallow shelf by a series of small islands (cays) and reefs except for a deep 50-km pass at its southeastern end and a much shallower one to the northeast (Stockhausen et al., 2000; Fig. 15-4). It is believed to be a much more closed system than that of the Florida Keys (Yeung and Lee, 2002). As a result of this isolation, the abundance of reef habitat, and

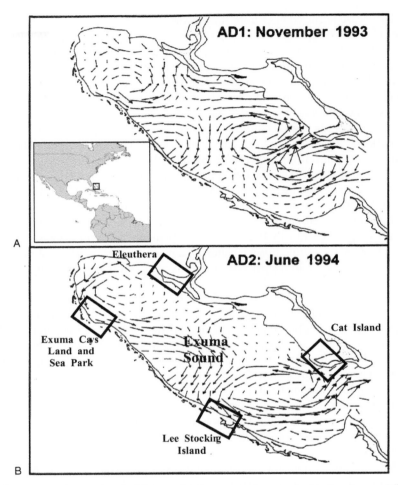

FIGURE 15-4. Exuma Sound, Bahamas, showing near-surface geostrophic flow in two different seasons. (A) Fall (November 1993). (B) Summer (June 1994). Study sites are indicated by boxes in B. Composite of Stockhausen et al. (2000, Fig. 2) and Lipcius et al. (2001, Fig. 1). Reprinted with permission.

the importance of spiny lobster as a fishery resource in the area, the Exuma Sound system provides an excellent opportunity to investigate metapopulation structure in marine systems.

The extremely long larval duration of this species makes establishing connectivity patterns difficult. For this reason, initial studies of this system sought to identify source and sink areas of the metapopulation based primarily on habitat quality as reflected by comparison between larval supply and juvenile and adult abundances as well as reproduction (Lipcius et al., 1997). Four sites around Exuma Sound were used for this study—Eleuthera in the northeast, Exuma Cays Land and Sea Park (ECLSP) in the northwest, Lee Stocking Island (LSI) in the southwest, and Cat Island in the southeast (Fig. 15-4B). Monitoring of larval supply indicated that Cat Island received far more larvae than the other three areas that were similar. Censuses indicated, however, that juvenile and adult densities were much lower at Cat Island than the other three sites. A look at conductivity, temperature, depth (CTD) profiles and drifter-derived current fields clearly indicates a twin set of gyres that are likely responsible for the high level of larval supply to Cat Island (Fig. 15-4A). From this information, Lipcius et al. (1997) concluded that Cat Island was likely serving as a sink within the metapopulation of Exuma Sound, because poor-quality habitat resulted in very high mortality rates for the postlarvae that were supplied there in such large quantities. It would seem that the other three sites would be serving, at least to some degree, as sources, because they are able to maintain high adult abundances (and therefore reproductive output) despite the relatively low rates of postlarval supply.

Stockhausen et al. (2000) used the results of this study to examine the issue of connectivity in this system and how the metapopulation structure affected conservation and fisheries using marine reserves. This study used a two-dimensional advection–diffusion model to describe larval movement within static flow fields provided by ship-based CTD/drifter studies and connected this adult production in different areas appropriately timed to the biology of *P. pargus*. Using this coupled model, they simulated reserves of various sizes located in proximity to the same four sites mentioned here under two slightly different hydrodynamic regimes (fall and summer, Figs. 15-4A and 15-4B respectively), and looked at the resulting changes in the larval production from reserves and the harvest of adults outside the reserves.

As expected with multiple variables, the results were complicated, but their chief findings were that (1) under the same hydrodynamic regime, reserves of similar size were equally good at increasing larval production regardless of where they were located, but this was not true for harvest rates; (2) the changes in larval production and adult harvest were very different for the same reserve design under the two different hydrodynamic scenarios; and (3) both the size and the location of the "optimal" reserve setup (for larval production and/or harvest) was different for each hydrodynamic regime.

So the nature of the connectivity pattern, as dictated by the hydrodynamic regime, very strongly influenced the ability of a reserve to supply larvae and eventually adults to areas outside the reserve. Furthermore, because of the highly spatial nature of the connectivity pattern, the ability to increase harvest outside reserves varied dramatically from one location to the next.

Building again on this information, Lipcius et al. (2001) used the same model to look at how well current and proposed marine reserves within this system could be expected to meet the goals of serving as sources for the rest of the metapopulation. Again, using the same four areas (where ECLSP is already a no-take reserve), they ran simulations to determine the likely larval supply to the rest of the system from these areas given the same static flow fields used by Stockhausen (2002), as well as the information about local density of adults and reproductive output. They estimated that fishing effort was about 47 to 98% lower in ECLSP than elsewhere and that, because of the circulation patterns, only ECLSP and Eleuthera supplied larvae to the rest of the system. Closed loop eddies near LSI and Cat Island resulted in retention and local settlement of the majority of larvae released at those two sites (Fig. 15-4). They also simulated making decisions about reserve location based solely on either adult abundance in an area or the amount of suitable habitat available. In both cases they found decisions based on flow information to be much more successful in identifying areas likely to be sources.

Spiny lobster in Exuma Sound exist in a highly spatially structured system but with enormous potential for connectivity resulting from the closed nature of the system and the very long planktonic larval duration of this species. These studies have pointed out the importance of considering the effect of local habitat as well as that of the connectivity afforded by hydrodynamic regimes to understanding the metapopulation-level consequences of conservation and management measures such as the establishment of marine reserves.

C. Red Sea Urchin (*Strongylocentrotus franciscanus*)

The red sea urchin is harvested along the length of the U.S. West Coast, but most of the take comes from the waters off California the roe is of commercial interest and the vast majority of the catch is shipped to Japanese markets (Andrew et al., 2002). The value of the fishery was estimated at $14.4 million (U.S.) in 1999 and catches of red sea urchin in California waters increased steadily since the beginning of the fishery in the early 1970s until the late 1980s (Andrew et al., 2002). Since the peak in 1988, at about 24,000 tons, the fishery has experienced a steady decline to catch levels of less than 10,000 tons in 1999 (Andrew et al., 2002). The fishery can be divided into two principal areas: southern and northern. Initial harvest occurred almost exclusively in areas south of Point Conception (southern) with primary collection grounds being the areas around the northern Channel Islands. In the mid 1980s, the northern fishery began to develop rapidly and peaked in 1988 when about 14,000 tons were taken (Andrew et al., 2002). Since then it has declined at a much faster rate than the southern fishery, and catches in recent years have been less than 1000 tons (Botsford et al., 1999).

It appears that the fisheries in these two regions are behaving quite distinctly. Although the northern fishery has experienced a very classic course for the discovery of a new fishery—initially increasing catch levels followed by rapid declines—the southern fishery, although also experiencing reductions, seems to be much more able to replenish itself and has thus not seen the collapse apparent in the northern region, despite relatively intense harvest. The combination of declining catches and the recent passage of conservation-oriented laws (the California Marine Life Management Act of 1998 and the Marine Life Protection Act) have generated greater interest in understanding better the large-scale population dynamics of the red sea urchin in California.

Adult red sea urchins typically inhabit shallow rocky subtidal shelf habitats (Botsford et al., 1994) and have relatively low mobility. Gonadal development and spawning follow a seasonal pattern that results in a period of peak spawning occurring typically sometime during the spring. The duration of the pelagic larval stage ranges from 7 to 19 weeks, at which point the larvae metamorphose and subsequently settle onto benthic habitat (Morgan et al., 2000). Tegner and Dayton (1977) have demonstrated a positive effect of adult density on the survival of new recruits mediated by the protection of these recruits under the spine canopy of adults. Generally speaking, settlement occurs year round, although it peaks in the summer months (Morgan et al., 2000). Settlement in the southern region tends to be relatively constant (Botsford et al., 1998), although influenced by basin-scale phenomena such as ENSO events (Botsford, 2001); recruitment in the northern region has been described as episodic and coincident with occasional upwelling relaxation (Morgan et al., 2000).

The physical setting of this metapopulation is that of the California Current system. At the largest scale the California Current system is the eastern arm of the Pacific basin anticyclonic gyre. It spans about 20° of latitude, with a net equatorward motion and a velocity of around 0.1 m/second. It is stronger and closer to shore in the summer. During the winter it moves offshore, making way for the poleward-flowing Davidson Current. In a process termed *the spring transition*, northward migration of the North Pacific high-pressure system in the spring reverses the typically northward winds and in so doing drives offshore Eckman transport and subsequent strong upwelling and southward flow along the coast. There is considerable interannual variability in the California Current system structure related to ENSO events as well as jets, eddies, and meanders, which develop as instabilities of the large-scale flow. Within California Current system–driven circulation, there are wind and freshwater outflow-driven patterns that result in more local variation to patterns of alongshore flow and upwelling. Finally, within these are small-scale coastal patterns forced primarily by waves, tides, and patchy winds (Botsford et al., 1994).

Given the life history characteristics of this species—isolated adult populations connected via the dispersal of larvae among them—investigations into the processes dictating its population biology have quite naturally gravitated toward metapopulation approaches. Through the use of metapopulation theory, valuable insights have been obtained about the processes at play that have been translated directly into management/conservation considerations, especially where the use of spatial closures such as marine reserves are concerned.

A study by Morgan et al. (2000) into factors driving the extreme variability in recruitment among subpopulations of urchins in northern California indicated a complicated but consistent relationship between recruitment and coastal circulation patterns. They found highest recruitment levels associated with areas receiving higher levels of onshore flow as a result of relaxations in upwelling. Because of the physics of the flow reversals, this resulted in areas to the north of major promontories typically having higher recruitment than those to the south. These processes indicate the possibility for retention of larvae on the scale of embayments despite strong offshore, equatorward mean advection, and in so doing can strongly influence the scale of dispersal and connectivity between subpopulations.

Botsford et al. (1994) summarized several modeling studies on the metapopulation dynamics of the red sea urchin with the goal of highlighting the ways in which physical oceanographic conditions can influence the stability, synchrony, and even the persistence of this metapopulation. They demonstrated that even in absence of any environmental forcing, the metapopulation structure itself could lead to fluctuations in local abundance, the frequency of which depends primarily on the rate of adult survival such that variations occur more often at low levels of survival. They found that the inclusion of deterministic larval dispersal among subpopulations would lead to asynchrony in these fluctuations, but that random environmental variability could serve to synchronize the system as long as it was on the scale of the whole coast. This effect was mediated by virtue of the influence of coastal circulation on larval transport time and, most specifically, competency period. Laboratory studies of the physiological response of larvae to different temperature and salinity regimes indicated that the observed characteristics and trends of the coastal circulation patterns could cause considerable variability in the duration of larval development, dramatically affecting mortality and thus connectivity within the system. Using three-dimensional circulation models with vertically migrating larvae, Botsford et al. (1994) found that natural variability in temperature, wind forcing, and the timing of the spring transition were able to explain adequately the observed variability in recruitment levels.

Botsford et al. (1994) point out that events such as ENSO, which occur on multiyear to decadal scales, can have dramatic effects on not only the coastal circulation patterns, but also the temperature and salinity of the coastal waters. This can result in alterations to the competency periods of larvae, which will in turn affect dispersal distances and general connectivity within the system. This can be especially important with regard to the establishment of refuges; simulations predict metapopulation extinction when the distance between refuges is greater than the dispersal distance. In addition, the accumulation of warm water along the U.S. West Coast that is associated with ENSO events can result in the destruction of the primary food base for urchins—giant kelp (*Macrocystis pyrifera*)—which will dramatically affect adult survival. As indicated earlier, because of the metapopulation structure of the system, reduction in survival of adults will increase the frequency of fluctuations from generational to more like annual timescales. Such changes, when not expected or understood, can wreak havoc on the management framework.

Additional studies have used the metapopulation approach to evaluate the usefulness of the mandated reserves for providing fisheries benefits from this system. Quinn et al. (1993) developed a model that included 24 isolated subpopulations distributed along the coast where larvae exchange occurred only with the immediate neighbor. This dispersal pattern would be a simple representation of the results of other studies indicating that dispersal distance can typically be on the scale of individual embayments with only small amounts of leakage to neighboring areas (Botsford et al., 1998). Their results indicated that when harvest rates are low, marine reserves will only reduce them further as they take away fishing grounds without providing any real benefit, because populations are very near carrying capacity anyway. However, at higher rates of harvest, reserves can increase the harvest rate but the spacing of those reserves becomes critical. If reserves are too far apart, not only are harvest rates low, but in many cases the population can actually go extinct. What exactly the minimum size will be relative to the harvest rate will depend on the specifics of dispersal as determined by the circulation pattern (Botsford et al., 1998).

The effect of reserves on metapopulation sustainability was shown by Botsford et al. (2001) to depend critically upon the fraction of natural larval settlement (FNLS) that remained after reserves were put in place. That is, the amount of larval settlement occurring after reserve establishment compared with the case in which no reserves had been established. They demonstrated that the FNLS depended critically on the reserve configuration in concert with dispersal distances. They point out that to maintain a value of FNLS adequate for the sustainability of many species will require either a very large proportion of coastline to be protected (>35%) or that individual reserves be larger than the dispersal distances of the resident species of interest. Additionally, because the FNLS is higher for organisms with shorter dispersal distances, it is possible that establishment of reserves will have effects on community composition (because spacing may favor persistence for some species more than others) or may even lead to genetic selection for shorter dispersal distances. These are all important and heretofore unrecognized considerations of protection of species such as red sea urchin through the use of marine reserves.

In a second modeling study focused specifically on evaluating whether reserves for red sea urchin in northern California would increase catch rates compared with the unregulated scenario, Botsford et al. (1999) identified the important role that uncertainty in model parameters could play. Specifically they identified a key parameter describing recruitment, the slope of the relationship between the number of larvae produced and those successfully settling as juveniles, and showed that equilibrium catch levels were very sensitive to it, where its value was near the origin (at low levels of larvae produced). When this value was high (steep slope), recruitment was strong and therefore relatively unaffected by harvest. In this case, reserves only served to diminish catch because they removed fishing opportunities. However, when this value was low (shallow slope), the establishment of reserves did increase the overall equilibrium catch.

Botsford et al. (1999) took this information and used various sources of empirical data to estimate the likelihood of different values of this key parameter. With this, they were able to perform a decision analysis to determine the most likely value for equilibrium catch given the determined distribution for the key parameter across different levels of reserve establishment (percent of area in reserves). They found that depending on whether you looked only at the short-term (50 years) or the long-term (equilibrium) reserves, incorporating 12 to 16% of the area gave the highest returns, which were in the range of 12 to 18% increases compared with the case of no reserve establishment. This analysis also showed that protection levels on both sides of these values resulted in lower returns that were in some cases actually worse than the case of no reserves. They also showed that there was an initial large drop in catch after the establishment of the reserves, but that this effect could be mitigated with a stepped introduction of reserves over time, although the build back to the long-term equilibrium value would then also be slower.

These studies have pointed out the importance of the metapopulation approach to understanding the processes that determine the distribution and abundance of the red sea urchin off California. They have also been able to explain the dramatic differences in recruitment reliability and subsequent fisheries performance between the northern and southern regions as a function of the large-, meso-, and small-scale circulation patterns that drive dispersal. They have identified the tendency for embayment-scale retention, highlighted how this may set the scale of dispersal, and demonstrated how such realizations affect the use of marine reserves to effectively conserve and manage this metapopulation both in terms of simple persistence as well as for the long-term benefit of the urchin fishery. The spatially explicit stock size information was used in combination with field data on size structure/recruitment and the understanding of the importance of coastal circulation in this system to identify areas that were overfished and others that seemed not to be (Botsford et al., 1998). Such spatially explicit information is invaluable to management and conservation and is currently being used in the marine reserve implementation process. None of these revelations would have become apparent without first treating the system as a metapopulation and seeking to understand both local subpopulation dynamics as well as the large-scale, physically driven connectivity between them.

D. LOGGERHEAD SEA TURTLES (*CARETTA CARETTA*)

Loggerhead sea turtles are considered threatened under the U.S. Endangered Species Act and their population demographics reflect the effects of nesting habitat loss, terrestrial and aquatic habitat degradation, as well as direct hunting and nontarget bycatch in several fisheries ranging over whole ocean basins. Specifically, beach modification and development have resulted in loss of nesting habitat, compromised incubation environments, and reduced hatchling production. These threats continue, but new threats include increases in egg incubation temperatures resulting from global warming (further skewing sex ratios that are influenced by incubation temperature) and loss of nesting habitat to rising sea levels on developed and armored beaches. In the marine system, accumulation of pollutants such as plastics, heavy metals, environmental estrogens, and oil products in pelagic nursery and demersal coastal habitats and feeding grounds as well as rapid degradation of marine and estuarine environments by nutrient runoff have the potential to reduce loggerhead populations further.

The natural history of the loggerhead sea turtle in the North Atlantic is reasonably well understood (Heppell et al., 2003). Many of its spatially distinct habitats are adjacent to rapidly growing human populations. Hence, we can often identify the multiple stressors that impact the turtles in different geographic regions. But these geographical as well as temporal effects (different size classes are in different habitats and have different susceptibilities to perturbations in their environment) have not been integrated into current population models used by managers. The most recent models of loggerhead population structure (Heppell et al., 2003) are still only incremental improvements on the stage-based model of Crouse et al. (1987).

Newly available data on spatial population structure suggests an approach emphasizing metapopulation structure, and current concerns about population size and reproduction in these spatial subpopulations suggest the need to consider sex ratio explicitly in future models. The North Atlantic loggerheads that nest in the southeastern United States have recently been documented to include at least five genetically distinct loggerhead nesting "subpopulations" (Encalada et al., 1998). These are described as the northern nesting subpopulation (turtles nesting from North Carolina southward to northeastern Florida), the south Florida nesting subpopulation (nests from approximately 29° N Lat. from the East Coast southward, through the Keys and northward to approximately the same latitude, near Sarasota), the Florida Panhandle subpopulation (perhaps 1000 nests/year documented at the Elgin Air Force Base and along the beaches of Panama City) and the Dry Tortugas population. The fifth population is the Yucatán subpopulation found nesting primarily on the Caribbean side of the Yucatán peninsula.

Unlike reef fishes, spiny lobsters, and red sea urchins, sea turtle dispersal occurs largely in the adult stage. Loggerhead hatchlings leave their nesting beaches after about 2 months of incubation and swim to the Gulf Stream, where they are entrained into the North Atlantic gyre. They spend about a decade at sea, after which they return to benthic foraging habitats in the U.S. coastal zone. Mature females develop distinct migratory routes between feeding grounds and nesting grounds (Polovina et al., 2000; Limpus and Limpus, 2003; Polovina et al., 2004). Male loggerheads probably also exploit predictable migratory routes, but wander among subpopulations more than females (much like green turtles and leatherbacks; Karl et al., 1992; P. Dutton, National Marine Fisheries Service (NMFS), La Jolla, California, unpublished). Current research on the nuclear DNA from these subpopulations suggests males move among nesting subpopulations. If males wander in sufficient numbers among nesting subpopulations, the population as a whole appears more robust, but the question of skewed sex ratios remains, and new two-sex, spatially explicit models will be necessary to address these issues. Particularly, these subpopulations differ dramatically in size and the numbers of clutches per season (fewer in the more northern latitudes), but differ little in clutch size. Hence, if hatching success and recruitment into the pelagic occurs in the same proportions across all subpopulations, each will make dramatically different contributions to the population as a whole.

Thus in the case of loggerheads, we are dealing with subpopulations defined primarily by breeding and to some extent foraging habitat areas with dispersal between subpopulations occurring via the wandering of adults in a manner that is most likely highly gender dependent. Latitudinal differences in reproductive output as well as sex ratios will lead to complicated metapopulation-level dynamics with potentially nonintuitive cause-and-effect relationships at the local level. The incorporation of such dynamics into conservation considerations will be critical to their success and therefore also to the long-term viability of the loggerhead sea turtle.

IV. SUMMARY

Many marine populations of concern to conservation biologists appear to have metapopulation structure driven either by larval dispersal among habitat patches or by juvenile dispersal and adult migration. To develop appropriate models to advise management, we need to model the entire life history of the species in question and to do so in the spatially explicit manner suggested by metapopulation theory. For those with larval transport it will be necessary to couple more mechanistically environmental variation and ocean physics to patterns of larval transport by using dispersal models to characterize the patterns of connectivity. Of key concern here is to establish the levels of variability in system connectivity and evaluate carefully how such variability interacts with local-scale demographic patterns to dictate both sub- and metapopulation-level dynamics. The reef fish and spiny lobster case studies have both pointed out the importance of considering both habitat and connectivity when seeking to understand local and metapopulation dynamics. We also must determine the relevant scale of the metapopulation—that at which it can be assumed to be effectively or truly closed. Metapopulation boundaries may be obvious and unchanging, as in those of Exuma Sound, Bahamas, for spiny lobster; or much more fluid and potentially variable, as would be the case for the same species in the Florida Keys.

For populations of species more controlled by migrations of juveniles and adults, like sea turtles, we need to understand better the migratory pathways and the risks these species encounter along their route. New models will need to be based on recently determined population structure in which females tend to return to nest in their region of origin, but males may wander among these subpopulations. The understanding that comes from this modeling will greatly aid our abilities to target conservation efforts in a more spatial and temporal manner in exactly the same way that the first stage-structured models allowed us to focus on the most sensitive segments of loggerhead life history.

It is no coincidence that the case studies presented here point to the same conclusion echoed in the ongoing evolution of place-based approaches to management of these systems—that is, the need to consider the dynamics of these systems at a larger scale for the sake of understanding them at the smaller scale. Management alternatives such as time–area closures, MPAs, and fully protected marine reserves have been proposed in reaction to this perceived need and, as the biology of the organisms themselves has shown, it is well justified. Although the biology has remained constant, advances in technology have given us the ability to look beyond the scale of the "open" local population and all the limitations of that demographic category. Through the application of metapopulation theory to marine systems, conservation biology is gaining a valuable tool in the effort to manage and conserve our dwindling marine resources.

REFERENCES

Allison, G.W., Gaines, S.D., Lubchenco, J., and Possingham, H.P. (2003). Enduring persistence of marine reserves: Catastrophes require adopting an insurance factor. *Ecological Applications.* 13, S8–S24.

Andrew, N.L., Agatsuma, Y., Ballesteros, E., Bazhin, A.G., Creaser, E.P., Barnes, D.K.A., Botsford, L.W., Bradbury, A., Campbell, A., Dixon, J.D., Einarsson, S., Gerring, P.K., Hebert, K., Hunter, M., Hur,

S.B., Johnson, C.R., Juinio-Menez, M.A., Kalvass, P., Miller, R.J., Moreno, C.A., Palleiro, J.S., Rivas, D., Robinson, S.M.L., Schroeter, S.C., Steneck, R.S., Vadas, R.S., Woodby, D.A., and Xiaoqi, Z. (2002). Status and management of world sea urchin fisheries. *Oceanography and Marine Biology.* **40**, 343–425.

Baum, J.K., and Myers, R.A. (2004). Shifting baselines and the decline of the pelagic sharks in the Gulf of Mexico. *Ecology Letters.* **7**, 135–145.

Baum, J.K., Myers, R.A., Kehler, D.G., Worm, B., Harley, S.J., and Doherty, P.A. (2003). Collapse and conservation of shark populations in the northwest Atlantic. *Science.* **299**, 389–392.

Botsford, L.W. (2001). Physical influences on recruitment to California Current invertebrate populations on multiple scales. *ICES Journal of Marine Science.* **58**, 1081–1091.

Botsford, L.W., Micheli, F., and Hastings, A. (2003). Principles for the design of marine reserves. *Ecological Applications.* **13**, S25–S31.

Botsford, L.W., Moloney, C.L., Hastings, A., Largier, J.L., Powell, T.M., Higgins, K., and Quinn, J.F. (1994). The influence of spatially and temporally varying oceanographic conditions on meroplanktonic metapopulations. *Deep Sea Research.* **41**, 107–145.

Botsford, L.W., Morgan, L.E., Lockwood, D.R., and Wilen, J.E. (1999). Marine reserves and management of the northern California red sea urchin fishery. *California Cooperative Oceanic Fisheries Investigations Reports.* **40**, 87–93.

Botsford, L.W., Wing, S.R., and Largier, J.L. (1998). Population dynamics and management implications of larval dispersal. *South African Journal of Marine Science.* **19**, 131–142.

Casey, J.M., and Myers, R.A. (1998). Near extinction of a large, widely distributed fish. *Science.* **281**, 690–692.

Cowen, R.K., Lwiza, K.M.M., Sponaugle, S., Paris, C.B., and Olson, D.B. (2000). Connectivity of marine populations: Open or closed? *Science.* **287**, 857–859.

Crouse, D.T., Crowder, L.B., and Caswell, H. (1987). A stage based population model for loggerhead sea turtles and implications for conservation. *Ecology.* **68**, 1412–1423.

Crowder, L.B., Lyman, S.J., Figueira, W.F., and Priddy, J. (2000). Source–sink population dynamics and the problem of siting marine reserves. *Bulletin of Marine Science.* **66**, 799–820.

Dayton, P.K., Thrush, S.F., Agardy, T., and Hofman, R.J. (1995). Environmental effects of marine fishing. *Aquatic Conservation Marine and Freshwater Ecosystems.* **5**, 205–232.

Doherty, P., and Fowler, T. (1994). An empirical test of recruitment limitation in a coral reef fish. *Science.* **263**, 935–939.

Encalada, S.E., Bjorndal, K.A., Bolten, A.B., Zurita, J.C., Schroeder, B., Possardt, E., Sears, C.J., and Bowen, B.W. (1998). Population structure of loggerhead turtle (*Caretta caretta*) nesting colonies in the Atlantic and Mediterranean as inferred from mitochondrial DNA control region sequences. *Marine Biology.* **130**, 567–575.

Figueira, W.F. (2002). *Metapopulation dynamics of coral reef fish: Understanding habitat, demography, and connectivity in source–sink systems.* PhD dissertation. Duke University, Durham, NC, USA.

Fogarty, M.J. (1998). Implications of larval dispersal and directed migration in American lobster stocks: Spatial structure and resilience. *Canadian Special Publication of Fisheries and Aquatic Sciences.* **125**, 273–283.

Gulland, J.A., ed. (1988). *Fish population dynamics* (2nd ed.). Chichester: John Wiley and Sons.

Halpern, B. (2002). The impact of marine reserves: Do reserves work and does reserve size matter? *Ecological Applications.* **13**, S117–S137.

Hanski, I. (1994). Patch-occupancy dynamics in fragmented landscapes. *Trends in Ecology and Evolution.* **9**, 131–135.

Hastings, A., and Botsford, L.W. (1999). Equivalence in yield from marine reserves and traditional fisheries management. *Science.* **284**, 1537–1538.

Heppell, S.S., Crowder, L.B., Crouse, D.T., Epperly, S.P., and Frazer, N.B. (2003). Population models of Atlantic loggerheads: Past, present, and future. In *Loggerhead sea turtles* (B.A. Bolten and B.E. Witherington, eds., pp. 225–273). Washington, DC: Smithsonian Institution Press.

Jackson, J.B.C., Kirby, M.X., Berger, W.H., Bjorndal, K.A., Botsford, L.W., Bourque, B.J., Bradbury, R.H., Cooke, R., Erlandson, J., Estes, J.A., Hughes, T.P., Kidwell, S., Lange, C.B., and Warner, R.R. (2001). Historical overfishing and the recent collapse of coastal ecosystems. *Science.* **293**, 629–638.

James, M.K., Armsworth, P.R., Mason, L.B., and Bode, L. (2002). The structure of reef fish metapopulations: Modelling larval dispersal and retention patterns. *Proceedings of the Royal Society of London Series B-Biological Sciences.* **269**, 2079–2086.

Jones, G.P., Milicich, M.J., Emslie, M.J., and Lunow, C. (1999). Self-recruitment in a coral reef fish population. *Nature.* **402**, 802–804.

Kamezaki, N. (2003). Loggerhead turtles nesting in Japan. In *Loggerhead sea turtles* (A.B. Bolten and B.E. Witherington, eds., pp. 210–217). Washington, DC: Smithsonian Institution Press.

Karl, S.A., Bowen, B.W., and Avise, J.C. (1992). Global population genetic structure and male-mediated gene flow in the green turtle (*Chelonia mydas*): RFLP analyses of anonymous nuclear loci. *Genetics.* **131**, 163–173.

Limpus, C.J., and Limpus, D.J. (2003). The loggerhead turtle, *Caretta caretta*, in the equatorial and southern Pacific Ocean: A species in decline. In *Loggerhead sea turtles* (A.B. Bolten and B.E. Witherington, eds., pp. 199–209). Washington, DC: Smithsonian Institution Press.

Lipcius, R.N., Crowder, L.B, and Morgan, L.E. (2005). Metapopulation structure and the enhancement of exploited species by marine reserves. In *Marine conservation biology: The science of maintaining the sea's biodiversity* (E.A. Norse and L.B. Crowder, eds., pp. 328–349). Washington, DC: Island Press.

Lipcius, R.N., Stockhausen, W.T., and Eggleston, D.B. (2001). Marine reserves for Caribbean spiny lobster: Empirical evaluation and theoretical metapopulation recruitment dynamics. *Marine and Freshwater Research.* **52**, 1589–1598.

Lipcius, R.N., Stockhausen, W.T., Eggleston, D.B., Marshall, Jr., L.S., and Hickey, B. (1997). Hydrodynamic decoupling of recruitment, habitat quality and adult abundance in the Caribbean spiny lobster: Source–sink dynamics? *Marine and Freshwater Research.* **48**, 807–815.

Lubchenco, J., Palumbi, S.R., Gaines, S.D., and Andelman, S. (2003). Plugging a hole in the ocean: The emerging science of marine reserves. *Ecological Applications.* **13**, S3–S7.

Menge, B.A., and Olson, A.M. (1990). Role of scale and environmental factors in regulation of community structure. *Trends in Ecology and Evolution.* **5(2)**, 52–57.

Mora, C., and Sale, P.F. (2002). Are populations of coral reef fish open or closed? *Trends in Ecology and Evolution.* **17**, 422–428.

Morgan, L.E., and Botsford, L.W. (2001). Managing with reserves: Modeling uncertainty in larval dispersal for a sea urchin fishery. In *Proceedings of the Symposium on Spatial Processes and Management of Marine Populations.* University of Alaska Sea Grant, Fairbanks, AK, pp. 667–684.

Morgan, L.E., Wing, S.R, Botsford, L.W, Lundquist, C.J., and Diehl, J.M. (2000). Spatial variability in red sea urchin (*Strongylocentrotus franciscanus*) recruitment in northern California. *Fisheries Oceanography.* **9**, 83–98.

Myers, R.A., and Worm, B. (2003). Rapid worldwide depletion of predatory fish communities. *Nature.* **423**, 280–283.

Myers, R.A., and Worm, B. (2004). Extinction, survival, or recovery of large predatory fishes. *Proceedings of the Royal Society of London Series B Biological Sciences.* **360(1453)**, 13–20.

Norse, E.A., Crowder, L.B., Gjerde, K., Hyrenbach, D., Roberts, C.M., Safina, C., and Soule, M.E. (2005). Place-based ecosystem management in the open ocean. In *Marine conservation biology: The science of maintaining the sea's biodiversity* (E.A. Norse and L.B. Crowder, eds., pp. 302–327). Washington, DC: Island Press.

Palumbi, S.R. (2003). Population genetics, demographic connectivity, and the design of marine reserves. *Ecological Applications.* **13**, S146–S158.

Pauly, D. (1995). Anecdotes and the shifting baseline syndrome in fisheries. *Trends in Ecology and Evolution.* **10**, 430.

Pauly, D., Christensen, A., Dalsgaard, J., Froese, R., and Torres, F.J. (1998). Fishing down marine food webs. *Science.* **279**, 860–863.
Pauly, D., and Maclean, J. (2003). *In a perfect ocean: The state of fisheries and ecosystems in the North Atlantic Ocean.* Washington, DC: Island Press.
Pew Oceans Commission. (2003). *America's living oceans: Charting a course for sea change.* Arlington, VA: Pew Oceans Commission.
Polovina, J.J., Balazs, G.H., Howell, E.A., Parker, D.M., Seki, M.P., and Dutton, P.H. (2004). Forage and migration habitat of loggerhead (*Caretta caretta*) and olive Ridley (*Lepidochelys olivacea*) sea turtles in the central North Pacific Ocean. *Fisheries Oceanography.* **13**, 36–51.
Polovina, J.J., Kobayashi, D.R., Parker, D.M., Seki, M.P., and Balazs, G.H. (2000). Turtles on the edge: Movement of loggerhead turtles (*Caretta caretta*) along oceanic fronts spanning longline fishing grounds in the central North Pacific, 1997–1998. *Fisheries Oceanography.* **9**, 1–13.
Pulliam, H.R. (1988). Sources, sinks, and population regulation. *American Naturalist.* **132**, 652–661.
Pulliam, H.R., and Danielson, B.J. (1991). Sources, sinks and habitat selection: A landscape perspective on population dynamics. *American Naturalist.* **137**, S50–S66.
Pulliam, H.R., Dunning, J.B.J., and Liu, J. (1992). Population dynamics in complex landscapes: A case study. *Ecological Applications.* **2**, 165–177.
Quinn, J.F., Wing, S.R., and Botsford, L.W. (1993). Harvest refugia in marine invertebrate fisheries: Models and applications to the Red Sea urchin, *Strongylocentrotus franciscanus. American Zoologist.* **33**, 537–550.
Roberts, C.M. (1997). Connectivity and management of Caribbean coral reefs. *Science.* **278**, 1454–1457.
Roberts, C.M. (1998). Sources, sinks, and the design of marine reserve networks. *Fisheries.* **23**, 16–19.
Roberts, C.M., Bohnsack, J.A., Gell, F., Hawkins, J.P., and Goodridge, R. (2001). Effects of marine reserves on adjacent fisheries. *Science.* **294**, 1920–1923.
Roberts, C.M., and Hawkins, J.P. (2000). *Fully-protected marine reserves: A guide.* Washington, DC: WWF Endangered Seas Campaign, Washington, DC (USA), and Environment Department, University of York, York (UK).
Roberts, C.M., and Polunin, N.V.C. (1991). Are marine reserves effective in management of reef fisheries? *Reviews in Fish Biology and Fisheries.* **1**, 65–91.
Roberts, C.M., and Polunin, N.V.C. (1993). Marine reserves: Simple solutions to managing complex fisheries? *Ambio.* **22**, 363–368.
Roughgarden, J., Gaines, S., and Possingham, H. (1988). Recruitment dynamics in complex life cycles. *Science.* **241**, 1460–1466.
Roughgarden, J., Iwasa, Y., and Baxter, C. (1985). Demographic theory for an open marine population with space-limited recruitment. *Ecology.* **66**, 54–67.
Russ, G.R., and Alcala, A.C. (1996). Do marine reserves export adult fish biomass? Evidence from Apo Island, Central Philippines. *Marine Ecology Progress Series.* **132**, 1–9.
Russ, G.R., and Alcala, A.C. (2004). Marine reserves: Rates and patterns of recovery and decline of predatory fish, 1983–2000. *Ecological Applications.* **13**(6), 1553–1565.
Sladek Nowlis, J.S., and Roberts, C.M. (1999). Fisheries benefits and optimal design of marine reserves. *Fishery Bulletin.* **97**, 604–616.
Spotila, J.R., Reina, R.R., Steyermark, A.C., Plotkin, A.C., and Paladino, F.V. (2000). Pacific leatherback turtles face extinction. *Nature.* **405**, 529–530.
Stockhausen, W.T., Lipcius, R.N., and Hickey, B.M. (2000). Joint effects of larval dispersal, population regulation, marine reserve design, and exploitation on production and recruitment in the Caribbean spiny lobster. *Bulletin of Marine Science.* **66**, 957–990.
Swearer, S.E., Shima, J.S., Hellberg, M.E., Thorrold, S.R., Jones, G.P., Robertson, D.R., Morgan, S.G., Selkoe, K.A., Ruiz, G.M., and Warner, R.R. (2002). Evidence of self-recruitment in demersal marine populations. *Bulletin of Marine Science.* **70**, 251–271.

Chapter 15 Metapopulation Ecology and Marine Conservation

Tegner, M.J., and Dayton, P.K. (1977). Sea urchin recruitment patterns and implications of commercial fishing. *Science*. **196**, 324–326.

Tuck, G.N., and Possingham, H.P. (2000). Marine protected areas for spatially structured exploited stocks. *Marine Ecology Progress Series*. **192**, 89–101.

United Nations. (2002). *World Summit on Sustainable Development: Plan of implementation*. New York: United Nations. Available at www.johannesburgsummit.org/html/documents/summit_docs/2309_planfinal.htm. Accessed 24 Sept. 2004.

U.S. Commission on Ocean Policy. (2004). *Preliminary report of the U.S. Commission on Ocean Policy governor's draft*. Washington, DC: U.S. Commission on Ocean Policy.

Watling, L., and Norse, E.A. (1998). Disturbance of the seabed by mobile fishing gear: A comparison to forest clearcutting. *Conservation Biology*. **12**, 1180–1197.

Yeung, C., and Lee, T.N. (2002). Larval transport and retention of the spiny lobster (*Panulirus argus*), in the coastal zone of the Florida Keys, USA. *Fisheries Oceanography*. **11**, 286–309.

CHAPTER **16**

The Future of Metapopulation Science in Marine Ecology

JACOB P. KRITZER and PETER F. SALE

I. Introduction
II. The Amphibiousness of Metapopulation Theory
   A. Commonalities between Marine and Terrestrial Ecology
   B. Differences between Marine and Terrestrial Systems
III. Where in the Sea is Metapopulation Theory Less Relevant?
   A. Highly Mobile Species
   B. Highly Isolated Populations
   C. Widely Dispersing Species
   D. The Special Case of Clonal Organisms
IV. Outstanding Research Questions
   A. Knowledge of Dispersal Pathways and Mechanisms
   B. Knowledge of Rates and Extents of Connectivity
   C. Marine Metacommunities
V. Spatially Explicit Management of Marine Fishery Resources
VI. Summary
   References

I. INTRODUCTION

The preceding pages have provided ample evidence that researchers are finding metapopulation approaches useful in their studies of marine populations. Is there evidence that metapopulation approaches will prove generally valuable, and should be generally recommended for ecological study of all marine populations? Can we anticipate that the rate of use of metapopulation approaches in marine ecology will continue to rise dramatically? More important, is there evidence that marine ecological studies are contributing in novel ways to the development of metapopulation science?

Metapopulation theory is most appropriate for organisms that are distributed as small populations scattered over a landscape, with some opportunities for connection through dispersal. How broadly does this describe marine populations? Chapters in this book demonstrate that such situations appear to be common for demersal, benthic, or sessile marine organisms in a broad variety of environments, particularly ones that contain topographic structures at scales that are appropriate for the organisms being studied. Coral reefs (Kritzer and Sale; Mumby and Dytham, this volume) are the epitome of spatially heterogeneous, patchy environments that support numerous small populations of the organisms that inhabit them. Rocky reefs (Gunderson and Vetter; Morgan and Shepherd, this volume) are comparable, although the degree of patchiness, particularly at the smallest scales, may be somewhat less. Hydrothermal vents (Neubert et al., this volume) are well isolated and ephemeral, and their populations must prevail by means of extensive dispersal, even if they do not function as units within metapopulations. Other substrata, such as the rocky intertidal (Johnson, this volume) or sea grass beds (Bell, this volume), provide less obviously patchy environments, yet even in these there is evidence that populations are patchily distributed, and metapopulation dynamics may well be involved.

In this chapter we first revisit the question of whether marine and terrestrial systems have such fundamentally different characteristics as to make metapopulation theory useful only on land, or at least to require drastically different constructs in the concepts and models (i.e., how amphibious is metapopulation ecology?). Not surprisingly (as evidenced by this book, and Kritzer and Sale, 2004), we conclude that metapopulation theory can be applied similarly to marine and terrestrial systems, with some structural differences, although not fundamental differences. But there are marine systems that possibly should not be seen as metapopulations, and we follow by reviewing those cases. Next, we provide a nonexhaustive list of key research topics that need to be addressed to build more fully an understanding of metapopulation structure and dynamics in the sea. We then consider the interaction between marine metapopulation ecology and management of living marine resources, before leaving with some brief concluding remarks.

II. THE AMPHIBIOUSNESS OF METAPOPULATION THEORY

A. Commonalities between Marine and Terrestrial Ecology

We return briefly to a question we addressed in Chapter 1, which is whether marine systems are sufficiently different from the terrestrial systems that provided the context for the development of metapopulation theory to call for an entirely separate set of ideas and approaches. And we reiterate our arguments from that introductory chapter.

First and foremost, the metapopulation concept can be used in several important ways that are equally applicable to marine and terrestrial systems. Even if the particular processes and dynamics being studied are somehow fundamentally different, application of the concept can help describe spatial processes and structures, provide a framework for asking research questions, and serve as a paradigm for conservation and population biology.

Second, there are certainly differences between marine and terrestrial systems in terms of, for example, the magnitude of dispersal rates (very high in the sea and lower on land) or the nature of the predominant anthropogenic impacts (overharvest at sea vs. habitat destruction on land). However, this does not mean that common ideas and a common concept cannot be used to understand both. After all, all populations are structured by the same four governing processes: birth, immigration, death, and emigration. No other processes can directly affect abundance, and those intraspecific processes such as growth, and interspecific processes such as competition, that do play roles operate through their effects on birth, immigration, death, and emigration. Still, while all population models will need to account for these four processes, specific models and questions can be expected to focus on different aspects of the particular system studied, and use very different parameter values, as dictated by the unique ecology of that system. In this way, specific studies may generate different types of insights. For example, models of fish populations will need to account for individual body size, because of its significant effects on fecundity (i.e., births), and protogynous fish species that are being exploited in a size-selected way have particular problems that are not nearly as significant for population models of most large mammal species.

Furthermore, the marine–terrestrial division is not a perfect dichotomy for separating taxa. Marine mammals have more in common with their terrestrial counterparts than they do with most fishes and marine invertebrates in terms of behavior, reproductive rates, and demography. Small, short-lived crustaceans (e.g., amphipods) are likewise probably more ecologically similar to terrestrial insects than they are to larger, longer lived crustaceans (e.g., lobsters). We might do well to think about types of metapopulations in terms of life history categories rather than whether the organisms live on land or in the sea.

B. Differences between Marine and Terrestrial Systems

There are some key differences between many marine and terrestrial metapopulations that might have important consequences for the way we construct models and hypotheses. One of these we touch on in Chapter 1 is more comprehensively addressed in this volume by Cynthia Jones in Chapter 4 on estuarine and diadromous fishes. Most marine metapopulations have evolved with a naturally patchy structure and reliance upon demographic connectivity among subpopulations ("naturally evolved" metapopulations, in Jones' terminology). In contrast, many terrestrial metapopulations appear to have been created by habitat loss and fragmentation, with demographic linkages a consequence of this fragmentation ("fragmentation-induced" metapopulations). Ecological characteristics such as stability and resilience (see Holling, 1973) might be fundamentally different for a system that evolved to be patchy and interconnected as opposed to one that has been recently restructured as such.

A second difference between many marine and terrestrial species is the nature of the dispersive stage that provides the demographic connectivity needed for metapopulation structure. On land, individuals dispersing among populations may be no different from those resident in habitat patches (e.g., small mammals; Smith and Gilpin, 1997), or they might have a unique form especially for the dispersal process (e.g., seeds designed to attach to passing animals or be carried on air currents; Murphy and Lovett–Doust, 2004). In either case, however, the purpose of the dispersive stage is to get from point A to point B. Although this is also the case for some marine taxa (e.g., sea grass spores; see Bell, this volume), it is not necessarily the case for many fishes and invertebrates. For these taxa, the pelagic larval stage is not only a dispersive stage, it is also an important developmental stage. Although dispersing individuals are developmentally static in most terrestrial species and certain marine species, most dispersing marine invertebrate and all fish larvae undergo profound changes in size, morphology, physiology, and behavior before recruiting to local populations.

Why is this important? A significant body of theory that is closely related to metapopulation theory addresses the evolution of dispersal, again with an implicit eye to terrestrial systems. This theory seeks to explain either mean dispersal rates/distances or the distribution of dispersal rates/distances in terms of the relative tradeoff between the risk of local extinction and the risk of leaving the natal patch (see, e.g., Poethke et al., 2003). Clearly, there is a risk in leaving a habitat patch that is known to be suitable in order to venture through inhospitable habitat with no certainty of ever reaching another suitable patch. However, at least some genetic predisposition to taking this risk among an individual's offspring might be worthwhile if there is a high probability of local extinction. Progeny (and therefore genes) will then be spread among multiple populations, hedging one's bets against any one population going extinct. There can also be incentives to disperse if a habitat patch is too crowded and resources are too hard to come by, resulting in density-dependent changes in demographic rates (Cadet et al., 2003). Again, there will be an evolutionary cost–benefit assessment, so to speak, of the costs of dispersing relative to the benefits of moving to a new population. By examining these tradeoffs, we can not only explain how various dispersal patterns have come to be, we can also glean important insights into the nature of the population processes that affect metapopulation dynamics.

This analysis of costs and benefits might be less useful for species with pelagic larvae if dispersal patterns have arisen not through selective pressure to spread offspring among populations but rather as an accidental outcome of particular developmental processes. If the length of the pelagic stage is determined by developmental needs, mediated by developmental constraints, species might have less than optimal control over their ability to determine ecologically optimal dispersal rates through evolutionary time. In other words, for many marine species, dispersal among local populations may happen because the organism is unable to avoid it, rather than because that species has evolved an optimal solution to a cost–benefit challenge. Of course, we know that larvae must eventually reach a habitat patch if they are to succeed, and that the relative advantages of different dispersal rates, as determined by the relative risks of staying close to home versus traveling farther afield, are likely to have shaped the evolution of marine larvae. For example, they might have evolved behaviors that promote local retention if there is a benefit in recruitment to the natal reef, but developmental needs that require a certain time in the pelagic realm. Still, the potential conflict between ecological needs and developmental needs might result in dispersal patterns that seem less than optimal when analyzed only according to the traditional components of dispersal theory. Doherty et al. (1985) addressed these issues implicitly in their model of dispersal success under different larval feeding strategies, and Strathmann et al. (2002) provide the first substantive discussion along these lines directed at marine organisms.

III. WHERE IN THE SEA IS METAPOPULATION THEORY LESS RELEVANT?

A. HIGHLY MOBILE SPECIES

Species for which the normal scale of movements and the size of home ranges are close to the scale over which interpopulation dispersal would occur are not likely to be strong candidates for a metapopulation approach. These will often be large pelagic species. For example, yellowfin tuna in the eastern Pacific show little genetic structure across the vast area managed by the Inter-America Tropical Tuna Commission, and likely represent a single stock (Uribe–Alcocer, 2003). In contrast, in the Mediterranean Sea, Atlantic bluefin tuna show some genetic differentiation between the eastern and western basins (Carlsson et al., 2004), so a simple two-population metapopulation structure might exist. However, most large pelagic species have sufficiently great movement potential to negate the possibility of metapopulation structure. Large shark species, such as the tiger shark and white shark, can have home ranges or seasonal migratory behaviors that span tens to hundreds of kilometers (Adams et al., 1994; Holland et al., 1999). The spatial structure characteristic of metapopulations will not be evident in such species.

Whales present an interesting case of where a metapopulation approach might be useful for a highly mobile species, although in a very nontraditional way. Movement patterns of most whale species clearly cover very large distances, and it would seem that considerable intermingling would occur, making formation of distinct subpopulations unlikely. However, across their extensive home ranges, most whale species separate into distinct "pods" composed primarily of females and calves in matrilineal groups. These groups, although not occupying distinctly identifiable habitat patches, could still be considered as something akin to subpopulations, although formed by social behavior rather than habitat. Males typically leave groups at maturity to live a solitary life, although there are exceptions when a male is at the center of a social group (Best et al., 2003). These males then join mature females for mating. Young females can also leave their matrilineal groups, groups can split, and groups can merge (Christal et al., 1998). Mating behavior by males and division or merging of groups represent mechanisms for linkages among "subpopulations," or social groups. There is genetic evidence that some degree of substructuring does occur (Lyrholm and Gyllensten, 1998), so there might be rates of exchange sufficient to separate animals into somewhat independent groups. These interactions among groups might then result in something analogous to metapopulation structure.

B. Highly Isolated Populations

The metapopulation approach is not suitable for populations with too little demographic connectivity among them. These need not necessarily be isolated geographically (i.e., with long distances separating populations), but rather can be isolated by behavior, hydrodynamics, or other mechanisms. To remain components of a single species, these populations will need to experience some exchange of individuals. But small amounts of dispersal, although evolutionarily significant, will not affect population or community dynamics over ecological timescales. Sinclair (1988) argues that some fish species have evolved strategies designed to preserve a high degree of integrity of local populations, with only very occasional emigration of "vagrants" into new populations. Species with so few vagrants that local population dynamics are unaffected should not be viewed as metapopulations.

Deep-sea corals require very specific conditions, typically found in deep-sea canyons or on seamounts occurring only occasionally across the vast area of the open ocean, and are often very poorly connected by larval dispersal (Koslow et al., 2000). These often long-lived, slow-growing, and highly sensitive species are also likely not structured as metapopulations.

Although the metapopulation concept has sometimes been used to examine apparently evolutionary questions, these tend to intersect closely with ecological questions. For example, one of Levins' original papers (Levins, 1970) examined how species persist (an evolutionary question) through local extinctions being offset by recolonization (an ecological question). By and large, a metapopulation framework is used to examine questions in ecology more so than evolution, and the framework will not be useful if the ecology of a given species results in minimal rates of dispersal. Such low rates of dispersal cannot have measurable impacts on local demography.

C. Widely Dispersing Species

Species that have very broad dispersal distances, or species with more modest dispersal patterns but with habitat patches existing in very close proximity, will experience too much interpopulation exchange to form partially independent subpopulations. Instead, local production of offspring will be completely decoupled from local recruitment, and a large, single population with complex internal spatial structure but without internal variation in demography will result. An example of this phenomenon may be the Caribbean spiny lobster, *Panulirus argus*. The larval duration for this species can be 6 to 12 months (Yeung and Lee, 2002). This apparently results in genetic homogeneity across the Caribbean, representative of a well-mixed population (Silberman et al., 1994). An important consequence of this is that management measures may need to be enacted on much larger spatial scales (Cochrane et al., 2004). In contrast, species with shorter dispersal distances, and therefore potentially with metapopulation structure, can be managed on much smaller scales as a result of the high reliance on local reproduction. Larval durations and behavior capabilities during the larval period are two general differences between reef fishes and crustaceans that might make the former generally more amenable to forming metapopulations than the latter; however, in both taxa, there are important interspecific variations in life history that cannot be neglected (reviewed by Sale and Kritzer, 2003). A metapopulation approach is more useful for study of some fish and some crustaceans than for others.

D. The Special Case of Clonal Organisms

A number of marine organisms are clonal, and potentially eternal as individuals. In such cases, the "need" for successful recruitment of new individuals is much reduced, except in cases when disturbances are prevalent and severe. Populations of such organisms may function effectively with very low rates of recruitment, and concomitantly low levels of connectivity among local populations. Whether metapopulation dynamics are important for these species, in an ecological sense and in the absence of severe anthropogenic impacts, is moot.

Corals are a good example of such species, but current trends in coral populations are potentially alarming, and the need is urgent to understand connectivity in these species. Global climate change, acting in consort with the many stresses that humanity imposes on coral populations, is believed responsible for the sudden increase in the occurrence of widespread bleaching events. During such events the corals expel their zooxanthellae (or the algae choose to leave) and the coral colonies are left bleached white in color. In cases where conditions favoring bleaching are not prolonged for too many days, the corals subsequently regain their symbionts and color, but in cases where bleaching events are lengthy, there is substantial mortality of the coral. In some cases, such as during the severe bleaching events of 1998 throughout the Indian Ocean, mortality is very extensive and over very broad areas of reef (e.g., McClanahan et al., 2004; Stobart et al., 2005). Under such circumstances the ability of the corals to recolonize from distant sources becomes paramount to any chance of natural recovery. Although populations of organisms such as corals may need the supplementation that metapopulation dynamics can provide only occasionally, its clear that knowledge of the capabilities of these organisms to disperse among populations can be important.

IV. OUTSTANDING RESEARCH QUESTIONS

A. Knowledge of Dispersal Pathways and Mechanisms

Uncovering the selective pressures determining dispersal patterns will help our understanding of why they came to be (see Section II.B), but our immediate need is to first accumulate far more data on those dispersal patterns irrespective of their origins. Our knowledge of dispersal pathways is limited in marine systems. However, because the great majority of dispersal takes place during larval life, hydrodynamics must play a major role both in defining pathways and in providing the mechanism primarily responsible for dispersal along them. Although knowledge of hydrodynamics is reasonably strong at larger spatial scales, our ability to predict details of hydrodynamics at scales of meters remains limited, particularly for water in close proximity to substrata. For certain species with inactive (kelp) or relatively inactive (corals) dispersive stages, hydrodynamics alone might be sufficient to predict most dispersal pathways. For others (most invertebrates and especially fishes), important sensory and behavioral abilities need to be factored in at some stage, potentially very early. There is considerable room for new research to determine the details of dispersal paths and the mechanisms responsible.

B. Knowledge of Rates and Extents of Connectivity

Dispersal alone does not equate to real demographic connectivity among populations. An individual that has moved from one population to another must then settle, recruit to the spawning population, and successfully reproduce to contribute most significantly to population dynamics (n.b., the individual can still have other trophic or competitive ecological effects even without reproducing). Currently there is very little information on the topic of connectivity in marine systems. Indeed, this is now recognized as a major gap in understanding of marine populations, and efforts are underway to gain this information in specific cases. The reasons for this gap are primarily because of the particular difficulties in tracking minute larval stages in the immense volume of the ocean, and then tracking the fate of individuals from different source populations after settlement. Furthermore, until the late 1980s there was little appreciation of the importance of larval dispersal for the dynamics of otherwise sedentary, demersal, or sessile populations. Yet, rates and extent of connectivity are crucial in determining whether a metapopulation structure actually exists for any case in which a set of neighboring, but separate, populations can be defined. High rates of dispersal and successful recruitment among the local groups can easily ensure homogeneity of demographic rates, converting the local groups into a single, but spatially subdivided, population (see Section III.C). Very low rates of dispersal among local groups, because they are too far apart relative to the dispersal distances of the larvae, will convert them into separate, isolated populations that simply do not interact in a demographically interesting manner, although they may still be interconnected genetically, and these interconnections may be important for their evolution (see Kritzer and Sale, 2004, for related discussion).

Recent advances in our ability to recognize origins of individual larvae or newly settled/recruited juveniles suggest that the next few years will be a time in which some solid baseline data on rates and extents of dispersal will be established. These baseline data will provide a perspective on the extent to which metapopulation concepts will be important in marine ecology.

C. Marine Metacommunities

One area in which marine ecologists may lead their terrestrial counterparts concerns the interplay between metapopulation dynamics and community ecology (indeed, Chapters 5, 9, and 14 herein represent major additions in this area). Marine systems have provided important insights into the processes structuring communities, arguably more so than they have into the processes structuring populations. This understanding was, for a long time, focused on within-patch dynamics; however, ideas have been developed that bring metapopulation processes into our perspective on community dynamics, either implicitly (e.g., Sale, 1991) or explicitly (Iwasa and Roughgarden, 1986). This perspective recognizes that having multiple sources of new recruits, many of which are located some distance away, will mean that continually varying numbers of a given species are being added to a given local population. One effect of this variability on local community structure can be to swamp the effects of local competitive interactions. In other words, local competition might occur, but it can be very difficult to predict precisely what its consequences will be because the numbers of the two or more species entering into competition can fluctuate so widely for reasons unrelated to the competition (Sale, 2004). As we learn more about the dispersal patterns of constituent species within different marine communities, we are likely to discover that different species are connected at different spatial scales. This will mean that a marine metacommunity will be comprised of a diverse mixture of marine metapopulations. See Chapter 14 by Ron Karlson for a much more comprehensive discussion of the interplay between metapopulation and community ecology in marine systems.

V. SPATIALLY EXPLICIT MANAGEMENT OF MARINE FISHERY RESOURCES

There is growing recognition that many marine organisms are distributed as local populations that do not travel far to reproduce, yet can be interconnected by dispersal of (typically) early life stages. As such, it makes sense to manage them in a spatially explicit way, and this has led some to an interest in metapopulation concepts. In studying metapopulation ecology, we will always face the logistic challenges of working on very large spatial scales. However, here science and management can assist one another by using the management process as a large-scale adaptive experiment to help generate the science needed to, in turn, to better inform management (Sale et al., 2005). The increase in the spatial scales of marine ecological research (see, for example, Fig. 2-1 in Chapter 2) coupled with the continued decline in many living marine resources has led many formerly "pure" marine ecologists into the management realm to take advantage of the larger scale at which management operates and to help rectify the problems (Sale, 2002).

Understanding marine metapopulation structure, where it exists, and designing appropriate management structures, will require additional scientific research. Sale et al. (2005) outlined five key research topics that will be needed to more effectively design networks of fishery reserves. Notably, three of these topics—larval dispersal patterns, adult and juvenile movement patterns, and hydrodynamics—all relate closely to interpopulation connectivity, and therefore are highly relevant to understanding metapopulation ecology.

Indeed, fishery reserves and other types of MPAs are management tools commonly linked with metapopulation concepts (see Crowder and Figueira, this volume).

How does marine population structure affect spatially explicit management? If local populations are organized into metapopulations, it will be important to structure management using MPAs of various types to reinforce, rather than to interfere with, that structure. If they are not organized this way, it is possible that suitable application of a spatially explicit management system could generate a metapopulation, possibly with benefits in terms of protecting populations from extinction risk or promoting productivity. Crowder et al. (2000) modeled a simple system that suggested that it made a considerable difference if MPAs were established over sites that were functioning as source populations or over ones that were functioning as sinks. Furthermore, this effect was modified by the overall intensity of fishing on the system.

Species that live in less patchy environments and have large home ranges (Section III.A) or species with dispersal patterns that cover broad distances (Section III.C) will be less amenable to management by protected areas of the sizes typically implemented. Conversely, species with very isolated populations (Section III.B) might seem suited to typical spatial management scales, but in fact protected populations will not be able to bolster unprotected populations. Understanding the spatial scale at which a given species' ecology is enacted is key to understanding the spatial scale at which management must take place and the structure of management units within a given area.

Although there has yet to develop a science of spatially explicit management that takes advantage of metapopulation ideas, the integration of metapopulation theory and spatially explicit management may be the area in which marine ecology will contribute most to the development or extension of the metapopulation paradigm.

VI. SUMMARY

Whether the study of marine systems in the metapopulation context is providing new insights to metapopulation theory at large is somewhat unclear. After all, the merging of marine ecology and metapopulation ecology is a relatively recent phenomenon (as outlined in Sale et al., this volume), and marine scientists are still coming to terms with the concept and its relevance (see Smedbol et al., 2002; Grimm et al., 2003; Kritzer and Sale, 2004; also see chapters herein). Nevertheless, adoption of the concept by marine ecologists has contributed at least in part to the expansion in the predominant spatial scales of marine ecological investigations, and to the increasing consideration and application of spatially explicit management strategies. Drawing upon metapopulation theory should continue to present new questions and guide new marine ecological research. Marine systems can at the very least increase the range of variation on the basic and universal set of population processes (i.e., births, immigration, deaths, emigration) represented in metapopulation studies, and marine ecologists can substantially widen the pool of scientists focused on metapopulation issues. If nothing else, drawing upon this common concept should encourage marine and terrestrial ecologists to learn about one another's work, interact more closely, and move toward more unified ideas in ecology.

REFERENCES

Adams, D.H., Mitchell, M.E., Parsons, G.R. (1994). Seasonal occurrence of the white shark, *Carcharodon carcharias*, in waters off the Florida west coast, with notes on its life history. *Mar. Fish. Rev.* **56**, 24–27.

Best, P.B., Schaeff, C.M., Reeb, D., Palsboell, P.J. (2003). Composition and possible function of social groupings of southern right whales in South African waters. *Behaviour.* **140**, 1469–1494.

Cadet, C., Ferriere, R., Metz, J.A.J., van Baalen, M. (2003). The evolution of dispersal under demographic stochasticity. *Am. Nat.* **162**, 427–441.

Carlsson, J., McDowell, J.R., Diaz-Jaimes, P., Carlsson, J.E.L., Boles, S.B., Gold, J.R., Graves, J.E. (2004). Microsatellite and mitochondrial DNA analyses of Atlantic bluefin tuna (*Thunnus thynnus thynnus*) population structure in the Mediterranean Sea. *Fish. Mol. Ecol.* **13**, 3345–3356.

Christal, J., Whitehead, H., Lettevall, E. (1998). Sperm whale social units: Variation and change. *Can. J. Zool.* **76**, 1431–1440.

Cochrane, K.L., Chakalall, B., Munro, G. (2004). The whole could be greater than the sum of the parts: The potential benefits of cooperative management of the Caribbean spiny lobster. In *Management of shared fish stocks* (A.I.L. Payne, C.M. O'Brien, S.I. Rogers, eds., pp. 223–239). London: Blackwell Publishing.

Crowder, L.B., Lyman, S.L., Figueira, W.F., Priddy, J. (2000). Source–sink population dynamics and the problem of siting marine reserves. *Bull. Mar. Sci.* **66**, 799–820.

Doherty, P.J., Williams, D.McB., and Sale. P.F. (1985). The adaptive significance of larval dispersal in coral reef fishes. *Env. Biol. Fish.* **12**, 81–90.

Grimm, V., Reise, K., Strasser, M. (2003). Marine metapopulations: A useful concept? *Helgoland Marine Research.* **56**, 222–228.

Holland, K.N., Wetherbee, B.M., Lowe, C.G., Meyer, C.G. (1999). Movements of tiger sharks (*Galeocerdo cuvier*) in coastal Hawaiian waters. *Mar. Biol.* **134**, 665–673.

Holling, C.S. (1973). Resilience and stability of ecological systems. *Ann. Rev. Ecol. Syst.* **4**, 1–23.

Iwasa, H., Roughgarden, J. (1986). Interspecific competition among metapopulations with space-limited subpopulations. *Theor. Popul. Biol.* **30**, 194–214.

Koslow, J.A., Boehlert, G.W., Gordon, J.D.M., Haedrich, R.L., Lorance, P., Parin, N. (2000). Continental slope and deep-sea fisheries: Implications for a fragile ecosystem. *ICES Journal of Marine Science.* **57**, 548–557.

Kritzer, J.P., Sale, P.F. (2004). Metapopulation ecology in the sea: From Levins' model to marine ecology and fisheries science. *Fish and Fisheries.* **5**, 131–140.

Levins, R. (1970). Extinction. In *Some mathematical problems in biology* (M. Desternhaber, ed., pp. 77–107). Providence, RI: American Mathematical Society.

Lyrholm, T., Gyllensten, U. (1998). Global matrilineal population structure in sperm whales as indicated by mitochondrial DNA sequences. *Proc. R. Soc. Lond. B.* **265**, 1679–1684.

McClanahan, T.R., Baird, A.H., Marshall, P.A., Toscano, M.A. (2004). Comparing bleaching and mortality responses of hard corals between southern Kenya and the Great Barrier Reef, Australia. *Mar. Pollut. Bull.* **48**, 327–335.

Murphy, H.T., Lovett–Doust, J. (2004). Context and connectivity in plant metapopulations and landscape mosaics: Does the matrix matter? *Oikos.* **105**, 3–14.

Poethke, H.J., Hovestadt, T., Mitesser, O. (2003). Local extinction and the evolution of dispersal rates: Causes and correlations. *Am. Nat.* **161**, 631–640.

Sale, P.F. (1991). Reef fish communities: Open nonequilibrial systems. In *The ecology of fishes on coral reefs* (P.F. Sale, ed., pp. 564–598). San Diego: Academic Press.

Sale, P.F. (2002). The science we need to develop for more effective management. In *Coral reef fishes: Dynamics and diversity in a complex ecosystem* (P.F. Sale, ed., pp. 361–376). San Diego: Academic Press.

Sale, P.F. (2004). Connectivity, recruitment variation, and the structure of reef fish communities. *Intergr. Comp. Biol.* **44**, 390–399.

Sale, P.F., Cowen, R.K., Danilowicz, B.S., Jones, G.P., Kritzer, J.P., Lindeman, K.C., Planes, S., Polunin, N.V., Russ, G.R., Sadovy, Y.J. (2005). Critical science gaps impede use of no-take fishery reserves. *Trends Ecol. Evol.* **20**, 74–80.

Sale, P.F., Kritzer, J.P. (2003). Determining the extent and spatial scale of population connectivity: Decapods and coral reef fishes compared. *Fish. Res.* **65**, 153–172.

Silberman, J.D., Sarver, S.K., Walsh, P.J. (1994). Mitochondrial DNA variation and population structure in the spiny lobster *Panulirus argus*. *Mar. Biol.* **120**, 601–608.

Sinclair, M. (1988). *Marine populations: An essay on population regulation and speciation.* Seattle: University of Washington Press.

Smedbol, R.K., McPherson, A., Hansen, M.M., Kenchington, E. (2002). Myths and moderation in marine "metapopulations?" *Fish and Fisheries.* **3**, 20–35.

Smith, A.T., Gilpin, M.E. (1997). Spatially correlated dynamics in a pika metapopulation. In *Metapopulation biology: Ecology, genetics and evolution* (I.A. Hanski and M.E. Gilpin, eds., pp. 407–428). San Diego: Academic Press.

Stobart, B., Teleki, K., Buckley, R., Downing, N., Callow, M. (2005). Coral recovery at Aldabra Atoll, Seychelles: Five years after the 1998 bleaching event. *Phil. Trans. A.* **363**, 251–255.

Strathmann, R.R., Hughes, T.P., Kuris, A.M., Lindeman, K.C., Morgan, S.G., Pandolfi, J.M., Warner, R.R. (2002). Evolution of local recruitment and its consequences for marine populations. *Bull. Mar. Sci.* **70**, 377–396.

Uribe-Alcocer, M. (2003). Allozyme and RAPD variation in the eastern Pacific yellowfin tuna (*Thunnus albacares*). *Fish. Bull.* **101**, 769–777.

Yeung, C., Lee, T.N. (2002). Larval transport and retention of the spiny lobster, *Panulirus argus*, in the coastal zone of the Florida Keys, USA. *Fish. Oceanogr.* **11**, 286–309.

INDEX

A

abalone
  adult habitat and spatial structure, 223–226
  fishing and management, 228–229
  Haliotis corrugata (pink abalone), 207, 211, 219, 221
  Haliotis cracherodii (black abalone), 207, 211, 219, 221
  Haliotis discus hannai (abalone), 207, 210, 211, 219, 221
  Haliotis diversicolor (abalone), 219, 221
  Haliotis fulgens (green abalone), 208, 211, 219, 221
  Haliotis iris (abalone), 207
  Haliotis kamtschatkana (northern abalone), 207, 210, 219
  Haliotis laevigata (greenlip abalone), 207, 210, 219, 221, 223
  Haliotis midae, 207, 219
  Haliotis roei, 207, 219, 221
  Haliotis rubra (blacklip abalone), 207, 210, 219, 221, 223
  Haliotis rufescens (red abalone), 207, 211, 219, 221
  Haliotis sorenseni (white abalone), 207
  larval dispersal and settlement, 209–212
    hydrodynamics and coastal topography, 211–212
    larval behavior at settlement, 212
    larval ecology and dispersal, 209–211
  life history of, 207–208
  population genetics of, 218–219
Acanthaster planci (crown-of-thorns starfish), 166

Acanthinucella spirata, 442
Acanthurus nigrofuscus (surgeonfish), 50, 56
acorn barnacle (*Balanus glandula*), 99, 298
acoustic Doppler current profilers (ADCPs), 372
Acropora, 36
Acropora cervicornis (staghorn coral), 171
Acropora longicyathus, 172
Acropora palmata (elkhorn coral), 167, 438, 441
Adalaria proxima (nudibranch), 256
Adams, D.J., 137
Adams, S., 46
Adams, T., 212
ADCPs (acoustic Doppler current profilers), 372
Aequipecten opercularis (scallop), 436
AFLPs (amplified fragment length polymorphisms), 437
Agaricia agaricites (lettuce coral), 170, 175
Agaricia humilis (lowrelief saucer coral), 175
algal overgrowth rate, 189–190
Allee effect, 229
Alosa alosa (alice shad), 135
Alosa fallax fallax (twaite shad), 135
Alosa sapidissima (American shad), 135
alosines, 135–137
Alvinella pompejana, 330
amplified fragment length polymorphisms (AFLPs), 437
analysis of variance (ANOVA), 190
Andrew, N.L., 212
Andrewartha, G.H., 3, 124
Anguilla anguilla (American eels), 130

ANOVA (analysis of variance), 190
anthropogenic extinctions, 53
Armsworth, P.R., 48, 418
Arnason, A.N., 129
atherinids, 138–139
Australian spiny lobster (*Panulirus cygnus*), 275
Avise, J.C., 445
Ayre, D.J., 171

B
Baker, J.L., 232
Balanophyllia elegans (solitary coral), 437, 439
Balanus glandula (acorn barnacle), 99, 298
Barents Sea, 303
barnacles
  Balanus glandula (acorn barnacle), 99, 298
  Tesseropera rosea (barnacle), 14–15
Barrett, S.C.H., 388
Barton, N.H., 432
Bascompte, J., 261
Baums, I.B., 441
Bayesian methods, 10
Beacham, T.D., 127
Bell, S.S., 393
Bergstrom, B.I., 302, 303
Bermingham, E., 35
Bernardi, G., 101
Berryman, A.A., 124
Berthelemy-Okazaki, N.J., 291
Beverton, R.J.H., 19, 141
BIDE (birth, immigration, death, and emigration), 462–463
biogeography and diversity
  biogeographic provinces and temperate reef fish communities, 78–79
  biogeological equilibrium, 339
  of deep-sea hydrothermal vents, 333–335
Birch, L.C., 3, 124, 462
Birkeland, C., 175
birth, immigration, death, and emigration (BIDE), 462–463
Black, R., 446
black abalone (*Haliotis cracherodii*), 207, 211, 219, 221
black smokers, 322
blue crab (*Calinectes sapidus*), 274, 285–292
Boehlert, G.W., 87, 88
Bohonak, A.J., 221
Bolden, S., 56
Borrell, Y., 441

Bosscher, H., 174
Botsford, L.W., 18, 230, 274, 308, 416, 418, 420, 424, 506, 507, 508
brachyuran crab, 215, 275–276
Breeman, A.M., 179
Breen, P.A., 223
Briers, R.A., 424
Briggs, J.C., 78
Brown, L.A., 18
Brown, L.D., 219, 232
brown king crab (*Lithodes aquispina*), 308
Bullock, J.M., 389
Burrows, M.T., 261
Burton, R.S., 220, 291
Bythell, J.C., 175

C
Caffey, H.M., 14–15, 17
CalCOFI ichthyoplankton survey, 88
Caley, M.J., 465
California Current system, 83, 214
Calinectes sapidus (blue crab), 274, 285–292
Cameron, R.A., 216
Campbell, M.L., 394
Camus, P.A., 124
Cancer magister (Dungeness crab), 293–300
Capricorn-Bunker group, 53
Carcinus maenas (green crab), 279
Caretta caretta (loggerhead sea turtles), 509–510
Caribbean seagrass (*Thalassia testudinum*), 402
Caribbean systems, 173
Carlson, H.R., 92
Carpenter, R.C., 185
Carr, M.H., 96
Carscadden, J.E., 136
Caulerpa (rhizophytic macroalgae), 392
Chaetodon plebius (butterflyfishes), 53
Charnov, E.L., 301
Chinocetes opilio (snow crab), 307
Chionoecetes bairdi (tanner crab), 308
Chromis (damselfishes), 36
clonal metapopulation dynamics, 463–464
clonal organisms, 523–524
Clupea harengus (Atlantic herring), 141
Coalescent models, 437
coastal decapods, 271–311
  American lobster (*Homarus americanus*), 279–285
  blue crab (*Callinectes sapidus*), 285–292

Index

Dungeness crab (Cancer magister), 293–300
  identifying decapod metapopulations, 277–279
  life histories of, 274–277
  overview, 271–274
  pink shrimp (Pandalus borealis), 300–305
Cobb, J.S., 307
colonization rate, 338
Colpophyllia natans (boulder brain coral), 175
Conacher, C.A., 390
conductivity, temperature, depth (CTD) profiles, 503
Connell, J.H., 14
conservation, *see* marine conservation
conservation and fisheries management, influences on marine ecology, 17–20
  marine protected areas, 19–20
  scale of study, 17–18
  spatial resolution, 18–19
conservation and population biology, 9–10
coral, *see also* hard corals
  Acropora cervicornis (staghorn coral), 171
  Acropora palmata (elkhorn coral), 167, 438, 441
  Agaricia agaricites (lettuce coral), 170, 175
  Agaricia humilis (lowrelief saucer coral), 175
  Balanophyllia elegans (solitary coral), 437, 439
  colonies, 15–16
  Colpophyllia natans (boulder brain coral), 175
  Diploria strigosa (symmetrical brian coral), 176
  Favia fragum (golfball coral), 170, 172, 436
  growth rate, 189–190
coral reef fishes, *see also names of specific coral reef fish*
  biology and ecology of, 42–55
    dispersal and connectivity, 51
    metapopulation dynamics, 51–55
    postsettlement life stages, 42–51
  as case studies, 495–499
  factors dissolving metapopulation structure, 55–60
    nursery habitats, 57–60
    spawning aggregations, 56–57
  metapopulation spatial structure, 34–42
    geographic extent, 34–35
    interpatch space, 41–42
    spatial subdivision, 35–41
    overview, 31–34

Coriolis effect, 330
Cornell, H.V., 475
Cowen, R.K., 35
crab
  brachyuran crab, 215, 275–276
  Calinectes sapidus (blue crab), 274, 285–292
  Cancer magister (Dungeness crab), 293–300
  Carcinus maenas (green crab), 279
  Chinocetes opilio (snow crab), 307
  Chionoecetes bairdi (tanner crab), 308
  Lithodes aquispina (brown king crab), 308
  Menippe mercenaria (stone crab), 274
  Paralithodes camtschaticus (red king crab), 308
  Petrolisthes cinctipes (porcelain crab), 297, 447
Craddock, C., 336
Crouse, D.T., 509
Crowder, L.B., 231, 527
crustaceans, 519
CTD (conductivity, temperature, depth) profiles, 503
Cymodocea nodosa, 391, 393
Cynoscion nebulosus (spotted sea trout), 138
Cynoscion regalis (weakfish), 138

D

Dadswell, M.J., 136
damselfishes, 36
  Pomacentrus amboinensis (ambon damselfish), 52, 496
  Stegastes partitus (bicolor damselfish), 498
Daume, S., 212
Davies, C.R., 47
Day, R.W., 223
Dayton, P.K., 14, 505
deep-sea hydrothermal vents, 321–346
  biogeography and diversity, 333–335
  metapopulation models for vent faunal diversity, 335–343
    facilitation, 341–343
    null model, 337–341
    overview, 321–327
  species interactions, 332–333
  vent systems as metapopulations, 328–332
    dispersal and colonization of, 328–332
    dynamics and distribution of vent habitat, 328
DeMartini, E.E., 20
demographic smoothing, 92

den Boer, P.J., 462
density-dependent recruitment, 299
density-independent larval demography, 48
depth, and temperate reef fish communities, 79–80
De Ruyter Van Steveninck, E.D., 179
Diadema antillarum (long-spined urchin), 179, 181, 185, 186, 442
diadromous fish, 119–146
  exemplifying metapopulation theory, 130–139
    alosines, 135–137
    atherinids, 138–139
    comparing salmonids and alosine herrings, 137
    salmonids, 131–135
    sciaenids, 137–138
  fisheries, value of metapopulation concept in understanding/managing, 139–146
    conservation of local populations, 140
    effect of demography on metapopulation management, 143–144
    and historical management of local populations, 140–142
    marine protected areas as spatial management tool, 144
    mixed-stock analysis, 142–143
    whether can be managed as metapopulations, 144–146
  mechanisms forming distinct populations, 125–126
  overview, 119–125
  tools to quantify migration in, 127–130
    artificial tags, 128–129
    genetics, 127–128
    natural tags with emphasis on otolith-geochemical tags, 129–130
Dictyota, 180, 185
Diploria strigosa (symmetrical brian coral), 176
dispersal
  and connectivity, of coral reef fishes, 51
  and population dynamics, 462–465
    background of, 462–463
    clonal metapopulation dynamics, 463–464
    population sources and sinks, 463
    recruitment limitation, 464–465
  and retention mechanisms, temperate rocky reef fishes, 84–85
  scale of, 460–462
distinct population segment (DPS), 71

diversity, of deep-sea hydrothermal vents, 333–335
Doherty, P.J., 17, 47, 54, 55, 521
Done, T.J., 176
Douglas, W.A., 42
DPS (distinct population segment), 71
Drengstig, A., 304
Duarte, C.M., 393
dungeness crab (*Cancer magister*), 293–300
Dupre, C., 401
Dybdahl, M.F., 444
dynamics, *see* marine metapopulation dynamics

E
East Pacific Rise, 329, 334–335
Ebert, T.A., 47, 214, 216
Echinometra, 220
Echinometra chloroticus, 220, 228
ECLSP (Exuma Cays Land and Sea Park), 503
ecology, marine and terrestrial
  commonalities between, 518–519
  differences between, 519–521
Edmands, S., 220
Ehrlen, J., 401
Ehrlich, P.R., 460
elkhorn coral (*Acropora palmata*), 167, 438, 441
Elliott, N.G., 219
El Niño events, 86–87
El Niño southern oscillation (ENSO) events, 355–356, 505
Elton, C.S., 121
Embiotoca jacksoni (black surfperch), 101
emigration, 126
ENSO (El Niño southern oscillation) events, 355–356, 505
Epifanio, C.E., 289
ESU (evolutionarily significant unit), 71
Evechinus chloroticus, 208, 216
evolutionarily significant unit (ESU), 71
extinction debt, 159
extinction-recolonization events, 52–53, 94
Exuma Cays Land and Sea Park (ECLSP), 503
Exuma Sound, 501–504

F
Faaborg, J., 472, 473
facilitation model, 342
Fain, S.R., 402
Fairweather, P.G., 20

Index

faulting, of temperate reef habitat, 74–76
Favia fragum (golfball coral), 170, 172, 436
Fevolden, S.E., 107, 304
Figueira, W.F., 497
Fisher, R.A., 477
fishing and management, effect on spatial variability, 228–233
  of abalone, 228–229
  and optimal harvesting of invertebrate metapopulations, 231–233
  of sea urchins, 229–230
fishing pressure, 11
Fleming, A., 223
FLEP (fraction lifetime egg production), 415–418
Florida Keys reef system., 497–501
FNLS (fraction of natural larval settlement), 507
Fogarty, M.J., 142, 281, 284
Fonseca, M.S., 393
Forrester, G.E., 36
Fowler, T., 54, 55
fraction lifetime egg production (FLEP), 415–418
fraction of natural larval settlement (FNLS), 507
fragmentation-induced metapopulations, 37, 120
Frank, K.T., 142, 143
Freckleton, R.P., 388, 389, 403
Fucus mosaic, 255–256
future of metapopulation science in marine ecology, 517–528
  amphibiousness of metapopulation theory, 518–521
    commonalities between marine and terrestrial ecology, 518–519
    differences between marine and terrestrial systems, 519–521
  areas of sea where metapopulation theory less relevant, 521–524
    highly isolated populations, 522–523
    highly mobile species, 521–522
    special case of clonal organisms, 523–524
    widely dispersing species, 523
  overview, 517–518
  research questions, 524–526
    knowledge of dispersal pathways and mechanisms, 524–525
    knowledge of rates and extents of connectivity, 525

  marine metacommunities, 525–526
  spatially explicit management of marine fishery resources, 526–527

G

Gadus morhua (northern cod), 107, 141
Gaggiotti, O.E., 7, 10, 445
Gaines, S.D., 260, 296
Gall, G.A.E., 132
Gause, G.F., 121
Gaylord, B., 260, 296, 371
Gell, F.R., 231
genetic homogeneity, 35
geological equilibrium, 339
George III Reef, 224–225
Gerber, L.R., 231
Gharrett, A.J., 126
giant kelp, 353–381
  connectivity among local populations, 375–377
  factors affecting colonization, 360
    life history constraints, 360–363
    modes of colonization, 363–365
    postsettlement processes, 366–367
    spore production, release, and competency, 365–366
  Macrocystis pyrifera, 80, 507
  overview, 353–355
  populations, dynamics of, 355–360
  spore dispersal, 367–375
    empirical estimates of spore dispersal, 368–371
    factors affecting colonization distance, 367–368
    modeled estimates of spore dispersal, 371–375
Gilpin, M.E., 70
glaciation, of temperate reef habitat, 76
Gobiidae, 45
Gobiodon (gobies), 36
Goodrich, D.M., 289
Gotelli, N.J., 358
Grant, W.S., 142
Grantham, B.A., 254
Great Barrier Reef, 36–37, 41, 58, 169, 173
Grimm, V., 5, 31, 32, 250, 251, 253, 262, 400
Gunderson, D.R., 94, 95
Gust, N., 48

H

Halichoeres bivittatus, 50
Halimeda (calcareous chlorophytes), 181
Haliotis, *see* abalone
Haliotis corrugata (pink abalone), 207, 211, 219, 221
Haliotis cracherodii (black abalone), 207, 211, 219, 221
Haliotis discus hannai (abalone), 207, 210, 211, 219, 221
Haliotis diversicolor (abalone), 219, 221
Haliotis fulgens (green abalone), 208, 211, 219, 221
Haliotis iris (abalone), 207
Haliotis kamtschatkana (northern abalone), 207, 210, 219
Haliotis laevigata (greenlip abalone), 207, 210, 219, 221, 223
Haliotis midae, 207, 219
Haliotis roei, 207, 219, 221
Haliotis rubra (blacklip abalone), 207, 210, 219, 221, 223
Haliotis rufescens (red abalone), 207, 211, 219, 221
Haliotis sorenseni (white abalone), 207
Halodule wrightii (seagrass), 393
Halophila decipiens (circumglobal seagrass), 396–398
Halophila johnsonii, 399–400
Halophila ovalis, 393
Hamilton, W.D., 463
Hammerstrom, K., 393
Hanski, I., 3, 5, 7, 10, 70, 158, 226, 248, 250, 251, 253, 262, 400
Hanson, J.M., 280
hard corals, 157–195
  existing models of dynamics on coral reefs, 166
  exploration of model behavior, 189–195
    effect of reduced hurricane frequency on reserve network, 194–195
    impact of hurricane frequency on local dynamics, 193–194
    interactions between initial coral cover, algal overgrowth rate, coral growth rate, and herbivory, 189–190
    recruitment scenario and overfishing of herbivores, 190–193
  initial conditions, sensitivity of model to, 186–189
  model structure and parameterization, 160–166
  overview, 158–160
  phase shifts in community structure: testing model, 185–186
  spatially structured coral reef communities, development of prototype model of, 167–185
    competition (and modeling dynamics of competitors), 178–181
    connectivity, 170–171
    growth, 174–175
    herbivory, 181–185
    mortality, 175–178
    recruitment, 171–174
    reproduction, 169–170
    scales, 167–168
Harper, J.L., 460
Harrison, P.L., 172
Harrold, C., 216
Harwell, M.C., 394
Hastings, A., 166, 274
Heincke, F., 141
Helix reef, 171
Hellberg, M.E., 130, 437
Herbivory, 183
herbivory, 189–190
herring
  Clupea harengus (Atlantic herring), 141
  salmonids and alosine herrings, comparing of, 137
Hessler, R.R., 332
Heterodontus francisci (horn shark), 101
highly isolated populations, 522–523
highly mobile species, 521–522
Hilborn, R., 49, 128
Hill, M.F., 422
Hixon, M.A., 47
Hjort, J., 16, 209
Holt, R.D., 459, 469
Holt, S.J., 19, 141
Homarus americanus (American lobster), 274, 275, 279–285
Horstedt, S.A., 302
Howard, D.F., 90
Howe, R.W., 231
Hubbell, S.P., 478, 480, 482
Huffaker, C.B., 121, 462
Hughes, T.P., 17, 54, 160, 171, 173, 178, 186
Hugueny, B., 475

Index

human impacts on temperate rocky reef fishes, 103–105
hurricanes
  effect of reduced frequency on reserve network, 194–195
  impact of frequency on local dynamics, 193–194
  incidences of partial-colony mortality induced by, 175–177
Husband, B.C., 388
hypothesis testing phase of marine ecology, 12

I

Idoine, J.S., 281
Iles, T.D., 141
immigration, 126, 332
Incze, L.S., 282
Indo-Pacific systems, 173
Inglis, G.J., 393
International Pacific Salmon Fisheries Commission, 133
Ivanov, B.G., 303
Iwasa, Y., 17, 218, 258

J

Jackson, J.B.C., 178, 461
James, M.K., 44, 496, 497
Jasus (spiny lobsters), 275
Jenkins, S.R., 262
Johnson, C.R., 166
Johnson, K.S., 333
Johnson, M.P., 253, 254, 255
Johnson, M.S., 446
Jolly, M.T., 445
Jompa, J., 179
Jones, Cynthia, 520
Jones, G.P., 496
Jones, Harden, 17
Juniper, S.K., 334, 336
juvenescence, 143

K

Kaplan, D.M., 420
Karlson, R.H., 463, 464, 475
Karlson, Ron, 526
kelp, *see also* giant kelp
  kelp bed experiment, 369
  kelp forests, 354
  Nereocystis luetkeana (bull kelp), 365
  Pterygophora californica (palm kelp), 366

Kendall, M.S., 390
Kiflawi, M., 50
Kordos, L.M., 291
Kritzer, J.P., 44, 46, 47, 70, 94, 96
Kunin, W.E., 263

L

Labyrinthula, 392
Labyrinthula zosterae, 392
Laminaria, 353
Lande, R., 425
landscape ecology, 42
Lanteigne, M., 280
Largier, J.R., 100, 170
Larson, R.J., 91
larval dispersal and settlement, 411–426
  of abalone, 209–212
  dispersal pattern, 416
  metapopulation persistence, 416–421
    consequences for distribution of meroplanktonic species, 420
    consequences for success of marine reserves, 421
  overview, 411–413
  review, 422–426
  role of variability, 421–422
  of sea urchins, 212–217
  single population persistence, 413–416
Lea, R.N., 91, 92
Leaman, B.M., 95
Lee Stocking Island (LSI), 503
Leggett, W.C., 136, 142, 143
Leiostomus xanthrus, 130
LEP (lifetime egg production), 414–415
Lessios, H.A., 442
Levin, D.A., 41
Levin, Richard, 7, 32
Levin, S.A., 247, 260
Levins, R., 3, 9, 10, 43, 45, 52, 70, 120, 121, 124, 158, 206, 249, 425
Lewis, R.I., 436
lifetime egg production (LEP), 414–415
Lima, M., 124
Limburg, K.E., 136
Lipcius, R., 288
Lipcius, R.N., 230, 231, 503
Lithodes aquispina (brown king crab), 308
Little, K.T., 289
Littorina fabalis, 255
Littorina saxatilis, 436

Lizard Island group, 44
Lobophora, 179–180, 185
Lobophora variegata, 179, 180, 185
lobsters
   Homarus americanus (American lobster), 274, 275, 279–285
   Jasus (spiny lobsters), 275
   Nephrops (clawed lobsters), 274, 275
   Nephrops norvegicus (Norway lobster), 274
   Panulirus (spiny lobsters), 275
   Panulirus argus (Caribbean spiny lobster), 275, 499–504, 523
   Panulirus cygnus (Australian spiny lobster), 275
Lockwood, D.R., 418, 422
loggerhead sea turtles (*Caretta caretta*), 509–510
Long Cay, 182
Loreau, M., 459, 482, 483
Loxechinus albus (sea urchins), 208, 213
LSI (Lee Stocking Island), 503
Lutjanus carponotatus (spanish flag snapper), 43–44, 46, 47
Lutz, R.A., 329
Lynx pardinus (Iberian lynx), 122
Lysy, 303, 304

M

MaCall, A.D., 231, 232
MacArthur, R.H., 3, 10, 159
Macrocystis, 355–356, 363, 366–368, 375
Macrocystis pyrifera, 80, 507
Macrocystis sporophytes, 366
Madigan, S.M., 210
Madracis mirabilis, 171
Magnuson-Stevens Fishery Conservation and Management Act, 19
major oceanographic domains, 83–84
Man, A., 20
Marba, N., 393
marine and terrestrial systems, differences and similarities of, 6–12
   applications to marine populations, 10–12
   describing actual spatial population processes and structures, 6–8
   framework for asking research questions, 8–9
   paradigm for population and conservation biology, 9–10
marine conservation, 491–511
   case studies, 495–510

      Caribbean spiny lobster (*Panulirus argus*), 499–504
      coral reef fishes, 495–499
      loggerhead sea turtles (*Caretta caretta*), 509–510
      red sea urchin (*Strongylocentrotus franciscanus*), 504–509
   overview, 491–493
   sources, sinks, and metapopulation dynamics, 494–495
marine ecology, conservation and fisheries management influences on, 17–20
   marine protected areas, 19–20
   scale of study, 17–18
   spatial resolution, 18–19
marine metapopulation dynamics, 431–449; *see also* metacommunities
   delineating populations, 434–438
   inferring of nonequilibrium population dynamics, 441–447
      mixing in plankton and genetics of larval cohorts, 445–447
      population extinction and recolonization, 442–444
      source–sink relationships, 444–445
   inferring patterns of connectivity, 438–441
   overview, 431–434
   island model and its limitations, 432–434
   subdivision in marine populations, 432
marine populations
   applications of metapopulation theory to, 10–12
   subdivisions in, 432
marine protected areas (MPAs), 19–20, 40, 144, 527
marine reserves, 421, 492
Martinez, I., 304
Massel, S.R., 176
May, R.M., 463
Mayr, E., 461
Mazzella, L., 402
McBride, R.S., 136
McCauley, D.E., 443
McCook, L.J., 179
McFarlane, G.A., 128
McQuinn, I.H., 19, 424
McShane, P.E., 209
Meesters, E.H., 174
Menidia menidia (Atlantic silverside), 138
Menippe mercenaria (stone crab), 274

Index

Merluccius productus (Pacific hake), 83
meroplanktonic species, 420
Mesoamerican Barrier Reef System, 58
mesopopulation, 36
metacommunities
   and local–regional species richness relationships, 472–477
      background, 472–474
      effects of transport processes and relative island position, 476–477
      relevance to, 474–476
   and relative species abundance patterns, 477–483
      background, 477–478
      and relative species abundance patterns, 478–482
      relevance to, 482–483
   relevance to, 465–466
   and species–area relationships, 466–472
      background, 466–468
      regional-scale differentiation, 470–471
      relevance to, 469–470
metapopulation dynamics
   of coral reef fishes, 51–55
      extinction–recolonization events, 52–53
      synchrony of population fluctuations, 53–55
   sinks and, 494–495
   sources and, 494–495
metapopulation ecology and marine conservation, 491–511
   case studies, 495–510
      Caribbean spiny lobster (*Panulirus argus*), 499–504
      coral reef fishes, 495–499
      loggerhead sea turtles (*Caretta caretta*), 509–510
      red sea urchin (*Strongylocentrotus franciscanus*), 504–509
   overview, 491–493
   sources, sinks, and metapopulation dynamics, 494–495
metapopulation theory
   amphibiousness of, 518–521
      commonalities between marine and terrestrial ecology, 518–519
      differences between marine and terrestrial systems, 519–521
   areas of sea where less relevant, 521–524
      highly isolated populations, 522–523
      highly mobile species, 521–522
      special case of clonal organisms, 523–524
      widely dispersing species, 523
   merging with marine ecology, establishing historical context, 3–22
      conservation and fisheries management influences on marine ecology, 17–20
      differences and similarities between marine and terrestrial systems, 6–12
      history and effects of predominant research questions in marine ecology, 12–17
      overview, 3–6
Micheli, F., 333
microsatellite DNA, 127
Mid-Atlantic Ridge, 328
migration matrices, 440
Minchin, D., 20
mitochondrial DNA (mtDNA), 99
Moberg, P.E., 220
Montastraea, 174
Montastraea annularis (boulder star coral), 167, 169, 174, 175, 176, 177, 187, 188
Mora, C., 461, 462
Morgan, L.E., 215, 216, 227, 230, 506
Morone saxatalis (anadromous striped bass), 137
Morse, D.E., 172, 175
Moser, H.G., 87, 88
Mouquet, N., 459, 481, 483
MPAs (marine protected areas), 19–20, 40, 144, 527
mtDNA, 138, 291
mtDNA (mitochondrial DNA), 99
Mullineaux, L.S., 332, 333
Mumby, P.J., 58
Murray, N.D., 219
Mycetophyllia, 169
Myers, R.A., 304, 305
Mytilus californianus (mussel-dominated shores), 250

N

Naime, C.E., 282
NAO (North Atlantic Oscillation), 281
Naso unicornis (brown unicornfish), 446
naturally evolved metapopulations, 37, 120
nearshore rockfishes (*Pteropodus* subgeneric group), 101
Nei, M., 443

neighborhood size (N_m), 218–219
Nephrops (clawed lobsters), 274, 275
Nephrops norvegicus (Norway lobster), 274
Nereocystis luetkeana (bull kelp), 365
network effect, 417
Newman, S.J., 44
nonequilibrium population dynamics, inferring of, 441–447
   mixing in plankton and genetics of larval cohorts, 445–447
   population extinction and recolonization, 442–444
   source–sink relationships, 444–445
North Atlantic Oscillation (NAO), 281
Northeast Pacific Ocean, 71
northern cod (Gadus morhua), 107, 141
North Pacific rocky reef fishes, population genetic studies in, 96–103
Norway lobster (Nephrops norvegicus), 274
Nucella (Thais) lamellosa, 255
nudibranch (Adalaria proxima), 256
null model, 337
nursery habitats, coral reef fishes, 57–60

O

oceanography, role of in metapopulation structuring, 84–85
Okazaki, R.K., 291
Olesen, B., 391
Olney, J.E., 136
OMZ (oxygen minimum zone), 104
Oncorhynchus gorbuscha (pink salmon), 131
Oncorhynchus tshawytscha (chinook salmon), 132
One Tree Reef, 53
Orth, R.J., 390, 394
otolith–geochemical tags, 129
otolith microincrements, 54
outbreeding depression, 126
overfishing of herbivores, 190–193
oxygen minimum zone (OMZ), 104

P

Pacific Decadal Oscillation (PDO), 86
Pagrus auratus (New Zealand snapper), 106
Paine, R.T., 247, 250, 465
Pandalus borealis (pink shrimp), 300–305
Panulirus (spiny lobsters), 275
Panulirus argus (Caribbean spiny lobster), 275, 499–504, 523

Panulirus cygnus (Australian spiny lobster), 275
Paracentrotus lividus, 208, 213
Paralithodes camtschaticus (red king crab), 308
patch models, 249–251, 336
Payri, C.E., 179
PDO (Pacific Decadal Oscillation), 86
Pearcy, W.G., 91
Pearse, J.S., 230
Pectinaria koreni, 445
Pedersen, O.P., 302, 303
pelagic zone, 12
persistence
   metapopulation, 416–421
      consequences for distribution of meroplanktonic species, 420
      consequences for success of marine reserves, 421
   single population, 413–416
Peterken, C.J., 390
Petersen, C.W., 49, 50
Petrolisthes cinctipes (porcelain crab), 297, 447
Pew Oceans Commission (2003), 492
philopatric model, 210
Phyllospadix scouleri, 399
pink shrimp (Pandalus borealis), 300–305
Planes, S., 446, 447
Plectropomus (grouper), 36
Plectropomus leopardus (bluedotted coral trout), 46, 55
Pogson, G.H., 107
Point Conception, 78
Polacheck, T., 20
Polunin, N.V.C., 20
Pomacentrus amboinensis (ambon damselfish), 52, 496
Pomacentrus mollucensis, 52
pool enrichment, 473
population
   and conservation biology, 9–10
   delineating of populations, 434–438
   extinction and recolonization, 442–444
   genetics, 218–221
      of abalone, 218–219
      of sea urchins, 219–221
   sources and sinks, 463
Porites astreoides (mustard hill coral), 169, 170, 175, 176
Porites lutea, 177
Porites porites, 171, 175
Posidonia australis, 394

Index

Posidonia coriacea, 394
Possingham, H.P., 231
Postelsia palmaeformis (sea palm), 250
postsettlement life stages, of coral reef fishes, 42–51
  behavior, 48–51
    movement, 49
    social and mating systems, 49–51
  demography, 44–48
    density dependence, 47–48
    life history patterns, 45–46
    spatial variation in demographic traits, 46–47
  population size, 43–44
Preece, A.L., 166
Preston, F.W., 477
Prince, J.D., 212, 223, 226, 229
Procaccini, C., 402
proportional sampling, 473
Pteropodus subgeneric group (nearshore rockfishes), 101
Pterygophora californica (palm kelp), 366
Puget Sound, 83, 98–99
Pulliam, H.R., 142, 463

Q

Queen Charlotte Sound, 98–99
Quinn, J.F., 507
Quinn, T.J., 126
Quinn, T.P., 131, 137

R

Raffaelli, D.G., 461
Ralph, P.J., 392
Ralston, S., 90
recolonization and population extinction, 442–444
recruitment, 207, 464–465
red sea urchins (*Strongylocentrotus franciscanus*), 208, 213, 227, 504–509
Reed, D.C., 96, 365
regional enrichment, 473
resource depletion zones, 460
restriction fragment length polymorphism (RFLP), 436
RFLP (restriction fragment length polymorphism), 436
rhizophytic macroalgae (*Caulerpa*), 392
Ricker, W.E., 130, 288

Ridgeia piscesae (vestimentiferan tubeworm), 334
Riftia, 330, 333
Riftia pachyptila (vent tubeworm), 329, 332
Robbins, B.D., 393
Roberts, C.M., 20, 231, 496
rocky intertidal invertebrates, potential for metapopulations within and among shores, 247–265
  measured scales of variability, 260–263
  metapopulations at different scales, 257–260
  overview, 247–248
  patch models, 249–251
  within-shore metapopulations, 251–257
Rogers–Bennett, L., 230
Rollon, R.N., 391
Rosenzweig, M.L., 467, 469, 474
Roughgarden, J., 15, 16, 17, 218, 258, 260, 475
Rousset, F., 100
Rowley, R.J., 216
Ruckelshaus, M., 444
Rugulo, L.J., 288
Russ, G.R., 54, 55
Russell, M.P., 214, 216, 227
Ruzzante, D.E., 436
Ryman, N., 145

S

Sale, P.E., 15, 16
Sale, P.F., 42, 70, 94, 96, 526
Salm, R.V., 20
salmonids, 131–135
  and alosine herrings, comparing of, exemplifying metapopulation theory, 137
  Oncorhynchus gorbuscha (pink salmon), 131
  Oncorhynchus tshawytscha (chinook salmon), 132
  Salmo salar (Atlantic salmon), 133
Salmo salar (Atlantic salmon), 133
Salmo trutta (Brown trout), 133
Sander, F., 174
Sand–Jensen, B., 391
Sargassum, 179, 185
Sasaki, R., 211, 218
scale of dispersal, 460–462
scale of study, as to marine ecology, 17–18
scallops (*Aequipecten opercularis*), 436
Scarus (parrotfish), 36

Schroeter, S.C., 216
Schwarz, C.J., 129
sciaenids, 137–138
seagrasses, 387–403
  examples of potential metapopulations, 394–400
    Halophila decipiens, 396–398
    Halophila johnsonii, 399–400
    Phyllospadix scouleri, 399
  overview, 388–389
  patches: colonization and extinction, 391–394
  reproduction, 389–391
  seed dormancy, 402–403
  spatial organization of seagrasses, 401
  spatial structure of population, 402
sea urchins
  adult habitat and spatial structure, 226–228
  Diadema antillarum (long-spined urchin), 179, 181, 185, 186, 442
  Echinometra, 220
  Echinometra chloroticus, 220, 228
  Evechinus chloroticus, 208, 216
  fishing and management, 229–230
  larval dispersal and settlement, 212–217
    hydrodynamics and coastal topography, 213–216
    larval behavior at settlement, 216–217
    larval ecology and dispersal, 212–213
  life history of, 208–209
  Loxechinus albus, 208, 213
  population genetics of, 219–221
  Strongylocentrotus, 207, 212, 215
  Strongylocentrotus droebachensis, 208, 213
  Strongylocentrotus franciscanus, 208, 213, 227
  Strongylocentrotus nudus, 208, 213
  Strongylocentrotus purpuratus (purple sea urchins), 105, 208, 213, 220
Sebastes, 85, 93, 103
Sebastes dalli (calico rockfish), 80
Sebastes helvomaculatus (rosethorn rockfish), 98
Sebastes jordani (shortbelly rockfish), 88
Sebastolobus alascanus (deepwater scorpaenid fishes), 96
Sebastolobus altivelis (deepwater scorpaenid fishes), 96, 97
seed dormancy, seagrasses, 402–403
Serpulidae, 323
Sevigny, 304
Shanks, A.L., 217, 278

Shapiro, D.Y., 445
Sheaves, M., 58, 59
Shepherd, J.G., 414
Shepherd, S.A., 18, 211, 218, 223, 228, 232
Short, F.T., 392
Shulman, M.J., 35
Shumway, S.E., 302
Siderastrea radians (lesser starlet coral), 170
Siderastrea siderea (massive starlet coral), 169, 175
Simberloff, D., 3, 10, 248, 400
Sinclair, M., 17, 141, 145, 435, 522
"sink" habitats, 494
sinks and population sources, 463
Sissenwine, M.P., 414
Skellam, J.G., 425
Slatkin, M., 438, 443
Smedbol, R.K., 5, 31, 424
Smidt, E., 302, 303
Smith, Maynard, 9
snapper
  Lutjanus carponotatus (spanish flag snapper), 43–44, 46, 47
  Pagrus auratus (New Zealand snapper), 106
snow crab (*Chinocetes opilio*), 307
Sork, V.L., 402
"source" habitats, 494
source–sink relationships, 444–445
Sparisoma viride, 182
spatially realistic metapopulation theory, 7
spatially structured coral reef communities, 34–42
  development of prototype model of, 167–185
    competition (and modeling dynamics of competitors), 178–181
    connectivity, 170–171
    growth, 174–175
    herbivory, 181–185
    mortality, 175–178
    recruitment, 171–174
    reproduction, 169–170
    scales, 167–168
  geographic extent, 34–35
  interpatch space, 41–42
  spatial subdivision, 35–41
spatial population processes and structures, 6–8
spatial resolution, as to marine ecology, 18–19

Index

spatial variability in adult distributions and demographics, 221–228
  abalone, 223–226
  sea urchins, 226–228
spawning aggregations, coral reef fishes, 56–57
spawning production ratio (SPR), 415
Spight, T.M., 255
spiny lobsters (*Jasus*), 275
spiny lobsters (*Panulirus*), 275
SPM (suspended particulate matter), 174
SPOM (stochastic patch-occupancy models), 7
spore dispersal, of giant kelp, 367–375
  empirical estimates of spore dispersal, 368–371
  factors affecting colonization distance, 367–368
  modeled estimates of spore dispersal, 371–375
spore production, release, and competency, 365–366
sporophylls spores, 365
SPR (spawning production ratio), 415
spring transition, 296, 505
Stanley, R.D., 91
Starr, R.M., 91
Stegastes partitus (bicolor damselfish), 498
stepping stone model, 440
Stiger, V., 179
stochastic patch-occupancy models (SPOM), 7
Stockhausen, W.T., 503, 504
stocks, 123
Stone, L., 166
Strathmann, R.R., 521
Strongylocentrotus, 207, 212, 215
Strongylocentrotus droebachensis, 208, 213
Strongylocentrotus franciscanus (red sea urchins), 208, 213, 227, 504–509
Strongylocentrotus nudus, 208, 213
Strongylocentrotus purpuratus (purple sea urchins), 105, 208, 213, 220
subduction, of temperate reef habitat, 74–76
suspended particulate matter (SPM), 174
Swearer, S.E., 35
Szmant, A.M., 169

T

tags, 128–130
  artificial, 128–129
  natural, 129–130
Tanner, J.E., 169, 179
Taylor, H.M., 463, 464
Tegner, M.J., 505
Teigsmark, G., 303
temperate reef communities, 105
temperate rocky reef fishes, 69–108
  climate, climate cycles, and historical metapopulation structuring, 85–87
  communities of, 78–80
    biogeographic provinces and, 78–79
    depth as master variable in, 79–80
    typical fish fauna of, 80
  empirical approaches to measuring dispersal and metapopulation structure, 94–96
  future directions for metapopulation studies of, 105–108
  human impacts, 103–105
  life history, role of, in metapopulation structuring, 87–93
    adults, 90–92
    early life history, 87–88
    juveniles, 88–90
    longevity, 92–93
  overview, 70–71
  population genetic studies in North Pacific rocky reef fishes, 96–103
  role of oceanography in metapopulation structuring, 80–85
    dispersal and retention mechanisms, 84–85
    major oceanographic domains, 83–84
  temperate reef habitat, geological processes and types and distribution of, 71–78
    fluvial and dynamic submarine erosive processes, 76–78
    glaciation, 76
    subduction, volcanism, and faulting, 74–76
Terborgh, J.W., 472, 473
terrestrial and marine systems, differences and similarities of, 6–12
  marine populations, applications of metapopulation theory to, 10–12
  population and conservation biology, 9–10
  research questions, 8–9
  spatial population processes and structures, 6–8
terrestrial metapopulation citations, 5
Tesseropera rosea (barnacle), 14–15
Tevnia, 333

Tevnia jerichonana (vestimentiferan tubeworm), 332
Thais (*Nucella*) *lamellosa*, 255
Thalassia hemphrichii, 391
Thalassia testudinum (Caribbean seagrass), 402
Thalassoma bifasciatum (Bluehead wrasses), 50, 52
Theory of Island Biogeography, The, 121
Thomas, C.D., 263
Thorpe, J.P., 436
Thorrold, S.R., 95, 137
Tigriopus brevicornis (rock pool copepods), 253
Tigriopus californicus (copepod), 253, 435, 444
Tilman, D., 159, 166
time–area closure, 492
Tomascik, T., 174
Toonen, R.J., 447
Tracey, M.L., 283
trout
   Cynoscion nebulosus (spotted sea trout), 138
   Plectropomus leopardus (bluedotted coral trout), 46, 55
   Salmo trutta (Brown trout), 133
Tsurumi, M., 335
Tuck, G.N., 231
Tunnicliffe, V., 334, 336
Turbinaria, 179, 185
Turnipseed, M., 335, 336
Tutschulte, T.C., 212

U
urchins, *see* sea urchins
U.S. Commission on Ocean Policy (2004), 492

V
Van Engel, W.A., 288
variability, role of, 421–422

Ventiella sulfuris, 330
vents, *see* deep-sea hydrothermal vents
Vidondo, B., 391
volcanism, of temperate reef habitat, 74–76

W
Wade, M.J., 443
Walters, C.J., 49
Waples, R.S., 120, 128, 142
Ward, S., 172
Warner, R.R., 49, 50, 54
Waterloo Bay, 225
Watkinson, A.R., 388, 389, 403
Webster, M.S., 47
West Florida Shelf, 397
Westrheim, S.J., 94
Whitlock, M.C., 432
widely dispersing species, 523
Wiens, J.A., 473
Wild, P.W., 296
Williams, D.M., 16, 44
Williams, I.D., 181, 182
Wilson, E.O., 3, 10, 159
Wilson, J., 145
Wing, S.R., 216, 227, 297
Wolanski, E., 166
Worm, B., 304, 305
Wright, S., 9
Wroblewski, J.W., 424

Z
Zeller, D.C., 56
Zhivotovsky, L.A., 126
Ziv, Y., 474
Zostera capricorni, 390
Zostera marina, 391, 392, 394, 402, 436
Zuñiga, G., 219

Printed and bound by CPI Group (UK) Ltd, Croydon, CR0 4YY
08/06/2025
01896870-0009